Molecular Autoimmunity

Molecular Autoimmunity

Edited by

MONCEF ZOUALI
Institut National de Santé et de Recherche Médicale (INSERM), Paris, France

A C.I.P. Catalogue record for this book is available from the Library of Congress.

ISBN 0-387-24533-2

©2005 Springer Science+Business Media, Inc.
All rights reserved. This work may not be translated or copied in whole or in part without the written permission of the publisher (Springer Science+Business Media, Inc., 233 Spring Street, New York, NY 10013, USA), except for brief excerpts in connection with reviews or scholarly analysis, Use in connection with any form of information storage and retrieval, electronic adaptation, computer software, or by similar or dissimilar methodology now known or hereafter developed is forbidden.
The use in this publicaiton of trade names, trademarks, service marks and similar terms, even if they are not identified as such, is not to be taken as an expression of opinion as to whether or not they are subject to proprietary rights.

Printed in the United States of America.

9 8 7 6 5 4 3 2 1

springeronline.com

Preface

A unique feature of the normal immune system is that, while being able to mount responses to virtually all antigens of the environment, it also exhibits tolerance to its own components, a property that prevents attack of the body's own tissues. At times, however, self-tolerance breaks down, and the immune system fails to recognize self-antigens and mounts a misguided immune attack against its own tissues, which culminates in autoimmune disease. Currently, a growing number of disorders affecting virtually all organs or tissues of the human body have a proven or a strongly suspected autoimmune etiology. Their prevalence is worldwide and their etiology remains under investigation. In the past few years, our understanding of autoimmunity has witnessed important advances. This volume commemorates the 100th anniversary of the discovery of the first human autoimmune disease by Julius Donath and Karl Landsteiner in 1904 in Vienna. It comprises a collection of papers that show some of the ways in which insight into autoimmunity is opening new avenues for understanding their etiology and for designing novel immunointervention strategies.

The first part of the book is concerned with innate immunity, a branch of the immune system mainly directed to recognition of invariant molecules of infectious agents. Most of them are essential for pathogen survival and are conserved and shared by groups of pathogens. The innate immune system is essential for the activation of the adaptive immune response, capable of coping with a high mutation rate and antigen heterogeneity of infectious agents, and generating a long-lasting immune memory. Addressing its role in induction, progression, and protection of myocarditis, a disease linked to adenovirus or coxsackievirus that accounts for approximately 25% of all heart failure in North America, Noel Rose *et al.* tackle brilliantly the daunting task of bringing together the many facets of innate immunity in autoimmunity. They discuss in detail four of the major components of the innate response found to contribute to disease susceptibility: the complement system, natural killer cells, macrophages/dendritic cells, and early-acting proinflammatory cytokines. Also important in the innate immunity branch are toll-like receptors (TLRs) present on a variety of cell types. Paul N. Moynagh reviews the crucial involvement of TLR9 in mediating the immunostimulatory effects of bacterial DNA, potentially leading to activation of B cells and production of autoantibodies independently of T cells. Another example of the interplay between innate immunity and autoimmunity is discussed in the paper from the Terry Du Clos laboratory. Here, members of the pentraxin family, a phylogenetically ancient, highly conserved component of the innate immune system, are shown to bind microbial determinants and autoantigens, and to have the potential to interact with the adaptive immune system through the complement system and Fcγ receptors. In studies of autoimmune type 1 diabetes, an autoimmune disease caused by T cell–mediated destruction of insulin-producing β cells in the pancreatic islets of Langerhans, Terry L. Delovitch and coworkers show how a subset of T cells act as regulators of both innate and adaptive immune responses. Since this cell population seems to be important in maintaining immune homeostasis, a further understanding of its role offers promise for the development of novel therapies for the prevention of diabetes.

The second part of this book focuses on genetic susceptibility. While early studies revealed that human autoimmune diseases require an inherited contribution, their genetics remains the focus of much investigation. Marta E. Alarcón-Riquelme discusses how the availability of the human genome sequence is playing an essential role in unraveling complex disease genetics, and how human genome scans are providing new discoveries. Most interesting is the observation that some of the genes identified are shared among various autoimmune diseases. In a search for factors that promote autoimmunity, Bruce Richardson's laboratory is exploring DNA methylation, an important determinant of chromatin structure that modifies gene expression through localized effects on the nucleosome polymers. Their article elegantly describes how the results can be used to predict functional, biochemical, and genetic alterations in T cells from patients with idiopathic lupus, and how failure to maintain T cell DNA methylation and chromatin structure contributes to human lupus. In myasthenia gravis, an organ-specific antibody-mediated autoimmune disease characterized by an immune response against the nicotinic acetylcholine receptor on the neuromuscular junction, the data described by Ann Kari Lefvert support the notion that the disease is polygenic, with subgroups of patients having different genetic backgrounds. Also polygenic is human lupus. C. Yung Yu and coworkers describe the strong association of complete C4A and C4B deficiencies with human lupus, providing support for the interpretation that C4A deficiency is a genetic risk factor for this disease.

Discussed in the third part of this volume are some potential triggers of autoimmunity that affect different organs. In rheumatic fever, a disease occurring as a delayed sequel of throat infection by *Streptococcus pyogenes* in 3–4 % of untreated children, Jorge Kalil and coworkers clearly show how molecular mimicry between streptococcal antigens and human heart tissue leads to rheumatic heart lesions. In this process, $CD4^+$ T lymphocytes are the major effectors of heart lesions, and several histocompatibility leucocyte antigen (HLA) class II molecules are associated with the disease worldwide, leading to multiple valvular lesions and/or mitral valve regurgitation. By contrast, in other disorders, no infectious agent has been identified. For example, celiac disease is an intestinal disorder caused by an inflammatory T cell response to gluten peptides bound to HLA-DQ2 or -DQ8, molecules with a preference for peptides that contain negatively charged amino acids. As described in the article from Frits Koning's laboratory, posttranslational modification of gluten is critical for the generation of a repertoire of T cell stimulatory peptides, an observation that may be relevant for other HLA-associated disease. The paper from Michael Hertl's laboratory discusses the pathogenic role of autoantibodies and the potential role of autoreactive T cells to desmogleins in pemphigus vulgaris. Aiming to develop antigen-specific immunotherapies, the authors put the emphasis on autoaggressive T cell epitopes and on a subset of T cells that may be critical in the maintenance/restoration of tolerance against desmogleins. Also unclear is the trigger involved in myasthenia gravis pathogenesis. Here, circumstantial evidence suggests a primary role of the thymus. Having established a model of intrathymic inflammation localized to the thymic medulla, Arnold I. Levinson *et al.* attempt to determine how intrathymic expression of the neuromuscular muscle type of acetylcholine receptors is involved in immunopathogenesis.

The fourth part of this volume is devoted to targets of autoimmunity. Paola Migliorini *et al.* focus on development of autoantibody-mediated nephritis. They review data indicating that distinct damage mechanisms probably coexist and play a role in the different phases of poststreptococcal nephritis, Goodpasture's syndrome, and systemic lupus nephritis. The contribution from Ansar Ahmed's laboratory lucidly addresses the role of hormonal factors in autoimmunity. Their effects have been demonstrated in many experimental settings. In humans, exposure to estrogens occurs through various sources, including physiological estrogens that vary during the lifetimes of women, pharmacological estrogens given for medical reasons and environmental estrogens, or endocrine-disrupting chemicals, (pesticides, plastic products, detergents, industrial by-products, municipal sewage–contaminated water that contains metabolites of estrogen-based contraceptive drugs). Nevertheless, their effects are complex and remain incompletely understood. Also important in deciphering autoimmunity are studies of the role of $CD4^+CD25^+$ regulatory T cells, discussed by Yi-chi M. Kong *et al.* in experimental murine autoimmune thyroiditis. Yet, our lack of understanding autoimmunity is perhaps best illustrated by the complexity of immune phenomena described in multiple sclerosis. This chronic inflammatory disease of the central nervous system represents one of the most common neurological diseases of young adults in developed countries. Its hallmarks include focal plaques of white matter demyelination, presence of autoreactive T cells in the blood of most patients, and autoreactive T cells and antibodies in the lesions. Arguing that the autoimmune responses in the affected patients are not invariably detrimental, but may even be beneficial, the provocative article from Hans Lassmann's laboratory challenges the "autoimmune hypothesis" of multiple sclerosis. Future work is required to provide a better understanding of the pathogenesis of this disease.

What could underlie the loss of tolerance in autoimmunity? As discussed in the fourth part of this volume, the reason may well relate to crippling of signaling pathways that govern the discriminatory potential of lymphocytes. Our immune system functions properly only because lymphocytes communicate with one another constantly. Recognizing molecules that are part of our body as selfantigens and distinguishing them from those that come from the external environment, lymphocytes instruct their relatives to attack invaders or produce growth factors or antibodies. This high-fidelity recognition is achieved through a network of intracellular communications wherein lymphocyte receptors are able to sense the nature of encountered molecules and to generate signals that are appropriately delivered to the internal machinery, allowing specific functional responses. As in other cells, the amount of signals generated is fine-tuned for optimal transmission, and kinases and phosphatases control most activities. The chapter on B cells sheds light on the biochemical and molecular aberrations that are responsible for the aberrant lupus B cell biology. Inactivation of genes encoding B cell signaling molecules leads to autoimmune phenomena, and crippled signaling pathways are detectable in the B lymphocytes from patients with systemic autoimmunity. Focusing on rheumatoid arthritis, Ana M. Blasini and Martín A. Rodríguez summarize how abnormalities in T cell responses seen in patients with systemic autoimmunity can be related to identifiable signaling abnormalities. In lupus too, T cells display diverse cellular and cytokine abnormalities. The paper form George C. Tsokos's laboratory elegantly describes biochemical abnormalities that underwrite the diverse T cell abnormalities in lupus. Here, the decreased T cell receptor–associated ζ chain in effector T cells is due mostly to increased degradation, rather than to decreased transcription. Lupus T cells also express increased amounts of the transcriptional repressor CREM that binds IL-2 promoter, thereby limiting its expression. Of further importance is the increased spontaneous aggregation of lipid rafts on the surface membrane of lupus T cells, an abnormality that may contribute to the well-established overexcitable T cell phenotype. Altered signaling in both B and T cells also might account for the aberrant rates of apoptosis in lupus. Koji Yasutomo argues that the resulting increased levels of free-circulating chromatin represent a potential source of antigen trigger in systemic autoimmunity.

Finally, the recent advances in the field of autoimmunity have given clinicians exciting new tools for diagnosing and treating autoimmune disorders. The final section of the book discusses state-of-the-art therapeutic intervention strategies. The rationale for B lymphocyte depletion therapy in autoimmune disorders stems from the paramount role of B cells in autoimmunity. Jonathan Edwards *et al*. discuss in detail its potential for clinical applications, the logistics employed, and the clinical results obtained with anti-CD20 antibody. While the precise mechanisms of action remain to be elucidated, an alternative to this therapy is based on study of B lymphocyte longevity. Following migration to the periphery, the selection and survival of B cells are controlled by a variety of signals. Longevity factors, such as B lymphocyte stimulator (BLyS), also called BAFF, TALL-1, THANK, or zTNF4, that support differentiation of selected B cells into mature long-lived B cells are critical in determining the capacity to mount protective immune responses and to generate deleterious autoimmune responses. Their vital role in survival and maturation of B cells is discussed by William Stohl. In experimental animals, treatment with BLyS/BAFF antagonists ameliorates disease progression and enhances survival. Since patients with lupus, rheumatoid arthritis, or systemic sclerosis overexpress this longevity factor, and because a phase I clinical trial in lupus patients with a neutralizing anti-BLyS monoclonal antibody has documented the safety and biological activity of this BLyS antagonist, additional phase I and phase II clinical trials with a variety of BLyS antagonists are currently under way. Another unifying theme in autoimmune diseases is the involvement of cytokines that play key roles throughout the whole course of the disease, from induction to effector functions. Hence, the control of autoimmunity by cytokine and anti-cytokine treatments represents a potential immunointervention strategy. The simplest approach, already in practice, is the specific inhibition of their action. As discussed in detail by Pierre Miossec, the use of TNFα inhibitors has provided clear evidence of the direct role of cytokines in complex inflammatory diseases. Another more physiological approach consists in stimulating endogenous regulatory mechanisms to restore an adequate balance. However, as Alan Tyndall and Paul Hasler point out, just as there is no consensual unifying mechanism in autoimmune diseases, there is no single successful treatment strategy. Most patients with severe autoimmune diseases are treated with a combination of glucocorticosteroids and immunosuppressive agents, but some either do not respond or require more toxic drugs to achieve or maintain clinical remission, and this subgroup poses a serious treatment dilemma. Rather than total eradication of clonal autoimmunity, hematopoietic stem cell transplantation techniques aim at resetting an imbalance in the complex immune network. The authors posit that this emerging alternative could be a viable option for selected autoimmune diseases patients. Currently, through an international collaboration, around 700 patients have received such treatment. The experience gained from the phase I and II clinical studies is sufficiently encouraging to be exploited in designing phase III randomized comparative trials in the major diseases.

Other potential immunointervention strategies have not reached the stage of clinical trials. In autoimmune uveitis, a disease that affects the inner eye of about 2% of the Western population, $CD4^+$ T helper$_1$ cells recruit inflammatory cells that can irreversibly destroy photoreceptors and neuronal tissue within the eye, leading to decreased vision or even blindness. Gerhild Wildner *et al.* describe several peptides mimicking a retinal autoantigen. Even though some of them are pathogenic in a rat model of experimental uveitis, they do not induce oral tolerance, thus indicating that pathogenic antigens are not obligatory oral tolerogens. The paper by Marc Monestier and coworker reviews the pathogenesis of atherosclerosis, with a particular emphasis on the role of the immune system. They also discuss studies that have addressed the importance of autoantibodies in this disease. Although their exact function is still not understood, manipulating humoral autoimmunity may represent a novel therapeutic or prophylactic approach in atherosclerosis. In their chapter, Silvia S. Pierangeli *et al.* review the molecular and intracellular pathways mediated by anti-phospholipid antibodies in platelets and endothelial cells that lead to thrombotic events. A better definition of the nature of the antiphospholipid antibody–target tissue interaction and the mechanism(s) by which these antibodies cause thrombosis may lead to devising new targeted treatment modalities. Finally, antigen-specific therapy represents a promising avenue for treating autoimmune diseases. It involves vaccination with autoantigens in a tolerogenic fashion, i.e., by nasal administration, oral feeding, and DNA vaccination, thought to induce regulatory T cells that produce anti-inflammatory cytokines. In the closing chapter, Matthias G. von Herrath and coworker focus on factors that influence the induction of autoantigen-specific regulatory T cells. In animal models, vaccination with autoantigens was successful in the prevention of autoimmune diseases, such as type 1 autoimmune diabetes and experimental allergic encephalomyelitis. In contrast, it has been more difficult to see an immediate benefit in human clinical trials.

Thus, the study of autoimmunity has penetrated several fields of medicine, such as neurology, cardiology, nephrology, endocrinology, gastroenterology, dermatology, and rheumatology. Integrating autoimmunity concepts with a variety of disorders, this book aims to provide both researchers and clinicians with a basic understanding of discoveries tangential to their own areas. As these advances push back the frontiers of our understanding of autoimmunity, it is likely that further studies of these and related pathways will provide means to tease apart some of the molecular strands involved in the complex interactions that culminate in autoaggressive immune reactions. Future insight into elucidating autoimmunity will have an impact on the pursuit of new and better designs of improved diagnosis and treatments.

October 2004   Moncef Zouali

Contents

Part I. Innate Immunity in Autoimmune Diseases

1. Innate Immunity in Experimental Autoimmune Myocarditis
Ziya Kaya and Noel R. Rose

1. Introduction	1
2. Experimental Models of Myocarditis in Mice	2
2.1. Coxsackievirus B3 (CB3)-Induced Autoimmune Myocarditis	2
2.2. Cardiac Myosin–Induced Autoimmune Myocarditis	2
2.3. Peptide-Induced Myocarditis	3
3. Susceptibility to Myocarditis	3
4. Mouse Genotype	4
5. Innate Immune System and Myocarditis	5
5.1. Complement and Myocarditis	6
5.2. NK Cells and Myocarditis	8
5.3. Cytokines and Myocarditis	8
5.4. Chemokines and Myocarditis	10
6. Conclusions	11
Acknowledgments	11
References	12

2. Toll-like Receptor 9 and Autoimmunity
Paul N. Moynagh

1. Introduction	17
2. TLRs as Receptors for Pathogen-Associated Molecules	17
3. TLR9 and the Immunostimulatory Effects of Bacterial DNA	18
4. TLR9 and Intracellular signaling	18
5. CpG Sequences in Self-DNA Trigger Autoantibody Production	20
6. TLR9 as a Target for Regulating RF Production	21
7. Concluding Remarks	22
Acknowledgments	22
References	22

3. C-Reactive Protein as a Regulator of Autoimmune Disease
Terry W. Du Clos and Carolyn Mold

1. Introduction	27
2. Structural Features of CRP	27
3. CRP as an Acute-Phase Reactant	28
4. CRP Interaction with Nuclear Antigens	29
5. CRP, SAP, and Nuclear Antigen Clearance	30
6. CRP Genetics and Autoimmunity	31
7. CRP Levels in Human SLE	31
8. CRP in Animal Models of Autoimmunity	32
9. CRP in Immune Complex Nephritis	33
10. CRP in Inflammation	34
11. Identification of FcγR as CRP Receptors	34
12. Role of FcγR in CRP Effects on Inflammation	35
13. Essential Role of IL-10 in Anti-inflammatory Activities of CRP	37
14. Current Perspective on CRP in Autoimmune Disease	37
References	38

4. NKT Cells and Autoimmune Type 1 Diabetes
Shabbir Hussain, Dalam Ly, Melany Wagner, and Terry L. Delovitch

1. Introduction	43
2. Type 1 Diabetes	44
3. NKT Cells	44
4. Role of iNKT Cells in the Pathogenesis of Type 1 Diabetes	45
4.1. iNKT Cell Deficiency and T1D	45
4.2. iNKT Cell Activation Induces Protection against Type 1 Diabetes	47
5. Future Directions	49
Acknowledgments	50
References	50

Part II. Genetics of Autoimmune Diseases

5. The Genetics of Human Autoimmune Diseases
Marta E. Alarcón-Riquelme

1. Introduction	55
2. Analysis of the Genetics of Complex Diseases	56
2.1. Linkage Analysis	56
2.2. Association Analysis	57
2.3. Combining Linkage and Association	58
3. Genetic Analysis in Autoimmunity	58
3.1. Genome Scans and Linkage Analysis in Autoimmune Diseases	58
3.2. Autoimmune Diabetes (T1D)	58
3.3. Multiple Sclerosis (MS)	59

3.4. Rheumatoid Arthritis (RA)	60
3.5. Crohn's Disease (CD) and Ulcerative Colitis (UC)	61
3.6. Systemic Lupus Erythematosus (SLE)	62
3.7. Genes Shared between Autoimmune Diseases	63
References	64

6. Failure to Maintain T Cell DNA Methylation and Chromatin Structure Contributes to Human Lupus

Donna Ray and Bruce Richardson

1. Introduction	69
2. DNA Methylation, Chromatin Structure, and Gene Expression	70
3. DNA Methylation and Drug-Induced Lupus	73
3.1. DNA Methylation and Autoimmunity	73
3.2. DNA Methylation and Drug-Induced Lupus	75
3.3. T Cell Genes Affected by DNA Methylation Inhibitors	76
4. Aberrant T cell DNA Methylation, Gene Expression, and Cellular Function in Idiopathic Lupus	77
4.1. DNA Methylation	77
4.2. Gene Expression and Cellular Function	78
5. Conclusions	80
References	81

7. Complement Components C4A and C4B in Human Lupus

Yan Yang, Erwin K. Chung, Karl Lhotta, Yee Ling Wu, Gloria C. Higgins, Robert M. Rennebohm, Lee A. Hebert, Daniel J. Birmingham, Brad H. Rovin, and C. Yung Yu

1. Introduction	85
2. Diversities of Complement Components C4A and C4B in Human Populations	86
2.1. Dichotomy in Gene Sizes, Polygenes, and RCCX Module Variants	86
2.2. Diversity of Human C4A and C4B Proteins	87
2.3. Genetic Determinants of C4 Plasma/Serum Protein Levels	89
3. Complete Deficiencies of C4A and C4B in SLE and Immune-Complex Diseases	90
3.1. Molecular Basis of Complete C4 Deficiency	90
3.2. Impairment of Immune Response in C4-Deficient Patients	91
4. Deficiencies of C4A or C4B in Human SLE	92
4.1. Low Complement Activity and C4 Protein Concentrations in SLE	92
4.2. Homozygous or "partial" Deficiency of C4A in SLE across multiple ethnic groups	92
4.3. Deficiency of C4B in SLE Patients from Spanish, Mexican, and Australian Aborigines	93
4.4. Partial Deficiencies versus Polygenic Variations of C4A and C4B	94
5. Concluding Remarks and Perspectives	94
Acknowledgments	95
References	96

8. Non-MHC Genetic Polymorphisms with Functional Importance for Human Myasthenia Gravis
Ann Kari Lefvert

1. Introduction 101
2. Pro- and Anti-inflammatory Cytokines in MG 102
   - 2.1. Association of MG to the High Secretory Alleles of TNF-α 102
   - 2.2. Functional Implications of the Association with the TNF-α-308 A2 Allele 103
   - 2.3. Association of MG to the high secretory Allele of IL-1β 103
   - 2.4. Functional Implications of the Association with the IL-1β TaqI RFLP A2 Allele 104
   - 2.5. Lack of Associations of MG to Genetic Variants of IL-4 and IL-6 105
   - 2.6. IL-10 Is Associated to MG with High Autoantibody Levels 105
3. The β2-Adrenergic Receptor in MG 105
4. The T Cell Receptor Cofactor CTLA-4 in MG 106
   - 4.1. Association to MG with Thymoma and Increased Activation of the Immune System 106
   - 4.2. Functional Correlates to the Genetic Variants of *Ctla-4* 107
   - 4.3. The C/T SNP at –318 108
   - 4.4. The A/G SNP in CDS1 108
   - 4.5. Promoter SNPs –1772 (C/T) and –1661 (A/G) 108
   - 4.6. CTLA-4 and Thymomas 109
   - 4.7. *Ctla-4* (AT)n Is Associated to ADCC 109
5. Conclusions 109
   - Acknowledgments 110
   - References 110

Part III. Triggers of the Autoimmune Attack

9. Rheumatic Heart Disease: Molecular Basis of Autoimmune Reactions Leading to Valvular Lesions
Luiza Guilherme, Kellen Faé, and Jorge Kalil

1. Introduction 115
2. The Etiopathogenic Agent: *Streptococcus Pyogenes* 116
3. Genetic Susceptibility 116
4. Molecular Mimicry and RF/RHD 118
   - 4.1. The Humoral Immune Response 119
   - 4.2. The Cellular Immune Response 119
   - 4.3. Humoral and Cellular Immune Responses Interface in RF/RHD 121
   - 4.4. T Cell Receptor (TCR) Usage 121
5. Cytokines 122
6. Animal Models 122
7. Conclusions 123
   - References 123

10. Autoimmunity against Desmogleins in Pemphigus Vulgaris
Christian Veldman and Michael Hertl

1. Introduction	127
2. Clinical Phenotype of Pemphigus Vulgaris	128
3. Epidemiology of Pemphigus and Association with HLA Class II Alleles	128
4. Pathogenesis of Pemphigus	128
5. Autoantibody Reactivity against Desmogleins	129
6. Autoreactive T Lymphocytes in Pemphigus	132
7. Regulatory T Lymphocytes in Pemphigus	134
8. Passive Animal Models of Pemphigus Vulgaris	135
9. Active Animal Model of Pemphigus Vulgaris	135
10. Conclusions	135
References	136

11. The Molecular Basis of Celiac Disease
Liesbeth Spaenij-Dekking and Frits Koning

1. Introduction	141
2. T Cell Recognition of Gluten Peptides	142
3. The Specificity of tTG Is Linked to Gluten Toxicity	145
4. Additional T cell Stimulatory Peptides in Barley, Rye, and Oats	145
5. The HLA Gene Dose Effect Is Linked to the Level of Gluten Presentation	146
6. Generation of Safer Foods for Patients	147
7. A Hypothesis for Disease Development	147
8. Future Research and Perspectives	148
Acknowledgments	149
References	149

12. Intrathymic Expression of Neuromuscular Acetylcholine Receptors and the Immunopathogenesis of Myasthenia Gravis
Arnold I. Levinson, Yi Zheng, Glen Gaulton, and Decheng Song

1. Introduction	151
2. Evidence Supporting the Role of the Thymus in MG Pathogenesis	152
3. Expression of Neuromuscular AChRs by Thymic Cells	153
4. The Thymus and Central Immune Tolerance	157
5. The Thymus and T Cell Trafficking	157
6. Development of an Experimental Model to Examine Peripheral T Cell Entry and Activation in the Thymus	158
7. Conclusions	160
Acknowledgments	160
References	161

Part IV. Targets of the Autoimmune Attack

13. Autoantibodies and Nephritis: Different Roads May Lead to Rome
Paola Migliorini, Consuelo Anzilotti, Laura Caponi, and Federico Pratesi

1. Introduction	165
2. Acute Poststreptococcal Glomerulonephritis	167
3. Goodpasture's Syndrome	169
4. Lupus Nephritis	170
5. Other Nephritogenic Autoantibodies	174
6. Conclusions	175
References	176

14. Estrogen, Interferon-Gamma, and Lupus
S. Ansar Ahmed and Ebru Karpuzoglu-Sahin

1. Introduction	181
2. Estrogen and Lupus: Human and Animal Studies	182
3. Mechanisms of Estrogen Effects on the Immune System	185
3.1. Estrogen Exerts Its Biological Effects on Cells by Both Estrogen Receptor–Dependent and –Independent Mechanisms	185
3.2. Estrogen Alterations of B cells	189
3.3. Estrogen Effects on Cytokines	189
4. IFNγ in SLE and Other Autoimmune Diseases	190
5. Conclusions	191
Acknowledgments	192
References	193

15. Extent of Regulatory T Cell Influence on Major Histocompatibility Complex Class II Gene Control of Susceptibility in Murine Autoimmune Thyroiditis
Yi-chi M. Kong, Gerald P. Morris, and Chella S. David

1. Introduction	197
2. Major Histocompatibility Complex (MHC) Class II Gene Control of Susceptibility	198
3. Establishment of $CD4^+$ T Cells as Mediators of Induced Resistance	199
3.1. Protection from EAT Induction by Elevating Circulatory Thyroglobulin Level	199
3.2. $CD4^+$ Regulatory T Cells as Mediators of Induced Resistance	200
3.3. Effect of Cytokines on $CD4^+$ Regulatory T Cell Induction and Function	200
4. CD25 Expression on $CD4^+$ Regulatory T Cells in Induced Resistance	201
4.1. Abrogation of Established Tolerance by $CD4^+CD25^+$ T Cell Depletion	201

4.2. Interference with CD4+CD25+ Regulatory T Cell Function by Cross-Linking TNFR Family Molecules	203
5. Naturally Existing CD4+CD25+ T Cells as Peripheral Barrier to Autoimmune Thyroiditis	204
6. T Cell Regulation and MHC Restriction	204
7. Conclusion	206
Acknowledgment	206
References	206

16. The Role of Autoimmunity in Multiple Sclerosis
Monika Bradl and Hans Lassmann

1. Introduction	209
2. The "Autoimmune Hypothesis" of MS	210
3. The Multiple Facets of Multiple Sclerosis	210
3.1. The Clinical Spectrum of MS	210
3.2. The Pathological Spectrum of MS	211
3.3. Evidence for T Cell–Mediated Autoimmunity	212
3.4. Evidence for B Cell– or Antibody-Mediated Autoimmunity	215
3.5. Evidence for Autoimmunity from Immunotherapies of MS	216
4. The Triggers for Autoimmune Reactions in MS Patients	216
4.1. Autoimmune Reactions Caused by a Defect in Immune Regulation	217
4.2. Autoimmune Reactions Caused by Infections	218
5. Protective Autoimmunity	220
6. What Remains of the "Autoimmune Hypothesis" of MS?	220
References	221

Part V. Immune Receptor Signaling Pathways

17. Crippled B Lymphocyte Signaling Checkpoints in Systemic Autoimmunity
Moncef Zouali

1. Introduction	227
2. B Lymphocytes Participate in Both Innate and Adaptive Immunity	228
3. The Critical Role of B Cells in Autoimmunity	229
4. B Cell Receptor–Mediated Signaling Checkpoints	230
5. Critical Regulators of B Cell Receptor Signaling	231
6. Negative Regulators of B Cell Receptor–Mediated Signal Transduction	234
7. Disrupted B Cell Signaling Pathways in Human Autoimmunity	237
8. Conclusions	239
References	239

18. Disrupted T Cell Receptor Signaling Pathways in Systemic Autoimmunity

Ana M. Blasini and Martín A. Rodríguez

1. Introduction 245
2. Signaling Pathways in T Cells 246
3. T Cell Signaling Abnormalities in Systemic Autoimmune Disease 250
   - 3.1. Signaling Abnormalities in Antigen-Presenting Cells and Autoimmune Disease 250
   - 3.2. Signaling Abnormalities in T Cells and Autoimmune Disease 252
4. Conclusions 255
   - Acknowledgments 257
   - References 257

19. Immune Cell Signaling and Gene Transcription in Human Systemic Lupus Erythematosus

Christina G. Katsiari and George C. Tsokos

1. Introduction 263
2. Altered Pattern of Tyrosine Phosphorylation and Calcium Responses 264
3. TCR ζ Chain Deficiency 265
   - 3.1. Impaired TCR ζ Chain Gene Transcription 266
   - 3.2. Impaired Translation and Posttranscription Events 266
   - 3.3. Impaired Posttranslational Functions 267
   - 3.4. Oxidative Stress 267
   - 3.5. Role of IFNγ 268
4. Mechanisms of Increased TCR/CD3-Mediated $[Ca^{2+}]_i$ Response in SLE T Cells 268
   - 4.1. FcRγ Chain Substitutes for Defective ζ Chain 268
   - 4.2. Altered Composition and Dynamics of Lipid Rafts 269
5. Protein Kinase A (PKA) Function 271
6. Regulation of Transcription Determines Interleukin 2 Deficiency in SLE T Cells 272
7. Conclusions 274
   - References 275

20. Accumulation of Self-Antigens in Systemic Lupus Erythematosus

Koji Yasutomo

1. Introduction 279
2. T Cell in Human Lupus 280
3. Antigen Clearance and Autoimmunity 281
   - 3.1. DNASE1-Deficient Patients: Gene Mutation and Clinical Features 281
   - 3.2. DNASE1-Deficient Patients: Laboratory Findings 282
   - 3.3. DNASE1-Deficient Patients: Effect on Autoreactivity 282
4. Defective Clearance of Self-Antigens in SLE 282
   - 4.1. Evidence from Knockout Mice 282

4.2. Mechanisms of Accumulation of Self-Antigens in SLE	284
4.3. Clearance of Self-Antigens as a Therapeutic Strategy	286
References	286

Part VI. Immunointervention Strategies

21. B Lymphocyte Depletion Therapy in Autoimmune Disorders: Chasing Trojan Horses
Jonathan C. W. Edwards, Geraldine Cambridge, and Maria J. Leandro

1. Introduction	291
2. Human Autoimmunity: An Abnormality of B Cell Function	291
2.1. A Brief History of Investigation of B and T Cell Autoreactivity in Human Autoantibody-Associated Diseases	292
2.2. Generation of Autoreactive T Cells	293
2.3. Autoantibodies as Effector Molecules	294
2.4. Autoantibodies as Trojan Horse Immunomodulators	294
3. Clinical Significance of the Trojan Horse Concept	297
3.1. Effector Mechanisms in RA	298
3.2. Logistics of B Cell Depletion	299
3.3. Anti-CD20 Therapeutic Agents	299
3.4. Rituximab	300
3.5. Efficacy	301
3.6. Failure of Seronegative Disease to Respond	302
3.7. Adverse Events Associated with BLyD	302
3.8. Repeated Cycles of B Cell Depletion	303
4. Do Data from BLyD Support the Trojan Horse Concept?	303
4.1. Autoantibody Levels Fall Selectively Compared with Antimicrobial and Total Immunoglobulin Levels	303
4.2. Total Immunoglobulin Levels May Fall after Repeat Cycles	304
4.3. Clinical Response Follows Serological Response, not B Cell Numbers	305
4.4. The Kinetics of Relapse Follow Autoantibody Rises Rather than B Cell Return	305
4.5. Why Are There Two Patterns of Relapse?	306
5. Conclusions	306
References	309

22. B lymphocyte Stimulator (BLyS) and Autoimmune Rheumatic Diseases
William Stohl

1. Introduction	313
2. BLyS and Its Receptors	313
2.1. General Biology	313

 2.2. *In Vivo* Deficiency of BLyS or Its Receptors 315
 2.3. Supranormal Levels of BLyS *In Vivo* 316
 2.4. APRIL and Its Relevance to BLyS 318
3. BLyS Antagonism as a Therapeutic Modality 319
 3.1. Mouse Models 319
 3.2. The Human Experience 320
 3.3. Which Patients Are Candidates for BLyS Antagonist Therapy? 321
 3.4. Concluding Comments 322
 References 322

23. Control and Induction of Autoimmunity by Cytokine and Anti-cytokine Treatments

Pierre Miossec

1. Introduction 329
2. TNFα and Its Receptors 330
3. Mode of Action of the Specific TNFα Inhibitors 331
4. The Local and Systematic Effects of TNFα Inhibition 332
5. Understanding the Side Effects of TNFα Inhibitors 335
6. Other Cytokine Inhibitors 336
7. Other Cytokines as Treatment Targets 337
8. Targeting One or More than One Cytokine 337
9. Understanding the Heterogeneity of the Response to TNFα Inhibitors 338
10. Autoimmune Manifestations with Cytokine Administration 338
11. Conclusions 341
 References 341

24. Hematopoietic Stem Cell Transplantation for the Treatment of Severe Autoimmune Diseases

Alan Tyndall and Paul Hasler

1. Introduction 347
2. Autoimmune Disease Mechanisms 348
3. Coincidental AD in Patients Receiving HSCT for Another Indication 349
4. Animal Models 350
5. Treatment of Human Autoimmune Disease with Hematopoietic Stem
 Cell Transplantation 351
6. Systemic Sclerosis (SSc) 353
7. Rheumatoid Arthritis 353
8. Juvenile Idiopathic Arthritis 354
9. Systemic Lupus Erythematosus 354
10. Prospective Randomized Controlled Clinical Trials 355
11. Open Issues 356
 11.1. Allogeneic HSCT 356
 11.2. Immune Reconstitution 357
 11.3. Ablative Therapy without HSCT 358
12. Conclusions 359
 References 359

25. Molecular Mimicry in Autoimmune Uveitis: From Pathogenesis to Therapy

Gerhild Wildner, Maria Diedrichs-Moehring, and Stephan R. Thurau

1. Introduction	365
2. Retinal Autoantigens and Mimicry Peptides	366
3. HLA Peptide B27PD in EAU	367
4. Pathogenic and Tolerogenic Epitopes of the Retinal Peptide PDSAg and Its Mimotope B27PD	368
5. Antigenic Mimicry of Retinal Autoantigen and Environmental Antigens	370
6. Treatment of Uveitis Patients with Oral Peptide B27PD	372
References	374

26. Molecular Pathogenesis of the Antiphospholipid Syndrome: Toward Novel Therapeutic Targets

Silvia S. Pierangeli, Mariano Vega-Ostertag, Azzudin E. Gharavi, and E. Nigel Harris

1. Introduction	377
2. Antiphospholipid Antibodies and Platelets	378
2.1. Effects of aPL on Platelets *In Vitro* and *In Vivo*	378
2.2. Hydroxychloroquine in aPL-mediated thrombosis	379
2.3. Intracellular Events in aPL-Mediated Platelet Activation	380
3. Antiphospholipid Antibodies and Endothelial Cells	381
3.1. Effects of aPL on Endothelial Cells	381
3.2. The Statins and Antiphospholipid Antibodies	385
3.3. Activation of the Complement Cascade and Antiphospholipid Antibodies	385
4. Conclusions	386
Acknowledgement	387
References	388

27. A Novel Approach to the Prevention of Atherosclerosis

Sun-Ah Kang and Marc Monestier

1. Introduction	393
2. Atherosclerosis	393
2.1. Lesion Initiation	394
2.2. Fatty Streak Formation	395
2.3. Fibrous Plaques	395
2.4. Plaque Rupture and Thrombosis	396
3. Immune Cells in Atherosclerosis	396
4. Cellular Immunity in Atherosclerosis	397
5. Humoral Immunity in Atherosclerosis	400
6. Vaccination or Immunoglobulin Administration in Atherosclerosis	401
7. Conclusions	402
References	403

28. Antigen-Specific Regulation of Autoimmunity
Amy E. Juedes and Matthias G. von Herrath

1. Introduction	407
2. Antigen-Specific Therapy	408
3. Antigen-Induced Regulatory T cells	408
4. Factors Involved in Treg Induction	410
4.1. Mechanisms of Protection	411
4.2. Application to Human Disease	412
5. Conclusions	414
References	414
Index	419

Contributors

Marta E. Alarcón-Riquelme
Department of Genetics and Pathology
Uppsala University
Dag Hammarsjkölds väg 20, 751 85 Uppsala
Sweden

S. Ansar Ahmed
Center for Molecular Medicine and Infectious
  Diseases
1410, Prices Fork Road
Virginia–Maryland Regional College
  of Veterinary Medicine
Virginia Tech
Blacksburg, Virginia 24061
United States

Consuelo Anzilotti
Clinical Immunology Unit
Department of Internal Medicine
Via Roma 67, 56126 Pisa
Italy

Daniel J. Birmingham
Department of Internal Medicine
The Ohio State University
700 Children's Drive
Columbus, Ohio 43205
United States

Ana M. Blasini
Centro Nacional de Enfermedades Reumáticas
Servicio de Reumatología
Hospital Universitario de Caracas, Caracas
Venezuela

Monika Bradl
Division of Neuroimmunology, Brain Research
  Institute
Medical University of Vienna
Spitalgasse 4, A-1090 Wien
Austria

Geraldine Cambridge
University College London Centre for
  Rheumatology
Arthur Stanley House, 40-50 Tottenham Street
London W1T 4NJ
United Kingdom

Laura Caponi
Department of Experimental Pathology
University of Pisa, Pisa
Italy

Erwin K. Chung
Center for Molecular and Human
  Genetics
Columbus Children's Research Institute
700 Children's Drive
Columbus, Ohio 43205
United States

Chella S. David
Department of Immunology, Mayo Clinic
Rochester, Minnesota 5590
United States

Terry L. Delovitch
Autoimmunity/Diabetes Group
Robarts Research Institute
London, Ontario N6G 2V4
Canada

Maria Diedrichs-Moehring
Section of Immunobiology, Department
  of Ophthalmology
Ludwig-Maximilians-University
Mathildenstrasse 8, 80336 Munich
Germany

Terry W. Du Clos
The Department of Veterans Affairs Medical
  Center and
The University of New Mexico School
  of Medicine
Department of Internal Medicine
Albuquerque, New Mexico 87108
United States

Jonathan C. W. Edwards
University College London Centre
  for Rheumatology
Arthur Stanley House
40-50 Tottenham Street
London W1T 4NJ
United Kingdom

Kellen Faé
Institute for Immunology Investigation
Millenium Institute, São Paulo
Brazil

Glen Gaulton
Department of Laboratory Medicine and
    Pathology
University of Pennsylvania School of Medicine
421 Curie Boulevard
Philadelphia, Pennsylvania 19104
United States

Azzudin E. Gharavi
Department of Medicine
Morehouse School of Medicine
720 Westview Drive SW
Atlanta, Georgia 30310-1495
United States

Luiza Guilherme
Heart Institute (InCor)
School of Medicine
University of São Paulo, São Paulo
Brazil

E. Nigel Harris
Office of the Dean
Morehouse School of Medicine
Atlanta, Georgia
United States

Paul Hasler
Department of Rheumatology
Kantonsspital, Aarau
Switzerland

Lee A. Hebert
Department of Internal Medicine
The Ohio State University
700 Children's Drive
Columbus, Ohio 43205
United States

Michael Hertl
Department of Dermatology
University of Erlangen
Hartmannstrasse 14, D-91054 Erlangen
Germany

Gloria C. Higgins
Department of Pediatrics
The Ohio State University
700 Children's Drive
Columbus, Ohio 43205
United States

Shabbir Hussain
Autoimmunity/Diabetes Group
Robarts Research Institute
London, Ontario N6G 2V4
Canada

Amy E. Juedes
Division of Developmental Immunology
La Jolla Institute for Allergy and Immunology
San Diego, California 92121
United States

Jorge Kalil
Clinical Immunology and Allergy
Department of Clinical Medicine
University of São Paulo
School of Medicine, São Paulo
Brazil

Sun-Ah Kang
Department of Microbiology and Immunology
Temple University School of Medicine
3400 North Broad Street
Philadelphia, Pennsylvania 19140
United States

Ebru Karpuzoglu-Sahin
Center For Molecular Medicine and Infectious
    Diseases
1410, Prices Fork Road
Virginia–Maryland Regional College of
    Veterinary Medicine
Virginia Tech
Blacksburg, Virginia 24061
United States

Christina G. Katsiari
Department of Cellular Injury
Walter Reed Army Institute of Research
Silver Spring, Maryland 20190
United States

Ziya Kaya
Department of Pathology and Feinstone
    Department of Molecular Microbiology and
    Immunology
The Johns Hopkins Medical Institutions
Baltimore, Maryland 21205
United States

Yi-chi M. Kong
Department of Immunology and Microbiology
Wayne State University School of Medicine
Detroit, Michigan 48201
United States

Frits Koning
Department of Immunohematology and Blood
    Transfusion
E3-Q, Leiden University Medical Center
PO Box 9600, 2300 RC Leiden
The Netherlands

Hans Lassmann
Division of Neuroimmunology
Brain Research Institute
Medical University of Vienna
Spitalgasse 4, A-1090 Wien
Austria

Contributors

Maria J. Leandro
University College London Centre for
  Rheumatology
Arthur Stanley House
40-50 Tottenham Street, London W1T 4NJ
United Kingdom

Ann Kari Lefvert
Immunological Research Laboratory
Center for Molecular Medicine and Department
  of Medicine
Karolinska Institutet, Karolinska Hospital
S-171 76 Stockholm
Sweden

Arnold I. Levinson
Section of Allergy and Immunology
University of Pennsylvania School
  of Medicine
Suite 1014, 421 Curie Boulevard
Philadelphia, Pennsylvania 19104
United States

Karl Lhotta
Division of Clinical Nephrology
Innsbruck University Hospital, Innsbruck
Austria

Dalam Ly
Department of Microbiology and
  Immunology
University of Western Ontario
London, Ontario N6A 5C1
Canada

Paola Migliorini
Clinical Immunology Unit
Department of Internal Medicine
Via Roma 67, 56126 Pisa
Italy

Pierre Miossec
Department of Immunology and Rheumatology
Hôpital Edouard Herriot
69437 Lyon Cedex 03
France

Carolyn Mold
Molecular Genetics and Microbiology
The University of New Mexico School of
  Medicine
Albuquerque, New Mexico 87108
United States

Marc Monestier
Department of Microbiology and
  Immunology
Temple University School of Medicine
3400 North Broad Street
Philadelphia, Pennsylvania 19140
United States

Gerald P. Morris
Department of Immunology and
  Microbiology
Wayne State University School
  of Medicine
Detroit, Michigan 48201
United States

Paul N. Moynagh
Department of Pharmacology
Conway Institute of Biomolecular and
  Biomedical Research
University College Dublin, Belfield,
  Dublin 4
Ireland

Silvia S. Pierangeli
Departments of Microbiology, Biochemistry
  and Immunology
Morehouse School of Medicine
720 Westview Drive SW
Atlanta, Georgia 30310-1495
United States

Federico Pratesi
Clinical Immunology Unit
Department of Internal Medicine
Via Roma 67, 56126, Pisa
Italy

Donna Ray
5310 Cancer Center and Geriatrics Center
  Building
1500 E. Medical Center Drive
Ann Arbor, Michigan 48109-0940
United States

Robert M. Rennebohm
Department of Pediatrics
The Ohio State University
700 Children's Drive
Columbus, Ohio 43205
United States

Bruce Richardson
Department of Medicine
University of Michigan
5310 Cancer Center and Geriatrics Center
  Building
1500 E. Medical Center Drive
Ann Arbor, Michigan 48109-0940
United States

Martín A. Rodríguez
Centro Nacional de Enfermedades
  Reumáticas
Servicio de Reumatología
Hospital Universitario de Caracas, Caracas
Venezuela

Contributors

Noel R. Rose
Department of Pathology and Feinstone
  Department of Molecular Microbiology
  and Immunology
The Johns Hopkins Medical Institutions
Baltimore, Maryland 21205
United States

Brad H. Rovin
Department of Internal Medicine
The Ohio State University
700 Children's Drive
Columbus, Ohio 43205
United States

Decheng Song
Section of Allergy and Immunology
University of Pennsylvania School
  of Medicine
Suite 1014, 421 Curie Boulevard
Philadelphia, Pennsylvania 19104
United States

Liesbeth Spaenij-Dekking
Department of Immunohematology and Blood
  Transfusion
E3-Q, Leiden University Medical Center
PO Box 9600, 2300 RC Leiden
The Netherlands

William Stohl
Division of Rheumatology
University of Southern California Keck
  School of Medicine
2011 Zonal Avenue HMR 711
Los Angeles, California 90033
United States

Stephan R. Thurau
Section of Immunobiology
Department of Ophthalmology
Ludwig-Maximilians-University
  Mathildenstrasse 8
80336 Munich
Germany

George C. Tsokos
Department of Cellular Injury
Walter Reed Army Institute of Research
Silver Spring, Maryland 20190
United States

Alan Tyndall
Department of Rheumatology
Felix-Platter Spital,
CH-4012 Basel
Switzerland

Mariano Vega-Ostertag
Departments of Microbiology, Biochemistry,
  and Immunology
Morehouse School of Medicine
720 Westview Drive SW
Atlanta, Georgia 30310-1495
United States

Christian Veldman
Department of Dermatology
University of Erlangen, Hartmannstrasse 14
D-91054 Erlangen
Germany

Matthias G. von Herrath
Division of Developmental Immunology
La Jolla Institute for Allergy and Immunology
San Diego, California 92121
United States

Melany Wagner
Department of Microbiology and Immunology
University of Western Ontario
London, Ontario N6A 5C1
Canada

Gerhild Wildner
Section of Immunobiology
Department of Ophthalmology
Ludwig-Maximilians-University
Mathildenstrasse 8, 80336 Munich
Germany

Yee Ling Wu
Center for Molecular and Human Genetics
Columbus Children's Research Institute
700 Children's Drive
Columbus, Ohio 43205
United States

Yan Yang
Center for Molecular and Human Genetics
Columbus Children's Research Institute
700 Children's Drive
Columbus, Ohio 43205
United States

Koji Yasutomo
Department of Immunology and Parasitology
Institute of Health Biosciences
The University of Tokushima Graduate School
3-18-15 Kuramoto
Tokushima 770-8503
Japan

C. Yung Yu
Center for Molecular and Human Genetics
Columbus Children's Research Institute
700 Children's Drive
Columbus, Ohio 43205
United States

Contributors

Yi Zheng
Section of Allergy and Immunology
University of Pennsylvania School of Medicine
Suite 1014, 421 Curie Boulevard
Philadelphia, Pennsylvania 19104
United States

Moncef Zouali
Unité d'Immunopathologie Humaine
INSERM U 430
15 rue de l'Ecole de Médecine
F-75006, Paris
France

Molecular Autoimmunity

1

Innate Immunity in Experimental Autoimmune Myocarditis

Ziya Kaya and Noel R. Rose

1. Introduction

A century has passed since the epic publication by Donath and Landsteiner (1904) on the pathogenesis of paroxysmal cold hemoglobunaria (PCH). The work provided the first hint that autoimmunity could be the cause of human disease. The concept remained fallow for half a century until improved immunologic methods and a broader view of the basis of the immune response validated the idea. Landsteiner associated PCH with his concurrent studies of syphilis, leading to the suggestion that infection may serve as the initiating factor for an autoimmune reaction. The idea became embedded in immunologic thought. Yet there are few firmly established examples of a human autoimmune disease caused by infection and little information about mechanisms by which infection might instigate such a pathologic autoimmune response. With the goal of elucidating the likely mechanisms, our group undertook a detailed study of one clear experimental model in mice of an autoimmune disease triggered by a viral infection, myocarditis (Rose *et al.*, 1988a).

Myocarditis accounts for approximately 25% of all heart failure in North America and is especially prevalent among young adults. Although most viral myocarditis patients recover, a few progress to chronic myocarditis and dilated cardiomyopathy (DCM), an often-fatal condition and a frequent reason for cardiac transplantation. The most common cause of myocarditis in the USA is infection with adenovirus or coxsackievirus. Progressive forms of myocarditis are characterized by the presence of cardiac myosin–specific autoantibodies (Caforio *et al.*, 2001). In this chapter, we review recent studies on the role of the innate immune system in induction, progression, and protection of the disease.

Ziya Kaya and Noel R. Rose • Department of Pathology and Feinstone Department of Molecular Microbiology and Immunology, The Johns Hopkins Medical Institutions, Baltimore, Maryland 21205.

Molecular Autoimmunity: In commemoration of the 100th anniversary of the first description of human autoimmune disease, edited by Moncef Zouali. Springer Science+Business Media, Inc., New York, 2005.

2. Experimental Models of Myocarditis in Mice

There is strong evidence that cardiac myosin is a dominant autoantigen in virus–induced myocarditis in mice (Neu et al., 1987a). The disease can be reproduced by immunization of susceptible strains of mice with cardiac myosin (Neu et al., 1987b). Myosin-induced myocarditis can be adoptively transferred by CD4+ T lymphocytes (Smith and Allen, 1991). In addition to T cells, passive administration of antimyosin monoclonal antibody was found to induce myocarditis in DBA/2 but not in BALB/c mice because of the presence of myosin or a myosin-like protein in the extracellular matrix of DBA/2 mice (Liao et al., 1995). Therefore, both antibody and T cells may contribute to the pathogenesis of inflammatory myocardial lesions. Gauntt et al. (1995) and Cunningham (2004) investigated the relationship between coxsackievirus and myosin and suggested that molecular mimicry between myosin and coxsackieviruses may play a role in myocarditis. Anti-coxsackievirus-neutralizing antibody produced myocardial inflammation in mice (Gauntt et al., 1995). On the other hand, Horwitz et al. (2000) presented evidence that virus-mediated damage to the heart is necessary for the induction of the autoimmune response, a finding that challenges the idea of molecular mimicry.

Studies to explore the inductive and the effector mechanisms involved in the development of experimental autoimmune myocarditis (EAM) implicate both innate and adaptive immune responses. Thus, important roles have been shown for autoreactive T cells (Smith and Allen, 1991), cardiac-specific autoantibodies (Neumann et al., 1991; Liao et al., 1995), various cytokines and chemokines (Afanasyeva and Rose, 2002a; Eriksson et al., 2003a, 2003b; Fairweather et al., 2003, 2004), natural killer (NK) cells (Fairweather et al., 2001, 2003), and the complement system (Kaya et al., 2001; Afanasyeva and Rose, 2002b) in the development of myocarditis.

2.1. Coxsackievirus B3 (CB3)–Induced Autoimmune Myocarditis

Our murine model of autoimmune myocarditis is based on genetic differences among inbred mouse strains in the immune response to CB3. In certain mouse strains, CB3-mediated myocarditis resolves into an early phase characterized by myocyte damage due to viral cytotoxicity and a late phase that is associated with the production of heart muscle–specific autoantibodies (Rose et al., 1988a). The later phase of CB3-induced heart disease can be mimicked by immunization of mice with purified murine cardiac myosin in the absence of viral infection, and experimental cardiac myosin–induced myocarditis has immunologic and histopathologic features that resemble postviral heart disease in mice and myocarditis in humans (Neu et al., 1987b). Thus, the myocarditis model offers a unique opportunity to study the factors contributing to the transition from a viral infection to an autoimmune disease.

2.2. Cardiac Myosin–Induced Autoimmune Myocarditis

Immunization of susceptible mice with cardiac myosin emulsified in complete Freund adjuvant induces myocarditis in mice with a peak of inflammation in the heart around day 21 (Afanasyeva *et al.*, 2001b). This inflammation is similar to that seen in the CB3-induced autoimmune myocarditis during the chronic phase. The immunization with cardiac myosin is linked with production of IgG1 autoantibodies to cardiac myosin and autoreactive CD4 T cells (Afanasyeva *et al.*, 2004).

2.3. Peptide-Induced Myocarditis

Unique epitopes within cardiac myosin have been described to produce myocarditis. Myocarditis can be induced in BALB/c mice by amino acid residues 736–1032 in cardiac myosin (Liao *et al.*, 1993), by amino acid residues 334–352, located in the S1 region of mouse cardiac myosin (Donermeyer *et al.*, 1995), or by acetylated amino acid residues 614–643 of rat cardiac myosin (Pummerer *et al.*, 1996). Myocarditis can also be induced in Lewis rats by amino acid residues 1070–1165 of porcine cardiac myosin (Inomata *et al.*, 1995), by amino acid residues 1107–1186 of the rat myosin heavy chain (Kohno *et al.*, 2002), and by acetylated amino acid residues 1539–1555 of rat cardiac myosin *alpha*-chain (Wegmann *et al.*, 1994).

3. Susceptibility to Myocarditis

Susceptibility to induction of EAM is dependent on the strain of mice (Table 1.1) (Fairweather *et al.*, 2003). The MHC class II haplotype (e.g., H-2a in highly susceptible A/J mice or H-2b in moderately susceptible A.BY mice) is an important genetic factor for disease susceptibility, but its effects are overshadowed by non-MHC traits (Neu *et al.*, 1987b; Rose *et al.*, 1988b). Thus, C57BL/6 (H-2s) mice are resistant to myosin-induced myocarditis, whereas A.SW mice are susceptible, even though they share the same H-2s haplotype. The same strains of mice susceptible to CB3-induced autoimmune myocarditis develop myocarditis following immunization with cardiac myosin. Susceptibility may be related to many genetic factors including target organ sensitivity or influences upon the immune response itself (Guler *et al.*, 2005).

Table 1.1. Susceptibility to Experimental Autoimmune Myocarditis

Mouse strain	Autoantibody titer	Myocarditis	
		Prevalence	Severity
A/J	++++	++++	++++
A.BY/SnJ	+++	+	++
A.CA/SnJ	++++	+++	+++
A.SW/SnJ	++++	++++	++++
B10.A/SgSnJ	+++	+	+
BALB/c	+++	+++	+++
C57BL/10J	++	−	−
C57BL/6J	++	−	−
DBA/2	+++	+++	+++

4. Mouse Genotype

To understand the multiple factors involved in the induction and progression of myocarditis, studies are under way to determine some of the non-H-2 genes that are involved in causing the inflammatory and autoimmune disease state in myocardium (Table 1.2). Autoimmune heart disease does not occur in mice lacking the Src family tyrosine kinase p56^{lck} or the tyrosine phosphatase CD45, which regulates the enzymatic activity of p56^{lck} (Liu et al., 2000). Mice lacking CD8 develop significantly more severe disease than their heterozygous littermates, suggesting that CD8 lymphocytes not only act as cytotoxic effector cells in autoimmunity but may also regulate disease severity (Penninger et al., 1993). T cell costimulatory molecule CD28-deficient animals develop significantly less severe disease and at lower prevalence than control littermates and have a defect in (Th2-related) IgG1 autoantibody production (Bachmaier et al.,

Table 1.2. Mouse Genotype—Impact on Prevalence and Severity of Myocarditis

	Myocarditis	
Genotype	Prevalence	Severity
p56$^{lck-/-}$	–	–
p56^{lck+}	++++	+++
CD45$^{-/-}$	–	–
CD45^{+}	++++	++++
CD4$^{-/-}$	–	–
CD8$^{-/-}$	++++	++++
CD28$^{-/-}$	+	++
CD28^{+}	++++	++++
TNF-Rp55$^{-/-}$	–	–
TNF-Rp55^{+}	++++	++++
hCD4TG	+	++
hCD4/DQ6TG	++++	++++
CR1/CR2$^{-/-}$	++	++
CR1/CR2^{+}	++++	++++
IFN-gamma$^{-/-}$	+++++	+++++
IFN-gamma^{+}	++++	++++
IL1R1$^{-/-}$	+	+
IL1R1^{+}	++++	++++
IL6$^{-/-}$	+	+
IL6^{+}	++++	++++
IL12R$^{-/-}$	+	+
IL12R^{+}	++++	++++
TLR4$^{-/-}$	+	+
TLR4^{+}	++++	++++
PD1$^{-/-}$	+++++	+++
PD1^{+}	+++	+++
STAT-4$^{-/-}$	+	+
STAT-4^{+}	++++	+++

1996). On the other hand, blocking the costimulatory molecule CTLA-4 enhanced myocarditis in susceptible A/J mice immunized with cardiac myosin and allowed induction of disease in resistant C57Bl6 mice (D. Cihakova et al., unpublished data). Recently, we showed that complement receptor CR1/CR2 knockout mice developed significantly less severe disease, demonstrating a critical effect of the complement system as one of the key components of the innate system in induction of the disease. Also, mice lacking CCR2 and CCR5 (chemokine receptors for MCP-1 and MIP-1*alpha*) develop less severe disease (Z. Kaya et al., unpublished data). Both chemokines play critical roles in recruitment of mononuclear cells to the inflammation site. Cytokines exert important effects in induction, progression, and manifestation of the disease. While IL-12R*beta*1 deficiency acting by reducing production of IL-1*beta* and IL-18 (Fairweather et al., 2003), IL-1-1R1 deficiency (Kaya et al., 2001), IL1-1R1 deficiency, and IL-6-deficiency (Eriksson et al., 2003a) resulted in significantly less severe inflammation in the heart, IFN-*gamma* deficiency exacerbates the severity of myocarditis. Other genes that have been reported to be involved in inflammatory myocarditis include Fas ligand that, when expressed in the heart, led to a mild inflammatory infiltrate, (Nelson et al., 2000), toll-like receptor 4 (Fairweather et al., 2001), and STAT-4 (Afanasyeva et al., 2001a). Mice lacking the negative immunoregulatory receptor PD-1 develop DCM with congestive heart failure and sudden death (Nishimura et al., 2001). Hearts from PD-1-deficient mice exhibited diffuse deposition of IgG on the surface of myocytes and antibody in the disease recognized a 33-kDa protein specific to heart tissues. It has been reported that this protein is troponin I and that DCM can be induced by treating wild mice with antibodies against troponin I (Okazaki et al., 2003).

Some structural genes are also involved in hereditary forms of cardiomyopathy. Mutations in genes encoding sarcomeric proteins including cardiac myosin heavy and light chains, cardiac tropomyosin, cardiac troponins, and myosin-binding protein C have been associated with familial hypertrophic cardiomyopathy (Leiden, 1997). Mutations in dystrophin, dystrophin-associated glycoproteins, and actin lead to DCM. Overexpression of calcineurin can, in transgenic mice, lead to cardiac hypertrophy and DCM. Also, abnormalities of the cytoskeletal proteins and the role of the coxsackieviral 2A protease in cleavage of dystrophin lead to DCM (Badorff et al., 2000, 1999).

Thus, the investigation of genes involved in susceptibility to myocarditis indicates that they are diverse and are related to several different mechanisms of pathogenesis.

5. Innate Immune System and Myocarditis

The innate immune response to pathogens has become a topic of renewed interest in recent years. In the past, innate immunity has been considered only to provide rapid, but incomplete, antimicrobial host defense until the slower, more definitive acquired immune response develops. However, recent research indicates that innate immunity critically impacts the subsequent development of the adaptive immune response. We focus our discussion on four major agents of the innate immune response: the complement system, NK cells, cytokines and chemokines. Other important elements of the innate immune response, such as dendritic cells, mast cells and toll-like receptors have been discussed elsewhere (Eriksson, 2004; Fairweather *et al.*, 2005; Afanasyeva, Georgakopoulos and Rose, 2004).

5.1. Complement and Myocarditis

The initiation of an adaptive immune response is modulated by humoral and cellular components of the innate immune system. The importance of complement, as one of the key components of the innate immune system, in mounting an immune response has been shown in a number of different studies (Carroll, 1998; Fearon and Carroll, 2000). An antigen can activate complement through either the alternative pathway or the classical pathway by interacting with natural or elicited antibodies. By either pathway, activation of complement leads to the generation of C3 split products, which in turn bind to complement receptors. In mice, complement receptor type 1 (CR1, also known as CD35) and type 2 (CR2, also known as CD21) are predominantly expressed on B cells, a subset of activated T cells, follicular dendritic cells, and activated granulocytes (Kinoshita *et al.*, 1988; Kaya *et al.*, 2001).

Interactions between complement components and infectious agents are important for further initiation of cellular responses and clearance of virus. On the other hand, complement components may also facilitate viral entry into host cells. CD55 has been identified as a receptor for entry of cardiovirulent CB3 into host cells (Kuhn, 1997; Martino *et al.*, 1998). The tyrosine kinase p56^{lck}, which is activated by cross-linking of the short consensus region 3 of CD55, was found to be required for efficient viral replication and the development of myocarditis (Liu *et al.*, 2000). Effective viral replication is clearly necessary for the induction of myocarditis, but the precise role for complement in the pathogenesis of myocarditis remains unclear.

We have previously demonstrated the importance of the complement system in the initiation of EAM (Kaya *et al.*, 2001). Depletion of C3 during disease induction prevented the development of autoimmune myocarditis. This effect could be mediated through any of the known complement receptors, including CR1, CR2, CR3, CR4, or the complement factor C3a receptor. However, blocking the complement receptors CR1 and CR2 and immunizing CR1/CR2 KO (knockout) mice also reduced myocarditis, demonstrating that the effect of C3 activation is mediated through these receptors. Membrane attack complex (MAC) formation (C5b-9) was not involved in the pathogenesis of autoimmune myocarditis in this study because A/J mice, which are deficient in C5, were used for these experiments. In contrast, C3 depletion during the later effector phase did not affect the disease severity or prevalence. Early C3 depletion, and CR1 and CR2 blockade, resulted in significantly decreased TNF-*alpha* and IL-1 production from splenocytes and decreased cardiac myosin–specific autoantibody production. CR1/CR2 KO mice also had decreased TNF-*alpha* production.

Additionally, IFN-*gamma* production was reduced and IL-10, an anti-inflammatory cytokine, was increased in CR1/CR2 KO mice. Overall, these findings suggest that C3, acting through the complement receptors CR1 and CR2, plays a regulatory role in the induction of autoimmune myocarditis and that its action results in increased activation of both B and T cells plus production of key proinflammatory cytokines, such as TNF-*alpha* and IL-1.

Impaired T cell activation as a result of reduced antigen presentation by macrophages and B cells may be the main cause for the reduction in myocarditis, since it is a T cell–dependent disease (Smith and Allen, 1991; Afanasyeva *et al.*, 2004). In addition, the complement system can act as an important non-antibody-mediated pathway for antigen uptake by B cells, efficiently directing antigens through the complement receptors CR1 and CR2 (Kerekes *et al.*, 1998). C3 deposition on antigen-presenting cells (APCs) lowers the activation threshold of antigen-specific T cells. Our results also suggest that CR1/CR2 receptors are necessary for optimal activation of B and T cells, since CR1/CR2 KO mice showed both lower expression of activation markers on both B and T cells and decreased amounts of T cell cytokines, such as IL-2, IL-4, and IFN-*gamma*, after antigen stimulation. In contrast, IL-10, an anti-inflammatory cytokine important for limiting inflammation in EAM (Kaya *et al.*, 2002), was significantly increased in the CR1/CR2 KO mice. These effects on T cells might be indirect through differential expression of the coreceptors B7.1 and B7.2 (ligands for CD28 and CTLA-4 on T cells) in CR1/CR2 KO and CR1/CR2 WT mice after antigen challenge or induced directly through an unknown mechanism provided by the CR1 and CR2 receptors on the subset of activated T cells. Interestingly, these CR1/CR2 receptors are limited to a subset of activated $CD44^{high}$, $CD62L^{low}$ T cells, but are not expressed on naive T cells (Kaya *et al.,* 2005). Recently it was reported that these receptors are limited to a subset of activated $CD4^+$ T cells (Pratt *et al.*, 2002). The expression of CR1/CR2 on activated T cells in mice is a finding that could provide fresh insight into the mechanism of the transition from innate to adaptive immune responses. Even though in few studies these receptors have been described in humans (Fischer *et al.*, 1991; Delibrias *et al.*, 1992, 1994), their exact function is unknown. Lately, Pratt *et al.* (2002) reported on a critical role for CR1/CR2 in a mouse model of kidney transplantation. B cells and follicular dendritic cells are known to express CR1/CR2 and their function on both these cells have been well studied (Fearon and Carroll, 2000). The role of CR1/CR2 in Ag-trapping, formation of germinal centers, B cell maturation, and long-term memory B cell development have been described (Carroll, 1998).

The C5b-9 terminal attack complex of complement (MAC) may also play a critical role in the pathogenesis of DCM (Zwaka *et al.*, 2002). It was demonstrated that C5b-9 accumulates in human myocardium in DCM. Its deposition significantly correlated with immunoglobulin deposition and myocardial expression of TNF-*alpha*. Further, *in vitro* C5b-9 attack on cardiac myocytes induced nuclear factor (NF)-kappaB activation as well as transcription, synthesis, and secretion of TNF-*alpha*. Thus it was concluded that chronic immunoglobulin-mediated complement activation in the myocardium may contribute in part to the progression of DCM via C5b-9-induced TNF-*alpha* expression in cardiac myocytes.

The role of the complement system in regulating the adaptive immune response has been demonstrated in a number of other studies. Complement is known to be important in the effector phase of several autoimmune diseases such as systemic lupus erythematosus (SLE) in humans (Morgan and Walport, 1991), type II collagen–induced arthritis in mice (Wang *et al.*, 1991), and experimental allergic encephalomyelitis (Davoust *et al.*, 1999).

5.2. NK Cells and Myocarditis

NK cells represent an important first line of defense because they are activated immediately after infection and their effector functions are not antigen-specific. They efficiently limit replication of a number of viruses including CB3 (Godeny and Gauntt, 1987). When APCs detect viral infection of host tissues they release cytokines and chemokines that attract NK cells to the site of infection (Godeny and Gauntt, 1986). NK cells rapidly produce cytokines and other products (e.g., IFN-*gamma*, perforin) before clonal expansion of T cells occurs and exert a critical function in early immune defense. NK cells and macrophages are present in the early infiltrate of the heart after CB3 infection, where they efficiently clear virus by releasing perforin to kill infected cells and by stimulating adaptive immunity via IFN-*gamma* production (Godeny and Gauntt, 1987). Thus, NK cells are believed to protect against the development of CB3-induced myocarditis by limiting viral replication. Recently it was demonstrated that although BALB/c mice have NK cells but no NK1.1$^+$ cells, they can primarily clear viral infection through the cytolytic activity of CD8$^+$ T cells. Depletion of NK1.1$^+$ cells from resistant C57BL/6 or BALB.B6-*Cmv1r* mice led to significantly increased myocarditis levels found in susceptible BALB/c mice. Furthermore, depletion of CD4$^+$, CD8$^+$, or both CD4$^+$ and CD8$^+$ T cells significantly reduced disease in susceptible BALB/c mice, although T cell depletion had little effect on myocarditis in resistant mice. These results suggest that the role of NK1.1 cells is more important in the development of myocarditis than other BALB/c genetic background genes, since the only difference between congenic BALB/c and wild type BALB/c mice is the NK cell gene complex region. The precise mechanisms of NK cells in the pathogenesis of myocarditis is still not known, but their rapid production of cytokines, such as IFN-*gamma*, may provide a clue.

5.3. Cytokines and Myocarditis

Pathogenic mechanisms in myocarditis can be related to specific cytokine production in a particular animal strain in both induction and progression of myocarditis. The profile of inflammatory mediators produced at different time points in the course of myocarditis has critical impact on the development of disease (Hill and Rose, 2001; Afanasyeva and Rose, 2004). Blocking IL-10 during the effector phase of myocarditis increased severity and retarded the reduction of disease, whereas blocking IL-10 during the induction phase had no effect (Kaya *et al.*, 2002). Cotreating genetically resistant mice with certain cytokines can make them susceptible to induction of myocarditis. For example, administration of either IL-1 or TNF-*alpha* promoted virus- and myosin-induced myocarditis in genetically resistant B10.A mice (Lane et al., 1992, 1993). The presence of myocarditis was associated with increased cardiac myosin–stimulated production of TNF-*alpha* (Lane et al., 1993). When A/J mice are infected with CB3 and treated with an IL-1 receptor antagonist, myocardial injury is diminished (Neumann et al., 1993). Furthermore, IL-1 receptor type 1–deficient IL-1R1$^{-/-}$ mice are protected from development of autoimmune myocarditis after immunization (Eriksson et al., 2003b). CD4$^+$ T cells from immunized IL-1R1$^{-/-}$ mice failed to transfer disease after injection into naive SCID mice. The activation of IL-1R1$^{-/-}$ CD4$^+$ T cells by dendritic cells was impaired in IL-1R1$^{-/-}$ mice. The release of TNF-*alpha*, IL-1, IL-6, and IL-12p70 was reduced in dendritic cells lacking the IL-1 receptor type 1. Indeed, injection of immature, antigen-loaded IL-1R1$^{+/+}$, but not IL-1R1$^{-/-}$, dendritic cells into IL-1R1$^{-/-}$ mice fully restored disease susceptibility. Further, overexpression of IL-1R antagonist in the mouse heart decreased myocardial inflammation in CB3 myocarditis. Thus, the authors concluded that IL-1R1 triggering is necessary for efficient activation of dendritic cells. This is conceivably a requirement for inducing an immune response and autoimmunity.

Cytokines are critical in controlling T cells responsive to self-antigens, sometimes by shifting the immune response toward a Th1 or a Th2 pattern. Recent studies suggest that both B and T cells are involved in polarized cytokine production and CD4$^+$, CD8$^+$ T cells, and NK and dendritic cells may also be involved in production of polarizing cytokines (Salazar-Mather et al., 1998). The Th1 response shifts the cytokine profile toward delayed hypersensitivity, macrophage activation, and a proinflammatory T cell response associated with IFN-*gamma* and IL-12, whereas the Th2 response is associated with B cell activation and humoral immunity, and IL-4, IL-5, IL-10, and IgE production. Th1 T cells secrete IL-2 and IFN-*gamma* that suppresses Th2 responses, whereas Th2 T cells secrete IL-4 and IL-10 that inhibit Th1 responses. EAM in A/J mice exhibits a Th2-like phenotype (Afanasyeva et al., 2001b). This was demonstrated by the histological picture of the heart lesions (eosinophils and giant cells) and by the humoral immune response (association with IgG1 response with disease and up-regulation of total IgE). The severity of disease could be reduced by blocking IL-4 with anti-IL-4 monoclonal antibody. It was associated with a shift from a Th2-like to a Th1-like phenotype represented by a reduction in CM-specific IgG1, an increase in CM-specific IgG2a, an abrogation of total IgE response, a decrease in IL-4, IL-5, IL-13, and an increase in IFN-*gamma* production *in vitro*.

IL-12 exerts a potent proinflammatory effect by stimulating Th1 responses (Afanasyeva et al., 2001a). This effect is believed to be mediated primarily through the activation of STAT4 and subsequent production of IFN-*gamma*. Both IL-12R*beta*1-deficient mice and STAT4-deficient mice were resistant to the induction of myocarditis. Treatment with exogenous IL-12 exacerbated disease. On the other hand, IFN-*gamma* deficiency was found to enhance EAM and treatment of mice with recombinant IFN-*gamma* suppressed the development of myocarditis. The disease-limiting effects of IFN-*gamma* may be explained by its ability to control the expansion of activated T lymphocytes. Thus, spleens from IFN-*gamma*-deficient mice immunized with cardiac myosin showed increased cellularity; greater numbers of $CD3^+$, $CD4^+$, $CD8^+$, and IL-2-producing cells; reduced early apoptosis; and heightened ability to produce cytokines on stimulation *in vitro* (Afanasyeva, 2005). In another study, transgenic mice expressing IFN-*gamma* in their pancreatic *beta* cells failed to develop CB3-induced myocarditis (Horwitz et al., 2000). The authors proposed that viral infection in the heart was subdued and cardiac myosin was not released from infected myocytes. The antiviral effect of IFN-*gamma* was proposed for the reduced viremia in the heart of the transgenic mice. This work challenges the concept of "molecular mimicry" in the CB3-induced autoimmune myocarditis model, and instead favors the idea of virus-mediated damage causing release of endogenous antigen.

5.4. Chemokines and Myocarditis

The cytotoxic action of leukocytes is a probable cause of the cardiac myocyte damage seen in chronic myocarditis and DCM. The migration and tissue infiltration of leukocytes is regulated by chemotactic cytokines. MCP-1 and MIP-1*alpha* are potent chemotactic factors for mononuclear cells. The inflammatory infiltrate observed in myocardial lesions in EAM consists of over 60% macrophages (Mac-1^+ cells).

Fuse et al. (2001a) reported increased mRNA of MCP-1 in the hearts of EAM rats from days 15 to 27. Also serum MCP-1 levels of the rats with EAM were significantly elevated. They found in a clinical study that serum levels of MCP-1 in patients with acute myocarditis at the time of admission were significantly elevated compared with those of healthy volunteers and serum MCP-1 levels of eight fatal cases were significantly higher than those cases who survived (Fuse et al., 2001b). Furthermore, DCM patients with severe left ventricular dysfunction showed a 2.35-fold higher MCP-1 messenger RNA expression when compared to DCM patients with less severe dysfunction. Positive immunohistochemical staining for MCP-1 was found in all seven patients with severe left ventricular dysfunction and was particularly distinct within the cardiac interstitium and there was a consistent trend toward a higher infiltration of inflammatory cells in DCM patients with lower ejection fraction (Lehman et al., 1998). To investigate the role of MCP-1 in myocarditis and cardiomyopathy Kolattukudy et al. (1998) created mice expressing the murine MCP-1(JE) gene under the control of the *alpha*-cardiac myosin heavy chain promoter. The mice showed targeted expression of MCP-1 transcripts and protein in the adult heart muscle and increased MCP-1 levels in the transgenic hearts with increased leukocyte infiltration into interstitium between cardiomyocytes. The infiltrate mainly comprised macrophages but not T cells. The presence of MCP-1 in the transgenic hearts did not induce cytokine production indicative of leukocyte activation. Echocardiographic analysis of 1-year-old mice that express MCP-1 in the myocardium and of age-matched controls revealed cardiac hypertrophy and dilation, increases in left ventricular mass, and systolic and diastolic left ventricular internal diameters. A significant decline in M-mode-shortening fraction showed depressed contractile function. Transgenic hearts were 65% heavier, and histological analysis showed moderate myocarditis, edema, and some fibrosis.

To further determine if mononuclear cells, chemokines MCP-1 or MIP-1*alpha*, and chemokine receptors CCR2 and CCR5 play a critical role in the pathogenesis of myocarditis, mononuclear cell activation and migration was inhibited in a set of studies to see if it would affect disease severity and prevalence in EAM. Blockade of MCP-1 or MIP-1*alpha* with monoclonal antibodies significantly reduced the severity of myocarditis (Kaya *et al.*, 2005). Similar results were obtained when CCR2-KO mice were immunized with cardiac myosin. In CCR2-KO mice not only the disease severity but also the prevalence of the disease was reduced (Kaya *et al.*, 2005). Experiments with MIP-1 KO and CCR5-KO are under way. Transfecting the mice before inducing EAM with a dominant-negative inhibitor of MCP-1 (7ND) gene significantly reduced the disease severity, decreased production of cardiac myosin-specific autoantibodies, especially of the IgG1 subclass, and resulted in a reduction in cardiac myosin–induced IL-1, IL-4 and in an increase in IFN-*gamma* and IL-10 cytokine production by splenocytes. These cytokines are known to regulate the development of autoimmune myocarditis (Kaya *et al.*, 2005). The MCP-1 (7ND) gene was used successfully in treatment of other diseases like nephritis (Shimizu *et al.*, 2003), renal fibrosis (Wada *et al.*, 2004), renal injury (Furuichi *et al.*, 2003; Shimizu *et al.*, 2003), cardiovascular diseases (Egashira, 2003; Kitamoto and Egashira, 2003), and pulmonary hypertension (Ikeda *et al.*, 2002). Thus, these experiments may open new pathways to the treatment of immune-mediated disease.

6. Conclusions

Autoimmune myocarditis can be induced in susceptible strains of mice by infection with coxsackievirus B3. The most prominent antibody elicited by the viral infection reacts with the cardiac isoform of myosin and immunization of susceptible mice with cardiac myosin replicates the autoimmune disease. A number of traits determine whether a particular strain of mice is susceptible to autoimmune myocarditis, but the critical decision is made early after infection during the innate immune response. Four of the major components of the innate response have been investigated and found to contribute to susceptibility: the complement system; NK cells; early-acting proinflammatory cytokines and chemokines.

Acknowledgments

The authors' research was supported by NIH grants HL-70729, HL-67290, and AI-51835.

References

Afanasyeva, M., Georgakopoulos, D., Belardi, D.F., Bedja D., Fairweather, D., Wang, Y., Kaya, Z., Gabrielson, K.L., Rodriguez, E.R., Caturegli, P., Kass, D.A., and Rose, N.R. (2005). Impaired up-regulation of CD25 on CD4$^+$ T cells in IFN-*gamma* knockout mice is associated with progression of myocarditis to heart failure. *Proc. Natl. Acad. Sci.*, **102**, 180–185.

Afanasyeva, M., Georgakopoulos, D., Belardi, D.F., Ramsundar, A.C., Barin, J.G., Kass, D.A., and Rose, N.R. (2004). Quantitative analysis of myocardial inflammation by flow cytometry in murine autoimmune myocarditis. *Am. J. Pathol.*, **164**, 807–815.

Afanasyeva, M., Georgakopoulos, D., and Rose, N.R. (2004). Autoimmune myocarditis: cellular mediators of cardiac dysfunction. *Autoimmun. Rev.*, **3**, 476–486.

Afanasyeva, M. and Rose, N.R. (2002a). Immune mediators in inflammatory heart disease: Insights from a mouse model. *Eur. Heart J.*, **4** (Suppl. 1), 131–146.

Afanasyeva, M. and Rose, N.R. (2002b). Cardiomyopathy is linked to complement activation. *Am. J. Pathol.*, **161**, 351–357.

Afanasyeva, M. and Rose, N.R. (2004). Viral infection and heart disease: autoimmune mechanisms. In: Y. Shoenfeld and N.R. Rose. (eds), *Infection and Autoimmunity*. Elsevier, Amsterdam. pp. 299–318.

Afanasyeva, M., Wang, Y., Kaya, Z., Park, S., Zilliox, M.J., Schofield, B.H., Hill, S.L., and Rose, N.R. (2001b). Experimental autoimmune myocarditis in A/J mice is an interleukin-4-dependent disease with a Th2 phenotype. *Am. J. Pathol.*, **159**, 193–203.

Afanasyeva, M., Wang, Y., Kaya, Z., Stafford, E.A., Dohmen, K.M., Sadighi Akha, A.A., and Rose, N.R. (2001a). Interleukin-12 receptor/STAT4 signaling is required for the development of autoimmune myocarditis in mice by an interferon-gamma-independent pathway. *Circulation*, **104**, 3145–3151.

Bachmaier, K., Pummerer, C., Shahinian, A., Ionescu, J., Neu, N., Mak, T.W., and Penninger, J.M. (1996). Induction of autoimmunity in the absence of CD28 costimulation. *J. Immunol.*, **157**, 1752–1757.

Badorff, C., Berkely, N., Mehrotra, S., Talhouk, J.W., Rhoads, R.E., and Knowlton, K.U. (2000). Enteroviral protease 2A directly cleaves dystrophin and is inhibited by a dystrophin-based substrate analogue. *J. Biol. Chem.*, **275**, 11191–11197.

Badorff, C., Lee, G.H., Lamphear, B.J., Martone, M.E., Campbell, K.P., Rhoads, R.E., and Knowlton, K.U. (1999). Enteroviral protease 2A cleaves dystrophin: Evidence of cytoskeletal disruption in an acquired cardiomyopathy. *Nat. Med.*, **5**, 320–326.

Caforio, A.L., Mahon, N.J., and Mckenna, W.J. (2001). Cardiac autoantibodies to myosin and other heart-specific autoantigens in myocarditis and dilated cardiomyopathy. *Autoimmunity*, **34**, 199–204.

Carroll, M.C. (1998). The role of complement and complement receptors in induction and regulation of immunity. *Annu. Rev. Immunol.*, **16**, 545–568.

Cunningham, M.W. (2004). T cell mimicry in inflammatory heart disease. *Mol. Immunol.*, **40**, 1121–1127.

Davoust, N., Nataf, S., Reiman, R., Holers, M.V., Campbell, I.L., and Barnum, S.R. (1999). Central nervous system-targeted expression of the complement inhibitor sCrry prevents experimental allergic encephalomyelitis. *J. Immunol.*, **163**, 6551–6556.

Delibrias, C.C., Fischer, E., Bismuth, G., and Kazatchkine, M.D. (1992). Expression, molecular association, and functions of C3 complement receptors CR1 (CD35) and CR2 (CD21) on the human T cell line HPB-ALL. *J. Immunol.*, **149**, 768–774.

Delibrias, C.C., Mouhoub, A., Fischer, E., and Kazatchkine, M.D. (1994). CR1(CD35) and CR2(CD21) complement C3 receptors are expressed on normal human thymocytes and mediate infection of thymocytes with opsonized human immunodeficiency virus. *Eur. J. Immunol.*, **24**, 2784–2788.

Donath, J. and Landsteiner, K. (1904). Ueber paroxysmale hemoglobinurie. *Z. Klin. Med.*, **58**, 173–189.

Donermeyer, D.L., Beisel, K.W., Allen, P.M., and Smith, S.C. (1995). Myocarditis-inducing epitope of myosin binds constitutively and stably to I-Ak on antigen-presenting cells in the heart. *J. Exp. Med.*, **182**, 1291–1300.

Egashira, K. (2003). Molecular mechanisms mediating inflammation in vascular disease: Special reference to monocyte chemoattractant protein-1. *Hypertension*, **41**, 834–841.

Eriksson, U., Kurrer, M.O., Schmitz, N., Marsch, S.C., Fontana, A., Eugster, H.P., and Kopf, M.U. (2003a). Interleukin-6-deficient mice resist development of autoimmune myocarditis associated with impaired upregulation of complement C3. *Circulation*, **107**, 320–325.

Eriksson, U., Kurrer, M.O., Sonderegger, I., and Iezzi, G. (2003b). Activation of dendritic cells through the interleukin 1 receptor 1 is critical for the induction of autoimmune myocarditis. *J. Exp. Med.*, **197**, 323–331.

Fairweather, D., Afanasyeva, M., and Rose, N.R. (2004). Cellular immunity: A role for cytokines. In A. Doria and P. Pauletto (eds), *Handbook of Systemic Autoimmune Diseases. Vol 1: The Heart in Systemic Autoimmune Diseases*. Elsevier, Amsterdam. pp. 3–17.

Fairweather, D., Frisancho-Kiss, S., Gatewood, S., Njoku, D., Steele, R., Barrett, M., and Rose, N.R. (2004). Mast cells and innate cytokines are associated with susceptibility to autoimmune heart disease following Coxsackievirus B3 infection. *Autoimmunity*, **37**, 131–145.

Fairweather, D., Kaya, Z., Shellam, G.R., Lawson, C.M., and Rose, N.R. (2001). From infection to autoimmunity. *J. Autoimmun.*, **16**, 175–186.

Fairweather, D., Yusung, S., Frisancho, S., Barrett, M., Gatewood, S., Steele, R., and Rose, N.R. (2003). IL-12 receptor beta 1 and toll-like receptor 4 increase IL-1 beta- and IL-18-associated myocarditis and coxsackievirus replication. *J. Immunol.*, **170**, 4731–4737.

Fearon, D.T. and Carroll, M.C. (2000). Regulation of B lymphocyte responses to foreign and self-antigens by the CD19/CD21 complex. *Annu. Rev. Immunol.*, **18**, 393–422.

Fischer, E., Delibrias, C., and Kazatchkine, M.D. (1991). Expression of CR2 (the C3dg/EBV receptor, CD21) on normal human peripheral blood T lymphocytes. *J. Immunol.*, **146**, 865–869.

Furuichi, K., Wada, T., Iwata, Y., Kitagawa, K., Kobayashi, K., Hashimoto, H., Ishiwata, Y., Tomosugi, N., Mukaida, N., Matsushima, K., Egashira, K., and Yokoyama, H. (2003). Gene therapy expressing amino-terminal truncated monocyte chemoattractant protein-1 prevents renal ischemia-reperfusion injury. *J. Am. Soc. Nephrol.*, **14**, 1066–1071.

Fuse, K., Kodama, M., Hanawa, H., Okura, Y., Ito, M., Shiono, T., Maruyama, S., Hirono, S., Kato, K., Watanabe, K., and Aizawa, Y. (2001a). Enhanced expression and production of monocyte chemoattractant protein-1 in myocarditis. *Clin. Exp. Immunol.*, **124**, 346–352.

Fuse, K., Kodama, M., Aizawa, Y., Yamaura, M., Tanabe, Y., Takahashi, K., Sakai, K., Miida, T., Oda, H., and Higuma, N. (2001b). Th1/Th2 balance alteration in the clinical course of a patient with acute viral myocarditis. *Circ. J.*, **65**, 1082–1084.

Gauntt, C.J., Arizpe, H.M., Higdon, A.L., Wood, H.J., Bowers, D.F., Rozek, M.M., and Crawley, R. (1995). Molecular mimicry, anti-coxsackievirus B3 neutralizing monoclonal antibodies, and myocarditis. *J. Immunol.*, **154**, 2983–2995.

Godeny, E.K. and Gauntt, C.J. (1986). Involvement of natural killer cells in coxsackievirus B3-induced murine myocarditis. *J. Immunol.*, **137**, 1695–1702.

Godeny, E.K. and Gauntt, C.J. (1987). Murine natural killer cells limit coxsackievirus B3 replication. *J. Immunol.*, **139**, 913–918.

Guler, M.L., Ligons, D.L., Wang, Y., Bianco, M., Broman, K.W., and Rose, N.R. (2005). Two autoimmune diabetes loci influencing T cell apoptosis control susceptibility to experimental autoimmune myocarditis. *J. Immunol.*, **174**, 2167–2173.

Hill, S.L. and Rose, N.R. (2001). The transition from viral to autoimmune myocarditis. *Autoimmunity*, **34**, 169–176.

Horwitz, M.S., La Cava, A., Fine, C., Rodriguez, E., Ilic, A., and Sarvetnick, N. (2000). Pancreatic expression of interferon-gamma protects mice from lethal coxsackievirus B3 infection and subsequent myocarditis. *Nat. Med.*, **6**, 631–632.

Ikeda, Y., Yonemitsu, Y., Kataoka, C., Kitamoto, S., Yamaoka, T., Nishida, K., Takeshita, A., Egashira, K., and Sueishi, K. (2002). Anti-monocyte chemoattractant protein-1 gene therapy attenuates pulmonary hypertension in rats. *Am. J. Heart Circ. Physiol.*, **283**, H2021–H2028.

Inomata, T., Hanawa, H., Miyanishi, T., Yajima, E., Nakayama, S., Maita, T., Kodama, M., Izumi, T., Shibata, A., and Abo, T. (1995). Localization of porcine cardiac myosin epitopes that induce experimental autoimmune myocarditis. *Circ. Res.*, **76**, 726–733.

Kaya, Z., Afanasyeva, M., Wang, Y., Dohmen, K.M., Schlichting, J., Tretter, T., Fairweather, D., Holers, V.M., and Rose, N.R. (2001). Contribution of the innate immune system to autoimmune myocarditis: A role for complement. *Nat. Immunol.*, **2**, 739–745.

Kaya, Z., Dohmen, K.M., Wang, Y., Schlichting, J., Afanasyeva, M., Leuschner, F., and Rose, N.R. (2002). Cutting edge: A critical role for IL-10 in induction of nasal tolerance in experimental autoimmune myocarditis. *J. Immunol.*, **168**, 1552–1556.

Kaya, Z., Göser, S., Öttl, R., Brodner, A., Dengler, T.J., Egashira, K., Kuziel, W.A., and Katus, H.A. (2005) (in press). Critical role for MCP-1 and MIP-1a in the induction of EAM and anti-MCP-1 gene therapy. *FASEB J.*, abstract.

Kaya, Z., Tretter, T., Schlichting, J., Leuschner, F., Afanasyeva, M., Katus, H.A., and Rose, N.R. (2005). Complement receptors regulate lipopolysaccharide-induced T-cell stimulation. *Immunology*, **114**, 493–498.

Kerekes, K., Prechl, J., Bajtay, Z., Jozsi, M., and Erdei, A. (1993). A further link between innate and adaptive immunity: C3 deposition on antigen-presenting cells enhances the proliferation of antigen-specific T cells. *Int. Immunol.*, **10**, 1923–1930.

Kinoshita, T., Takeda, J., Hong, K., Kozono, H., Sakai, H., and Inoue, K. (1988). Monoclonal antibodies to mouse complement receptor type 1 (CR1). Their use in a distribution study showing that mouse erythrocytes and platelets are CR1-negative. *J. Immunol.*, **140**, 3066–3072.

Kitamoto, S. and Egashira, K. (2003). Anti-monocyte chemoattractant protein-1 gene therapy for cardiovascular diseases. *Exp. Rev. Card. Ther.*, **1**, 393–400.

Kohno, K., Takagaki, Y., Nakajima, Y., and Izumi, T. (2000). Advantage of recombinant technology for the identification of cardiac myosin epitope of severe autoimmune myocarditis in Lewis rats. *Jpn. Heart J.*, **41**, 67–77.

Kolattukudy, P.E., Quach, T., Bergese, S., Breckenridge, S., Hensley, J., Altschuld, R., Gordillo, G., Klenotic, S., Orosz, C., and Parker-Thornburg, J. (1998). Myocarditis induced by targeted expression of the MCP-1 gene in murine cardiac muscle. *Am. J. Pathol.*, **152**, 101–111.

Kuhn, R.J. (1997). Identification and biology of cellular receptors for the coxsackie B viruses group. *Curr. Top. Microbiol. Immunol.*, **223**, 209–226.

Lane, J.R., Neumann, D.A., Lafond-Walker, A., Herskowitz, A., and Rose, N.R. (1992). Interleukin 1 or tumor necrosis factor can promote coxsackie B3-induced myocarditis in resistant B10. A mice. *J. Exp. Med.*, **175**, 1123–1129.

Lane, J.R., Neumann, D.A., Lafond-Walker, A., Herskowitz, A., and Rose, N.R. (1993). Role of IL-1 and tumor necrosis factor in coxsackie virus-induced autoimmune myocarditis. *J. Immunol.*, **151**, 1682–1690.

Lehmann, M.H., Kuhnert, H., Muller, S., and Sigusch, H.H. (1998). Monocyte chemoattractant protein 1 (MCP-1) gene expression in dilated cardiomyopathy. *Cytokine*, **10**, 739–746.

Leiden, J.M. (1997). The genetics of dilated cardiomyopathy—emerging clues to the puzzle. *New Engl. J. Med.*, **337**, 1080–1081.

Liao, L., Sindhwani, R., Leinwand, L., Diamond, B., and Factor, S. (1993). Cardiac alpha-myosin heavy chains differ in their induction of myocarditis. Identification of pathogenic epitopes. *J. Clin. Invest.*, **92**, 2877–2882.

Liao, L., Sindhwani, R., Rojkind, M., Factor, S., Leinwand, L., and Diamond, B. (1995). Antibody-mediated autoimmune myocarditis depends on genetically determined target organ sensitivity. *J. Exp. Med.*, **181**, 1123–1131.

Liu, P., Aitken, K., Kong, Y.Y., Opavsky, M.A., Martino, T., Dawood, F., Wen, W.H., Kozieradzki, I., Bachmaier, K., Straus, D., Mak, T.W., and Penninger, J.M. (2000). The tyrosine kinase p56^{lck} is essential in coxsackievirus B3-mediated heart disease. *Nat. Med.*, **6**, 429–434.

Martino, T.A., Petric, M., Brown, M., Aitken, K., Gauntt, C.J., Richardson, C.D., Chow, L.H., and Liu, P.P. (1998). Cardiovirulent coxsackieviruses and the decay-accelerating factor (CD55) receptor. *Virology*, **244**, 302–314.

Morgan, B.P. and Walport, M.J. (1991). Complement deficiency and disease. *Immunol. Today*, **12**, 301–306.

Nelson, D.P., Setser, E., Hall, D.G., Schwartz, S.M., Hewitt, T., Klevitsky, R., Osinska, H., Bellgrau, D., Duke, R.C., and Robbins, J. (2000). Proinflammatory consequences of transgenic fas ligand expression in the heart. *J. Clin. Invest.*, **105**, 1199–1208.

Neu, N., Beisel, K.W., Traystman, M.D., Rose, N.R., and Craig, S.W. (1987a). Autoantibodies specific for the cardiac myosin isoform are found in mice susceptible to coxsackievirus B_3-induced myocarditis. *J. Immunol.*, **138**, 2488–2492.

Neu, N., Rose, N.R., Beisel, K.W., Herskowitz, A., Gurri-Glass, G., and Craig, S.W. (1987b). Cardiac myosin induces myocarditis in genetically predisposed mice. *J. Immunol.*, **139**, 3630–3636.

Neumann, D.A., Lane, J.R., LaFond-Walker, A., Allen, G.S., Frondoza, C., Herskowitz, A., and Rose, N.R. (1991). Elution of autoantibodies from the hearts of coxsackie-virus-infected mice. *Eur. Heart J.*, **12** (Suppl. D), 113–116.

Neumann, D.A., Lane, J.R., Allen, G.S., Herskowitz, A., and Rose, N.R. (1993). Viral myocarditis leading to cardiomyopathy: Do cytokines contribute to pathogenesis? *Clin. Immunol. Immunopathol.*, **68**, 181–190.

Nishimura, H., Okazaki, T., Tanaka, Y., Nakatani, K., Hara, M., Matsumori, A., Sasayama, S., Mizoguchi, A., Hiai, H., Minato, N., and Honjo, T. (2001). Autoimmune dilated cardiomyopathy in PD-1 receptor-deficient mice. *Science*, **291**, 319–322.

Okazaki, T., Tanaka, Y., Nishio, R., Mitsuiye, T., Mizoguchi, A., Wang, J., Ishida, M., Hiai, H., Matsumori, A., Minato, N., and Honjo, T. (2003). Autoantibodies against cardiac troponin I are responsible for dilated cardiomyopathy in PD-1-deficient mice. *Nat. Med.*, **9**, 1477–1483.

Penninger, J.M., Neu, N., Timms, E., Wallace, V.A., Koh, D.R., Kishihara, K., Pummerer, C., and Mak, T.W. (1993). The induction of experimental autoimmune myocarditis in mice lacking CD4 or CD8 molecules [corrected]. *J. Exp. Med.*, **178**, 1837–1842.

Pratt, J.R., Basheer, S.A., and Sacks, S.H. (2002). Local synthesis of complement component C3 regulates acute renal transplant rejection. *Nat. Med.*, **8**, 582–587.

Pummerer, C.L., Luze, K., Grassl, G., Bachmaier, K., Offner, F., Burrell, S.K., Lenz, D.M., Zamborelli, T.J., Penninger, J.M., and Neu, N. (1996). Identification of cardiac myosin peptides capable of inducing autoimmune myocarditis in BALB/c mice. *J. Clin. Invest.*, **97**, 2057–2062.

Rose, N.R., Herskowitz, A., Neumann, D.A., and Neu, N. (1988a). Autoimmune myocarditis: A paradigm of post-infection autoimmune disease. *Immunol. Today*, **9**, 117–119.

Rose, N.R., Neumann, D.A., Herskowitz, A., Traystman, M.D., and Beisel, K.W. (1988b). Genetics of susceptibility to viral myocarditis in mice. *Pathol. Immunopathol. Res.*, **7**, 266–278.

Salazar-Mather, T.P., Orange, J.S., and Biron, C.A. (1998). Early murine cytomegalovirus (MCMV) infection induces liver natural killer (NK) cell inflammation and protection through macrophage inflammatory protein 1alpha (MIP-1alpha)-dependent pathways. *J. Exp. Med.*, **187**, 1–14.

Shimizu, H., Maruyama, S., Yuzawa, Y., Kato, T., Miki, Y., Suzuki, S., Sato, W., Morita, Y., Maruyama, H., Egashira, K., and Matsuo, S. (2003). Anti-monocyte chemoattractant protein-1 gene therapy attenuates renal injury induced by protein-overload proteinuria. *J. Am. Soc. Nephrol.*, **14**, 1496–1505.

Smith, S.C. and Allen, P.M. (1991). Myosin-induced acute myocarditis is a T cell-mediated disease. *J. Immunol.*, **147**, 2141–2147.

Wada, T., Furuichi, K., Sakai, N., Iwata, Y., Kitagawa, K., Ishida, Y., Kondo, T., Hashimoto, H., Ishiwata, Y., Mukaida, N., Tomosugi, N., Matsushima, K., Egashira, K., and Yokoyama, H. (2004). Gene therapy via blockade of monocyte chemoattractant protein-1 for renal fibrosis. *J. Am. Soc. Nephrol.*, **15**, 940–948.

Wang, Y.C., Herskowitz, A., Gu, L.B., Kanter, K., Lattouf, O., Sell, K.W., and Ahmed-Ansari, A. (1991). Influence of cytokines and immunosuppressive drugs on major histocompatibility complex class I/II expression by human cardiac myocytes in vitro. *Hum. Immunol.*, **31**, 123–133.

Wegmann, K.W., Zhao, W., Griffin, A.C., and Hickey, W.F. (1994). Identification of myocardiogenic peptides derived from cardiac myosin capable of inducing experimental allergic myocarditis in the Lewis rat. The utility of a class II binding motif in selecting self-reactive peptides. *J. Immunol.*, **153**, 892–900.

Zwaka, T.P., Manolov, D., Ozdemir, C., Marx, N., Kaya, Z., Kochs, M., Hoher, M., Hombach, V., and Torzewski, J. (2002). Complement and dilated cardiomyopathy: A role of sublytic terminal complement complex-induced tumor necrosis factor-alpha synthesis in cardiac myocytes. *Am. J. Pathol.*, **161**, 449–457.

2

Toll-like Receptor 9 and Autoimmunity

Paul N. Moynagh

1. Introduction

Most studies that have explored the molecular and cellular basis to the generation of systemic autoimmune diseases have focused on the role played by the specific immune system. This is hardly surprising since systemic autoimmune diseases such as systemic lupus erythematosus (SLE) and rheumatoid arthritis (RA) are associated with the production of autoantibodies. The latter include rheumatoid factor (RF), an autoantibody that is directed toward normal antibodies, antinuclear antibodies, and antibodies to DNA. However, in the last 2 years the innate immune system has jumped to the fore as a key contributor to these diseases. This has been due mainly to a pioneering report showing the key role played by the innate immune system in promoting the production of RF in a manner independent of help from T cells (Leadbetter et al., 2002). This process is facilitated by a member of the toll-like receptor (TLR) family, namely TLR9. This chapter initially overviews the crucial involvement of TLR9 in mediating the immunostimulatory effects of bacterial DNA. It then describes a role for TLR9 in recognizing self-DNA leading to the activation of B cells and production of autoantibodies. The intracellular signaling pathway employed by TLR9 is also discussed and its value as a therapeutic target for the design of novel strategies for treating autoimmune diseases is emphasized.

2. TLRs as Receptors for Pathogen-Associated Molecules

Human TLRs play crucial roles at the host–pathogen interface due to their capacity to recognize pathogen-associated molecules. Many of the TLRs have defined functions in the innate immune system. Thus TLR2 recognizes peptidoglycan and bacterial lipoprotein from Gram-positive bacteria (Aliprantis et al., 1999; Takeuchi et al., 1999), TLR3 mediates responses to double-stranded RNA (Alexopoulou et al., 2001), TLR4 is involved in recognition of Gram-negative

---

Paul N. Moynagh • Department of Pharmacology, Conway Institute of Biomolecular and Biomedical Research, University College Dublin, Belfield, Dublin 4, Ireland.

Molecular Autoimmunity: In commemoration of the 100th anniversary of the first description of human autoimmune disease, edited by Moncef Zouali. Springer Science+Business Media, Inc., New York, 2005.

lipopolysaccharide (LPS) (Poltorak *et al.*, 1998; Chow *et al.*, 1999; Hoshino *et al.*, 1999; Qureshi *et al.*, 1999; Takeuchi *et al.*, 1999), TLR5 recognizes bacterial flagellin (Hayashi *et al.*, 2001), TLR7 and TLR8 sense single-stranded RNA (Diebold *et al.*, 2004; Heil *et al.*, 2004), and TLR9 functions as a receptor for bacterial DNA (Hemmi *et al.*, 2000). Some TLRs, such as TLR2 and TLR6, also show functional cooperativity (Ozinsky *et al.*, 2000). The engagement of TLRs by pathogenic components results in induction of specific gene expression profiles that are suited to ensuring efficient removal and destruction of the invading pathogen.

3. TLR9 and the Immunostimulatory Effects of Bacterial DNA

TLR9 acts as a recognition system for bacterial DNA containing unmethylated CpG motifs. Since these motifs are quite rare and predominantly methylated in vertebrate DNA, TLR9 provides a means to distinguish between self and nonself. The expression of TLR9 in humans is mainly restricted to plasmacytoid dendritic cells (Krug *et al.*, 2001) and B cells (Bauer *et al.*, 2001). Its engagement in dendritic cells by unmethylated bacterial CpG motifs synergizes with CD40 ligand to induce high levels of IL-12 that facilitates activation of Th1 cells (Krug *et al.*, 2001). In addition, CpG motifs induce high-level expression of costimulatory molecules in plasmacytoid dendritic cells and this ensures strong activation of allogeneic T cells (Hartmann *et al.*, 1999).

CpG motifs are extremely strong stimulators for B cells, activating them to enter the G1 phase of the cell cycle (Krieg *et al.*, 1995). Indeed, relatively low concentrations of CpG DNA can synergize with the B cell receptor (BCR) and cause a 10-fold increase in B cell proliferation and antigen-specific antibody secretion. Furthermore, CpG DNA can produce antiapoptotic effects in B cells. Thus CpG DNA can inhibit the proapoptotic capacity of BCR ligation in B cell lines (Yi and Krieg, 1998a) and can promote survival of primary B cells in culture by blocking their spontaneous apoptosis (Yi *et al.*, 1998a). The ability of CpG DNA to promote sustained activation of the prosurvival transcription factor NFκB is a major contributor to these antiapoptotic effects (Yi *et al.*, 1998a, 1999; Yi and Krieg, 1998a). In addition to enhancing survival and proliferation of B cells and facilitating increased secretion of antigen-specific antibodies, CpG DNA can also induce the expression of costimulatory molecules in B cells (Krieg *et al.*, 1995) and thus increase their efficacy with respect to antigen presentation and T cell activation. Since the induction of many of these costimulatory molecules is dependent on NFκB (Medzhitov *et al.*, 1997), this transcription factor emerges as a key player in mediating the biological effects of CpG DNA. Consequently, the intracellular signaling pathways employed by TLR9 in mediating activation of NFκB in response to CpG DNA has become the focus for much research.

4. TLR9 and Intracellular signaling

Cells display DNA-binding proteins on their surface, but lack selectivity in recognizing specific sequences (Krieg *et al.*, 1995). This study also demonstrated that cell uptake of CpG DNA is required to produce its effects in B cells. While the mechanism of uptake is incompletely understood, internalized CpG DNA has been localized to the endosome and studies suggest that this is the site where intracellular signaling pathways are initiated by the DNA. Thus agents such as chloroquine, which interfere with endosomal acidification and/or maturation, block the signaling pathways (Hacker *et al.*, 1998; Yi and Krieg, 1998b; Bauer *et al.*, 2001) and immunostimulatory effects (Yi *et al.*, 1998b) of CpG DNA. In contrast, other pathogen-associated molecules such as LPS, which are recognized by specific cell surface receptors, are unaffected by chloroquine (Yi *et al.*, 1998b). TLR9 was subsequently shown to be the specific receptor for CpG DNA since the dendritic and B cells of mice genetically deficient in TLR9 are unresponsive to CpG (Hemmi *et al.*, 2000). Furthermore, some evidence suggests that TLR9 is localized to the endosomes where it may be able to physically interact with the internalized CpG DNA (Hemmi *et al.*, 2000; Takeshita *et al.*, 2001).

The proximal signaling events subsequent to TLR9 engagement by CpG have been well characterized. Like most other TLRs, activation of TLR9 recruits the adaptor molecule Myd88 (Medzhitov *et al.*, 1998). Indeed, CpG DNA, TLR9, and Myd88 colocalize in late endosomes and the initiation of signaling is dependent on endosome maturation (Takeshita *et al.*, 2001; Ahmad-Nejad *et al.*, 2002). Myd88 subsequently recruits and activates members of the IL-1 receptor–associated kinase (IRAK) family (Muzio *et al.*, 1997; Wesche *et al.*, 1997; Kobayashi *et al.*, 2002; Li *et al.*, 2002; Suzuki *et al.*, 2002). The IRAK-Myd88 association triggers hyperphosphorylation of IRAK by itself (Cao *et al.*, 1996) and/or by other additional kinases (Li *et al.*, 1999), leading to its dissociation from Myd88 and its interaction with and activation of the downstream adaptor TNF receptor–associated factor 6 (TRAF-6) (Burns *et al.*, 2000). The latter is a ubiquitin ligase that activates TGFβ-activating kinase (TAK1) (Ninomiya-Tsuji *et al.*, 1999). Activated TAK1 promotes downstream activation of the IκB-kinases (IKK), IKKα and IKKβ, that form a large multiprotein complex with a scaffold protein called NEMO (IKKγ). Of the two active IKK isoforms, IKKβ appears to be the more important for CpG signaling (Chu *et al.*, 2000). It affects phosphorylation of members of the inhibitory IκB family (IκB-α and IκB-β) that normally sequester NFκB in an inactive form in the cytosol (Yi and Krieg, 1998a). The phosphorylation of the IκB proteins represents a signal for polyubiquitination followed by their degradation via the 26 S proteosome and this allows for translocation of NFκB to the nucleus, where it activates a plethora of genes encoding proinflammatory proteins and costimulatory molecules (Medzhitov *et al.*, 1997; O'Neill, 2002). Interestingly, the activation of NFκB is also a requisite for CpG-induced protection of B cells against apoptosis (Yi and Krieg, 1998a; Yi *et al.*, 1999). Such activation of NFκB is sustained in nature and usually associated with prolonged disappearance of the inhibitory IκB-β protein (Bourke *et al.*, 2000).

In addition to NFκB activation, CpG motifs can also engage TLR9 to stimulate mitogen-activated protein kinase (MAPK) signaling cascades and activate multiple transcription factors, including AP-1. Thus CpG DNA induces phosphorylation of p38 and c-Jun N-terminal kinase (JNK) in B cells (Yi and Krieg, 1998b). Interestingly, while CD40 utilizes many of the same signaling molecules and activates the same pathways as CpG in B cells (Brady *et al.*, 2000, 2001), the activation of MAPKs by CpG is slower in onset but longer in duration (Yi and Krieg, 1998b). The sustained activation of both NFκB and MAPK pathways by CpG in B cells suggests that these intracellular pathways are long-lived in response to CpG and not subject to acute downregulation. Whilst the mechanisms by which the MAPKs are activated by TLR9 are incompletely understood, upstream regulators have been identified. Such regulators also play integral roles in mediating activation of NFκB. Thus TAK1 can activate the MAPKKs MKK3/6 and MKK4, which in turn activate p38 and JNK, respectively (Ninomiya-Tsuji *et al.*, 1999). The activation of the MAPK pathways by CpG motifs in B cells contributes significantly to their biological responsiveness to CpG since inhibitors of these pathways block CpG-induced secretion of cytokines by B cells (Yi and Krieg, 1998b).

The overall effects of CpG DNA on B cells are thus mediated via TLR9 activation of the NFκB and MAPK pathways. As stated previously, TLR9 and BCR show strong functional synergy in activating B cells. It is worth noting that a recent report has provided a molecular basis for this synergy by demonstrating that their signaling pathways converge at the level of NFκB and p38/JNK (Yi *et al.*, 2003).

5. CpG Sequences in Self-DNA Trigger Autoantibody Production

For many years vertebrate self-DNA was considered to be immunologically inert. However, a report in 2002 demonstrated that immune complexes containing self-DNA activate RF-specific B cells (Leadbetter *et al.*, 2002). This study employed a genetically engineered mouse strain in which most B cells express a BCR with low affinity for self-IgG2a. The affinity is insufficient to trigger activation of the B cells. However, other workers had previously shown that when the mice are crossed with a strain susceptible to autoimmune diseases, such as RA and SLE, the self-IgG2a becomes a powerful activator of B cells and stimulates B cell proliferation and secretion of high levels of circulating RF autoantibodies (Wang and Shlomchik, 1999). Leadbetter *et al.* (2002) went on to show that self-IgG2a accumulates in complexes with self-DNA in the bloodstream of the autoimmune mice. It is likely that the self-DNA is released during physiological and/or pathological cell death. In this model the synergy between the signals originating from BCR activation by self-IgG2a and TLR9 activation by self-DNA is sufficiently powerful to provoke strong B cell activation. Furthermore, the BCR is likely to facilitate the uptake of self-DNA and its ultimate delivery to endosomal TLR9 (Viglianti *et al.*, 2003). However, as stated previously, TLR9 acts as a receptor for DNA containing unmethylated CpG motifs, and mammalian DNA tends to be methylated. This questions the role of TLR9 in recognizing self-DNA. However, mammalian DNA is not exclusively methylated and indeed methylation is restricted to 70–80% of the CpG motifs (Bird, 1987). The remaining 20–30% of unmethylated CpG motifs may be responsible for self-DNA activating TLR9. This notion is supported by two findings. First, DNA methylation status is reduced in cells from animals and humans with autoimmune disease (Richardson *et al.*, 1990). Second, drugs such as azacytidine that inhibit CpG methylation cause an autoimmune disease with features of lupus (Yung *et al.*, 1995). Krieg (2002) has put forward an alternative model wherein cross-linking of the BCR may be sufficiently extensive to reduce the specificity of TLR9 recognition, allowing its activation by self-DNA (Goeckeritz *et al.*, 1999). Interestingly, Krieg (2002) also suggests activating immune complexes may result from an earlier loss of B cell tolerance to DNA-associated self-antigens. This is consistent with the clinical finding that most autoantibodies in SLE are directed against nuclear antigens that bind DNA (Krieg, 2002). Since high levels of RF against IgG are measured in patients suffering from RA and SLE, therapeutic potential may lie in designing novel approaches to control the production of RFs. The key role played by TLR9 in promoting RF production makes this pathway an ideal target for therapeutic intervention.

6. TLR9 as a Target for Regulating RF Production

Since TLR9 acts as a costimulus for autoreactive B cells, and because endosomal maturation and acidification is required for its signaling, a molecular basis is now at hand to explain the therapeutic effects of chloroquine in systemic autoimmune diseases. Chloroquine interferes with endosomal acidification and maturation and blocks all CpG-TLR9 signaling pathways without inhibiting other B cell activators (Hacker *et al.*, 1998; Yi and Krieg, 1998b; Yi *et al.*, 1998b; Bauer *et al.*, 2001). Chloroquine and other similarly acting agents have now been shown to block immune complex–induced activation of autoreactive B cells (Leadbetter *et al.*, 2002) and this is likely to underlie the therapeutic efficacy of chloroquine. This highlights the value of targeting the TLR9 pathway in the design of novel therapeutics for the treatment of autoimmune diseases. Valuable clues in this regard may be provided by identifying physiological mechanisms for controlling the TLR9 pathway. A recent report has shown that continuous self-antigen signaling via the BCR MAPK pathway inhibits CpG DNA–induced plasma cell differentiation and controls TLR9-mediated autoantibody production (Rui *et al.*, 2003). It is also worth noting the presence of a number of endogenous regulators of TLR signaling pathways. SIGIRR is a membrane protein that negatively regulates TLR signaling (Wald *et al.*, 2003). Myd88S, a splice variant of Myd88, inhibits Myd88-dependent pathways (Janssens *et al.*, 2003). IRAK-M is a member of the IRAK family and acts as a negative regulator of TLR signaling pathways (Kobayashi *et al.*, 2002). Furthermore, the phosphorylation of IRAK-1 during TLR signaling ultimately leads to its degradation by proteosomes and this may represent a regulatory mechanism by which cells become desensitized after prolonged activation of TLRs (Yamin and Miller, 1997; Li *et al.*, 2000). Finally, the antiapoptotic protein A20 acts to terminate signaling pathways triggered by TLRs (O'Reilly and Moynagh, 2003; Boone *et al.*, 2004; Gon *et al.*, 2004). The existence of these numerous braking systems on TLR signaling pathways emphasizes the importance of tightly regulating such pathways. Some of these mechanisms may offer future opportunity to develop novel strategies to dampen TLR9 signaling and to control the activity of autoreactive B cells in autoimmune disease. Interestingly, as stated previously, the temporal activation of TLR9 signaling pathways in B cells tends to be sustained and many of the above autoregulatory mechanisms may not be at play in these cells. The identification of means to promote these mechanisms may be of immense value in learning to regulate TLR9 signaling and autoantibody production in B cells.

7. Concluding Remarks

The discovery that TLR9 can promote the production of autoantibodies by synergistically activating B cells in conjunction with BCR raises the intriguing possibility that self-DNA is immunostimulatory in a T cell–independent manner. Furthermore, TLRs can no longer be regarded as recognition systems in the innate immune system that shows exclusive selectivity for pathogen-associated molecules. TLR9 compromises the capacity for absolute distinction between self and nonself. Our current molecular appreciation of the TLR9 signaling pathway provides hope for therapeutic manipulation and the design of novel strategies for treating autoimmune diseases.

Acknowledgments

This publication has emanated from research conducted with the financial support of Science Foundation Ireland. Dr. Paul N. Moynagh is also funded by Health Research Board of Ireland and Higher Education Authority of Ireland.

References

Ahmad-Nejad, P., Hacker, H., Rutz, M., Bauer, S., Vabulas, R.M., and Wagner, H. (2002). Bacterial CpG-DNA and lipopolysaccharides activate toll-like receptors at distinct cellular compartments. *Eur. J. Immunol.*, **32**, 1958–1968.

Alexopoulou, L., Holt, A.C., Medzhitov, R., and Flavell, R.A. (2001). Recognition of double-stranded RNA and activation of NF-kappaB by toll-like receptor 3. *Nature*, **413**, 732–738.

Aliprantis, A.O., Yang, R.B., Mark, M.R., Suggett, S., Devaux, B., Radolf, J.D., Klimpel, G.R., Godowski, P., and Zychlinsky, A. (1999). Cell activation and apoptosis by bacterial lipoproteins through toll-like receptor-2. *Science*, **285**, 736–739.

Bauer, S., Kirschning, C.J., Hacker, H., Redecke, V., Hausmann, S., Akira, S., Wagner, H., and Lipford, G.B. (2001). Human TLR9 confers responsiveness to bacterial DNA via species-specific CpG motif recognition. *Proc. Natl. Acad. Sci. USA*, **98**, 9237–9242.

Bird, A.P. (1987). CpG islands as gene markers in the vertebrate nucleus. *Trends Genet.*, **3**, 342–347.

Boone, D.L., Turer, E.E., Lee, E.G., Ahmad, R.C., Wheeler, M.T., Tsui, C., Hurley, P., Chien, M., Chai, S., Hitotsumatsu, O., McNally, E., Pickart, C., and Ma, A. (2004). The ubiquitin-modifying enzyme A20 is required for termination of toll-like receptor responses. *Nat. Immunol.*, **5**, 1052–1060.

Bourke, E., Kennedy, E.J., and Moynagh, P.N. (2000). Loss of Ikappa B-beta is associated with prolonged NF-kappa B activity in human glial cells. *J. Biol. Chem.*, **275**, 39996–40002.

Brady, K., Fitzgerald, S., Ingvarsson, S., Borrebaeck, C.A., and Moynagh, P.N. (2001). CD40 employs p38 MAP kinase in IgE isotype switching. *Biochem. Biophys. Res. Commun.*, **289**, 276–281.

Brady, K., Fitzgerald, S., and Moynagh, P.N. (2000). Tumour-necrosis-factor-receptor-associated factor 6, NF-kappaB-inducing kinase and IkappaB kinases mediate IgE isotype switching in response to CD40. *Biochem. J.*, **350**, 735–740.

Burns, K., Clatworthy, J., Martin, L., Martinon, F., Plumpton, C., Maschera, B., Lewis, A., Ray, K., Tschopp, J., and Volpe, F. (2000). Tollip, a new component of the IL-1RI pathway, links IRAK to the IL-1 receptor. *Nat. Cell Biol.*, **2**, 346–351.

Cao, Z., Henzel, W.J., and Gao, X. (1996). IRAK: A kinase associated with the interleukin-1 receptor. *Science*, **271**, 1128–1131.

Chow, J.C., Young, D.W., Golenbock, D.T., Christ, W.J., and Gusovsky, F. (1999). Toll-like receptor-4 mediates lipopolysaccharide-induced signal transduction. *J. Biol. Chem.*, **274**, 10689–10692.

Chu, W., Gong, X., Li, Z., Takabayashi, K., Ouyang, H., Chen, Y., Lois, A., Chen, D.J., Li, G.C., Karin, M., and Raz, E. (2000). DNA-PKcs is required for activation of innate immunity by immunostimulatory DNA. *Cell*, **103**, 909–918.

Diebold, S.S., Kaisho, T., Hemmi, H., Akira, S., and Reis e Sousa, C. (2004). Innate antiviral responses by means of TLR7-mediated recognition of single-stranded RNA. *Science*, **303**, 1529–1531.

Goeckeritz, B.E., Flora, M., Witherspoon, K., Vos, Q., Lees, A., Dennis, G.J., Pisetsky, D.S., Klinman, D.M., Snapper, C.M., and Mond, J.J. (1999). Multivalent cross-linking of membrane Ig sensitizes murine B cells to a broader spectrum of CpG-containing oligodeoxynucleotide motifs, including their methylated counterparts, for stimulation of proliferation and Ig secretion. *Int. Immunol.*, **11**, 1693–1700.

Gon, Y., Asai, Y., Hashimoto, S., Mizumura, K., Jibiki, I., Machino, T., Ra, C., and Horie, T. (2004). A20 inhibits toll-like receptor 2- and 4-mediated interleukin-8 synthesis in airway epithelial cells. *Am. J. Respir. Cell Mol. Biol.*, **31**, 330–336.

Hacker, H., Mischak, H., Miethke, T., Liptay, S., Schmid, R., Sparwasser, T., Heeg, K., Lipford, G.B., and Wagner, H. (1998). CpG-DNA-specific activation of antigen-presenting cells requires stress kinase activity and is preceded by non-specific endocytosis and endosomal maturation. *EMBO J.*, **17**, 6230–6240.

Hartmann, G., Weiner, G.J., and Krieg, A.M. (1999). CpG DNA: A potent signal for growth, activation, and maturation of human dendritic cells. *Proc. Natl. Acad. Sci. USA*, **96**, 9305–9310.

Hayashi, F., Smith, K.D., Ozinsky, A., Hawn, T.R., Yi, E.C., Goodlett, D.R., Eng, J.K., Akira, S., Underhill, D.M., and Aderem, A. (2001). The innate immune response to bacterial flagellin is mediated by toll-like receptor 5. *Nature*, **410**, 1099–1103.

Heil, F., Hemmi, H., Hochrein, H., Ampenberger, F., Kirschning, C., Akira, S., Lipford, G., Wagner, H., and Bauer, S. (2004). Species-specific recognition of single-stranded RNA via toll-like receptor 7 and 8. *Science*, **303**, 1526–1529.

Hemmi, H., Takeuchi, O., Kawai, T., Kaisho, T., Sato, S., Sanjo, H., Matsumoto, M., Hoshino, K., Wagner, H., Takeda, K., and Akira, S. (2000). A toll-like receptor recognizes bacterial DNA. *Nature*, **408**, 740–745.

Hoshino, K., Takeuchi, O., Kawai, T., Sanjo, H., Ogawa, T., Takeda, Y., Takeda, K., and Akira, S. (1999). Cutting edge: Toll-like receptor 4 (TLR4)-deficient mice are hyporesponsive to lipopolysaccharide: Evidence for TLR4 as the Lps gene product. *J. Immunol.*, **162**, 3749–3752.

Janssens, S., Burns, K., Vercammen, E., Tschopp, J., and Beyaert, R. (2003). MyD88S, a splice variant of MyD88, differentially modulates NF-kappaB- and AP-1-dependent gene expression. *FEBS Lett.*, **548**, 103–107.

Kobayashi, K., Hernandez, L.D., Galan, J.E., Janeway, C.A., Jr., Medzhitov, R., and Flavell, R.A. (2002). IRAK-M is a negative regulator of toll-like receptor signaling. *Cell*, **110**, 191–202.

Krieg, A.M. (2002). A role for toll in autoimmunity. *Nat. Immunol.*, **3**, 423–424.

Krieg, A.M., Yi, A.K., Matson, S., Waldschmidt, T.J., Bishop, G.A., Teasdale, R., Koretzky, G.A., and Klinman, D.M. (1995). CpG motifs in bacterial DNA trigger direct B-cell activation. *Nature*, **374**, 546–549.

Krug, A., Towarowski, A., Britsch, S., Rothenfusser, S., Hornung, V., Bals, R., Giese, T., Engelmann, H., Endres, S., Krieg, A.M., and Hartmann, G. (2001). Toll-like receptor expression reveals CpG DNA as a unique microbial stimulus for plasmacytoid dendritic cells which synergizes with CD40 ligand to induce high amounts of IL-12. *Eur. J. Immunol.*, **31**, 3026–3037.

Leadbetter, E.A., Rifkin, I.R., Hohlbaum, A.M., Beaudette, B.C., Shlomchik, M.J., and Marshak-Rothstein, A. (2002). Chromatin–IgG complexes activate B cells by dual engagement of IgM and toll-like receptors. *Nature*, **416**, 603–607.

Li, L., Cousart, S., Hu, J., and McCall, C.E. (2000). Characterization of interleukin-1 receptor-associated kinase in normal and endotoxin-tolerant cells. *J. Biol. Chem.*, **275**, 23340–23345.

Li, S., Strelow, A., Fontana, E.J., and Wesche, H. (2002). IRAK-4: A novel member of the IRAK family with the properties of an IRAK-kinase. *Proc. Natl. Acad. Sci. USA*, **99**, 5567–5572.

Li, X., Commane, M., Burns, C., Vithalani, K., Cao, Z., and Stark, G.R. (1999). Mutant cells that do not respond to interleukin-1 (IL-1) reveal a novel role for IL-1 receptor-associated kinase. *Mol. Cell. Biol.*, **19**, 4643–4652.

Medzhitov, R., Preston-Hurlburt, P., and Janeway, C.A., Jr. (1997). A human homologue of the Drosophila toll protein signals activation of adaptive immunity. *Nature*, **388**, 394–397.

Medzhitov, R., Preston-Hurlburt, P., Kopp, E., Stadlen, A., Chen, C., Ghosh, S., and Janeway, C.A., Jr. (1998). MyD88 is an adaptor protein in the hToll/IL-1 receptor family signaling pathways. *Mol. Cell*, **2**, 253–258.

Muzio, M., Ni, J., Feng, P., and Dixit, V.M. (1997). IRAK (Pelle) family member IRAK-2 and MyD88 as proximal mediators of IL-1 signaling. *Science*, **278**, 1612–1615.

Ninomiya-Tsuji, J., Kishimoto, K., Hiyama, A., Inoue, J., Cao, Z., and Matsumoto, K. (1999). The kinase TAK1 can activate the NIK-I kappaB as well as the MAP kinase cascade in the IL-1 signaling pathway. *Nature*, **398**, 252–256.

O'Neill, L.A. (2002). Signal transduction pathways activated by the IL-1 receptor/toll-like receptor superfamily. *Curr. Top. Microbiol. Immunol.*, **270**, 47–61.

O'Reilly, S.M. and Moynagh, P.N. (2003). Regulation of toll-like receptor 4 signaling by A20 zinc finger protein. *Biochem. Biophys. Res. Commun.*, **303**, 586–593.

Ozinsky, A., Underhill, D.M., Fontenot, J.D., Hajjar, A.M., Smith, K.D., Wilson, C.B., Schroeder, L., and Aderem, A. (2000). The repertoire for pattern recognition of pathogens by the innate immune system is defined by cooperation between toll-like receptors. *Proc. Natl. Acad. Sci. USA*, **97**, 13766–13771.

Poltorak, A., He, X., Smirnova, I., Liu, M.Y., Van Huffel, C., Du, X., Birdwell, D., Alejos, E., Silva, M., Galanos, C., Freudenberg, M., Ricciardi-Castagnoli, P., Layton, B., and Beutler, B. (1998). Defective LPS signaling in C3H/HeJ and C57BL/10ScCr mice: Mutations in Tlr4 gene. *Science*, **282**, 2085–2088.

Qureshi, S.T., Lariviere, L., Leveque, G., Clermont, S., Moore, K.J., Gros, P., and Malo, D. (1999). Endotoxin-tolerant mice have mutations in toll-like receptor 4 (Tlr4). *J. Exp. Med.*, **189**, 615–625.

Richardson, B., Scheinbart, L., Strahler, J., Gross, L., Hanash, S., and Johnson, M. (1990). Evidence for impaired T cell DNA methylation in systemic lupus erythematosus and rheumatoid arthritis. *Arthritis Rheumatol.*, **33**, 1665–1673.

Rui, L., Vinuesa, C.G., Blasioli, J., and Goodnow, C.C. (2003). Resistance to CpG DNA-induced autoimmunity through tolerogenic B cell antigen receptor ERK signaling. *Nat. Immunol.*, **4**, 594–600.

Suzuki, N., Suzuki, S., Duncan, G.S., Millar, D.G., Wada, T., Mirtsos, C., Takada, H., Wakeham, A., Itie, A., Li, S., Penninger, J.M., Wesche, H., Ohashi, P.S., Mak, T.W., and Yeh, W.C. (2002). Severe impairment of interleukin-1 and toll-like receptor signaling in mice lacking IRAK-4. *Nature*, **416**, 750–756.

Takeshita, F., Leifer, C.A., Gursel, I., Ishii, K.J., Takeshita, S., Gursel, M., and Klinman, D.M. (2001). Cutting edge: Role of toll-like receptor 9 in CpG DNA-induced activation of human cells. *J. Immunol.*, **167**, 3555–3558.

Takeuchi, O., Hoshino, K., Kawai, T., Sanjo, H., Takada, H., Ogawa, T., Takeda, K., and Akira, S. (1999). Differential roles of TLR2 and TLR4 in recognition of Gram-negative and Gram-positive bacterial cell wall components. *Immunity*, **11**, 443–451.

Viglianti, G.A., Lau, C.M., Hanley, T.M., Miko, B.A., Shlomchik, M.J., and Marshak-Rothstein, A. (2003). Activation of autoreactive B cells by CpG dsDNA. *Immunity*, **19**, 837–847.

Wald, D., Qin, J., Zhao, Z., Qian, Y., Naramura, M., Tian, L., Towne, J., Sims, J.E., Stark, G.R., and Li, X. (2003). SIGIRR, a negative regulator of toll-like receptor-interleukin 1 receptor signaling. *Nat. Immunol.*, **4**, 920–927.

Wang, H. and Shlomchik, M.J. (1999). Autoantigen-specific B cell activation in Fas-deficient rheumatoid factor immunoglobulin transgenic mice. *J. Exp. Med.*, **190**, 639–649.

Wesche, H., Henzel, W.J., Shillinglaw, W., Li, S., and Cao, Z. (1997). MyD88: An adapter that recruits IRAK to the IL-1 receptor complex. *Immunity*, **7**, 837–847.

Yamin, T.T. and Miller, D.K. (1997). The interleukin-1 receptor-associated kinase is degraded by proteosomes following its phosphorylation. *J. Biol. Chem.*, **272**, 21540–21547.

Yi, A.K., Chang, M., Peckham, D.W., Krieg, A.M., and Ashman, R.F. (1998a). CpG oligodeoxyribonucleotides rescue mature spleen B cells from spontaneous apoptosis and promote cell cycle entry. *J. Immunol.*, **160**, 5898–5906.

Yi, A.K. and Krieg, A.M. (1998a). CpG DNA rescue from anti-IgM-induced WEHI-231 B lymphoma apoptosis via modulation of I kappa B alpha and I kappa B beta and sustained activation of nuclear factor-kappa B/c-Rel. *J. Immunol.*, **160**, 1240–1245.

Yi, A.K. and Krieg, A.M. (1998b). Rapid induction of mitogen-activated protein kinases by immune stimulatory CpG DNA. *J. Immunol.*, **161**, 4493–4497.

Yi, A.K., Peckham, D.W., Ashman, R.F., and Krieg, A.M. (1999). CpG DNA rescues B cells from apoptosis by activating NFkappaB and preventing mitochondrial membrane potential disruption via a chloroquine-sensitive pathway. *Int. Immunol.*, **11**, 2015–2024.

Yi, A.K., Tuetken, R., Redford, T., Waldschmidt, M., Kirsch, J., and Krieg, A.M. (1998b). CpG motifs in bacterial DNA activate leukocytes through the pH-dependent generation of reactive oxygen species. *J. Immunol.*, **160**, 4755–4761.

Yi, A.K., Yoon, J.G., and Krieg, A.M. (2003). Convergence of CpG DNA- and BCR-mediated signals at the c-jun-N-terminal kinase and NF-kB activation pathways; regulation by mitogen-activated protein kinases. *Int. Immunol.*, **15**, 577–591.

Yung, R.L., Quddus, J., Chrisp, C.E., Johnson, K.J., and Richardson, B.C. (1995). Mechanism of drug-induced lupus. I. Cloned Th2 cells modified with DNA methylation inhibitors in vitro cause autoimmunity in vivo. *J. Immunol.*, **154**, 3025–3035.

3

C-Reactive Protein as a Regulator of Autoimmune Disease

Terry W. Du Clos and Carolyn Mold

1. Introduction

C-reactive protein (CRP) is a phylogenetically ancient, highly conserved component of the innate immune system. The pentraxin family, which includes CRP and serum amyloid P (SAP) component, is represented in all vertebrate species studied as well as in several invertebrates. Throughout evolution, pentraxins have retained similar amino acid sequence, structure, and calcium-dependent ligand-binding sites. CRP was identified and named for its ability to precipitate the C-polysaccharide of *Streptococcus pneumoniae* (Tillett and Francis, 1930) to which it binds through phosphocholine (PC) residues (Volanakis and Kaplan, 1971). CRP and SAP bind to microbial determinants, as well as to potential autoantigens, and interact with the adaptive immune system through the complement system and Fcγ receptors (FcγR). The general properties of CRP have been reviewed recently (Du Clos, 2000; Volanakis, 2001; Du Clos and Mold, 2003) and will only be described briefly here. The focus of the current review will be our current understanding of the properties of CRP that affect autoimmune disease.

2. Structural Features of CRP

CRP is a pentameric protein composed of five identical subunits in planar configuration (Thompson *et al.*, 1999) (Figure 3.1). Each CRP subunit contains a ligand-binding site on one face. Binding to PC and related ligands requires Ca^{++} ions, which interact with the phosphate moiety of PC. The opposite face of CRP is responsible for its binding to C1q (Agrawal *et al.*, 2001) and FcγR (Marnell *et al.*, 1995). The interactions of CRP with complement and with FcγR account

---

Terry W. Du Clos • The Department of Veterans Affairs Medical Center and the Department of Internal Medicine, University of New Mexico School of Medicine, Albuquerque, New Mexico 87108.
Carolyn Mold • The Department of Molecular Genetics and Microbiology, University of New Mexico School of Medicine, Albuquerque, New Mexico 87131.

Molecular Autoimmunity: In commemoration of the 100th anniversary of the first description of human autoimmune disease, edited by Moncef Zouali. Springer Science+Business Media, Inc., New York, 2005.

for many of its biological activities including host defense against infection, phagocytosis, and regulation of inflammation. The regulation of CRP synthesis as an acute-phase reactant and the ability of CRP to bind to damaged cells and nuclear antigens also contribute to its role in autoimmunity.

3. CRP as an Acute-Phase Reactant

A striking feature of CRP and a key to its function is the regulation of its synthesis (reviewed in Ballou and Kushner, 1992; Volanakis, 2001). CRP is the prototypic acute-phase reactant in man. CRP synthesis is rapidly induced in hepatocytes by interleukin-6 (IL-6) and other cytokines causing baseline serum levels of <1 µg/ml to rapidly increase to several hundred micrograms per milliliter after an acute stimulus. CRP is also rapidly cleared from the circulation and its serum concentration is a sensitive indicator of inflammation in rheumatic and infectious diseases (Du Clos, 2000).

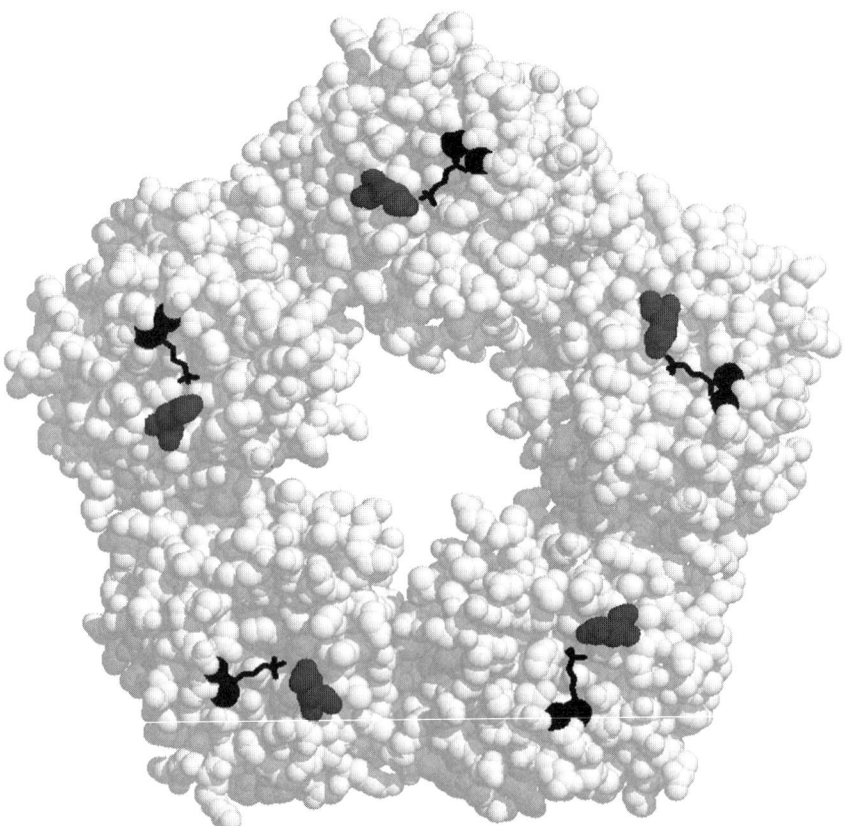

Figure 3.1A. C-reactive protein (CRP) is a cyclic pentamer. Space-fill diagram of the ligand-binding face of CRP showing phosphocholine (PC) interacting with the bound calcium ions and glutamic acid at position 88 (Thompson *et al.*, 1999).

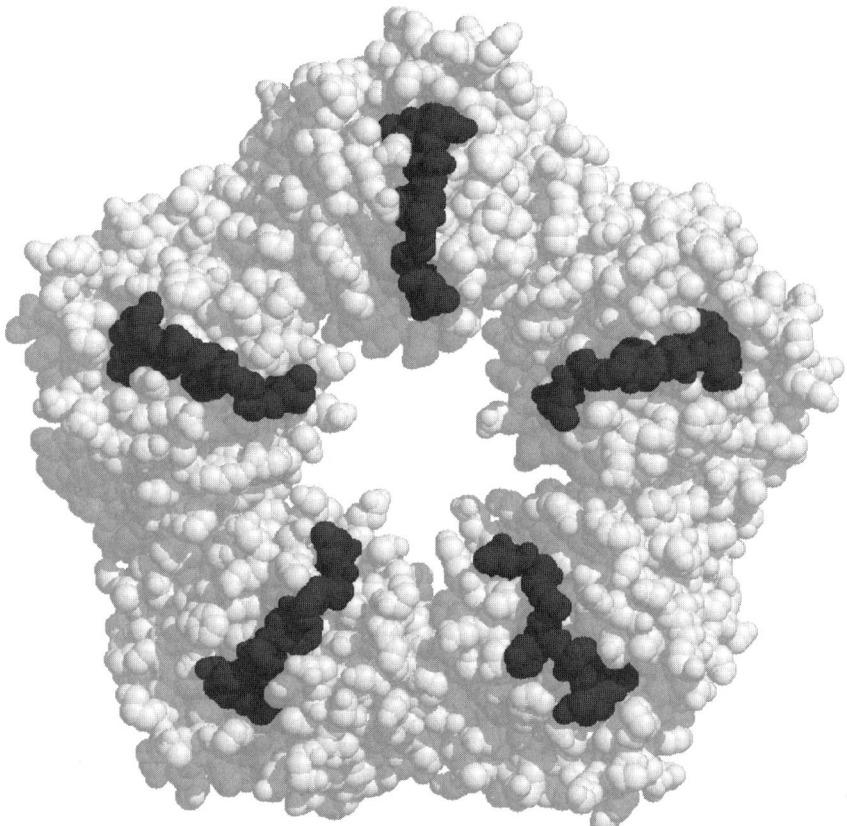

Figure 3.1B. *Continued* Space-fill diagram of the opposite face of CRP showing the region homologous to the FcγR-binding site on IgG (amino acids 175–185). Several residues in this region affect CRP binding to FcγR.

More recently even low elevations of CRP have been shown to have predictive value in assessing cardiovascular risk (Ridker, 2003). Although the use of CRP as a predictive marker for cardiovascular events has been supported by many studies, controversy remains as to whether elevated CRP contributes directly to cardiovascular disease or is merely a very sensitive indicator of inflammation. A recent study using apolipoprotein E–deficient mice transgenic for human CRP provided direct evidence that CRP may contribute to the development of aortic lesions (Paul *et al.*, 2004). CRP binds to lipoproteins, phospholipids, and apoptotic cells present in atherosclerotic lesions and may interact locally with complement or with macrophages through FcγR (Zwaka *et al.*, 2001).

4. CRP Interaction with Nuclear Antigens

The localization of CRP to sites of tissue injury was recognized early (Kushner and Kaplan, 1961). CRP-binding sites were believed to be membrane phospholipids on damaged cells. CRP binds to the PC head group of phosphatidylcholine in cell membranes damaged by detergent or complement (Narkates and Volanakis, 1982; Li *et al.*, 1994).

The first indication that CRP also binds to nuclear antigens was the report by Gitlin *et al.* (1977) of intense staining for CRP detected by immunofluorescence in the nuclei of cells from rheumatoid arthritis synovium. Robey *et al.* (1984) later reported localization of fluorescently labeled CRP to the nuclei of damaged fibroblasts and attributed this to an interaction of CRP with chromatin. On further investigation, we found that CRP actively localizes to the nuclei of cells following microinjection (Du Clos *et al.*, 1990) where it binds to small nuclear ribonucleoprotein (snRNP) particles (Figure 3.2, see color insert) (Du Clos, 1989). SAP also has a nuclear localization signal, but binds to chromatin rather than to snRNP in intact cells (Du Clos *et al.*, 1990; Pepys *et al.*, 1994). We further investigated the binding of CRP to isolated nuclear antigens, and found that CRP binds through its PC-binding sites to a conserved peptide motif on histones (H1, H2A, and H2B), isolated chromatin and nucleosomes, and the snRNP proteins, SmD and the U1 70K protein (reviewed in Du Clos, 1996). These findings led to studies on the possible role of CRP in the clearance of nuclear antigens, which are prominent targets of the autoimmune response in systemic lupus erythematosus (SLE).

5. CRP, SAP, and Nuclear Antigen Clearance

The ability of CRP to promote phagocytosis and clearance of its ligands and to bind to nuclear antigens suggested a role in removal of potential autoantigens such as chromatin and snRNP. In an initial study, CRP injection did not have a major effect on the rate of clearance of mononucleosomes from the circulation of BALB/c mice (Du Clos *et al.*, 1994). However, CRP effects on chromatin clearance were seen in a later study using H1-stripped chromatin and taking into account the presence of SAP (Burlingame *et al.*, 1996). SAP binds to DNA, histones, and chromatin (Pepys and Butler, 1987; Hicks *et al.*, 1992). The clearance of H1-stripped chromatin was markedly slower in mice with high serum SAP levels (BALB/c) compared to mice with low serum SAP levels (C57BL/10). Induction of an acute-phase response in C57BL/10 mice increased SAP levels and decreased both the rate of chromatin clearance and the localization of chromatin in the kidney. Coinjection of either SAP or CRP with chromatin into C57BL/10 mice also decreased chromatin clearance. Recently the rate of clearance of long native chromatin was found to be accelerated in SAP$^{-/-}$ mice (Bickerstaff *et al.*, 1999). These findings suggested that CRP and SAP produce major changes in the rate and path of chromatin clearance that could affect its presentation to the immune system.

Many investigators believe that the autoantibody response in SLE is driven by endogenous nuclear antigens released from dying cells (Walport, 2000). This hypothesis is based on the characteristics of the autoantibody response to chromatin in SLE (Burlingame *et al.*, 1993, 1994), on the identification of novel epitopes on autoantigens exposed on apoptotic cells (Casciola-Rosen *et al.*, 1994), and on the association of deficiencies in apoptotic cell clearance with the

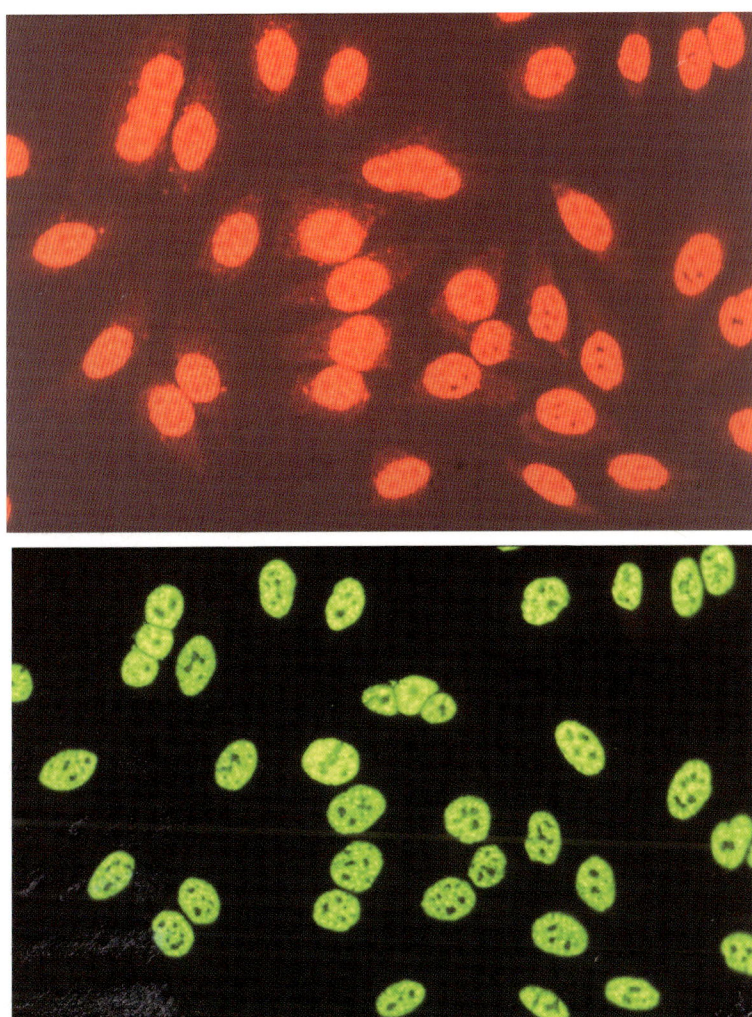

Figure 3.2. C-reactive protein (CRP) binding to small nuclear ribonucleoprotein (snRNP) in cell nuclei. Top: Antinuclear antibody slides prepared with fixed HEp2 cells were incubated with CRP, mAb to CRP and rhodamine-anti-mouse IgG. Bottom: The pattern of staining was identical to that seen with serum from a patient with anti-RNP antibody and FITC-anti-human IgG (Du Clos, 1989).

development of autoimmunity (Carroll, 2000; Walport, 2000). In addition to their effects on clearance of chromatin, both CRP and SAP also bind to apoptotic cells (Gershov et al., 2000; Familian et al., 2001). CRP increases the uptake of apoptotic cells by macrophages through complement-dependent opsonization (Gershov et al., 2000), and both CRP and SAP enhance FcγR-dependent phagocytosis of apoptotic cells (Mold et al., 2002a). Gershov et al. (2000) reported that opsonization of apoptotic cells by CRP and complement has anti-inflammatory effects.

6. CRP Genetics and Autoimmunity

If CRP and SAP play important roles in the clearance of nuclear antigens then deficiency of these proteins is expected to predispose to autoimmune disease. There are no known deficiencies of either CRP or SAP in man. However, a recent study of SLE families has suggested that genetically determined low baseline levels of CRP might be associated with development of SLE. Two polymorphisms at the CRP locus that affected basal levels of CRP were identified. One of these polymorphisms, associated with reduced CRP expression, was linked with SLE and antinuclear autoantibody production (Russell et al., 2004).

Other polymorphisms of CRP that affect baseline levels have been described, but have not thus far been associated with risk of autoimmune disease (Szalai et al., 2002b; Zee and Ridker, 2002; Brull et al., 2003). Polymorphisms of cytokine genes such as IL-6 and IL-1 may also influence levels of CRP in the blood (Berger et al., 2002; Vickers et al., 2002).

In mouse, unlike in man, SAP is a major acute-phase reactant and CRP is found only at low levels (Pepys et al., 1979; Whitehead et al., 1990). Two groups who independently produced SAP$^{-/-}$ mice by targeted gene deletion have studied the role of SAP in autoimmunity. In each case a greater percentage of the SAP-deficient mice produced antinuclear antibodies compared to the SAP$^{+/+}$ mice, and for one strain, SAP$^{-/-}$ was associated with nephritis in female animals (Bickerstaff et al., 1999; Soma et al., 2001).

7. CRP Levels in Human SLE

In addition to genetic effects on CRP synthesis, there is evidence for altered induction of CRP synthesis in human SLE. A number of studies have concluded that CRP levels are abnormally low in comparison to other measures of inflammation during flares of SLE (Becker et al., 1980). Despite these low CRP levels during the course of SLE, the same patients may produce a vigorous CRP response to infection, leading to the suggestion that elevated CRP levels in lupus may be used as an indicator of infection or serositis (Zein et al., 1979; ter Borg et al., 1990). The reason for the low CRP response during the course of SLE is unknown, but it has been reported that in contrast to other inflammatory diseases, in SLE serum CRP and IL-6 levels do not correlate (Gabay et al., 1993).

8. CRP in Animal Models of Autoimmunity

We, and others, have taken advantage of the low levels of CRP in mouse, and the ability of human CRP to interact with mouse complement and FcγR to study human CRP as a therapeutic agent in mouse models of infection and autoimmunity (Mold et al., 1981, 2000b; Du Clos et al., 1994; Szalai et al., 2002a, 2003). Several studies have tested CRP as a therapeutic agent in the (NZB × NZW)F$_1$ female mouse model (NZB/W). These mice develop antinuclear antibodies and proteinuria at about 20 weeks of age and then undergo a progressive increase in proteinuria, which is associated with renal failure, severe glomerulonephritis, and 50% mortality by about 34 weeks of age (Theofilopoulos and Dixon, 1985). The first study of CRP in the treatment of SLE was made in our laboratory using NZB/W mice. Our hypothesis was that CRP might provide protection from immunization with nuclear autoantigens. We injected chromatin into NZB/W mice to accelerate disease and examined the effect of adding CRP with the antigen (Du Clos et al., 1994). Chromatin injection did in fact accelerate disease, and mice treated with CRP and chromatin survived significantly longer than mice treated with chromatin alone. Consistent with the hypothesis that CRP could prevent immunization with chromatin, antibody responses to histones and DNA were transiently suppressed in the CRP-treated mice. However, antibody responses to an irrelevant antigen were also transiently suppressed, suggesting a more general effect of CRP on the immune system.

Szalai et al. (2002a, 2003) have developed mouse strains that express human CRP as an acute-phase reactant from a transgene, and used these mice to study the effect of CRP on autoimmune disease. These investigators first examined experimental autoimmune encephalomyelitis, an induced autoimmune disease, in CRP transgenic and control mice (Szalai et al., 2002a). CRP transgenic mice had delayed onset of disease in response to immunization with an encephalitogenic peptide. This effect of CRP was associated with the ability of CRP to inhibit proliferation and production of inflammatory cytokines and chemokines. CRP also increased production of the anti-inflammatory cytokine IL-10 in cultures containing antigen-presenting cells, encephalitogenic T cells, and the peptide antigen.

More recently, these investigators have bred the human CRP transgene onto the NZB/W background and studied the development of SLE (Szalai et al., 2003). The results confirmed the protective effect of CRP in this SLE model. Despite the expression of <1 μg/ml of human CRP, CRP transgenic mice developed proteinuria later and survived longer than controls. The CRP transgenic mice did not have decreased anti-dsDNA antibodies. CRP transgenic mice had delayed accumulation of immunoglobulin in glomeruli with greater mesangial localization.

We recently initiated a new series of studies using a single injection of a high concentration of CRP (200 μg/mouse) into NZB/W mice either before or after the onset of renal disease (Rodriguez et al., 2004). NZB/W mice that received CRP treatment at 18 weeks of age had a marked delay in the onset of renal disease and significantly prolonged survival. Untreated NZB/W mice developed 3+ proteinuria at a median age of 26.5 weeks compared to 42.5 weeks for the CRP-treated mice ($p < 0.0001$). CRP treatment extended survival to a median age of 49 weeks compared to 36 weeks for untreated NZB/W mice ($p < 0.005$). These effects on proteinuria and survival are greater than those observed previously, and were not associated with differences in anti-dsDNA or antinuclear antibodies in CRP-treated mice. In the same experiment, NZB/W mice that had developed 4+ proteinuria (> 500 mg/dl) were injected with 200 μg of CRP at 30 weeks of age to determine whether CRP would affect ongoing renal disease. This late CRP treatment resulted in a rapid decrease in proteinuria within 2 days of treatment. The mice became completely free of measurable proteinuria and significant proteinuria did not return until 12 weeks after CRP treatment. Interestingly, mice that were treated with CRP at 30 weeks of age again developed proteinuria at a similar age as the mice treated with CRP at 18 weeks of age and had a similar median survival age of 46 weeks ($p < 0.005$ vs. untreated). The ability of CRP to treat ongoing renal disease in NZB/W mice suggests that the protective effect of CRP in SLE is independent of nuclear antigen processing. Further, there is no evidence from the latter studies that CRP inhibits autoantibody responses.

The MRL/lpr mouse is a rapid onset model of autoimmune disease accelerated by homozygosity of the lpr gene, which encodes a mutant form of the death receptor Fas. On the autoimmune MRL background, homozygosity of the lpr gene causes a more rapid development of SLE. In addition to antinuclear antibodies and glomerulonephritis, MRL/lpr mice develop lymphadenopathy and vasculitic skin lesions (Theofilopoulos and Dixon, 1985). Using the single high dose CRP treatment, we have observed delayed onset of proteinuria in the MRL/lpr model similar to the results in NZB/W mice. In MRL/lpr mice, CRP treatment at 6 weeks of age also delayed the development of anti-dsDNA antibodies and lymphadenopathy. Injection of CRP after the onset of renal disease reversed proteinuria and prolonged survival, but did not affect anti-dsDNA levels.

9. CRP in Immune Complex Nephritis

The most striking effects of CRP in mouse models of SLE were the prevention and reversal of proteinuria in two different mouse strains. The original hypothesis that CRP can prevent the presentation of nuclear antigens to the immune system would predict that CRP would be effective before disease development and that CRP treatment would be associated with decreased autoantibody responses to these antigens. However, CRP treatment does not consistently suppress autoantibody levels, and treatment after disease development is rapidly effective. These findings suggest that although CRP may alter processing and presentation of antigens from dead and dying cells, its primary effect in these SLE models is directed at the inflammatory response in the end organ, the kidney. To test this idea, we studied the protective effect of CRP in nephrotoxic nephritis (NTN), a non-autoimmune model of immune complex–mediated glomerulonephritis. To induce NTN, mice are injected with rabbit IgG in adjuvant followed 1 week later by rabbit antibody to mouse glomerular antigens (anti-GBM). The result is the rapid development of proteinuria, with pathological changes in the glomeruli over a period of several weeks that include mesangial hypercellularity, inflammatory infiltrates, fibrin deposition, and crescent formation. The development of these later changes is dependent on FcγRI and FcγRIII (Tarzi et al., 2003), and in particular the influx into the lesions of macrophages expressing these FcγR (Tarzi et al., 2002). Since NTN and lupus nephritis are both immune complex–mediated and FcγR-dependent (Clynes et al., 1998) inflammatory diseases, we postulated that any effect of CRP on renal disease in SLE would also be seen in NTN. We did in fact observe that treatment of mice with CRP at the time of anti-GBM administration prevented the development of proteinuria and pathologic changes in the kidneys (Rodriguez et al., 2004). In addition, CRP treatment 9 days after anti-GBM reversed proteinuria and significantly decreased the pathologic changes in the kidneys within 2 days. These findings support the hypothesis that it is the anti-inflammatory role of CRP in these diseases that is of primary importance.

10. CRP in Inflammation

The ability of CRP to suppress acute inflammation has been well documented in studies using mice transgenic for rabbit CRP developed by Samols et al. In these mice the level of CRP expression can be regulated by diet. These investigators have examined several models of inflammation in these mice after dietary manipulation. In systemic inflammation, mice expressing high levels of CRP were protected from lethal shock induced by endotoxin (LPS), platelet-activating factor (PAF), or the combination of tumor necrosis factor (TNF) and IL-1, but not TNF alone (Xia and Samols, 1997). These results suggest that CRP may regulate the inflammatory cytokine cascade or the response to it. Transgenic mice expressing high levels of CRP and challenged by intratracheal administration of chemotactic agents (C5a, FMLP, leukotriene B_4, or IL-8) showed less neutrophil influx and less protein in bronchoalveolar lavage than mice expressing low levels of CRP (Heuertz et al., 1994; Ahmed et al., 1996). The mechanisms of these effects were not well understood, although a direct inhibitory effect of CRP on neutrophil chemotaxis was proposed to account for protection from alveolitis (Heuertz et al., 1999).

11. Identification of FcγR as CRP Receptors

To clarify the mechanism(s) for the diverse activities reported for CRP in phagocytosis, inflammation, and cytokine responses, it was essential to identify the receptor for CRP on leukocytes. Multiple investigations over the years by Mortensen and others had established specific binding of CRP to monocytes, macrophages, myeloid cell lines, neutrophils, and platelets. In many cases this binding was inhibited by aggregated IgG, leading to the proposal that CRP interacted with FcγR (Mortensen and Duszkiewicz, 1977). However, based on the inability of mAb to

FcγR to inhibit CRP binding, the lack of CRP inhibition of IgG binding, and other experiments, these same investigators later proposed a unique CRP receptor on mouse and human monocytes (Zahedi *et al.*, 1989; Tebo and Mortensen, 1990).

The cloning of individual human FcγR and the development of FcγR$^{-/-}$ mice made possible the identification by our laboratory of FcγRI and FcγRII as the leukocyte receptors for CRP in both man and mouse. We first determined by immunoprecipitation and cross-linking that CRP binds to the high-affinity receptor for IgG, FcγRI (Crowell *et al.*, 1991). This was confirmed using cells transfected with FcγRI (Marnell *et al.*, 1995). However, cells that lacked or displayed low levels of FcγRI such as neutrophils, platelets, and the K562 cell line all bound CRP, indicating that there must be a second CRP receptor. We turned our attention to FcγRIIa, because of a strong correlation between expression of FcγRIIa by various cells and the ability to bind CRP. We demonstrated specific binding of CRP to COS cells transfected with human FcγRIIa. CRP binding to FcγRIIa-transfected COS cells and K562 cells, which express FcγRIIa, was inhibited by aggregated IgG at nearly identical concentrations (Bharadwaj *et al.*, 1999). The identification of FcγRIIa as a CRP receptor was confirmed by functional studies showing FcγR-dependent phagocytosis of CRP-opsonized zymosan by human neutrophils (Bharadwaj *et al.*, 2001). In surface plasmon resonance studies by Bodman-Smith *et al.* (2002), using immobilized extracellular domains of FcγRI and soluble CRP, the binding affinity of CRP for FcγRI was 10-fold higher than that of IgG. These investigators have also demonstrated phagocytosis of CRP-opsonized erythrocytes through human FcγRI and FcγRIIa in transfected COS cells (Bodman-Smith *et al.*, 2002, 2004). Other investigators have confirmed CRP-mediated signaling through FcγRI (Han *et al.*, 2004) and FcγRIIa on human myeloid cell lines (Chi *et al.*, 2002).

To identify CRP receptors in the mouse, we used strains of mice genetically deficient in FcγR. In short, these studies demonstrated that CRP completely failed to bind to cells from mice lacking all three FcγRs (Stein *et al.*, 2000). Mice lacking only FcγRII bound less CRP than controls and mice lacking FcγRI and FcγRIII also had decreased binding. There was no evidence that FcγRIII$^{-/-}$ mice had altered CRP binding and natural killer (NK) cells that express only FcγRIII did not bind CRP. Moreover, the phagocytosis of CRP-opsonized zymosan by mouse macrophages was dependent on FcγRI (Mold *et al.*, 2001). Thus, we concluded that in the mouse, as in the human, FcγRI and FcγRII function as the receptors for CRP. Mouse FcγR differ from human receptors in that FcγRII in man is found in three forms, two with activating signaling motifs (ITAMs) and one with an inhibitory signaling motif (ITIM). In the mouse only inhibitory forms of FcγRIIb are found (Ravetch and Bolland, 2001).

12. Role of FcγR in CRP Effects on Inflammation

The identification of FcγRI and FcγRIIb as the receptors for CRP in the mouse provided the means to determine whether these receptors were important for the regulation of the inflammatory response. We injected human CRP into normal and FcγR$^{-/-}$ mice and challenged with high-dose LPS (Mold *et al.*, 2002c) (Figure 3.3). In normal mice, CRP increased survival following LPS. CRP treatment of normal mice also increased the IL-10 response to LPS, both *in vivo* and in bone marrow macrophage cultures. IL-10 is an anti-inflammatory cytokine that has protective effects in LPS shock. CRP did not protect FcR γ-chain$^{-/-}$ mice and did not enhance the IL-10 response to LPS either *in vivo* or in macrophage cultures. We assume that CRP is acting through FcγRI, because CRP binds to this γ-chain-associated receptor (Stein *et al.*, 2000). Mosser *et al.* have described enhanced IL-10 synthesis in macrophages cultures stimulated with LPS and IgG immune complexes (Sutterwala *et al.*, 1998; Gerber and Mosser, 2001).

In addition to IL-10, CRP binding to FcγR may induce synthesis of other anti-inflammatory cytokines by macrophages. Tilg *et al.* (1993) demonstrated increased IL-1 receptor antagonist (IL-1RA) synthesis in human peripheral blood monocytes responding to LPS. Cross-linking FcγR is an effective stimulus for IL1-RA production by monocytes and macrophages, and IL1-RA is a potent inhibitor of the proinflammatory effects of IL-1 (Arend *et al.*, 1998).

FcγRIIb is a regulatory receptor found on B lymphocytes, monocytes, macrophages, neutrophils, mast cells, and basophils (Ravetch and Lanier, 2000). Cross-linking of FcγRIIb results in recruitment of phosphatases that inhibit the activation pathways of ITAM-containing receptors such as the B cell antigen

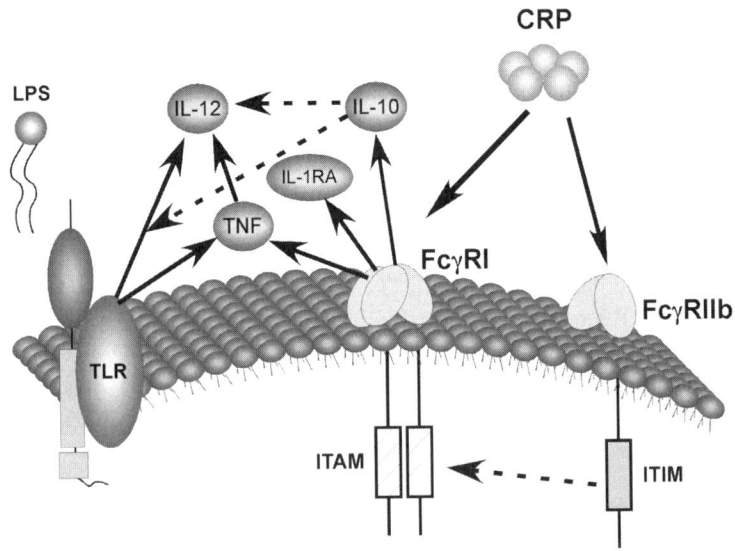

Figure 3.3. Model for the role of FcγR in C-reactive protein (CRP)–mediated protection against endotoxin challenge. Normal mice injected with a lethal dose of LPS were protected by CRP injection. This protection required the FcR γ-chain, used by FcγRI for signaling, and was associated with an enhanced IL-10 response. CRP induction of IL-10 synthesis through FcγRI is proposed to downregulate the inflammatory cytokine response to LPS and protect mice from lethality. CRP increased lethality as well as the TNF, IL-12, and Il-10 responses in FcγRIIb-deficient mice injected with LPS. This indicates an important regulatory role for CRP binding to FcγRIIb (Mold *et al.*, 2002c).

receptor, FcγRI, FcγRIII, and FcεRI. Since CRP binds to FcγRIIb, we tested the role of this receptor in CRP-mediated protection from LPS shock (Mold et al., 2002c). Surprisingly, CRP treatment increased the lethality of LPS in FcγRIIb-deficient mice. This increased lethality was associated with increased production of the proinflammatory cytokines TNF and IL-12 in mice injected with CRP and low-dose LPS. CRP did not affect levels of TNF or IL-12 in wild-type mice, indicating a regulatory role for CRP binding to FcγRIIb. Figure 3.3 shows diagrammatically the two regulatory pathways for CRP in systemic inflammation.

13. Essential Role of IL-10 in Anti-inflammatory Activities of CRP

IL-10 is a cytokine with potent anti-inflammatory activity (Moore et al., 2001). IL-10 inhibits the production of inflammatory cytokines by macrophages, NK cells, and Th1 cells and thus prevents excessive damage to the host following infection or exposure to microbial products such as LPS. The induction of acute-phase protein synthesis by IL-6 is part of the innate inflammatory response to microbial determinants. Elevated CRP concentrations will be found at the time when it is crucial to regulate potentially harmful cytokines such as TNF, IL-12, IL-1, IL-18, and IFN-γ. Thus, the ability of CRP to stimulate IL-10 synthesis may help control the innate immune response, preventing damage to the host. If this regulatory pathway is important in autoimmune inflammatory disease, then the ability of CRP to prevent and reverse nephritis is expected to require IL-10 and FcγRI. We tested the ability of CRP to protect IL-10-deficient mice in the NTN model. CRP had no effect in either preventing or reversing proteinuria in IL-10-deficient mice injected with anti-GBM (Rodriguez et al., 2004). This indicates that the short-term anti-inflammatory activity of CRP may be mediated by IL-10. Studies are under way to determine whether FcγRI is required for CRP protection in NTN.

14. Current Perspective on CRP in Autoimmune Disease

CRP is a mediator of the innate immune system, which binds to damaged cells and nuclear antigens at the cell surface and in the nucleus. When an acute-phase response is initiated by tissue injury, infection, or inflammation, CRP synthesis is stimulated and serum CRP localizes at the damaged tissue site. CRP also activates the complement cascade and helps to initiate and direct the adaptive immune response. Furthermore, CRP is capable of initiating phagocytosis and modulating cytokine responses through interactions with the FcγR. One function of these interactions with nuclear antigens is altered clearance of nuclear antigens. In addition, CRP and complement may increase uptake of dying cells and enhance TGFβ synthesis by macrophages (Gershov et al., 2000). Another important function of CRP is likely to be the generation of anti-inflammatory cytokines like IL-10 and IL-1RA. This response is dependent on interaction with FcγRI, which is expressed at increased levels on phagocytic cells exposed to interferon-γ. The timing of peak CRP synthesis 24–48 hr after an inflammatory stimulus is expected to coincide with increased receptor expression and to occur at an appropriate time for regulating the innate cytokine response. The net effect of CRP on inflammatory responses and autoimmune diseases is to decrease tissue injury.

References

Agrawal, A., Shrive, A.K., Greenhough, T.J., and Volanakis, J.E. (2001). Topology and structure of the C1q-binding site on C-reactive protein. *J. Immunol.*, **166**, 3998–4004.

Ahmed, N., Thorley, R., Xia, D., Samols, D., and Webster, R.O. (1996). Transgenic mice expressing rabbit C-reactive protein exhibit diminished chemotactic factor-induced alveolitis. *Am. J. Respir. Crit. Care Med.*, **153**, 1141–1147.

Arend, W.P., Malyak, M., Guthridge, C.J., and Gabay, C. (1998). Interleukin-1 receptor antagonist: Role in biology. *Annu. Rev. Immunol.*, **16**, 27–55.

Ballou, S.P. and Kushner, I. (1992). C-reactive protein and the acute phase response. *Adv. Intern. Med.*, **37**, 313–336.

Becker, G.J., Waldburger, M., Hughes, G.R., and Pepys, M.B. (1980). Value of serum C-reactive protein measurement in the investigation of fever in systemic lupus erythematosus. *Ann. Rheum. Dis.*, **39**, 50–52.

Berger, P., McConnell, J.P., Nunn, M., Kornman, KS., Sorrell, J., Stephenson, K., and Duff, G.W. (2002). C-reactive protein levels are influenced by common IL-1 gene variations. *Cytokine*, **17**, 171–174.

Bharadwaj, D., Mold, C., Markham, E., and Du Clos, T.W. (2001). Serum amyloid P component binds to Fcγ receptors and opsonizes particles for phagocytosis. *J. Immunol.*, **166**, 6735–6741.

Bharadwaj, D., Stein, M.P., Volzer, M., Mold, C., and Du Clos, T.W. (1999). The major receptor for C-reactive protein on leukocytes is Fcγ receptor II. *J. Exp. Med.*, **190**, 585–590.

Bickerstaff, M.C., Botto, M., Hutchinson, W.L., Herbert, J., Tennent, G.A., Bybee, A., Mitchell, D.A., Cook, H.T., Butler, P.J., Walport, M.J., and Pepys, M.B. (1999). Serum amyloid P component controls chromatin degradation and prevents antinuclear autoimmunity. *Nat. Med.*, **5**, 694–697.

Bodman-Smith, K.B., Gregory, R.E., Harrison, P.T., and Raynes, J.G. (2004). FcγRIIa expression with FcγRI results in C-reactive protein- and IgG-mediated phagocytosis. *J. Leukoc. Biol.*, **75**, 1029–1035.

Bodman-Smith, K.B., Melendez, A.J., Campbell, I., Harrison, P.T., Allen, J.M., and Raynes, J.G. (2002). C-reactive protein-mediated phagocytosis and phospholipase D signaling through the high-affinity receptor for immunoglobulin G (FcγRI). *Immunology*, **107**, 252–260.

Brull, D.J., Serrano, N., Zito, F., Jones, L., Montgomery, H.E., Rumley, A., Sharma, P., Lowe, G.D., World, M.J., Humphries, S.E., and Hingorani, A.D. (2003). Human CRP gene polymorphism influences CRP levels: Implications for the prediction and pathogenesis of coronary heart disease. *Arterioscler. Thromb. Vasc. Biol.*, **23**, 2063–2069.

Burlingame, R.W., Boey, M.L., Starkebaum, G., and Rubin, R.L. (1994). The central role of chromatin in autoimmune responses to histones and DNA in systemic lupus erythematosus. *J. Clin. Invest.*, **94**, 184–192.

Burlingame, R.W., Rubin, R.L., Balderas, R.S., and Theofilopoulos, A.N. (1993). Genesis and evolution of antichromatin autoantibodies in murine lupus implicates T-dependent immunization with self antigen. *J. Clin. Invest.*, **91**, 1687–1696.

Burlingame, R.W., Volzer, M.A., Harris, J., and Du Clos, T.W. (1996). The effect of acute phase proteins on clearance of chromatin from the circulation of normal mice. *J. Immunol.*, **156**, 4783–4788.

Carroll, M.C. (2000). A protective role for innate immunity in autoimmune disease. *Clin. Immunol.*, **95**, S30–S38.

Casciola-Rosen, L.A., Anhalt, G., and Rosen, A. (1994). Autoantigens targeted in systemic lupus erythematosus are clustered in two populations of surface structures on apoptotic keratinocytes. *J. Exp. Med.*, **179**, 1317–1330.

Chi, M., Tridandapani, S., Zhong, W., Coggeshall, K.M., and Mortensen, R.F. (2002). C-reactive protein induces signaling through FcγRIIa on HL-60 granulocytes. *J. Immunol.*, **168**, 1413–1418.

Clynes, R., Dumitru, C., and Ravetch, J.V. (1998). Uncoupling of immune complex formation and kidney damage in autoimmune glomerulonephritis. *Science*, **279**, 1052–1054.

Crowell, R.E., Du Clos, T.W., Montoya, G., Heaphy, E., and Mold, C. (1991). C-reactive protein receptors on the human monocytic cell line U-937. Evidence for additional binding to Fc gamma RI. *J. Immunol.*, **147**, 3445–3451.

Du Clos, T.W. (1989). C-reactive protein reacts with the U1 small nuclear ribonucleoprotein. *J. Immunol.*, **143**, 2553–2559.

Du Clos, T.W. (1996). The interaction of C-reactive protein and serum amyloid P component with nuclear antigens. *Mol. Biol. Rep.*, **23**, 253–260.

Du Clos, T.W. (2000). Function of C-reactive protein. *Ann. Med.*, **32**, 274–278.

Du Clos, T.W. and Mold, C. (2003). C-reactive protein: Structure, synthesis and function. In S.Y. Wong and G. Arsequell (eds), *Immunobiology of Carbohydrates*. Kluwer Academic/Plenum Publishers, New York. pp. 39–55.

Du Clos, T.W., Mold, C., and Stump, R.F. (1990). Identification of a polypeptide sequence that mediates nuclear localization of the acute phase protein C-reactive protein. *J. Immunol.*, **145**, 3869–3875.

Du Clos, T.W., Zlock, L.T., Hicks, P.S., and Mold, C. (1994). Decreased autoantibody levels and enhanced survival of (NZB × NZW) F1 mice treated with C-reactive protein. *Clin. Immunol. Immunopathol.*, **70**, 22–27.

Familian, A., Zwart, B., Huisman, H.G., Rensink, I., Roem, D., Hordijk, P.L., Aarden, L.A., and Hack, C.E. (2001). Chromatin-independent binding of serum amyloid P component to apoptotic cells. *J. Immunol.*, **167**, 647–654.

Gabay, C., Roux-Lombard, P., de Moerloose, P., Dayer, J.M., Vischer, T., and Guerne, P.A. (1993). Absence of correlation between interleukin 6 and C-reactive protein blood levels in systemic lupus erythematosus compared with rheumatoid arthritis. *J. Rheumatol.*, **20**, 815–821.

Gerber, J.S. and Mosser, D.M. (2001). Reversing lipopolysaccharide toxicity by ligating the macrophage Fcγ receptors. *J. Immunol.*, **166**, 6861–6868.

Gershov, D., Kim, S., Brot, N., and Elkon, K.B. (2000). C-reactive protein binds to apoptotic cells, protects the cells from assembly of the terminal complement components, and sustains an antiinflammatory innate immune response: Implications for systemic autoimmunity. *J. Exp. Med.*, **192**, 1353–1363.

Gitlin, J.D., Gitlin, J.I., and Gitlin, D. (1977). Localization of C-reactive protein in synovium of patients with rheumatoid arthritis. *Arthritis Rheum.*, **20**, 1491–1499.

Han, K.H., Hong, K.H., Park, J.H., Ko, J., Kang, D.H., Choi, K.J., Hong, M.K., Park, S.W., and Park, S.J. (2004). C-reactive protein promotes monocyte chemoattractant protein-1-mediated chemotaxis through upregulating CC chemokine receptor 2 expression in human monocytes. *Circulation*, 2566–2571.

Heuertz, R.M., Dongyuan, X., Samols, D., and Webster, R.O. (1994). Inhibition of C5a des Arg-induced neutrophil alveolitis in transgenic mice expressing C-reactive protein. *Am. J. Physiol.*, **266**, L649–L654.

Heuertz, R.M., Tricomi, S.M., Ezekieli, U.R., and Webster, R.O. (1999). C-reactive protein inhibits chemotactic peptide-induced p38 mitogen-activated protein kinase activity and human neutrophil movement. *J. Biol. Chem.*, **275**, 17968–17974.

Hicks, P.S., Saunero-Nava, L., Du Clos, T.W., and Mold, C. (1992). Serum amyloid P component binds to histones and activates the classical complement pathway. *J. Immunol.*, **149**, 3689–3694.

Kushner, I. and Kaplan, M.H. (1961). Studies of acute phase protein. I. An immunohistochemical method for the localization of Cx-reactive protein in rabbits: Association with necrosis in inflammatory lesions. *J. Exp. Med.*, **114**, 961–973.

Li, Y.P., Mold, C., and Du Clos, T.W. (1994). Sublytic complement attack exposes C-reactive protein binding sites on cell membranes. *J. Immunol.*, **152**, 2995–3005.

Marnell, L.L., Mold, C., Volzer, M.A., Burlingame, R.W., and Du Clos, T.W. (1995). C-reactive protein binds to Fc gamma RI in transfected COS cells. *J. Immunol.*, **155**, 2185–2193.

Mold, C., Baca, R., and Du Clos, T.W. (2002a). Serum amyloid P component and C-reactive protein opsonize apoptotic cells for phagocytosis through Fcγ receptors. *J. Autoimmunity*, **19**, 147–154.

Mold, C., Gresham, H.D., and Du Clos, T.W. (2001). Serum amyloid P component and C-reactive protein mediate phagocytosis through murine FcγRs. *J. Immunol.*, **166**, 1200–1205.

Mold, C., Nakayama, S., Holzer, T.J., Gewurz, H., and Du Clos, T.W. (1981). C-reactive protein is protective against *Streptococcus pneumoniae* infection in mice. *J. Exp. Med.*, **154**, 1703–1708.

Mold, C., Rodic-Polic, B., and Du Clos, T.W. (2002b). Protection from *Streptococcus pneumoniae* infection by C-reactive protein and natural antibody requires complement but not Fcγ receptors. *J. Immunol.*, **168**, 6375–6381.

Mold, C., Rodriguez, W., Rodic-Polic, B., and Du Clos, T.W. (2002c). C-reactive protein mediates protection from lipopolysaccharide through interactions with FcγR. *J. Immunol.*, **169**, 7019–7025.

Moore, K.W., Malefyt, R.D.W., Coffman, R.L., and O'Garra, A. (2001). Interleukin-10 and the Interleukin-10 receptor. *Annu. Rev. Immunol.*, **19**, 683–765.

Mortensen, R.F. and Duszkiewicz, J.A. (1977). Mediation of CRP-dependent phagocytosis through mouse macrophage Fc-receptors. *J. Immunol.*, **119**, 1611–1616.

Narkates, A.J. and Volanakis, J.E. (1982). C-reactive protein binding specificities: Artificial and natural phospholipid bilayers. *Ann. N Y Acad. Sci.*, **389**, 172–181.

Paul, A., Ko, K.W., Li, L., Yechoor, V., McCrory, M.A., Szalai, A.J., and Chan, L. (2004). C-reactive protein accelerates the progression of atherosclerosis in apolipoprotein E–deficient mice. *Circulation*, **109**, 647–655.

Pepys, M.B., Baltz, M.L., Gomer, K., Davis, J.S., and Doenhoff, M. (1979). Serum amyloid P component is an acute phase reactant in mouse. *Nature*, **278**, 259–261.

Pepys, M.B., Booth, S.E., Tennent, G.A., Butler, P.J.G., and Williams, D.G. (1994). Binding of pentraxins to different nuclear structures: C-reactive protein binds to small nuclear ribonucleoprotein particles, serum amyloid P component binds to chromatin and nucleoli. *Clin. Exp. Immunol.*, **97**, 152–157.

Pepys, M.B., and Butler, P.J.G. (1987). Serum amyloid P component is the major calcium-dependent specific DNA binding protein of the serum. *Biochem. Biophys. Res. Commun.*, **148**, 308–313.

Ravetch, J.V. and Bolland, S. (2001). IgG Fc receptors. *Annu. Rev. Immunol.*, **19**, 275–290.

Ravetch, J.V. and Lanier, L.L. (2000). Immune inhibitory receptors. *Science*, **290**, 84–89.

Ridker, P.M. (2003). Clinical application of C-reactive protein for cardiovascular disease detection and prevention. *Circulation*, **107**, 363–369.

Robey, F.A., Jones, K.D., Tanaka, T., and Liu, T.-Y. (1984). Binding of C-reactive protein to chromatin and nucleosome core particles. A possible physiological role of C-reactive protein. *J. Biol. Chem.*, **259**, 7311–7316.

Rodriguez, W., Mold, C., Kataranovski, M., Marnell, L., Hutt, J., and Du Clos, T.W. (2004). C-reactive protein reverses ongoing nephritis in autoimmune mice. *Arthritis Rheum.*, **52**, 642–650

Russell, A.I., Cunninghame Graham, D.S., Shepherd, C., Roberton, C.A., Whittaker, J., Meeks, J., Powell, R.J., Isenberg, D.A., Walport, M.J., and Vyse, T.J. (2004). Polymorphism at the C-reactive protein locus influences gene expression and predisposes to systemic lupus erythematosus. *Hum. Mol. Genet.*, **13**, 137–147.

Soma, M., Tamaoki, T., Kawano, H., Ito, S., Sakamoto, M., Okada, Y., Ozaki, Y., Kanba, S., Hamada, Y., Ishihara, T., and Maeda, S. (2001). Mice lacking serum amyloid P component do not necessarily develop severe autoimmune disease. *Biochem. Biophys. Res. Commun.*, **286**, 200–205.

Stein, M.P., Mold, C., and Du Clos, T.W. (2000). C-reactive protein binding to murine leukocytes requires Fcγ receptors. *J. Immunol.*, **164**, 1514–1520.

Sutterwala, F.S., Noel, G.J., Salgame, P., and Mosser, D.M. (1998). Reversal of proinflammatory responses by ligating the macrophage Fcγ receptor type I. *J. Exp. Med.*, **188**, 217–222.

Szalai, A.J., McCrory, M.A., Cooper, G.S., Wu, J., and Kimberly, R.P. (2002b). Association between baseline levels of C-reactive protein (CRP) and a dinucleotide repeat polymorphism in the intron of the CRP gene. *Genes Immun.*, **3**, 14–19.

Szalai, A.J., Nataf, S., Hu, X.-Z., and Barnum, S.R. (2002a). Experimental allergic encephalomyelitis is inhibited in transgenic mice expressing human C-reactive protein. *J. Immunol.*, **168**, 5792–5797.

Szalai, A.J., Weaver, C.T., McCrory, M.A., van Ginkel, F.W., Reiman, R.M., Kearney, J.F., Marion, T.N., and Volanakis, J.E. (2003). Delayed lupus onset in (NZB × NZW)F1 mice expressing a human C-reactive protein transgene. *Arthritis Rheum.*, **48**, 1602–1611.

Tarzi, R.M., Davies, K.A., Claassens, J.W., Verbeek, J.S., Walport, M.J., and Cook, H.T. (2003). Both Fcγ receptor I and Fcγ receptor III mediate disease in accelerated nephrotoxic nephritis. *Am. J. Pathol.*, **162**, 1677–1683.

Tarzi, R.M., Davies, K.A., Robson, M.G., Fossati-Jimack, L., Saito, T., Walport, M.J., and Cook, H.T. (2002). Nephrotoxic nephritis is mediated by Fcγ receptors on circulating leukocytes and not intrinsic renal cells. *Kidney Int.*, **62**, 2087–2096.

Tebo, J.M. and Mortensen, R.F. (1990). Characterization and isolation of a C-reactive protein receptor from the human monocytic cell line U-937. *J. Immunol.*, **144**, 231–238.

ter Borg, E.J., Horst, G., Limburg, P.C., van Rijswijk, M.H., and Kallenberg, C.G. (1990). C-reactive protein levels during disease exacerbations and infections in systemic lupus erythematosus: A prospective longitudinal study. *J. Rheumatol.*, **17**, 1642–1648.

Theofilopoulos, A.N. and Dixon, F.J. (1985). Murine models of systemic lupus erythematosus. *Adv. Immunol.*, **37**, 269–391.

Thompson, D., Pepys, M.B., and Wood, S.P. (1999). The physiological structure of human C-reactive protein and its complex with phosphocholine. *Structure*, **7**, 169–177.

Tilg, H., Vannier, E., Vachino, G., Dinarello, C.A., and Mier, J.W. (1993). Antiinflammatory properties of hepatic acute phase proteins: Preferential induction of Interleukin 1 (IL-1) receptor antagonist over IL-1b synthesis by human peripheral blood mononuclear cells. *J. Exp. Med.*, **178**, 1629–1636.

Tillett, W.S. and Francis, T., Jr. (1930). Serological reactions in pneumonia with a non-protein fraction of pneumococcus. *J. Exp. Med.*, **52**, 561–571.

Vickers, M.A., Green, F.R., Terry, C., Mayosi, B.M., Julier, C., Lathrop, M., Ratcliffe, P.J., Watkins, H.C., and Keavney, B. (2002). Genotype at a promoter polymorphism of the interleukin-6 gene is associated with baseline levels of plasma C-reactive protein. *Cardiovasc. Res.*, **53**, 1029–1034.

Volanakis, J.E. (2001). Human C-reactive protein: Expression, structure, and function. *Mol. Immunol.*, **38**, 189–197.

Volanakis, J.E. and Kaplan, M.H. (1971). Specificity of C-reactive protein for choline phosphate residues of pneumococcal C-polysaccharide. *Proc. Soc. Exp. Biol. Med.*, **136**, 612–614.

Walport, M.J. (2000). Lupus, DNase and defective disposal of cellular debris. *Nat. Genet.*, **25**, 135–136.

Whitehead, A.S., Zahedi, K., Rits, M., Mortensen, R.F., and Lelias, J.M. (1990). Mouse C-reactive protein generation of cDNA clones, structural analysis, and induction of mRNA during inflammation. *Biochem. J.*, **266**, 283–290.

Xia, D. and Samols, D. (1997). Transgenic mice expressing rabbit C-reactive protein are resistant to endotoxemia. *Proc. Natl. Acad. Sci. USA*, **94**, 2575–2580.

Zahedi, K., Tebo, J.M., Siripont, J., Klimo, G.F., and Mortensen, R.F. (1989). Binding of human C-reactive protein to mouse macrophages is mediated by distinct receptors. *J. Immunol.*, **142**, 2384–2392.

Zee, R.Y. and Ridker, P.M. (2002). Polymorphism in the human C-reactive protein (CRP) gene, plasma concentrations of CRP, and the risk of future arterial thrombosis. *Atherosclerosis*, **162**, 217–219.

Zein, N., Ganuza, C., and Kushner, I. (1979). Significance of serum C-reactive protein elevation in patients with systemic lupus erythematosus. *Arthritis Rheum.*, **22**, 7–12.

Zwaka, T.P., Hombach, V., and Torzewski, J. (2001). C-reactive protein-mediated low density lipoprotein uptake by macrophages: Implications for atherosclerosis. *Circulation*, **103**, 1194–1197.

4

NKT Cells and Autoimmune Type 1 Diabetes

Shabbir Hussain, Dalam Ly, Melany Wagner, and Terry L. Delovitch

1. Introduction

Natural killer T (NKT) cells are a small subset of T cells that regulate both innate and adaptive immune responses. An NKT cell is defined by the coexpression of an $\alpha\beta$ T cell receptor (TCR) and the NK cell markers DX5 and NK1.1. The identification of an NKT cell specific ligand, α-galactosylceramide (α-GalCer), and the generation of α-GalCer-loaded CD1d tetramers has enabled the tracking and functional analysis of a CD1d-restricted invariant natural killer T (iNKT) cell subset. iNKT cells represent the majority of the total NKT cell population. Based upon the expression of cell surface markers and the binding of CD1d-loaded tetramers, a numerical deficiency of iNKT cells is detectable in both nonobese diabetic (NOD) mice and humans with autoimmune type 1 diabetes (T1D). Interestingly, iNKT cells from NOD mice and humans with T1D also show a functional deficiency in iNKT cells, such as an impaired polarization toward a Th2-like immune response. Restoring the numerical and/or functional deficiency of iNKT cells in NOD mice either by treatment with α-GalCer, transgenic induction of Vα14-Jα18 expression, or transgenic expression of CD1d in NOD islets under the control of the human insulin promoter confers protection from T1D in these mice. Recently, considerable progress has been made in understanding the developmental biology and function of iNKT cells. In this chapter, we focus on the protective role of iNKT cells against T1D, with an emphasis on the iNKT cell deficiency, defects in iNKT cell development, and the mechanisms that may be involved in iNKT cell activation–dependent protection against T1D.

Shabbir Hussain • Autoimmunity/Diabetes Group, Robarts Research Institute, London, ON N6G 2V4, Canada. **Dalam Ly, Melany Wagner, and Terry L. Delovitch** • Autoimmunity/Diabetes Group, Robarts Research Institute, London, ON N6G 2V4, Canada and Department of Microbiology & Immunology, University of Western Ontario, London, ON N6A 5C1, Canada.

Molecular Autoimmunity: In commemoration of the 100th anniversary of the first description of human autoimmune disease, edited by Moncef Zouali. Springer Science+Business Media, Inc., New York, 2005.

2. Type 1 Diabetes

T1D is an autoimmune disease caused by the T cell–mediated destruction of insulin-producing β cells in the pancreatic islets of Langerhans (Tisch and McDevitt, 1996; Delovitch and Singh, 1997). T1D pathology begins with the development of peri-insulitis as a result of recruitment of a heterogeneous population of leukocytes comprising T cells, B cells, macrophages (Mφ), and dendritic cells (DCs) to the perivascular region of the islets, which slowly progresses to an invasive and destructive insulitis (Pankewycz et al., 1991; Dahlen et al., 1998; Hussain et al., 2004). The period of insulitis varies between individuals (years in humans, months in rodents) before it finally progresses to overt T1D manifested by hyperglycemia (Delovitch and Singh, 1997; Mathis et al., 2001). Both $CD4^+$ and $CD8^+$ T cells are required for islet β cell destruction (Christianson et al., 1993; Delovitch and Singh, 1997). $CD8^+$ T cells function as effector cells but also require help from $CD4^+$ T cells (Christianson et al., 1993; Serreze et al., 1994; Wicker et al., 1994). The presence of T cells, B cells, Mφ, and DCs in islets suggests that collaborative actions of these immune cells are required for the destruction of islet β cells and onset of T1D.

The NOD mouse is an extensively studied animal model of human T1D (Atkinson and Leiter, 1999). NOD mice spontaneously develop T1D as a result of immune dysregulation and exhibit immunopathology similar to the human disease (Delovitch and Singh, 1997). Immune dysregulation in NOD mice occurs as a result of defective antigen presentation, impaired deletion and/or suppression of autoreactive T cells (Delovitch and Singh, 1997), and deficiencies in subsets of regulatory T (T_{reg}) cells including $CD4^+CD25^+$ T_{reg} cells (Salomon et al., 2000; Wu et al., 2002) and CD1d-restricted iNKT cells (Baxter et al., 1997; Yang et al., 2001; Laloux et al., 2002) (Figure 4.1). Although iNKT cells represent a small T cell subset, their potent immunoregulatory activity is sufficient to protect NOD mice from T1D (Hong et al., 2001; Naumov et al., 2001; Sharif et al., 2001; Wang et al., 2001; Beaudoin et al., 2002).

3. NKT Cells

An NKT cell is defined by the coexpression of an αβ TCR and NK cell markers, such as DX5 and NK1.1 (Gombert et al., 1996; Hammond et al., 2001; Yang et al., 2001). $NK1.1^+$ NKT cells were originally identified in the mature $CD4^-CD8^-$ double-negative (DN) αβ T cell subpopulation (Godfrey et al., 2000). Subsequently, $CD4^+$ or $CD8^+$ subsets of NKT cells were identified (MacDonald, 2002). The discovery of the NKT cell ligand α-GalCer and the MHC class I–like molecule CD1d led to the generation of α-GalCer-loaded CD1d tetramers (Kawano et al., 1999; Benlagha et al., 2000; Matsuda et al., 2000), which allowed investigators to identify the CD1d-restricted iNKT cell subset. iNKT cells express a specific invariant TCR α gene rearrangement: Vα14-Jα18 and Vβ8.2/2/7 in mice and Vα24-Jα15 and Vβ11 in humans (Wilson and Delovitch, 2003). Although the majority of iNKT cells express NK1.1, the specific binding of

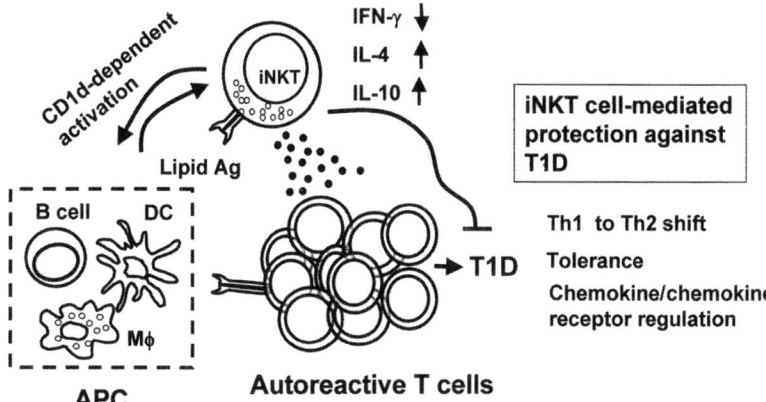

Figure 4.1. Invariant natural killer T (iNKT) cell activation corrects immune dysregulation and protects nonobese diabetic (NOD) mice from type 1 diabetes (T1D). Defective antigen presentation together with suboptimal costimulation by antigen-presenting cell (APC) costimulatory molecules causes the generation and accumulation of islet β cell autoreactive T cells. Uncontrolled generation and accumulation of these autoreactive T cells induces islet β cell destruction and onset of T1D. Upon activation by α-galactosylceramide (α-GalCer), iNKT cells secrete low levels of Th1 cytokine (IFN-γ) and increased levels of Th2 cytokines (IL-4, IL-10) that polarizes the immune response toward a Th2 phenotype. α-GalCer administration to NOD mice also induces tolerance via interaction of iNKT cells with other cell types and regulation of chemokine/chemokine receptor gene expression. DC, dendritic cell; Mϕ, macrophage; APC, antigen-presenting cell.

α-GalCer-loaded tetramers (Tet$^+$) can also identify an NK1.1$^-$ subset of iNKT cells (Kronenberg and Gapin, 2002). NK1.1$^-$Tet$^+$ iNKT cells are immature precursors to the NK1.1$^+$Tet$^+$ iNKT cell subset (Benlagha et al., 2002; Pellicci et al., 2002). Upon exit from the thymus into the periphery, the majority of these Tet$^+$ iNKT cells acquire NK1.1 on their surface (Pellicci et al., 2002).

iNKT cells recognize glycolipid antigen when presented in the context of CD1d (Brigl et al., 2003; Taniguchi et al., 2003). Although iNKT cells react to endogenous self-ligands (Brigl et al., 2003), a strong stimulator of these cells is α-GalCer, a glycolipid derived from marine sponge (Fuji et al., 2000). iNKT cells represent the majority of the total NKT cell population, and in NOD mice protection from T1D has been associated specifically with the Tet$^+$NK1.1$^{+/-}$ subset of NKT cells (Hong et al., 2001; Sharif et al., 2001).

4. Role of iNKT Cells in the Pathogenesis of Type 1 Diabetes

4.1. iNKT Cell Deficiency and T1D

NOD mice exhibit functional and numerical deficiencies of iNKT cells (Baxter et al., 1997; Hong et al., 2001; Naumov et al., 2001; Sharif et al., 2001; Yang et al., 2001). The correction of these deficiencies either by α-GalCer-stimulation (Hong et al., 2001; Naumov et al., 2001; Sharif et al., 2001), transgenic induction of Vα14-Jα281 expression (Lehuen et al., 1998; Laloux et al., 2001), or transgenic expression of CD1d in NOD islets (Falcone et al., 2004) protects these mice from T1D. The most severe deficiency in iNKT cell numbers in NOD mice occurs in the thymus and spleen (Gombert et al., 1996; Baxter et al., 1997; Poulton et al., 2001). CD1d-restricted NKT cells from NOD mice also show reduced secretion of IL-4 after TCR cross-linking and α-GalCer-stimulation (Sharif et al., 2001). Thus, decreased IL-4 secretion by NOD T cells following TCR cross-linking (Arreaza et al., 1997), coupled with a numerical deficiency of IL-4-secreting iNKT cells (Hong et al., 2001; Sharif et al., 2001), may be an important factor that exacerbates T1D onset in NOD mice.

A deficiency in iNKT cells has been demonstrated in humans with T1D. In twins discordant for T1D, the iNKT cell population defined by the invariant TCR Vα24JαQ and Vβ11 expression was shown to be deficient in numbers and production of IL-4 in the diabetic twins (Wilson et al., 1998). However, other investigators (Lee et al., 2002; Oikawa et al., 2002) have failed to detect an iNKT cell deficiency in T1D patients. The reason(s) for the discrepancy between these studies is not known but may result from the different reagents and cell populations used in the studies. Nonetheless, an increased frequency of iNKT cells in the peripheral blood of T1D patients is associated with recent T1D onset rather than established disease (Oikawa et al., 2003). However, a recent study (Berzins et al., 2004) found normal or even elevated iNKT cell numbers in peripheral blood of NOD mice, suggesting that circulating iNKT frequencies may not reflect lymphoid tissue deficiencies in iNKT cells. Although the iNKT cell frequency in peripheral blood may not be a reliable tool for T1D diagnosis, the immunomodulatory capacity of iNKT cells may still offer much potential for future therapies.

An iNKT cell deficiency arising from a block in iNKT cell development after thymectomy on day 3 causes gastritis in (BALB/c × C57BL6) F_1 mice (Hammond et al., 1998). This suggests that significant iNKT cell development occurs early in life to prevent autoimmunity. Previously, an NKT cell deficiency was reported as early as 3 weeks of age in NOD mice (Gombert et al., 1996). Interestingly, our real time PCR analyses of Vα14Jα18 expression indicate that this iNKT cell deficiency may be present even earlier at 1 week of age in an NOD thymus (our unpublished data), a time point considered to be the immunological equivalent of a newborn human (Adkins et al., 2004). Our finding raises the possibility that an iNKT cell deficiency may be present at birth and may predispose NOD mice to the development of T1D. Precursors in the iNKT cell pathway, e.g., $CD44^{low}$ iNKT or DP thymocytes, need to be examined for the expression of Vα14Jα18 TCR to determine the earliest stage at which the iNKT cell deficiency occurs in an NOD thymus (Gapin et al., 2001; Benlagha et al., 2002).

NOD mice possess a defect in central tolerance, as engraftment of thymic epithelia from NOD mice into athymic C57BL/6 mice induces insulitis in diabetes-resistant C57BL/6 mice (Thomas-Vaslin et al., 1997). In addition to thymic epithelia, other unknown factor(s) may also contribute to the iNKT cell deficiency in NOD mice as intrathymic cotransfer of thymic precursors in a 1:1 ratio from NOD and AKR mice into NOD.Scid mice can restore the frequency of iNKT cells in an NOD.Scid thymus (Yang et al., 2001). Thus, the factor(s) responsible for a defect in central tolerance and an iNKT cell deficiency in the NOD thymus requires further investigation.

4.2. iNKT Cell Activation Induces Protection against Type 1 Diabetes

The discovery of synthetic glycolipids specific for CD1d-restricted T cells has greatly increased our understanding of iNKT cell biology and has identified possible avenues for iNKT cell–targeted therapies. The unique ability of iNKT cells to secrete immunoregulatory cytokines upon activation has become an important research area especially in the context of disease models. In addition to maintaining tumor and infectious immunity (Wilson and Delovitch, 2003), iNKT cells are important in the regulation of susceptibility to several autoimmune diseases, including systemic lupus erythematosus, rheumatoid arthritis, multiple sclerosis, and T1D (Taniguchi et al., 2003; Wilson and Delovitch, 2003). Recent progress in our understanding of iNKT cell activation–induced protection from T1D has identified several possible mechanisms of protection. These include a shift in the immune response toward a Th2-like phenotype, iNKT cell/antigen-presenting cell (APC) interaction–dependent tolerance induction and the regulation of chemokine/chemokine receptor expression and interaction (Hong et al., 2001; Naumov et al., 2001; Sharif et al., 2001; Beaudoin et al., 2002; Mi et al., 2003). These mechanisms are illustrated in Figure 4.1, and are discussed below in light of recent findings.

4.2.1. iNKT Cell Activation Polarizes the Immune Response Toward a Th2-like Phenotype

We, and others, have previously shown that administration of α-GalCer using a multi-low-dose treatment protocol can significantly protect NOD mice from spontaneous and cyclophosphamide-induced T1D (Hong et al., 2001; Naumov et al., 2001; Sharif et al., 2001). iNKT cell activation by multi-low-dose α-GalCer administration preferentially enhances IL-4 and IL-10 production by NOD spleen cells, while a single dose induces tumor immunity by enhancing a Th1 response (Cui et al., 1997; Toura et al., 1999). We also observed the upregulation of IL-10 receptor expression by cDNA microarray analyses of α-GalCer-treated NOD spleen mRNA. While the significance of this elevated IL-10 receptor expression is presently unknown, it may indicate the enhanced utilization of IL-10 produced in response to α-GalCer. Our further investigation of the relative contribution of IL-4 and IL-10 in protection against T1D conducted using NOD.IL-4$^{-/-}$ and NOD.IL-10$^{-/-}$ mice revealed that NOD.IL-10$^{-/-}$, but not NOD.IL-4$^{-/-}$ mice, are protected against T1D following α-GalCer administration, demonstrating that IL-4 but not IL-10 is required to induce protection (Mi et al., 2004). These results highlight the important contribution of IL-4 in the iNKT cell–mediated protection against T1D in NOD mice.

In addition to the antigen dose, the type of antigen administered is also important to enhance the polarization of an immune response toward a Th2 phenotype. For instance, compared to α-GalCer treatment, β-GalCer (C12) stimulation failed to enhance cell proliferation and cytokine release from iNKT cells (Ortaldo et al., 2004). Both glycolipids, however, were able to deplete iNKT cells in vivo. Furthermore, β-GalCer (C12) stimulation did not induce transactivation of other cells, demonstrating the ability of iNKT cells to discriminate antigens. Interestingly, another glycolipid antigen, designated as OCH, a sphingosine-truncated analog of α-GalCer preferentially induced IL-4 production by iNKT cells and reduced IFN-γ secretion (Miyamoto et al., 2001). This preferential bias of IL-4 secretion in response to the OCH glycolipid may be beneficial for the treatment of various autoimmune diseases (Chiba et al., 2004; Oki et al., 2004; Yamamura et al., 2004).

4.2.2. iNKT Cell–Antigen-Presenting Cell Interaction Induces Tolerance

Studies have suggested that iNKT cell activation leads to the transactivation or modulation of other cell types, including T cells, B cells, NK cells, and DCs (Sharif et al., 2002; Wilson and Delovitch, 2003). As CD1d is expressed on all types of APCs (Pulendran et al., 1997; Sonoda and Stein-Streilein, 2002), the effect of transactivation may depend on the APC subset presenting the antigen. In the iNKT cell–dependent anterior chamber-associated immune deviation (ACAID) model of tolerance induction, $CD1d^+$ spleen marginal zone (MZ) B cells promote tolerance by enhancing the development of antigen-specific T_{reg} cells (Sonoda and Stein-Streilein, 2002). The exact mechanism of iNKT cell–dependent increase in T_{reg} development is not known, but may be due to an increase in IL-10 secretion by activated iNKT cells (Sonoda et al., 2001).

DCs also interact with iNKT cells. Repeated administration of α-GalCer to NOD mice preferentially induces the accumulation of $CD11c^+CD8\alpha^-$ tolerogenic DC in the pancreatic-draining lymph nodes (PLNs) of these mice. Furthermore, PLN-derived DCs from α-GalCer-treated NOD mice respond poorly to LPS-induced secretion of IL-12, which elicits a Th1-type response (Naumov et al., 2001). This suggests that accumulation of low IL-12-producing tolerogenic DC in the PLN of α-GalCer-treated NOD mice may favor the development of a protective Th2 environment.

4.2.3. iNKT Cell Activation Modulates Cytokine/Cytokine Receptor and Chemokine/Chemokine Receptor Expression

Chemokines and their receptors play an important role in the pathogenesis of T1D due to their ability to recruit leukocytes to islets. An increased ratio of macrophage inflammatory protein-1α (MIP-1α):MIP-1β in the pancreas of NOD mice relative to diabetes-resistant NOR mice correlates with destructive insulitis and progression to T1D (Cameron et al., 2000). MIP-1α is now termed CCL3 and MIP-1β is termed CCL4 (Zlotnik and Yoshie, 2000). Systemic treatment of NOD mice with IL-4 decreases the CCL3:CCL4 ratio and protects them from T1D (Cameron et al., 2000). Interestingly, iNKT cell activation by α-GalCer increases both IL-4 production and CCL4 mRNA expression in spleen cells of NOD mice. Furthermore, antibody neutralization of CCL4 abrogates IL-4-mediated protection from T1D in these mice (Sharif et al., 2001; Mi et al., 2003). *In vitro* studies indicate that CCL4 recruits CD4+CD25+ T_{reg} cells (Bystry et al., 2001). However, whether protection from T1D in α-GalCer- and IL-4-treated NOD mice is associated with recruitment of CD4+CD25+ T_{reg} cells remains to be determined.

Our cDNA microarray analyses revealed that in contrast to increased CCL4 gene expression, IL-16 mRNA expression was significantly decreased in the spleen of α-GalCer-treated NOD mice (Mi et al., 2003). IL-16 is a proinflammatory cytokine with chemoattractant activity. Interestingly, neutralization of IL-16 by a mouse anti-human IL-16 antibody protects NOD mice from T1D, and this effect can be abrogated upon coinjection of anti-IL-16 with an anti-CCL4-neutralizing antibody treatment (Mi et al., 2003). IL-16 is produced by several cell types, including CD4+ and CD8+ T cells, B cells, and DCs (Cruikshank et al., 2000; Kaser et al., 2000). IL-16 binds to its CD4 receptor and preferentially recruits CD4+ Th1 cells (Lynch et al., 2003). The binding of IL-16 to CD4 also desensitizes the ability of CCL4 to signal through its receptor CCR5 (Mashikian et al., 1999). Thus, neutralization of IL-16 by an anti-IL-16 antibody may block this desensitization and enable CCL4 to signal normally to CCR5.

We also found a decreased expression of stromal cell–derived factor-1 (SDF-1) mRNA in the spleens of α-GalCer-treated NOD mice (Mi et al., 2003). SDF-1 binds to its receptor CXCR4, and thus recruits both T cells and B cells (Bleul et al., 1996). Neutralization of SDF-1 in NOD mice by administration of an anti-mouse SDF-1 polyclonal antibody protects them from T1D by decreasing the number of B cells that migrate to the spleen (Matin et al., 2002). The expression of CCL21 (6 Ckine), CCR2, CCR3, and CXCR4 are also reduced in the spleen of α-GalCer-treated NOD mice (Mi et al., 2003). In contrast, the mRNA transcripts of CCL4, eotaxin (CCL24), IFN-γ-inducible protein 10 (IP10; CXCL10), and monokine induced by IFN-γ (MIG; CXCL9) are increased in the spleens of α-GalCer-treated NOD mice (Mi et al., 2003). Together, these studies suggest that iNKT cell activation by α-GalCer regulates the expression of several chemokines and chemokine receptors. Elucidation of the mechanism(s) of regulation of chemokines and chemokine receptor expression following α-GalCer stimulation of iNKT cells requires further experimentation, and may yield new avenues for therapeutic intervention.

5. Future Directions

Despite considerable efforts, the precise mechanism of iNKT cell–mediated protection of NOD mice from T1D is still not completely understood. Studies to date indicate that polarization of an immune response toward a Th2-like phenotype may be involved and be accompanied by the α-GalCer-induced recruitment of tolerogenic DCs to the PLN. While previous studies identified an iNKT cell deficiency in NOD mice as early as 3 weeks after birth, our more recent findings indicate that this deficiency may occur during fetal development and be present at birth. Precursors common to conventional T cells and iNKT cells, e.g., $CD44^{low}$

iNKT or DP thymocytes, need to be examined for the expression of Vα14Jα18 TCR to determine the precise stage at which the iNKT cell deficiency occurs in the NOD thymus. Emphasis should also be given to identify the subset of APC that presents α-GalCer and stimulates the production of Th1 or Th2 cytokines. Furthermore, identification of the natural self-ligands that regulate iNKT cell activity in various tissues and the role of iNKT cell activation in maintaining immune homeostasis offers promise for the development of novel therapies for the prevention of T1D.

Acknowledgments

This work was supported by grants from the Canadian Institutes of Health Research (CIHR), Juvenile Diabetes Research Foundation (JDRF), Canadian Diabetes Association (CDA) in honor of the late Olive I. Moore and the Ontario Research and Development Challenge Fund (ORDCF). S. Hussain is recipient of a postdoctoral fellowship from the CDA in the honor of the late Flora I. Nichol. T. L. Delovitch is the Sheldon H. Weinstein Scientist in Diabetes at the Robarts Research Institute and University of Western Ontario.

References

Adkins, B., Leclerc, C., and Marshall-Clarke, S. (2004). Neonatal adaptive immunity comes of age. *Nat. Rev. Immunol.*, **4**, 553–564.

Arreaza, G.A., Cameron, M.J., Jaramillo, A., Gill, B.M., Hardy, D., Laupland, K.B., Rapoport, M.J., Zucker, P., Chakrabarti, S., Chensue, S.W., Qin, H.Y., Singh, B., and Delovitch, T.L. (1997). Neonatal activation of CD28 signaling overcomes T cell anergy and prevents autoimmune diabetes by an IL-4-dependent mechanism. *J. Clin. Invest.*, **100**, 2243–2253.

Atkinson, M.A. and Leiter, E.H. (1999). The NOD mouse model of type 1 diabetes: As good as it gets? *Nat. Med.*, **5**, 601–604.

Baxter, A.G., Kinder, S.J., Hammond, K.J., Scollay, R., and Godfrey, D.I. (1997). Association between alphabetaTCR+CD4−CD8− T-cell deficiency and IDDM in NOD/Lt mice. *Diabetes*, **46**, 572–582.

Beaudoin, L., Laloux, V., Novak, J., Lucas, B., and Lehuen, A. (2002). NKT cells inhibit the onset of diabetes by impairing the development of pathogenic T cells specific for pancreatic beta cells. *Immunity*, **17**, 725–736.

Benlagha, K., Kyin, T., Beavis, A., Teyton, L., and Bendelac, A. (2002). A thymic precursor to the NKT cell lineage. *Science*, **296**, 553–555.

Benlagha, K., Weiss, A., Beavis, A., Teyton, L., and Bendelac, A. (2000). In vivo identification of glycolipid antigen-specific T cells using fluorescent CD1d tetramers. *J. Exp. Med.*, **191**, 1895–1903.

Berzins, S.P., Kyparissoudis, K., Pellicci, D.G., Hammond, K.J., Sidobre, S., Baxter, A., Smyth, M.J., Kronenberg, M., and Godfrey, D.I. (2004). Systemic NKT cell deficiency in NOD mice is not detected in peripheral blood: Implications for human studies. *Immunol. Cell Biol.*, **82**, 247–252.

Bleul, C.C., Fuhlbrigge, R.C., Casasnovas, J.M., Aiuti, A., and Springer, T.A. (1996). A highly efficacious lymphocyte chemoattractant, stromal cell-derived factor-1 (SDF-1). *J. Exp. Med.*, **184**, 1101–1109.

Brigl, M., Bry, L., Kent, S.C., Gumperz, J.E., and Brenner, M.B. (2003). Mechanism of CD1d-restricted natural killer T cell activation during microbial infection. *Nat. Immunol.*, **4**, 1230–1237.

Bystry, R.S., Aluvihare, V., Welch, K.A., Kallikourdis, M., and Betz, A.G. (2001). B cells and professional APCs recruit regulatory T cells via CCL4. *Nat. Immunol.*, **2**, 1126–1132.

Cameron, M.J., Arreaza, G.A., Grattan, M., Meagher, C., Sharif, S., Burdick, M.D., Strieter, R.M., Cook, D.N., and Delovitch, T.L. (2000). Differential expression of CC chemokines and the CCR5 receptor in the pancreas is associated with progression to type 1 diabetes. *J. Immunol.*, **165**, 1102–1110.

Chiba, A., Oki, S., Miyamoto, K., Hashimoto, H., Yamamura, T., and Miyake, S. (2004). Suppression of collagen-induced arthritis by natural killer T cell activation with OCH, a sphingosine-truncated analog of alpha-galactosylceramide. *Arthritis Rheum.*, **50**, 305–313.

Christianson, S.W., Shultz, L.D., and Leiter, E.H. (1993). Adoptive transfer of diabetes into immunodeficient NOD-scid/scid mice. Relative contributions of $CD4^+$ and $CD8^+$ T-cells from diabetic versus prediabetic NOD.NON-Thy-1a donors. *Diabetes*, **42**, 44–55.

Cruikshank, W.W., Kornfeld, H., and Center, D.M. (2000). Interleukin-16. *J. Leukoc. Biol.*, **67**, 757–766.

Cui, J., Shin, T., Kawano, T., Sato, H., Kondo, E., Toura, I., Kaneko, Y., Koseki, H., Kanno, M., and Taniguchi, M. (1997). Requirement for Valpha14 NKT cells in IL-12-mediated rejection of tumors. *Science*, **278**, 1623–1626.

Dahlen, E., Dawe, K., Ohlsson, L., and Hedlund, G. (1998). Dendritic cells and macrophages are the first and major producers of TNF-alpha in pancreatic islets in the nonobese diabetic mouse. *J. Immunol.*, **160**, 3585–3593.

Delovitch, T.L. and Singh, B. (1997). The nonobese diabetic mouse as a model of autoimmune diabetes: Immune dysregulation gets the NOD. *Immunity*, **7**, 727–738.

Falcone, M., Facciotti, F., Ghidoli, N., Monti, P., Olivieri, S., Zaccagnino, L., Bonifacio, E., Casorati, G., Sanvito, F., and Sarvetnick, N. (2004). Up-regulation of CD1d expression restores the immunoregulatory function of NKT cells and prevents autoimmune diabetes in nonobese diabetic mice. *J. Immunol.*, **172**, 5908–5916.

Fuji, N., Ueda, Y., Fujiwara, H., Toh, T., Yoshimura, T., and Yamagishi, H. (2000). Antitumor effect of alpha-galactosylceramide (KRN7000) on spontaneous hepatic metastases requires endogenous interleukin-12 in the liver. *Clin. Cancer Res.*, **6**, 3380–3387.

Gapin, L., Matsuda, J.L., Surh, C.D., and Kronenberg, M. (2001). NKT cells derive from double-positive thymocytes that are positively selected by CD1d. *Nat. Immunol.*, **2**, 971–978.

Godfrey, D.I., Hammond, K.J., Poulton, L.D., Smyth, M.J., and Baxter, A.G. (2000). NKT cells: Facts, functions and fallacies. *Immunol. Today*, **21**, 573–583.

Gombert, J.M., Herbelin, A., Tancrede-Bohin, E., Dy, M., Carnaud, C., and Bach, J.F. (1996). Early quantitative and functional deficiency of $NK1^+$-like thymocytes in the NOD mouse. *Eur. J. Immunol.*, **26**, 2989–2998.

Hammond, K.J., Cain, W., van Driel, I., and Godfrey, D. (1998). Three day neonatal thymectomy selectively depletes $NK1.1^+$ T cells. *Int. Immunol.*, **10**, 1491–1499.

Hammond, K.J., Pellicci, D.G., Poulton, L.D., Naidenko, O.V., Scalzo, A.A., Baxter, A.G., and Godfrey, D.I. (2001). CD1d-restricted NKT cells: An interstrain comparison. *J. Immunol.*, **167**, 1164–1173.

Hong, S., Wilson, M.T., Serizawa, I., Wu, L., Singh, N., Naidenko, O.V., Miura, T., Haba, T., Scherer, D.C., Wei, J., Kronenberg, M., Koezuka, Y., and Van Kaer, L. (2001). The natural killer T-cell ligand alpha-galactosylceramide prevents autoimmune diabetes in non-obese diabetic mice. *Nat. Med.*, **7**, 1052–1056.

Hussain, S., Salojin, K.V., and Delovitch, T.L. (2004). Hyperresponsiveness, resistance to B-cell receptor-dependent activation-induced cell death, and accumulation of hyperactivated B-cells in islets is associated with the onset of insulitis but not type 1 diabetes. *Diabetes*, **53**, 2003–2011.

Kaser, A., Dunzendorfer, S., Offner, F.A., Ludwiczek, O., Enrich, B., Koch, R.O., Cruikshank, W.W., Wiedermann, C.J., and Tilg, H. (2000). B lymphocyte-derived IL-16 attracts dendritic cells and Th cells. *J. Immunol.*, **165**, 2474–2480.

Kawano, T., Tanaka, Y., Shimizu, E., Kaneko, Y., Kamata, N., Sato, H., Osada, H., Sekiya, S., Nakayama, T., and Taniguchi, M. (1999). A novel recognition motif of human NKT antigen receptor for a glycolipid ligand. *Int. Immunol.*, **11**, 881–887.

Kronenberg, M. and Gapin, L. (2002). The unconventional lifestyle of NKT cells. *Nat. Rev. Immunol.*, **2**, 557–568.

Laloux, V., Beaudoin, L., Jeske, D., Carnaud, C., and Lehuen, A. (2001). NKT cell-induced protection against diabetes in Valpha14-Jalpha281 transgenic nonobese diabetic mice is associated with a Th2 shift circumscribed regionally to the islets and functionally to islet autoantigen. *J. Immunol.*, **166**, 3749–3756.

Laloux, V., Beaudoin, L., Ronet, C., and Lehuen, A. (2002). Phenotypic and functional differences between NKT cells colonizing splanchnic and peripheral lymph nodes. *J. Immunol.*, **168**, 3251–3258.

Lee, P.T., Putnam, A., Benlagha, K., Teyton, L., Gottlieb, P.A., and Bendelac, A. (2002). Testing the NKT cell hypothesis of human IDDM pathogenesis. *J. Clin. Invest.*, **110**, 793–800.

Lehuen, A., Lantz, O., Beaudoin, L., Laloux, V., Carnaud, C., Bendelac, A., Bach, J.F., and Monteiro, R.C. (1998). Overexpression of natural killer T cells protects Valpha14-Jalpha281 transgenic nonobese diabetic mice against diabetes. *J. Exp. Med.*, **188**, 1831–1839.

Lynch, E.A., Heijens, C.A., Horst, N.F., Center, D.M., and Cruikshank, W.W. (2003). Cutting edge: IL-16/CD4 preferentially induces Th1 cell migration: Requirement of CCR5. *J. Immunol.*, **171**, 4965–4968.

MacDonald, H.R. (2002). Development and selection of NKT cells. *Curr. Opin. Immunol.*, **14**, 250–254.

Mashikian, M.V., Ryan, T.C., Seman, A., Brazer, W., Center, D.M., and Cruikshank, W.W. (1999). Reciprocal desensitization of CCR5 and CD4 is mediated by IL-16 and macrophage inflammatory protein-1 beta, respectively. *J. Immunol.*, **163**, 3123–3130.

Mathis, D., Vence, L., and Benoist, C. (2001). Beta-cell death during progression to diabetes. *Nature*, **414**, 792–798.

Matin, K., Salam, M.A., Akhter, J., Hanada, N., and Senpuku, H. (2002). Role of stromal cell-derived factor-1 in the development of autoimmune diseases in non-obese diabetic mice, *Immunology*, **107**, 222–232.

Matsuda, J.L., Naidenko, O.V., Gapin, L., Nakayama, T., Taniguchi, M., Wang, C.R., Koezuka, Y., and Kronenberg, M. (2000). Tracking the response of natural killer T cells to a glycolipid antigen using CD1d tetramers. *J. Exp. Med.*, **192**, 741–754.

Mi, Q.S., Ly, D., Zucker, P., McGarry, M., and Delovitch, T.L. (2004). Interleukin-4 but not interleukin-10 protects against spontaneous and recurrent type 1 diabetes by activated CD1d-restricted invariant natural killer T-cells. *Diabetes*, **53**, 1303–1310.

Mi, Q.S., Meagher, C., and Delovitch, T.L. (2003). CD1d-restricted NKT regulatory cells: Functional genomic analyses provide new insights into the mechanisms of protection against Type 1 diabetes. *Novartis Found. Symp.*, **252**, 146–160.

Miyamoto, K., Miyake, S., and Yamamura, T. (2001). A synthetic glycolipid prevents autoimmune encephalomyelitis by inducing T_H2 bias of natural killer T cells. *Nature*, **413**, 531–534.

Naumov, Y.N., Bahjat, K.S., Gausling, R., Abraham, R., Exley, M.A., Koezuka, Y., Balk, S.B., Strominger, J.L., Clare-Salzer, M., and Wilson, S.B. (2001). Activation of CD1d-restricted T cells protects NOD mice from developing diabetes by regulating dendritic cell subsets. *Proc. Natl. Acad. Sci. USA*, **98**, 13838–13843.

Oikawa, Y., Shimada, A., Yamada, S., Motohashi, Y., Nakagawa, Y., Irie, J., Maruyama, T., and Saruta, T. (2002). High frequency of Valpha24(+) Vbeta11(+) T-cells observed in type 1 diabetes. *Diabetes Care*, **25**, 1818–1823.

Oikawa, Y., Shimada, A., Yamada, S., Motohashi, Y., Nakagawa, Y., Irie, J., Maruyama, T., and Saruta, T. (2003). NKT cell frequency in Japanese type 1 diabetes. *Ann. NY Acad. Sci.*, **1005**, 230–232.

Oki, S., Chiba, A., Yamamura, T., and Miyake, S. (2004). The clinical implication and molecular mechanism of preferential IL-4 production by modified glycolipid-stimulated NKT cells. *J. Clin. Invest.*, **113**, 1631–1640.

Ortaldo, J.R., Young, H.A., Winkler-Pickett, R.T., Bere, E.W., Jr., Murphy, W.J., and Wiltrout, R.H. (2004). Dissociation of NKT stimulation, cytokine induction, and NK activation *in vivo* by the use of distinct TCR-binding ceramides. *J. Immunol.*, **172**, 943–953.

Pankewycz, O., Strom, T.B., and Rubin-Kelley, V.E. (1991). Islet-infiltrating T cell clones from non-obese diabetic mice that promote or prevent accelerated onset of diabetes. *Eur. J. Immunol.*, **21**, 873–879.

Pellicci, D.G., Hammond, K.J., Uldrich, A.P., Baxter, A.G., Smyth, M.J., and Godfrey, D.I. (2002). A natural killer T (NKT) cell developmental pathway involving a thymus-dependent NK1.1(−) CD4(+) CD1d-dependent precursor stage. *J. Exp. Med.*, **195**, 835–844.

Poulton, L.D., Smyth, M.J., Hawke, C.G., Silveira, P., Shepherd, D., Naidenko, O.V., Godfrey, D.I., and Baxter, A.G. (2001). Cytometric and functional analyses of NK and NKT cell deficiencies in NOD mice. *Int. Immunol.*, **13**, 887–896.

Pulendran, B., Lingappa, J., Kennedy, M.K., Smith, J., Teepe, M., Rudensky, A., Maliszewski, C.R., and Maraskovsky, E. (1997). Developmental pathways of dendritic cells *in vivo*: Distinct function, phenotype, and localization of dendritic cell subsets in FLT3 ligand-treated mice. *J. Immunol.*, **159**, 2222–2231.

Salomon, B., Lenschow, D.J., Rhee, L., Ashourian, N., Singh, B., Sharpe, A., and Bluestone, J.A. (2000). B7/CD28 costimulation is essential for the homeostasis of the $CD4^+CD25^+$ immunoregulatory T cells that control autoimmune diabetes. *Immunity*, **12**, 431–440.

Serreze, D.V., Leiter, E.H., Christianson, G.J., Greiner, D., and Roopenian, D.C. (1994). Major histocompatibility complex class 1-deficient NOD-B2m null mice are diabetes and insulitis resistant. *Diabetes*, **43**, 505–509.

Sharif, S., Arreaza, G.A., Zucker, P., Mi, Q.S., and Delovitch, T.L. (2002). Regulation of autoimmune disease by natural killer T cells. *J. Mol. Med.*, **80**, 290–300.

Sharif, S., Arreaza, G.A., Zucker, P., Mi, Q.S., Sondhi, J., Naidenko, O.V., Kronenberg, M., Koezuka, Y., Delovitch, T.L., Gombert, J.M., Leite-De-Moraes, M., Gouarin, C., Zhu, R., Hameg, A., Nakayama, T., Taniguchi, M., Lepault, F., Lehuen, A., Bach, J.F., and Herbelin, A. (2001). Activation of natural killer T cells by alpha-galactosylceramide treatment prevents the onset and recurrence of autoimmune type 1 diabetes. *Nat. Med.*, **7**, 1057–1062.

Sonoda, K.H., Faunce, D.E., Taniguchi, M., Exley, M., Balk, S., and Stein-Streilein, J. (2001). NKT cell-derived IL-10 is essential for the differentiation of antigen-specific T regulatory cells in systemic tolerance. *J. Immunol.*, **166**, 42–50.

Sonoda, K.H. and Stein-Streilein, J. (2002). CD1d on antigen-transporting APC and splenic marginal zone B cells promotes NKT cell-dependent tolerance. *Eur. J. Immunol.*, **32**, 848–857.

Taniguchi, M., Harada, M., Kojo, S., Nakayama, T., and Wakao, H. (2003). The regulatory role of Valpha14 NKT cells in innate and acquired immune response. *Annu. Rev. Immunol.*, **21**, 483–513.

Thomas-Vaslin, V., Damotte, D., Coltey, M., Le Douarin, N.M., Coutinho, A., and Salaun, J. (1997). Abnormal T cell selection on NOD thymic epithelium is sufficient to induce autoimmune manifestations in C57BL/6 athymic nude mice. *Proc. Natl. Acad. Sci. USA*, **94**, 4598–4603

Tisch, R. and McDevitt, H. (1996). Insulin-dependent diabetes mellitus. *Cell*, **85**, 291–297.

Toura, I., Kawano, T., Akutsu, Y., Nakayama, T., Ochiai, T., and Taniguchi, M. (1999). Cutting edge: Inhibition of experimental tumor metastasis by dendritic cells pulsed with alpha-galactosylceramide. *J. Immunol.*, **163**, 2387–2391.

Wang, B., Geng, Y.B., and Wang, C.R. (2001). CD1-restricted NKT cells protect nonobese diabetic mice from developing diabetes. *J. Exp. Med.*, **194**, 313–320.

Wicker, L.S., Leiter, E.H., Todd, J.A., Renjilian, R.J., Peterson, E., Fischer, P.A., Podolin, P.L., Zijlstra, M., Jaenisch, R., and Peterson, L.B. (1994). Beta 2-microglobulin-deficient NOD mice do not develop insulitis or diabetes. *Diabetes*, **43**, 500–504.

Wilson, S.B. and Delovitch, T.L. (2003). Janus-like role of regulatory iNKT cells in autoimmune disease and tumour immunity. *Nat. Rev. Immunol.*, **3**, 211–222.

Wilson, S.B., Kent, S.C., Patton, K.T., Orban, T., Jackson, R.A., Exley, M., Porcelli, S., Schatz, D.A., Atkinson, M.A., Balk, S.P., Strominger, J.L., and Hafler, D.A. (1998). Extreme Th1 bias of invariant Valpha24JalphaQ T cells in type 1 diabetes. *Nature*, **391**, 177–181.

Wu, A.J., Hua, H., Munson, S.H., and McDevitt, H.O. (2002). Tumor necrosis factor-alpha regulation of $CD4^+CD25^+$ T cell levels in NOD mice. *Proc. Natl. Acad. Sci. USA*, **99**, 12287–12292.

Yamamura, T., Miyamoto, K., Illes, Z., Pal, E., Araki, M., and Miyake, S. (2004). NKT cell-stimulating synthetic glycolipids as potential therapeutics for autoimmune disease. *Curr. Top. Med. Chem.*, **4**, 561–567.

Yang, Y., Bao, M., and Yoon, J.W. (2001). Intrinsic defects in the T-cell lineage results in natural killer T-cell deficiency and the development of diabetes in the nonobese diabetic mouse. *Diabetes*, **50**, 2691–2699.

Zlotnik, A. and Yoshie, O. (2000). Chemokines: A new classification system and their role in immunity. *Immunity*, **12**, 121–127.

5

The Genetics of Human Autoimmune Diseases

Marta E. Alarcón-Riquelme

1. Introduction

Most diseases, if not all, affecting human populations are determined by genes and by the surrounding environment. The extent to which genes are important depends on several factors such as penetrance, genetic interactions, and gene–environment interactions. Even infectious diseases have an important genetic component that may determine the extent our immune system can respond to infection and neutralize it (Lander and Schork, 1994).

Complex diseases are characterized by the requirement of many genes and environmental factors. Genetically, the genes may be interacting for their effect to be observed or the genetic effect may be the result of the sum of several genes. The mutations found in genes involved in complex diseases are usually relatively common polymorphisms that are not selected against because either they have been useful during human evolution for unknown reasons or simply because they do not normally pose any lethal effect as long as the combination with other genes or the environment does not occur.

Identification of genes for complex diseases is complicated by the fact that the genes involved may be different in different families or in different ethnic groups (genetic heterogeneity). Even if the same gene is found to be involved, allelic heterogeneity may also play an important role (Terwilliger and Goring, 2000).

The identification of the genes involved in complex diseases is of interest because it may lead us to the understanding of disease mechanisms. The disease pathways in which the genes exert their effects may also lead to the discovery of new therapeutic targets. It may also change our understanding on the interactions between genes and environment.

Most autoimmune diseases are complex and the view that genetic factors might predispose to autoimmune disorders has come through several pieces of

---

Marta E. Alarcón-Riquelme • Department of Genetics and Pathology, Uppsala University, Dag Hammarsjkölds väg 20, 751 85, Uppsala, Sweden.

Molecular Autoimmunity: In commemoration of the 100th anniversary of the first description of human autoimmune disease, edited by Moncef Zouali. Springer Science+Business Media, Inc., New York, 2005.

evidence. First, studies done in twins have shown a higher concordance rate in identical twins than in nonidentical twins (Arnett and Shulman, 1976; Ebers *et al*., 1995). Second, studies in families showed a tendency of the diseases to aggregate in families. A common way in complex disease genetics to "measure" familial aggregation is to use the λ value. Familial aggregation has to be calculated for a given family relatedness and the one most commonly used is sib ship. Thus, the λ_S value is the risk of disease in sibs divided by the population prevalence of the disease. Even if the λ_S value does not distinguish between genetics and environment within families, it does give an idea of the genetic contribution and it is generally assumed that the higher the λ_S value, the larger the genetic component of the disease. Hence, the identification of the genes and polymorphisms involved in that particular disease may be easier (Risch, 1990a,b). This chapter discusses how the availability of the human genome sequence has played an essential role in the recent advances in unraveling complex disease genetics, and how human genome scans have provided some recent new discoveries. Most interesting is the understanding that some of the genes identified are shared between the various autoimmune diseases, allowing us to investigate further the pathogenic mechanisms.

2. Analysis of the Genetics of Complex Diseases

2.1. Linkage Analysis

The advances in genome technology, including arrays and electrophoresis, and the availability of the human genome sequence have played an essential role in the recent advances in complex disease genetics. The initial phases of the study of complex diseases were initiated when microsatellites were identified as extremely useful polymorphic markers that could be mapped to specific locations within the genome and used for genetic linkage analysis in families. In order to perform linkage analysis the collection and recruitment of families was done by several groups and high-throughput genotyping was begun. Geneticists brought up several issues at the time regarding the types of families and analyses required, mainly because of the assumption that complex diseases do not follow an inheritance pattern and methods for linkage analysis were designed, of which nonparametric score methods became the most popular (Lathrop *et al*., 1984; Lander and Botstein, 1986; Lander *et al*., 1987; Lander and Schork, 1994).

Multiplex families, with two or more affected individuals, were recruited, frequently with two affected sibs. It is important to note that such type of families will bias the identification of genes and polymorphic disease alleles to those showing predominantly a recessive pattern of inheritance. Even if complex diseases do not follow a defined Mendelian inheritance mode, the alleles at disease genes do (Lander and Schork, 1994).

Methods for the identification of genetic linkage or of association in the presence of linkage have been widely used. In some cases where recruitment of the families has included extended pedigrees, parametric methods have been used. The main difference between nonparametric and parametric methods is the requirement in the latter of an inclusion of a set of parameters such as the estimated disease gene frequency (calculated from the disease prevalence in the population under study), allele frequencies of the markers to use (that can be determined from the family material that is used), penetrance values (that can be set flexibly with high and low prevalence for testing), and the mode of inheritance (that can also be tested for dominant or recessive). Nonparametric analysis is model-free in theory. Because it does not use healthy relatives or several generations of individuals, model-free analysis is less powerful in detecting linkage. The most common model-free analysis is the sib-pair analysis (Durner *et al.*, 1992; Greenberg *et al.*, 1998; Abreu *et al.*, 1999; Goedken *et al.*, 2000).

Linkage analysis is used to define, the distance between two markers being the presence of disease phenotype, one of the markers for which linkage is tested against. Therefore linkage analysis assumes the probable identification of a disease locus based on the correlation between genotype and phenotype. It is calculated using the logarithm of the odds (LOD) score that is considered significant when the LOD score reaches a value of 3.3. This corresponds to a p value of 4.9×10^{-5}, representing a 5% chance to obtain a LOD score in a genome scan. Usually the interval that is detected is very large, reaching 20–30 centiMorgans (cM) (Ott, 1979, 1989b; Ott and Bhat, 1999).

2.2. Association Analysis

When there is a hypothesis about the genes involved in disease susceptibility, a candidate gene approach is used. Polymorphisms are identified and analyzed using two possible types of materials, trio families, that is, families comprising the patient and the parents. These families are very useful to define haplotypes. The tests used to analyze association are based on transmission of the alleles from the parents to the disease-affected offspring, where the nontransmitted alleles are the control alleles. This design is however less powerful than a simple case-control design, in particular if the most popular test (transmission disequilibrium test, TDT) is used (Matise, 1995; Todd, 1995). The TDT uses only alleles transmitted from heterozygous parents. An alternative is the haplotype relative risk (HRR) test that simply uses alleles transmitted also from homozygous parents, increasing the power slightly (Ott, 1989a). Multiplex pedigrees can also be used in association analysis together with the pedigree disequilibrium test (PDT) (Abecasis *et al.*, 2000). One can also simply randomly choose a nuclear family from the pedigree to include in the association test and augment the power of the test in a trio study by including the trios available from multiplex families.

The case-control approach, even with its caveats, is still the most powerful approach from a statistical point of view. Cases and controls matched by ethnicity, age, and sex are the most common and simple type of approach for the study of candidate genes through genetic association or for replicating an association detected in families. The rationale of the approach is based on the fact that gene alleles are in linkage disequilibrium and that patients share ancestral haplotypes. The major drawback of such studies is the introduction of false-positive results (and false negatives) because of population stratification due to population admixture. There are several ways of avoiding population stratification, one being the use of carefully matched close population controls; an alternative is to use a large number of randomly selected markers across the genome that should not differ significantly in the allele frequencies between both groups (Cardon and Bell, 2001; Cardon and Palmer, 2003). Haplotypes can be estimated from genotype data in cases and controls, but estimation algorithms usually create nonexistent haplotypes, so care should be taken.

2.3. Combining Linkage and Association

Genome scans and linkage analysis are followed by genetic association analyses of the genes and haplotypes found in the intervals identified. The close characterization of single-nucleotide polymorphisms (SNPs) across the genome allows one to choose the best SNPs that will serve as tags and will identify the haplotypes within a given region of the genome (Cardon and Abecasis, 2003). With the new high-throughput genotyping methods available, hundreds of SNPs can be genotyped fairly quickly for thousands of individuals. Therefore, studies that begin with linkage analysis are now on the phase of identifying the associated haplotypes and closing the genes within these haplotypes that will become candidates for disease susceptibility. In some cases these genes have already been identified.

3. Genetic Analysis in Autoimmunity

3.1. Genome Scans and Linkage Analysis in Autoimmune Diseases

During the last 10 years, several groups have collected multiplex families of the various autoimmune diseases for genome screening with microsatellites and linkage analysis. The very first report of linkage analysis came from the field of type 1 diabetes (T1D) mellitus (Davies *et al.*, 1994) followed by multiple sclerosis (MS) (Ebers *et al.*, 1996), rheumatoid arthritis (RA) (Cornelis *et al.*, 1998), and systemic lupus erythematosus (SLE) (Gaffney *et al.*, 1998; Moser *et al.*, 1998). I limit this review to these diseases, adding only Crohn's disease (CD) and ulcerative colitis (UC).

3.2. Autoimmune Diabetes (T1D)

As mentioned above, the first genome scan performed with multiplex families was on this disease. It identified 20 potential susceptibility regions of which the locus for insulin was already known (Davies *et al.*, 1994). A second genome scan did not allow replication of most of the regions identified in the first scan (Concannon *et al.*, 1998). This led to skepticism regarding genome scans, probably because of the inflated expectations promoted initially by some investigators.

Later on, through careful fine mapping, it was possible to pinpoint a variable number tandem repeat (VNTR) as the causative polymorphism within the insulin gene that represented the *IDDM2* locus. The VNTR was classified in types

I–III depending on the size of the repeats. It was also found that class III alleles could associate with low expression levels of insulin despite the fact that these alleles have previously shown a protective effect. In Caucasians, the insulin gene VNTR alleles divide into two discrete size classes. Class I alleles (26–63 repeats) predispose in a recessive way to T1D, while class III alleles (140 to more than 200 repeats) are dominantly protective. The protective effect may be explained by higher levels of class III VNTR-associated insulin mRNA in thymus such that elevated levels of preproinsulin protein enhance immune tolerance to preproinsulin, a key autoantigen in T1D pathogenesis. The paternal effect (preferential transmission of a given allele from the father) was observed only when the father's untransmitted allele was class III. Therefore, the hypothesis has been that in individuals carrying the class I VNTR alleles the low expression of proinsulin in the thymus leads to lack of induction of immune tolerance (Undlien *et al.*, 1995; Bennett *et al.*, 1996, 1997; Vafiadis *et al.*, 1997; Ahmed *et al.*, 1999).

A second important region of linkage identified by the genome scans was at 2q33. Recently, an extremely thorough analysis of the region, including the genes *ICOS* and *CTLA4*, was performed. A common allelic variation was found to be genetically associated not only with T1D but also with Graves' disease and autoimmune hyperthyroidism correlated with the presence of a molecule producing lower levels of a soluble alternative splice form of *CTLA4* lacking the CD80/CD86 ligand-binding domain, providing an interesting mechanistic explanation of the genetic association with *CTLA4* (Nistico *et al.*, 1996; Hill *et al.*, 2000; Ueda *et al.*, 2003). In short, the insulin gene VNTR and the splice form of the *CTLA4* gene have with certainty been identified as susceptibility genes for T1D. Various other IDDM loci were detected in the genome scans performed to date and the genes contributing to the effects are being identified.

3.3. Multiple Sclerosis

Just as with T1D, full-genome scans have been performed for MS (Ebers *et al.*, 1996). In the initial genome scan, possible loci were detected on chromosome 6p21 and 5p and evidence was later provided on 5p with 21 Finnish MS families (Kuokkanen *et al.*, 1996). A full-genome scan also performed in the Finnish families identified a region in chromosome 17q22–24 (Kuokkanen *et al.*, 1997) also revealed in a genome scan from the UK and in a limited scan in Scandinavian families and in a whole Nordic linkage analysis. However the 5p region has not been replicated (Oturai *et al.*, 1999; Akesson *et al.*, 2002). A new region was found in chromosome 19q13, in particular the 19q13.1 subregion when a larger set of Finnish families were stratified for HLA-DR15, known to be strongly associated with MS (Reunanen *et al.*, 2002), as seen with HLA and several other autoimmune diseases (Ligers *et al.*, 2001; Masterman and Hillert, 2002). An interaction between markers at HLA-DR15 and the *CTLA4* gene has been described in MS (Alizadeh *et al.*, 2003). The *CTLA4* gene has been an interesting candidate as linkage has also been found to 2q33, where this gene lies. It appears that the identification of genes for MS is more difficult than for other autoimmune diseases. An interesting candidate gene, *CD45* (*PTPRC*) was found to have a splice variant associated with MS in German individuals, but this association was not replicated in several other studies (Jacobsen et al., 2002; Gomez-Lira et al., 2003; Sabouri et al., 2003). In short, the most consistent genomic region containing a potential gene for MS susceptibility is the 17q22 region, but no gene has been identified to date though some interesting observations have been done (Nelissen et al., 2000; Saarela et al., 2002; Chen et al., 2004).

3.4. Rheumatoid Arthritis

Several genome scans have been produced for RA in France (mainly south European) (Cornelis et al., 1998) and the UK (MacKay et al., 2002), and two screens from the NARAC Consortium in the US focused on European Americans and erosive arthritis (Jawaheer et al., 2001, 2003). As for the other autoimmune diseases, the HLA region at 6p21 showed the strongest linkage signal in all studies performed to date. The results from the various scans were rather consistent for non-HLA regions in spite of the apparent nonreplication observed initially. The most outstanding novel region that was found in the French genome screen was located at 18q22 as well as at 3q13. When the first NARAC genome scan appeared the results were quite different from those identified in the French screen. The main regions identified were at chromosomes 1, 4, 12, 16, and 17, but the second NARAC screen identified chromosome 18q21 and new loci at 9p22 and 10q21. The combined analysis supported regions from both scans. Regions most influenced by using the HLA-DRβ alleles as covariates (shared epitope alleles such as DRβ1*0401) were in chromosomes 1p, 1q, and 18q (Gregersen et al., 1987; Jawaheer et al., 2002).

The results of the UK genome scan were rather disappointing as the only region reaching nominal significance was the HLA while all non-HLA regions did not. However, the study results got support from the previous NARAC and French genome screens in six regions. To date, no gene has been identified through the direct use of the multiplex families, but the NARAC material has been used in alternative approaches, as will be described further below.

Work performed by Japanese groups has used a completely different approach and has identified at least three new interesting and potential genes involved in RA, *PADI4*, *RUNX1*, and *SLC22A4* (*OCTN1*). The approach consisted in the analysis of a dense map of SNPs in regions selected for having been found in human linkage studies or containing interesting candidate genes. The two regions were in 1p36 and in 5q31, the latter being that of the cytokine gene family. The former study identified the *PADI4* gene (peptidylarginine deiminase type 4), an enzyme involved in citrullination, a process thought to be defective in RA where antibodies against citrullinated peptides are major biomarkers of the disease (Suzuki et al., 2003). Whether *PADI4* is the gene for RA is as yet uncertain as the results have not been replicated (Barton et al., 2004).

The second gene identified by the Japanese groups was after the analysis of the 5q31 region with a very dense map of SNPs (Tokuhiro et al., 2003). This analysis identified the *SLC22A4* gene, an organic cation transporter member of the OCTN family that was recently found also associated with CD (Peltekova *et al.*, 2004). The mutation found within the first intron of this gene results in the disruption of a binding site for the runt-domain family of regulatory proteins known as RUNX, in particular RUNX1. A similar disruption was previously identified in SLE (described below). Thus the Japanese approach led to the discovery of *PADI4*, *SLC22A4* (*OCTN1*), and preliminarily *RUNX1*.

Interestingly, another candidate gene, *TNFRII*, has been found associated in familial RA in Europeans but not in sporadic cases (Dieude *et al.*, 2002). However, clinically, no differences have been described between sporadic and familial cases of RA. The possibility exists that several genes are responsible for disease susceptibility in this cytogenetic region.

Most recently, the genotyping of 87 SNPs that were considered functional (showing amino acid changes) was performed in a case-control design in RA. This analysis led to the identification of an amino acid change in codon 620 of the protein tyrosine phosphatase *PTPN22*, also known as LYP (Begovich *et al.*, 2004), an enzyme that dephosphorylates and inactivates antigen-induced T cell activation. The change converted an arginine to tryptophan and disrupted the binding of the SH3 domain of LYP to the kinase Csk. This variant was found previously to be associated with T1D (Bottini *et al.*, 2004).

3.5. Crohn's Disease and Ulcerative Colitis

As for the other autoimmune diseases, genome scans have been performed for both CD and UC (Rioux *et al.*, 2000). The best results have been obtained with the former while linkages remain elusive for UC. Some of the most important results were found for chromosomes 5q31–33, 3p, and 6p (Rioux *et al.*, 2001) as well as for chromosome 12 in a separate scan and in chromosome 16 in a French study (Lesage *et al.*, 2000). Importantly, the first breakthrough in gene identification in autoimmune diseases and complex diseases came from identification of mutations in the *CARD15* gene in CD (Hugot *et al.*, 2001; Ogura *et al.*, 2001). *CARD15*, or *NOD2*, is a leucine-rich domain-containing protein that has an inhibitory role and acts as a receptor for intracellular pathogens. Three main variants were identified in three independent low-frequency haplotypes, thus challenging the common disease–common variant hypothesis. Weak association of *CARD15* has been found with UC and epistasis involving 5q31 and *CARD15*.

The next region to be studied, 5q31, was recently fine-mapped and as a surprise the results pinned down to two related genes, *SLC22A4* (*OCTN1*) and *SLC22A5* (*OCTN2*) (Peltekova *et al.*, 2004). As noted previously, variants of the *SLC22A4* gene were identified in susceptibility to RA. Two mutations were identified in these genes in CD and it remains uncertain if the main contributor to susceptibility is *SLC22A4* or *SLC22A5*. Recently, it has been observed that *OCTN1* is importantly expressed in lymphoid tissues, while *OCTN2* is mainly expressed in kidney and is a proven carnitine transporter involved in carnitine deficiency (Xuan *et al.*, 2003; Yamada *et al.*, 2004). The mutation found in *OCTN1* was a missense substitution, while that identified in *OCTN2* was a G-C transversion in the promoter, leading to changes in gene expression (Peltekova *et al.*, 2004). However, o gene has been identified for UC.

3.6. Systemic Lupus Erythematosus

The prototype of systemic autoimmune disease, SLE, has been extensively studied and four groups have performed genome screenings for various populations (Gaffney et al., 1998; Moser et al., 1998; Shai et al., 1999; Lindqvist et al., 2000). The disease has several features that make the genetic component even more important than for the other autoimmune diseases. There are animal models that develop the disease spontaneously. Conversely, the genetic component for SLE is not dominated by the HLA with the exception of some populations where HLA DR3 has been implicated. However, DR3 is in linkage disequilibrium with a null allele of C4, making interpretations difficult (Arnett, 1985; Howard et al., 1986). One of the genome scans in Caucasians detected the MHC region as an important susceptibility locus. A similar finding was observed in Caucasian Swedes, but not in other ethnic groups (Gaffney et al., 1998; Lindqvist et al., 2000).

The most consistent loci identified for SLE were found in chromosome 1 (1p36, 1q23, 1q31, and 1q42–44), but the most significant were detected in chromosomes 2q37, 4p13, 4p15, and 16q12 (Gaffney et al., 1998; Moser et al., 1998; Shai et al., 1999; Lindqvist et al., 2000). Other minor loci have been identified by several groups and replicated as well, but remain with minor effects (Gaffney et al., 2000; Tsao, 2000). Some of the strongest linkages were found in families of Icelandic and Swedish origins (Lindqvist et al., 2000). Recently, a genome screen in Finnish families identified and confirmed loci in 14q and 6q observed in previous scans (Koskenmies et al., 2004). Another locus identified recently in families of Argentine–European origin is on chromosome 17q12 (Johansson et al., 2004). As expected, high degree of heterogeneity has been found between populations.

The region 1q23 contains as major candidate genes the low-affinity receptors for the Fc portion of immunoglobulins or FcγR. Earlier studies had been performed on these genes, because their products are involved in immune complex clearance (Salmon et al., 1996). The main association has been found for FcγRIIIA, with a clearly functional polymorphism (valine to phenylalanine substitution) affecting immunoglobulin-binding affinity, but association to FcγRIIA has also been observed (Seligman et al., 2001; Zuniga et al., 2001). However, the functional polymorphism of FcγRIIA (histidine to arginine substitution) has also been associated. Recently it was proposed that variants in both genes are required for susceptibility and act functionally as if it would be a compound heterozygosity situation (Magnusson et al., 2004). In other populations, in particular Asians, the polymorphism found in the FcγRIIB gene that could affect the transmembrane region (Li et al., 2003) is associated with SLE.

Most of the loci on chromosome 1 have been found also to be quantitative trait loci in animal models of SLE although importance of the FcG receptors has not been confirmed in mice. Instead, these genes confer susceptibility in animals deficient for the FcγRIIB crossed with lupus susceptible mice (Bolland et al., 2002).

A breakthrough in SLE has been the identification of variation in the *PDCD1* gene coding for the immunoreceptor PD-1, a 50-kDa protein expressed on the surface of various early lymphoid tissues and activated T and B cells (Prokunina et al., 2002). It is involved in inhibition of T cell activation and in thymic tolerance. It had been previously shown that a knockout model for PD-1 developed a lupus-like disease, and the gene became an important candidate when linkage was identified to 2q37.3 in sets of Nordic multiplex families with SLE, where the gene for PD-1, *PDCD1*, had been mapped (Magnusson et al., 2000). The variant identified to be associated segregated in a single haplotype and disrupted the binding site for the runt-related silencer *RUNX1*, the same identified later for RA, and even psoriasis (Helms et al., 2003; Tokuhiro et al., 2003). Several studies have now replicated the genetic association initially found in lupus and have also identified association of the same polymorphism with RA and T1D in Europeans (Nielsen et al., 2003; Lin et al., 2004, 2004; Prokunina et al., 2004a,b). Association with RA was identified with a second polymorphism in *PDCD1* in the Chinese (Lin et al., 2004).

3.7. Genes Shared between Autoimmune Diseases

The overlap of clinical manifestations, the presence of similar autoantibody specificities in several diseases, and the common immunologic features suggest that genetic susceptibility may be shared among various autoimmune diseases as it was suspected in earlier studies. Now that the genes are beginning to be identified, it is indeed confirmed that this is the case. For example, a splicing mutation in *CTLA4* has been found associated with T1D as well as with SLE. *PDCD1* also has been associated with RA and T1D, and, as mentioned previously, *SLC22A4* has been associated both with CD and with RA. The coding and functional variant of *PTPN22* was first identified in T1D and then independently discovered in RA. Association was also recently identified with SLE (Kyogoku et al., 2004). However, the gene most commonly shared among autoimmune diseases has been the HLA DRβ1 (Table 5.1). Does this gene modify the clinical picture of the autoimmune diseases depending on the variation? Why do we identify several genetic variants being shared among clinically distinct autoimmune diseases? "How does the combination of susceptibility variation lead to disease" is becoming a question with a higher probability to be answered.

Table 5.1. Genes Shared between Autoimmune Diseases

Genes	Diseases
HLADRβ1	RA/MS/SLE/T1D
PDCD1	SLE/RA/T1D
CTLA4	T1D/SLE
SLC22A4	RA/Crohn's disease
PTPN22	T1D/RA/SLE

References

Abecasis, G.R., Cookson, W.O., and Cardon, L.R. (2000). Pedigree tests of transmission disequilibrium. *Eur. J. Hum. Genet.*, **8**, 545–551.

Abreu, P.C., Greenberg, D.A., and Hodge, S.E. (1999). Direct power comparisons between simple LOD scores and NPL scores for linkage analysis in complex diseases. *Am. J. Hum. Genet.*, **65**, 847–857.

Ahmed, S., *et al.* (1999). INS VNTR allelic variation and dynamic insulin secretion in healthy adult non-diabetic Caucasian subjects. *Diabet. Med.*, **16**, 910–917.

Akesson, E., *et al.* (2002). A genome-wide screen for linkage in Nordic sib-pairs with multiple sclerosis. *Genes Immun.*, **3**, 279–285.

Alizadeh, M., *et al.* (2003). Genetic interaction of CTLA-4 with HLA-DR15 in multiple sclerosis patients. *Ann. Neurol.*, **54**, 119–122.

Arnett, F.C. (1985). HLA and genetic predisposition to lupus erythematosus and other dermatologic disorders. *J. Am. Acad. Dermatol.*, **13**, 472–481.

Arnett, F.C. and Shulman, L.E. (1976). Studies in familial systemic lupus erythematosus. *Medicine (Baltimore)*, **55**, 313–322.

Barton, A., *et al.* (2004). A functional haplotype of the PADI4 gene associated with rheumatoid arthritis in a Japanese population is not associated in a United Kingdom population. *Arthritis Rheum.*, **50**, 1117–1121.

Begovich, A.B., *et al.* (2004). A missense single-nucleotide polymorphism in a gene encoding a protein tyrosine phosphatase (PTPN22) is associated with rheumatoid arthritis. *Am. J. Hum. Genet.*, **75**, 330–337.

Bennett, S.T., *et al.* (1996). IDDM2-VNTR-encoded susceptibility to type 1 diabetes: Dominant protection and parental transmission of alleles of the insulin gene-linked minisatellite locus. *J. Autoimmun.*, **9**, 415–421.

Bennett, S.T., *et al.* (1997). Insulin VNTR allele-specific effect in type 1 diabetes depends on identity of untransmitted paternal allele. The IMDIAB Group. *Nat. Genet.*, **17**, 350–352.

Bolland, S., *et al.* (2002). Genetic modifiers of systemic lupus erythematosus in FcgammaRIIB(−/−) mice. *J. Exp. Med.*, **195**, 1167–1174.

Bottini, N., *et al.* (2004). A functional variant of lymphoid tyrosine phosphatase is associated with type I diabetes. *Nat. Genet.*, **36**, 337–338.

Cardon, L.R. and Abecasis, G.R. (2003). Using haplotype blocks to map human complex trait loci. *Trends Genet.*, **19**, 135–140.

Cardon, L.R. and Bell, J.I. (2001). Association study designs for complex diseases. *Nat. Rev. Genet.*, **2**, 91–99.

Cardon, L.R. and Palmer, L.J. (2003). Population stratification and spurious allelic association. *Lancet*, **361**, 598–604.

Chen, D.C., *et al.* (2004). Segmental duplications flank the multiple sclerosis locus on chromosome 17q. *Genome Res.*, **14**, 1483–1492.

Concannon, P., *et al.* (1998). A second-generation screen of the human genome for susceptibility to insulin-dependent diabetes mellitus. *Nat. Genet.*, **19**, 292–296.

Cornelis, F., *et al.* (1998). New susceptibility locus for rheumatoid arthritis suggested by a genome-wide linkage study. *Proc. Natl. Acad. Sci. USA*, **95**, 10746–10750.

Davies, J.L., *et al.* (1994). A genome-wide search for human type 1 diabetes susceptibility genes. *Nature*, **371**, 130–136.

Dieude, P., *et al.* (2002). Association between tumor necrosis factor receptor II and familial, but not sporadic, rheumatoid arthritis: Evidence for genetic heterogeneity. *Arthritis Rheum.*, **46**, 2039–2044.

Durner, M., *et al.* (1992). Inter- and intrafamilial heterogeneity: Effective sampling strategies and comparison of analysis methods. *Am. J. Hum. Genet.*, **51**, 859–870.

Ebers, G.C., *et al.* (1995). A genetic basis for familial aggregation in multiple sclerosis. Canadian Collaborative Study Group. *Nature*, **377**, 150–151.

Ebers, G.C., *et al.* (1996). A full genome search in multiple sclerosis. *Nat. Genet.*, **13**, 472–476.

Gaffney, P.M., *et al.* (1998). A genome-wide search for susceptibility genes in human systemic lupus erythematosus sib-pair families. *Proc. Natl. Acad. Sci. USA*, **95**, 14875–14879.

Gaffney, P.M., et al. (2000). Genome screening in human systemic lupus erythematosus: Results from a second Minnesota cohort and combined analyses of 187 sib-pair families. *Am. J. Hum. Genet.*, **66**, 547–556.

Goedken, R., et al. (2000). Drawbacks of GENEHUNTER for larger pedigrees: Application to panic disorder. *Am. J. Med. Genet.*, **96**, 781–783.

Gomez-Lira, M., et al. (2003). CD45 and multiple sclerosis: The exon 4 C77G polymorphism (additional studies and meta-analysis) and new markers. *J. Neuroimmunol.*, **140**, 216–221.

Greenberg, D.A., et al. (1998). The power to detect linkage in complex disease by means of simple LOD-score analyses. *Am. J. Hum. Genet.*, **63**, 870–879.

Gregersen, P.K., et al. (1987). The shared epitope hypothesis. An approach to understanding the molecular genetics of susceptibility to rheumatoid arthritis. *Arthritis Rheum.*, **30**, 1205–1213.

Helms, C., et al. (2003). A putative RUNX1 binding site variant between SLC9A3R1 and NAT9 is associated with susceptibility to psoriasis. *Nat. Genet.*, **35**, 349–356.

Hill, N.J., et al. (2000). NOD Idd5 locus controls insulitis and diabetes and overlaps the orthologous CTLA4/IDDM12 and NRAMP1 loci in humans. *Diabetes*, **49**, 1744–1747.

Howard, P.F., et al. (1986). Relationship between C4 null genes, HLA-D region antigens, and genetic susceptibility to systemic lupus erythematosus in Caucasian and black Americans. *Am. J. Med.*, **81**, 187–193.

Hugot, J.P., et al. (2001). Association of NOD2 leucine-rich repeat variants with susceptibility to Crohn's disease. *Nature*, **411**, 599–603.

Jacobsen, M., et al. (2002). A novel mutation in PTPRC interferes with splicing and alters the structure of the human CD45 molecule. *Immunogenetics*, **54**, 158–163.

Jawaheer, D., et al. (2001). A genomewide screen in multiplex rheumatoid arthritis families suggests genetic overlap with other autoimmune diseases. *Am. J. Hum. Genet.*, **68**, 927–936.

Jawaheer, D., et al. (2002). Dissecting the genetic complexity of the association between human leukocyte antigens and rheumatoid arthritis. *Am. J. Hum. Genet.*, **71**, 585–594.

Jawaheer, D., et al. (2003). Screening the genome for rheumatoid arthritis susceptibility genes: A replication study and combined analysis of 512 multicase families. *Arthritis Rheum.*, **48**, 906–916.

Johansson, C.M., et al. (2004). Chromosome 17p12–q11 harbors susceptibility loci for systemic lupus erythematosus. *Hum. Genet.*, **115**, 230–238.

Koskenmies, S., et al. (2004). Haplotype associations define target regions for susceptibility loci in systemic lupus erythematosus. *Eur. J. Hum. Genet.*, **12**, 489–494.

Kuokkanen, S., et al. (1996). A putative vulnerability locus to multiple sclerosis maps to 5p14-p12 in a region syntenic to the murine locus Eae2. *Nat. Genet.*, **13**, 477–480.

Kuokkanen, S., et al. (1997). Genomewide scan of multiple sclerosis in Finnish multiplex families. *Am. J. Hum. Genet.*, **61**, 1379–1387.

Kyogoku, C., et al. (2004). Genetic association of the R620W polymorphism of protein tyrosine phosphatase PTPN22 with human SLE. *Am. J. Hum. Genet.*, **75**, 504–507.

Lander, E.S. and Botstein, D. (1986). Mapping complex genetic traits in humans: New methods using a complete RFLP linkage map. *Cold Spring Harb. Symp. Quant. Biol.*, **51** Pt 1, 49–62.

Lander, E.S. and Schork, N.J. (1994). Genetic dissection of complex traits. *Science*, **265**, 2037–2048.

Lander, E.S., et al. (1987). MAPMAKER: An interactive computer package for constructing primary genetic linkage maps of experimental and natural populations. *Genomics*, **1**, 174–181.

Lathrop, G.M., et al. (1984). Strategies for multilocus linkage analysis in humans. *Proc. Natl. Acad. Sci. USA*, **81**, 3443–3446.

Lesage, S., et al. (2000). Genetic analyses of chromosome 12 loci in Crohn's disease. *Gut*, **47**, 787–791.

Li, X., et al. (2003). A novel polymorphism in the Fcgamma receptor IIB (CD32B) transmembrane region alters receptor signaling. *Arthritis Rheum.*, **48**, 3242–3252.

Ligers, A., et al. (2001). Evidence of linkage with HLA-DR in DRB1*15-negative families with multiple sclerosis. *Am. J. Hum. Genet.*, **69**, 900–903.

Lin, S.C., et al. (2004). Association of a programmed death 1 gene polymorphism with the development of rheumatoid arthritis, but not systemic lupus erythematosus. *Arthritis Rheum.*, **50**, 770–775.

Lindqvist, A.K., et al. (2000). A susceptibility locus for human systemic lupus erythematosus (hSLE1) on chromosome 2q. *J. Autoimmun.*, **14**, 169–178.
MacKay, K., et al. (2002). Whole-genome linkage analysis of rheumatoid arthritis susceptibility loci in 252 affected sibling pairs in the United Kingdom. *Arthritis Rheum.*, **46**, 632–639.
Magnusson, V., et al. (2000). Fine mapping of the SLEB2 locus involved in susceptibility to systemic lupus erythematosus. *Genomics*, **70**, 307–314.
Magnusson, V., Johanneson, B., Lima, G., Odeberg, J., Alarcón-Segovia, D., Alarcón-Riquelme, M.E., and the SLE Genetics Collaboration Group. (2004). Both risk alleles for FcgRIIA and FcgRIIIA are involved in susceptibility for SLE. A unifying hypothesis. *Genes Immun.*, **5**, 130–137.
Masterman, T. and Hillert, J. (2002). HLA-DR15 and age at onset in multiple sclerosis. *Eur. J. Neurol.*, **9**, 179–180.
Matise, T.C. (1995). Genome scanning for complex disease genes using the transmission/disequilibrium test and haplotype-based haplotype relative risk. *Genet. Epidemiol.*, **12**, 641–645.
Moser, K.L., et al. (1998). Genome scan of human systemic lupus erythematosus: Evidence for linkage on chromosome 1q in African-American pedigrees. *Proc. Natl. Acad. Sci. USA*, **95**, 14869–14874.
Nelissen, I., et al. (2000). PECAM1, MPO and PRKAR1A at chromosome 17q21–q24 and susceptibility for multiple sclerosis in Sweden and Sardinia. *J. Neuroimmunol.*, **108**, 153–159.
Nielsen, C., et al. (2003). Association of a putative regulatory polymorphism in the PD-1 gene with susceptibility to type 1 diabetes. *Tissue Antigens*, **62**, 492–497.
Nistico, L., et al. (1996). The CTLA-4 gene region of chromosome 2q33 is linked to, and associated with, type 1 diabetes. Belgian Diabetes Registry. *Hum. Mol. Genet.*, **5**, 1075–1080.
Ogura, Y., et al. (2001). A frameshift mutation in NOD2 associated with susceptibility to Crohn's disease. *Nature*, **411**, 603–606.
Ott, J. (1979). Genetic linkage studies in man. *Transplant. Proc.*, **11**, 1689–1691.
Ott, J. (1989a). Statistical properties of the haplotype relative risk. *Genet. Epidemiol.*, **6**, 127–130.
Ott, J. (1989b). Computer-simulation methods in human linkage analysis. *Proc. Natl. Acad. Sci. USA*, **86**, 4175–4178.
Ott, J. and Bhat, A. (1999). Linkage analysis in heterogeneous and complex traits. *Eur. Child Adolesc. Psychiatry*, **8** (Suppl. 3), 43–46.
Oturai, A., et al. (1999). Linkage and association analysis of susceptibility regions on chromosomes 5 and 6 in 106 Scandinavian sibling pair families with multiple sclerosis. *Ann. Neurol.*, **46**, 612–616.
Peltekova, V.D., et al. (2004). Functional variants of OCTN cation transporter genes are associated with Crohn's disease. *Nat. Genet.*, **36**, 471–475.
Prokunina, L., et al. (2002). A regulatory polymorphism in PDCD1 is associated with susceptibility to systemic lupus erythematosus in humans. *Nat. Genet.*, **32**, 666–669.
Prokunina, L., et al. (2004). The systemic lupus erythematosus-associated PDCD1 polymorphism PD1.3A in lupus nephritis. *Arthritis Rheum.*, **50**, 327–328.
Prokunina, L.P.L., Bennet, A., De Faire, U., Wiman, B., Prince, J., Alfredsson, L., Klareskog, L., and Alarcón-Riquelme, M.E. (2004). Association of the PD1.3 A allele of the PDCD1 gene in patients with rheumatoid arthritis negative for rheumatoid factor and the shared epitope. *Arthritis Rheum.*, **50**, 1770–1773.
Reunanen, K., et al. (2002). Chromosome 19q13 and multiple sclerosis susceptibility in Finland: A linkage and two-stage association study. *J. Neuroimmunol.*, **126**, 134–142.
Rioux, J.D., et al. (2000). Genomewide search in Canadian families with inflammatory bowel disease reveals two novel susceptibility loci. *Am. J. Hum. Genet.*, **66**, 1863–1870.
Rioux, J.D., et al. (2001). Genetic variation in the 5q31 cytokine gene cluster confers susceptibility to Crohn's disease. *Nat. Genet.*, **29**, 223–228.
Risch, N. (1990a). Linkage strategies for genetically complex traits. I. Multilocus models. *Am. J. Hum. Genet.*, **46**, 222–228.
Risch, N. (1990b). Linkage strategies for genetically complex traits. II. The power of affected relative pairs. *Am. J. Hum. Genet.*, **46**, 229–241.
Saarela, J., et al. (2002). Fine mapping of a multiple sclerosis locus to 2.5 Mb on chromosome 17q22–q24. *Hum. Mol. Genet.*, **11**, 2257–2267.

Sabouri, A.H., et al. (2003). A C77G point mutation in CD45 exon 4, which is associated with the development of multiple sclerosis and increased susceptibility to HIV-1 infection, is undetectable in Japanese population. *Eur. J. Neurol.*, **10**, 737–739.

Salmon, J.E., et al. (1996). Fc gamma RIIA alleles are heritable risk factors for lupus nephritis in African Americans. *J. Clin. Invest.*, **97**, 1348–1354.

Seligman, V.A., et al. (2001). The Fcgamma receptor IIIA-158F allele is a major risk factor for the development of lupus nephritis among Caucasians but not non-Caucasians. *Arthritis Rheum.*, **44**, 618–625.

Shai, R., et al. (1999). Genome-wide screen for systemic lupus erythematosus susceptibility genes in multiplex families. *Hum. Mol. Genet.*, **8**, 639–644.

Suzuki, A., et al. (2003). Functional haplotypes of PADI4, encoding citrulinating enzyme peptidyl-arginine deiminase 4, are associated with rheumatoid arthritis. *Nat. Genet.*, **34**, 395–402.

Terwilliger, J.D. and Goring, H.H. (2000). Gene mapping in the 20th and 21st centuries: Statistical methods, data analysis, and experimental design. *Hum. Biol.*, **72**, 63–132.

Todd, J.A. (1995). Genetic analysis of type 1 diabetes using whole genome approaches. *Proc. Natl. Acad. Sci. USA*, **92**, 8560–8565.

Tokuhiro, S., et al. (2003). An intronic SNP in a RUNX1 binding site of SLC22A4, encoding an organic cation transporter, is associated with rheumatoid arthritis. *Nat. Genet.*, **35**, 341–348.

Tsao, B.P. (2000). Lupus susceptibility genes on human chromosome 1. *Int. Rev. Immunol.*, **19**, 319–334.

Ueda, H., et al. (2003). Association of the T-cell regulatory gene CTLA4 with susceptibility to autoimmune disease. *Nature*, **423**, 506–511.

Undlien, D.E., et al. (1995). Insulin gene region-encoded susceptibility to IDDM maps upstream of the insulin gene. *Diabetes*, **44**, 620–625.

Vafiadis, P., et al. (1997). Insulin expression in human thymus is modulated by INS VNTR alleles at the IDDM2 locus. *Nat. Genet.*, **15**, 289–292.

Xuan, W., et al. (2003). Characterization of organic cation/carnitine transporter family in human sperm. *Biochem. Biophys. Res. Commun.*, **306**, 121–128.

Yamada, R., et al. (2004). SLC22A4 and RUNX1: Identification of RA susceptible genes. *J. Mol. Med.*, **82**, 558–564.

Zuniga, R., et al. (2001). Low-binding alleles of Fcgamma receptor types IIA and IIIA are inherited independently and are associated with systemic lupus erythematosus in Hispanic patients. *Arthritis Rheum.*, **44**, 361–367.

6

Failure to Maintain T Cell DNA Methylation and Chromatin Structure Contributes to Human Lupus

Donna Ray and Bruce Richardson

1. Introduction

Systemic lupus erythematosus (SLE) is a chronic relapsing autoimmune disease that predominately affects women, mainly in their childbearing years. The incidence of lupus has been on the rise over the past few decades with an estimated half to three quarters of a million people having the disease in the USA (Gladman, 2004). Lupus affects the skin, serous membranes (pleura, pericardium, and peritoneum), bones, joints, the gastrointestinal system, the cardiovascular system, the brain, and the clotting system. The clinical manifestations of lupus vary widely, ranging from mild to severe, and lupus is sometimes fatal. Pathologically SLE is characterized by tissue damage resulting from antibody and complement-fixing immune complex deposition as well as abnormalities in T cell function. A better understanding of the mechanisms causing these immune abnormalities may lead to improvements in the therapy of human lupus.

While the specific etiology of SLE remains unknown, genetic and environmental factors contribute significantly to disease development. Associations between specific major histocompatibility complex (MHC) alleles and the development of idiopathic lupus are well described (Goldstein and Arnett, 1987), supporting genetic associations. Similarly, somewhat more than 5% of lupus cases are familial (Tsao and Grossman, 2001), and a number of loci predisposing to autoimmunity have been identified in these families. Twin studies show that concordance rates within monozygotic twins, ranging from 24% to 58%, are significantly higher compared with 2% to 5% in dizygotic twins (Kyttaris and Tsokos, 2003). However, the departure from 100% concordance in identical twins indicates a requirement for other factors for the development of the disease, and may include environmental triggers.

Donna Ray and Bruce Richardson • Department of Medicine, University of Michigan, Michigan 48109-0940.

Molecular Autoimmunity: In commemoration of the 100th anniversary of the first description of human autoimmune disease, edited by Moncef Zouali. Springer Science+Business Media, Inc., New York, 2005.

Many agents have been associated with lupus-like syndromes (Mongey and Hess, 2002),These include dietary factors, a variety of chemicals, certain metals, infectious agents, UV radiation, and various drugs. Though many of these factors may be capable of inducing autoimmune disease, they most likely represent examples of inciting agents in genetically predisposed individuals (Hess, 1995).

Drug-induced lupus (DIL) is perhaps one of the best examples of an autoimmune disease caused by an exogenous agent in genetically predisposed individuals. Over 100 drugs have been implicated in causing DIL (Yung and Richardson, 1994, 2003). Procainamide and hydralazine, the two drugs most often associated with DIL, have been studied most extensively. Genetic factors associated with procainamide- and hydralazine-induced lupus include female gender, slow acetylator status, and perhaps certain MHC alleles (Yung and Richardson, 1994). The fact that structurally unrelated drugs can induce a similar clinical syndrome suggests that more than one mechanism may be involved in the induction of lupus-like autoimmunity. It is also likely that a single drug may be acting by more than one mechanism. Nonetheless, the fact that certain drugs can induce a lupus-like autoimmune disease in genetically predisposed individuals suggests that DIL may represent a potentially important model for the study of idiopathic lupus. Understanding the pathogenesis and underlying mechanisms of DIL may help to elucidate the mechanisms involved in idiopathic SLE.

Our group has provided evidence that inhibition of T cell DNA methylation may be one of the mechanisms by which procainamide and hydralazine cause DIL, and used this model to predict mechanisms causing autoimmunity in idiopathic lupus. Using murine models, we found that T cells treated with the DNA methylation inhibitor 5-azacytidine (5-azaC) cause a lupus-like disease, that procainamide and hydralazine are DNA methylation inhibitors, and that T cells treated with procainamide and hydralazine induce an autoimmune disease identical to that caused by 5-azaC (Quddus *et al.*, 1993; Yung *et al.*, 1995, 1996, 1997). Subsequent studies demonstrated that DNA hypomethylation may also contribute to idiopathic human lupus (Richardson *et al.*, 1990; Deng *et al.*, 2001; Lu *et al.*, 2002).

DNA methylation is an important determinant of chromatin structure, and modifies gene expression through localized effects on the nucleosome polymers. Sections 2 and 3 review the relationship between DNA methylation, chromatin structure, and gene expression, and the role that DNA methylation and chromatin structure may play in DIL. We then describe how these results can be used to predict functional, biochemical, and genetic alterations in T cells from patients with idiopathic lupus.

2. DNA Methylation, Chromatin Structure, and Gene Expression

DNA methylation is the only postsynthetic modification of mammalian DNA. DNA methylation refers to the methylation of deoxycytosine (dC) bases at the 5 position to form deoxymethylcytosine (d^mC) (Adams and Burdon, 1985). Methylated dC residues are nearly always found in 5′-cytosine guanine-3′ (CG) dinucleotides. In mammals most (~70–80%) CG pairs are methylated (Ehrlich and Wang, 1981). The exceptions occur in regions of DNA rich in CG pairs, referred to as CpG islands. CpG islands are found in or near promoters of active genes and account for most of the nonmethylated CG dinucleotides. Approximately half of human and murine genes have CpG islands (Antequera and Bird, 1993).

A correlation between DNA methylation and gene expression was first proposed in the 1970s (Holliday and Pugh, 1975). Since then, numerous reports have supported this relationship. In general, hypomethylation of CpG islands, as well as the promoters of those genes lacking CpG islands, correlates with gene expression, while methylation results in transcriptional suppression (Attwood et al., 2002). DNA methylation patterns are established during ontogeny and are involved in suppressing expression of genes unnecessary or potentially detrimental to the function of the mature cells. Following differentiation, methylation patterns are maintained by replication during mitosis.

DNA methylation is mediated by a family of structurally related enzymes termed DNA methyltransferases. DNA (cytosine-5-)-methyltransferase 1 (Dnmt1), the first DNA methyltransferase identified, is responsible for replicating methylation patterns during mitosis (Attwood et al., 2002), a process referred to as maintenance methylation. Dnmt1 associates with proteins binding the replication fork, including proliferating cell nuclear antigen (PCNA) (Chuang et al., 1997), where it preferentially recognizes hemimethylated CpG dinucleotides in the newly synthesized double-stranded (ds) DNA. If the parent strand is methylated, it catalyzes the transfer of a methyl group from S-adenosylmethionine (SAM) to the unmethylated cytosine residue in the newly synthesized daughter strand, producing 5-methylcytosine and S-adenosylhomocysteine and replicating methylation patterns (Figure 6.1). Because of this recognition of hemimethylated DNA, Dnmt1 is referred to as a maintenance methyltransferase (Yoder et al., 1997).

The methylation of previously unmethylated sequences, such as occurs during differentiation, is mediated by DNA methyltransferases 3a and 3b (Dnmt3a and Dnmt3b, respectively). These enzymes are capable of methylating both unmethylated and hemimethylated DNA and are referred to as de novo methyltransferases (Figure 6.1). DNA methyltransferase 2 (Dnmt2) has been identified by sequence homology but lacks methyltransferase activity in vitro, and its biological function remains unknown (Okano et al., 1998a, 1998b).

DNA methylation is important in differentiation, X-chromosome inactivation, genomic imprinting, and suppression of parasitic DNA (Riggs, 1985; Stoger et al., 1993; Bestor, 1998). The importance of DNA methylation in differentiation is evidenced by the fact that homozygous deletion of Dnmt1, Dnmt3a, or Dnmt3b results in death during embryonic development or in the early neonatal period (Li et al., 1992; Okano et al., 1999). Several genetic disorders have been attributed to defects in DNA methylation, further attesting to its importance. Mutations in Dnmt3b are associated with the immunodeficiency, centromere instability, and facial anomalies (ICF) syndrome, a rare autosomal recessive disorder (Hansen et al., 1999). Furthermore, Rett syndrome, one of the most common causes of mental retardation afflicting young girls, is caused by

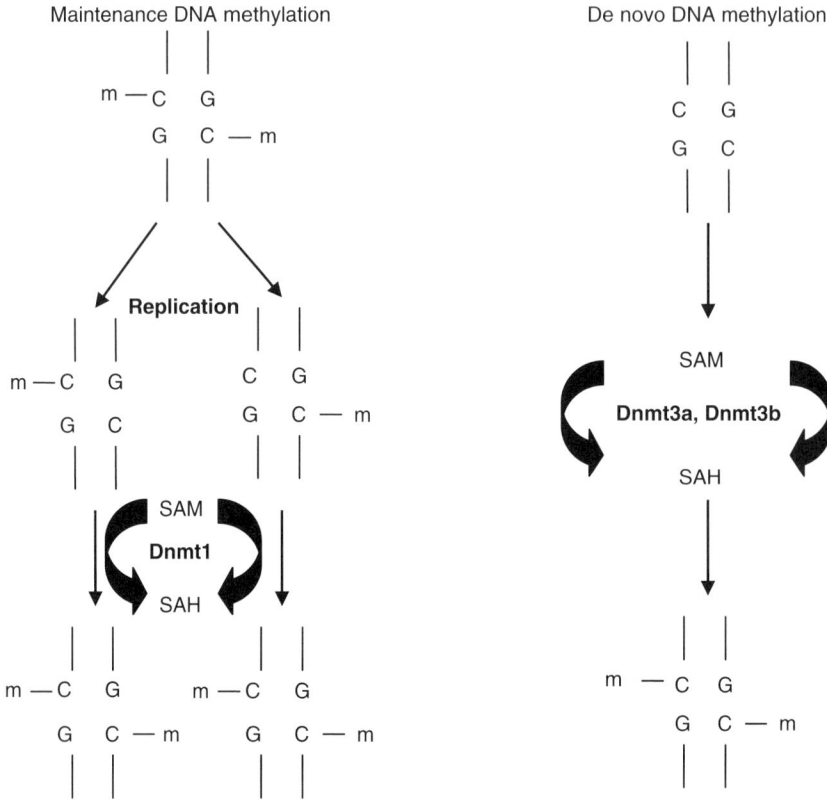

Figure 6.1. Maintenance and *de novo* DNA methylation. In maintenance methylation, DNA methyltransferase 1 (Dnmt1) binds the replication fork where it recognizes hemimethylated DNA and catalyzes the transfer of the methyl group from S-adenosylmethionine (SAM) to deoxycytosine in the newly synthesized DNA strand, producing methylcytosine (m-C) and S-adenosylhomocysteine. In *de novo* methylation DNA methyltransferases 3a and 3b (Dnmt3a and Dnmt3b, respectively) symmetrically methylate previously unmethylated regions of DNA by catalyzing a similar transmethylation reaction.

mutations in the methylcytosine-binding protein MeCP2, discussed below (Amir *et al.*, 1999). Abnormal methylation patterns have been associated with malignancies as well: hypomethylation of proto-oncogenes and hypermethylation of tumor suppressor genes may contribute to cellular transformation (Baylin and Herman, 2000).

Several mechanisms have been proposed by which DNA methylation might modify gene expression. Early studies demonstrated that the methylation of CG dinucleotides can prevent the binding of some transcription factors such as AP-2 to their recognition sequences (Comb and Goodman, 1990), and that members of the methylcytosine-binding protein family may prevent the binding of transcription factors by binding the methylated sequences (Yu *et al.*, 2000). Sp1 is a transcription factor that may be affected by this mechanism (Clark *et al.*, 1997).

Failure to Maintain T Cell DNA Methylation

More recent studies, though, demonstrate that DNA methylation acts synergistically with epigenetic processes such as histone deacetylation to modify chromatin structure and suppress gene transcription. Methylcytosine-binding proteins, such as MBD1, MBD2, and MeCP2, bind chromatin remodeling complexes containing histone deacetylases (HDACs), which remove acetyl groups from adjacent histones, condensing the nearby chromatin into an inactive configuration inaccessible to transcription factors (Attwood et al., 2002). This may be the most important mechanism by which DNA methylation suppresses gene expression, and is illustrated in Figure 6.2.

3. DNA Methylation and Drug-Induced Lupus

3.1. DNA Methylation and Autoimmunity

Abnormalities in DNA methylation are implicated in the pathogenesis of DIL. Our group and others have examined the effects of DNA methylation on T cell function and gene expression using the DNA methylation inhibitor 5-azaC. 5-Azacytidine is incorporated into newly synthesized DNA, where it covalently binds DNA methyltransferases during the S-phase cytosine transmethylation reaction, depleting intracellular pools of the enzymes and demethylating the daughter cells (Glover and Leyland-Jones, 1987).

Our group reported that cloned and polyclonal human and murine $CD4^+$ T cells become responsive to normally subthreshold stimuli when treated with

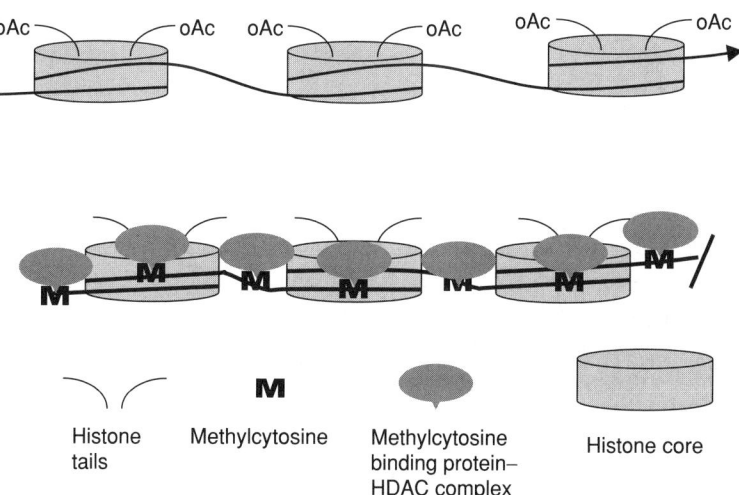

Figure 6.2. DNA methylation and chromatin structure. Chromatin is formed from nucleosome polymers, each consisting of DNA wrapped twice around a histone core. Transcriptionally permissive chromatin is characterized by unmethylated DNA and acetylated (oAc) histone tails. In contrast, transcriptionally nonpermissive chromatin is characterized by methylated DNA, the binding of chromatin inactivation complexes containing methylcytosine-binding proteins and histone deacetylases (HDACs), and unacetylated histone tails.

5-azaC, including self-class II MHC determinants without specific antigens (Richardson, 1986; Richardson et al., 1986, 1990; Quddus et al., 1993; Yung et al., 1995). This autoreactivity has been confirmed in mitogen-stimulated primary CD4+ human T cells (Richardson et al., 1990) and in a panel of cloned tetanus toxoid–specific CD4+ human T cell lines (Richardson, 1986; Richardson et al., 1992), as well as in mitogen-stimulated and alloreactive primary CD4+ murine T cells (Quddus et al., 1993), and conalbumin and pigeon cytochrome c–specific cloned murine CD4+ T cells (Yung et al., 1995, 2001). The autoreactivity is due, at least in part, to overexpression of the adhesion molecule LFA-1 (CD11a/CD18) by the 5-azaC-treated cells, since T cells transfected with LFA-1 demonstrate an identical autoreactivity (Richardson et al., 1994; Yung et al., 1996). 5-Azacytidine appears to increase LFA-1 expression by demethylating sequences 5′ to the CD11a (*ITGAL*) gene, causing increases in CD11a mRNA (Richardson et al., 1992; Lu et al., 2002).

Adoptive transfer of T cells made autoreactive by treatment with DNA methylation inhibitors causes a lupus-like disease in murine models. In initial studies, unirradiated female DBA/2 mice were injected intravenously with syngeneic CD4+ T cells made autoreactive with 5-azaC. Mice receiving treated, but not untreated, T cells, developed hematuria, proteinuria, and red cell casts beginning 7–21 days after each injection and lasting 7–14 days. A proliferative glomerulonephritis with IgG, IgM, and IgA immune complex deposition, as well as IgG deposition at the dermal–epidermal junction resembling the lupus band test, was observed upon histologic analysis. The mice also developed anti-DNA antibodies (Quddus et al., 1993). These *in vivo* effects are likely due to the effector functions expressed by the autoreactive cells, since viable cells are required for disease induction (Quddus et al., 1993). However, since the number of potential effector mechanisms expressed by polyclonal CD4+ cells is large, the cloned, conalbumin-reactive Th2 cell line D10.G4.1 (D10) isolated from AKR mice and AE7 cells, a cloned Th1 line from B10.A mice, were used to test the ability of demethylated Th1 and Th2 cells to induce autoimmunity (Yung et al., 1995, 2001). 5-Azacytidine-treated D10 and AE7 cells both induced anti-dsDNA antibodies in syngeneic recipients. The mice receiving D10 cells also developed lymphocytic interstitial pneumonitis, an immune complex glomerulonephritis, liver disease resembling primary biliary cirrhosis, and central nervous system (CNS) changes resembling CNS lupus (Yung et al., 1995), while the mice receiving AE7 cells did not. These differences may be due to functional differences between the Th1 and the Th2 cells or to host genetic differences (Yung et al., 1995, 2001).

The effector functions of the autoreactive cells were studied *in vitro*. The autoreactive T cells promote B cell immunoglobulin secretion and spontaneously kill autologous or syngeneic macrophages (Mø). For example, polyclonal human CD4+ T cells treated with DNA methylation inhibitors and cultured with autologous B lymphocytes induce differentiation of the B cells into IgG-secreting cells, without the addition of antigen or mitogen, possibly due, in part, to secretion of cytokines with B cell costimulatory properties (Richardson et al., 1990). More recently we reported that overexpression of the B cell costimulatory molecule

CD70 by drug-treated CD4+ T cells may also contribute to B cell overstimulation and subsequent immunoglobulin overproduction (Oelke et al., 2004).

Macrophage killing by the hypomethylated, autoreactive CD4+ T cells may also contribute to the development of autoimmunity. Normally, antigen-presenting Mø die by apoptosis after activating CD4+ T cells (Richardson et al., 1993). 5-Azacytidine-treated, autoreactive CD4+ T cells respond to autologous Mø without specific antigen, promiscuously killing the stimulating Mø (Quddus et al., 1993; Yung et al., 1995). This Mø killing may contribute to lupus-like autoimmunity by two mechanisms. Mevorach et al. (1998) reported that i.v. injection of apoptotic thymocytes induces anti-DNA antibodies in normal mice, suggesting that overproduction of apoptotic material from the dying Mø may contribute to the development of anti-DNA antibodies in the hypomethylation model. Others have reported that mice genetically deficient in any one of a number of molecules important for clearing apoptotic material will also develop anti-DNA antibodies and a lupus-like disease (reviewed in Walport, 2000). Since Mø are the cells responsible for removing apoptotic material, inducing Mø apoptosis would increase release of antigenic apoptotic nucleosome fragments as well as decrease their clearance, potentially synergizing in the stimulus for anti-DNA antibodies.

3.2. DNA Methylation and Drug-Induced Lupus

The studies summarized above demonstrate that treating normal CD4+ T cells with drugs that inhibit DNA methylation is sufficient to cause a lupus-like disease. This suggests that some lupus-inducing drugs could be DNA methylation inhibitors. Since procainamide and hydralazine are the two drugs most strongly associated with DIL (Yung and Richardson, 1994), we studied the effects of these drugs on T cell DNA methylation. Treatment of human CD4+ T cells with procainamide, hydralazine, or 5-azaC results in a decrease of total T cell genomic d^mC content as measured by reverse-phase high-performance liquid chromatography (HPLC), with 5-azaC being the most potent (Cornacchia et al., 1988). This demonstrates that both the lupus-inducing drugs are DNA methylation inhibitors. Subsequent studies demonstrated that procainamide and hydralazine treatment of human or murine CD4+ T cells causes DNA hypomethylation, LFA-1 overexpression, and autoreactivity *in vitro*, and adoptive transfer of the autoreactive murine cells into syngeneic recipients causes a lupus-like disease identical to 5-azaC-treated T cells (Cornacchia et al., 1988; Yung et al., 1997). More recently we reported that 5-azaC and procainamide demethylate the same sequences in the human CD11a promoter (Lu et al., 2002).

The mechanisms by which these drugs inhibit DNA methylation has been explored. Procainamide is a competitive inhibitor of some, but not all, nuclear DNA methyltransferase activity (Scheinbart et al., 1991). In contrast, hydralazine has no direct effect on DNA methyltransferase enzyme activity, but appears to prevent activation of DNA methyltransferase genes during mitosis. Treatment of T cell receptor (TCR)–stimulated or phorbol myristate acetate (PMA)–stimulated human T cells with hydralazine or a MEK inhibitor decreases ERK phosphorylation by 70–90%, indicating that hydralazine is a selective ERK (Ras-MAPK) signaling pathway inhibitor. Furthermore, inhibiting ERK pathway signaling with either hydralazine or MEK inhibitors, such as PD98059 or U0126, prevents upregulation of Dnmt1 and Dnmt3a in stimulated T cells by a mechanism analogous to MEK inhibitors. This subsequently leads to a decrease in the levels of these enzymes, thereby resulting in DNA hypomethylation (Deng et al., 2003).

The possibility that decreased ERK pathway signaling contributes to hydralazine-induced lupus was further confirmed by demonstrating that murine T cells treated with U0126, a MEK inhibitor, overexpress LFA-1 and become autoreactive similar to T cells treated with hydralazine or DNA methyltransferase inhibitors. Furthermore, injecting U0126-treated T cells into syngeneic mice induces anti-DNA antibodies similar to T cells treated with hydralazine or procainamide, further confirming that decreased T cell ERK pathway signaling can contribute to the development of lupus-like autoimmunity (Deng et al., 2003).

3.3. T Cell Genes Affected by DNA Methylation Inhibitors

To identify methylation-sensitive T cell genes, monoclonal antibody and oligonucleotide array–based approaches were used to compare gene expression in untreated and 5-azaC-treated human T cells. More than 100 genes increased expression following 5-azaC treatment. Genes of potential interest to autoimmunity include CD11a (Richardson et al., 1992), as noted above, and also perforin (Lu et al., 2003) and CD70 (Oelke et al., 2004). CD11a is a subunit of the β2-integrin LFA-1 (CD11a/CD18), which has adhesive and costimulatory functions (Kaplan et al., 2000). Bisulfite DNA sequencing, a technique used to precisely quantitate methylated and unmethylated dC bases, was used to determine the methylation status of the CD11a promoter and flanking regions in normal T cells, and T cells treated with DNA methylation inhibitors (Lu et al., 2002). The 1200-bp region 5′ to the transcription start site was found to demethylate in primary T cells treated with 5-azaC or procainamide, and regional, or "patch," methylation of the affected sequences in reporter constructs suppressed CD11a promoter function. These findings suggest that methylation is important in suppressing CD11a expression. Importantly, as noted above, overexpressing LFA-1 by transfection of human and murine CD4$^+$ T cells causes MHC-specific T cell autoreactivity that is identical to the autoreactivity caused by DNA methylation inhibitors (Yung et al., 1996). Furthermore, adoptive transfer of LFA-1-transfected murine CD4$^+$ T cells causes a lupus-like disease resembling that caused by DNA methylation inhibitors (Yung et al., 1996). These results indicate an important role for T cell LFA-1 overexpression in autoreactivity *in vitro* and autoimmunity *in vivo*.

Perforin is a cytotoxic molecule normally expressed in NK cells and a subset of cytotoxic CD8$^+$ T cells. However, CD4$^+$ T cells treated with 5-azaC aberrantly express perforin, and perforin expression is increased in CD8$^+$ T cells treated with 5-azaC. Bisulfite sequencing of the perforin promoter and upstream enhancer indicates that the methylation status of the region linking the enhancer and promoter correlates with perforin expression, and that methylation of this region suppresses promoter function in reporter constructs. Interestingly, hypomethylation of this region may alter chromatin structure around the enhancer, because demethylation of this region in CD8$^+$ T cells correlates with hypersensitivity of the enhancer to DNase1 digestion (Lu *et al.*, 2003). Finally, perforin contributes to macrophage killing by 5-azaC-treated T cells, suggesting a potential role in autoimmunity (Lu *et al.*, 2003).

CD70, a costimulatory ligand for B cell CD27, is a member of the tumor necrosis factor (TNF) family that is expressed on activated CD4$^+$ T cells (Lens *et al.*, 1998). We recently reported that CD4$^+$ T cells treated with 5-azaC, procainamide, hydralazine, and ERK pathway inhibitors overexpress CD70 and overstimulate B cell IgG production, and that the increased IgG synthesis is abrogated by anti-CD70 (Oelke *et al.*, 2004), suggesting another mechanism by which demethylated T cells might contribute to autoantibody production.

Together, these studies demonstrate that DNA methylation inhibitors including hydralazine and procainamide can increase T cell expression of LFA-1, perforin, and CD70, and that the drug-treated cells can cause a lupus-like disease *in vivo*. Autoreactivity, caused by LFA-1 overexpression, together with macrophage killing mediated in part by aberrant perforin overexpression and increased B cell costimulatory signals mediated in part from CD70 overexpression, together could contribute to the development of autoantibodies in DIL. We therefore determined if similar changes in T cell DNA methylation and gene expression also occur in patients with idiopathic lupus.

4. Aberrant T Cell DNA Methylation, Gene Expression, and Cellular Function in Idiopathic Lupus

4.1. DNA Methylation

Initial studies compared total genomic d^{m}C content in T cells from patients with inactive and active lupus with controls using reverse-phase HPLC. These studies demonstrated that T cells from patients with active lupus have decreased genomic d^{m}C levels as well as decreased levels of DNA methyltransferase activity. Intracellular pools of SAM and S-adenosylhomocysteine, both regulators of transmethylation reactions, were normal (Richardson *et al.*, 1990). Subsequent studies demonstrated decreased Dnmt1 mRNA levels, implicating decreased maintenance DNA methylation in the DNA hypomethylation (Richardson *et al.*, 1990; Deng *et al.*, 2001). Dnmt3a, but not Dnmt3b, transcripts were also decreased in lupus T cells (C. Deng and B. Richardson, unpublished data). Mechanisms contributing to the decreased DNA methyltransferase expression and consequent DNA hypomethylation were sought.

Others have reported that signaling is abnormal in lupus T cells (Kammer *et al.*, 2002), and our group reported that Dnmt1 and Dnmt3a levels are regulated in part by the JNK and ERK pathways (Deng *et al.*, 2003). Furthermore, inhibiting either ERK or JNK signaling in proliferating cells demethylates DNA (Deng *et al.*, 1998). We therefore examined JNK and ERK pathway signaling in lupus T cells. JNK pathway signaling was found to be intact. ERK pathway signaling, however, is decreased in T cells from patients with active lupus, with levels of Dnmt1 mRNA decreased to the same degree as in T cells treated with ERK pathway inhibitors (Deng et al., 2001). Since hydralazine and MEK inhibitors decrease Dnmt1 and Dnmt3a expression, demethylate DNA, and because T cells treated with these drugs induce autoimmunity (Deng et al., 2003), decreased T cell ERK pathway signaling may contribute to DNA hypomethylation and autoimmunity in idiopathic lupus by similar mechanisms.

4.2. Gene Expression and Cellular Function

As noted above, CD11a, perforin, and CD70 have been identified as methylation-sensitive T cell genes that may contribute to the pathogenesis of DIL. Evidence for overexpression of these genes was therefore sought in T cells from lupus patients.

A subset of T cells was found to express high levels of LFA-1 in lupus patients compared with controls, which included normal individuals, patients with active infections, and patients with multiple sclerosis (Richardson et al., 1992). The degree of overexpression correlated directly with clinical disease activity. In addition, the cells overexpressing LFA-1 were found to be autoreactive and spontaneously lysed autologous, but not allogeneic, macrophages, similar to T cells treated with DNA hypomethylating agents (Richardson et al., 1992). Bisulfite sequencing of the CD11a promoter and flanking regions in T cells from active lupus patients showed that the sequences demethylated are the same as those demethylated in T cells treated with 5-azaC and procainamide (Lu et al., 2002). Figure 6.3 compares the effects of 5-azaC and lupus on CD11a promoter methylation in T cells.

Perforin is also overexpressed in lupus T cells. $CD4^+$ T cells from patients with active, but not inactive, lupus aberrantly express perforin, and the abnormal expression is related to demethylation of the same sequences suppressing perforin transcription in primary $CD4^+$ T cells, which are demethylated by DNA methylation inhibitors (Kaplan et al., 2004) (Figure 6.4). Furthermore, the perforin inhibitor concanamycin A blocks autologous monocyte killing by $CD4^+$ lupus T cells, suggesting that aberrant perforin expression in $CD4^+$ lupus T cells may contribute to autoreactive monocyte/macrophage killing (Kaplan et al., 2004).

CD70 was recently found to be overexpressed in lupus T cells as well. $CD4^+$ SLE T cells overexpress CD70 and stimulate B cell IgG production, similar to T cells treated with DNA methylation inhibitors (Oelke et al., 2004), or transfected with CD70 (Kobata et al., 1995). The abnormal T cell–dependent IgG secretion that characterizes lupus B cells is inhibited by anti-CD70 (Liossis and Tsokos, 1999), as predicted by the in vitro model. Preliminary work has shown that DNA methylation inhibitors demethylate a region ~500 bp 5′ to the CD70 transcription start site and that $CD4^+$ T cells from patients with active lupus exhibit hypomethylation in the same region (Q. Lu and B. Richardson, unpublished data).

Together, these studies demonstrate that lupus T cells aberrantly overexpress LFA-1, perforin, and CD70, that the same genes are overexpressed in the hypomethylation model, and that the same sequences are demethylated as in T cells treated with DNA methyltransferase or ERK pathway inhibitors. These observations strongly suggest that similar mechanisms may be contributing to overexpression of these genes in the DNA hypomethylation model and in lupus.

Failure to Maintain T Cell DNA Methylation

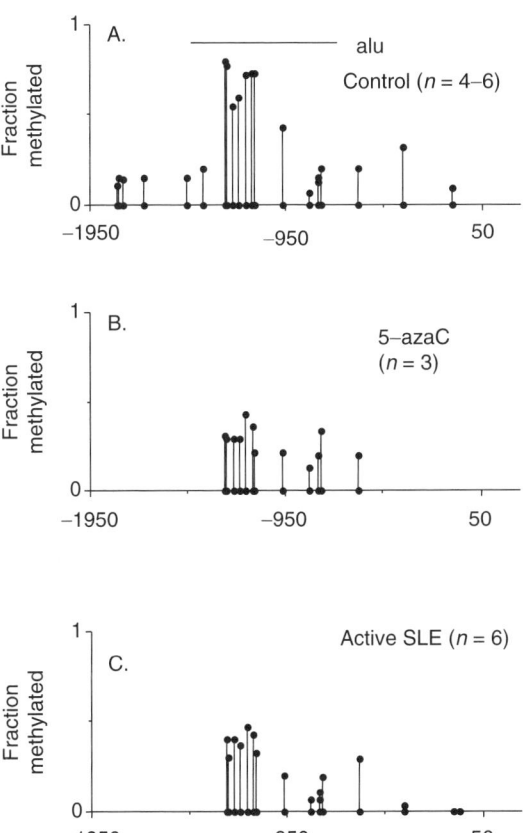

Figure 6.3. CD11a promoter methylation. (A) T cell DNA was isolated from four to six healthy donors, treated with sodium bisulfite, and the indicated region amplified in overlapping fragments. Five cloned fragments from each section were then sequenced from each donor, and the fraction methylated for each CpG pair averaged across the five cloned and sequenced fragments. The average fraction methylated is shown on the y-axis, and the location of each CG pair is shown on the x-axis, numbered relative to the transcription start site at 0. The region from the beginning (−1950) to bp −1262 represents the mean of five fragments from each of four donors, the region from −1261 to −68 the mean of five fragments from each of six donors, and the region from −68 to the end the mean of five fragments from each of four donors. A region containing *Alu* elements is denoted by the horizontal line. (B). T cells were isolated from three healthy donors, stimulated with PHA, treated with 5-azacytidine (5-azaC), DNA isolated, and bisulfite sequencing performed as in panel A. PHA stimulation alone had no effect on the methylation pattern (not shown). (C) T cell DNA from six patients with active (SLE Disease Activity Index (SLEDAI) > 6) lupus was analyzed as in panel A. The methylation pattern resembles that seen in the 5-azaC-treated cells. Patients with inactive lupus did not have demethylation of this region (not shown).

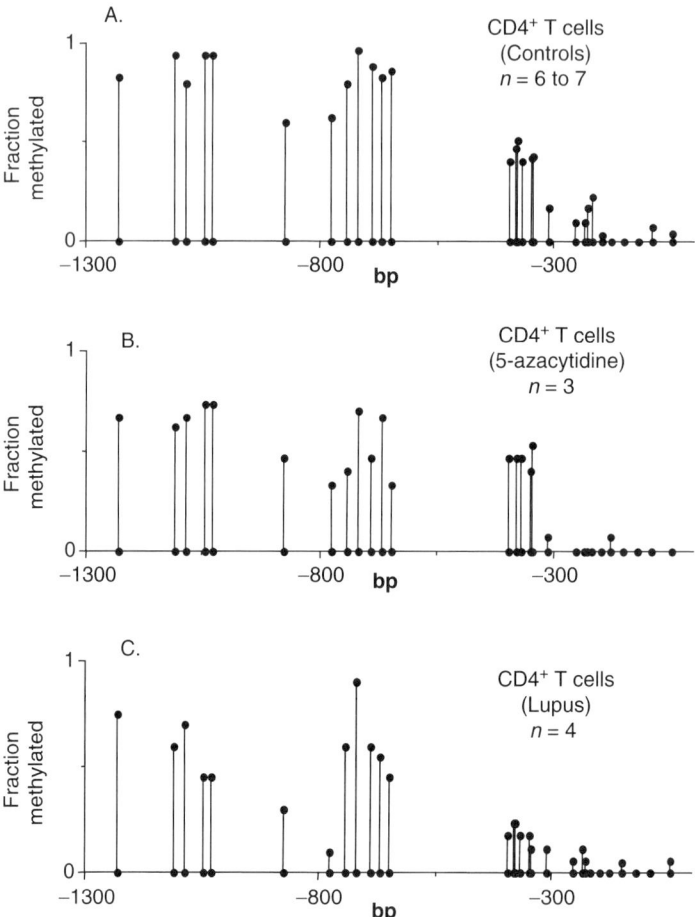

Figure 6.4. Perforin promoter methylation. (A) The methylation of the perforin promoter was determined as in Figure 6.3A, using CD4+ T cells from six to seven healthy donors, and amplified in two fragments. The *y*-axis represents the average methylation of five cloned and sequenced fragments from each segment amplified from each of the six to seven donors. The *x*-axis represents each CG pair in the region shown, numbered relative to the transcription start site. (B) CD4+ T cells from three healthy donors were stimulated with PHA, treated with 5-azacytidine (5-azaC), and methylation determined as in Figure 6.3B. The results thus represent the mean of 15 determinations per point. Again, PHA stimulation alone had no effect on the methylation pattern (not shown). (C) T cell DNA from four patients with active (SLE Disease Activity Index (SLEDAI) 8–12) lupus was analyzed as in Figure 6.3C. The methylation is decreased and the pattern resembles that seen in the 5-azaC-treated cells. Again, patients with inactive lupus did not have demethylation of this region (not shown).

5. Conclusions

The studies summarized in this chapter demonstrate that the failure to maintain DNA methylation patterns in mature CD4+ T cells causes aberrant expression of several methylation-sensitive genes, including LFA-1, perforin, CD70, and likely others, and that their overexpression alters T cell function, promoting autoreactivity, monocyte/macrophage killing, and B cell overstimulation. Furthermore, T cells experimentally demethylated with DNA methyltransferase or ERK signaling pathway inhibitors cause a lupus-like disease in murine models. Procainamide and hydralazine are DNA methylation inhibitors and cause a lupus-like disease in genetically susceptible individuals. Patients with idiopathic lupus have hypomethylated DNA, overexpress the same genes due to the same changes in DNA methylation patterns as in the methylation inhibition model, and demonstrate identical changes in CD4$^+$ T cell function including autoreactive, perforin-mediated monocyte killing and B cell overstimulation. Thus, similar changes in DNA methylation and chromatin structure likely contribute to the pathogenesis of autoimmunity in the DNA hypomethylation model as in idiopathic lupus. The DNA hypomethylation model may also provide an approach to predict additional aberrantly expressed genes in human lupus T cells, since CD11a, perforin, and CD70 were predicted by this model. Finally, these studies also suggest that environmental agents may act by mechanisms analogous to those seen in DIL, triggering changes in chromatin structure and affecting gene expression through signaling inhibition or direct DNA methyltransferase inhibition. Clearly, there is a fundamental role for a failure to maintain DNA methylation patterns and chromatin structure in this disease.

References

Adams, R.L.P. and Burdon, R.H. (1985). DNA methylation in the cell. In A. Rich (ed.), *Molecular Biology of DNA Methylation*. Springer-Verlag, New York. pp. 9–18.

Amir, R.E., Van den Veyver, I.B., Wan, M., Tran, C.Q., Francke, U., and Zoghbi, H.Y. (1999). Rett syndrome is caused by mutations in X-linked MECP2, encoding methyl-CpG-binding protein 2. *Nat. Genet.*, **23**, 185–188.

Antequera, F. and Bird, A. (1993). Number of CpG islands and genes in human and mouse. *Proc. Natl. Acad. Sci. U S A*, **90**, 11995–11999.

Attwood, J.T., Yung, R.L., and Richardson, B.C. (2002). DNA methylation and the regulation of gene transcription. *Cell. Mol. Life Sci.*, **59**, 241–257.

Baylin, S.B. and Herman, J.G. (2000). DNA hypermethylation in tumorigenesis: Epigenetics joins genetics. *Trends Genet.*, **16**, 168–174.

Bestor, T.H. (1998). The host defence function of genomic methylation patterns. *Novartis Foundation Symposium*, **214**, 187–195; discussion 195–199, 228–232.

Chuang, L.S., Ian, H.I., Koh, T.W., Ng, H.H., Xu, G., and Li, B.F. (1997). Human DNA-(cytosine-5) methyltransferase-PCNA complex as a target for p21WAF1. *Science*, **277**, 1996–2000.

Clark, S.J., Harrison, J., and Molloy, P.L. (1997). Sp1 binding is inhibited by (m)Cp(m)CpG methylation. *Gene*, **195**, 67–171.

Comb, M. and Goodman, H.M. (1990). CpG methylation inhibits proenkephalin gene expression and binding of the transcription factor AP-2. *Nucleic Acids Res.*, **18**, 3975–3982.

Cornacchia, E., Golbus, J., Maybaum, J., Strahler, J., Hanash, S., and Richardson, B. (1988). Hydralazine and procainamide inhibit T cell DNA methylation and induce autoreactivity. *J. Immunol.*, **140**, 2197–2200.

Deng, C., Kaplan, M.J., Yang, J., Ray, D., Zhang, Z., McCune, W.J., Hanash, S.M., and Richardson, B.C. (2001). Decreased Ras-mitogen-activated protein kinase signaling may cause DNA hypomethylation in T lymphocytes from lupus patients. *Arthritis Rheum.*, **44**, 397–407.

Deng, C., Lu, Q., Zhang, Z., Rao, T., Attwood, J., Yung, R., and Richardson, B. (2003). Hydralazine may induce autoimmunity by inhibiting extracellular signal-regulated kinase pathway signaling. *Arthritis Rheum.*, **48**, 746–756.

Deng, C., Yang, J., Scott, J., Hanash, S., and Richardson, B.C. (1998). Role of the Ras-MAPK signaling pathway in the DNA methyltransferase response to DNA hypomethylation. *Biol. Chem.*, **379**, 1113–1120.

Ehrlich, M. and Wang, R.Y. (1981). 5-Methylcytosine in eukaryotic DNA. *Science*, **212**, 1350–1357.

Gladman, D. (2004). Epidemiology of systemic lupus erythematosus. In R.G. Lahita (ed.), *Systemic Lupus Erythematosus*. Elsevier Academic, New York. pp. 697–715.

Glover, A.B. and Leyland-Jones, B. (1987). Biochemistry of azacitidine: A review. *Cancer Treatm. Rep.*, **71**, 959–964.

Goldstein, R. and Arnett, F.C. (1987). The genetics of rheumatic disease in man. *Rheum. Dis. Clin. N Am.*, **13**, 487–510.

Hansen, R.S., Wijmenga, C., Luo, P., Stanek, A.M., Canfield, T.K., Weemaes, C.M., and Gartler, S.M. (1999). The DNMT3B DNA methyltransferase gene is mutated in the ICF immunodeficiency syndrome. *Proc. Natl. Acad. Sci. U S A*, **96**, 14412–14417.

Hess, E.V. (1995). Environmental lupus syndromes. *Br. J. Rheum.*, **34**, 597–599.

Holliday, R. and Pugh, J.E. (1975). DNA modification mechanisms and gene activity during development. *Science*, **187**, 226–232.

Kammer, G.M., Perl, A., Richardson, B.C., and Tsokos, G.C. (2002). Abnormal T cell signal transduction in systemic lupus erythematosus. *Arthritis Rheum.*, **46**, 1139–1154.

Kaplan, M.J., Beretta, L., Yung, R.L., and Richardson, B.C. (2000). LFA-1 overexpression and T cell autoreactivity: Mechanisms. *Immunol. Invest.*, **29**, 427–442.

Kaplan, M.J., Lu, Q., Wu, A., Attwood, J., and Richardson, B. (2004). Demethylation of promoter regulatory elements contributes to perforin overexpression in CD4+ lupus T cells. *J. Immunol.*, **172**, 3652–3661.

Kobata, T., Jacquot, S., Kozlowski, S., Agematsu, K., Schlossman, S.F., and Morimoto, C. (1995). CD27–CD70 interactions regulate B-cell activation by T cells. *Proc. Natl. Acad. Sci. U S A*, **92**, 11249–11253.

Kyttaris, V. and Tsokos, G. (2003). Uncovering the genetics of systemic lupus erythematosus: Implications for therapy. *Am. J. Pharmacogenomics*, **3**, 193–202.

Lens, S.M., Tesselaar, K., van Oers, M.H., and van Lier, R.A. (1998). Control of lymphocyte function through CD27–CD70 interactions. *Semin. Immunol.*, **10**, 491–499.

Li, E., Bestor, T.H., and Jaenisch, R. (1992). Targeted mutation of the DNA methyltransferase gene results in embryonic lethality. *Cell*, **69**, 915–926.

Liossis, S. and Tsokos, G. (1999). B cells in systemic lupus erythematosus. In G. Kammer and G. Tsokos (eds), *Lupus. Molecular and Cellular Pathogenesis*. Humana Press Inc, Totowa, New Jersey. pp. 167–180.

Lu, Q., Kaplan, M., Ray, D., Ray, D., Zacharek, S., Gutsch, D., and Richardson, B. (2002). Demethylation of ITGAL (CD11a) regulatory sequences in systemic lupus erythematosus. *Arthritis Rheum.*, **46**, 1282–1291.

Lu, Q., Wu, A., Ray, D., Deng, C., Attwood, J., Hanash, S., Pipkin, M., Lichtenheld, M., and Richardson, B. (2003). DNA methylation and chromatin structure regulate T cell perforin gene expression. *J. Immunol.*, **170**, 5124–5132.

Mevorach, D., Zhou, J.L., Song, X., and Elkon, K.B. (1998). Systemic exposure to irradiated apoptotic cells induces autoantibody production. *J. Exp. Med.*, **188**, 387–392.

Mongey, A.-B. and Hess, E.V. (2002). The role of environment in systemic lupus erythematosus and associated disorders. In D.J. Wallace and B.H. Hahn (eds), *Dubois's Lupus Erythematosus*. Lippincott Williams and Wilkins, Philadelphia. pp. 33–64.

Oelke, K., Lu, Q., Richardson, D., Wu, A., Deng, C., Hanash, S., and Richardson, B. (2004). Overexpression of CD70 and overstimulation of IgG synthesis by lupus T cells and T cells treated with DNA methylation inhibitors. *Arthritis Rheum.*, **50**, 1850–1860.

Okano, M., Bell, D.W., Haber, D.A., and Li, E. (1999). DNA methyltransferases Dnmt3a and Dnmt3b are essential for de novo methylation and mammalian development. *Cell*, **99**, 247–257.

Okano, M., Xie, S., and Li, E. (1998a). Dnmt2 is not required for de novo and maintenance methylation of viral DNA in embryonic stem cells. *Nucleic Acids Res.*, **26**, 2536–2540.

Okano, M., Xie, S., and Li, E. (1998b). Cloning and characterization of a family of novel mammalian DNA (cytosine-5) methyltransferases. *Nat. Genet.*, **19**, 219–220.

Quddus, J., Johnson, K.J., Gavalchin, J., Amento, E.P., Chrisp, C.E., Yung, R.L., and Richardson, B.C. (1993). Treating activated CD4+ T cells with either of two distinct DNA methyltransferase inhibitors, 5-azacytidine or procainamide, is sufficient to cause a lupus-like disease in syngeneic mice. *J. Clin. Invest.*, **92**, 38–53.

Richardson, B. (1986). Effect of an inhibitor of DNA methylation on T cells. II. 5-Azacytidine induces self-reactivity in antigen-specific T4+ cells. *Hum. Immunol.*, **17**, 456–470.

Richardson, B., Kahn, L., Lovett, E.J., and Hudson, J. (1986). Effect of an inhibitor of DNA methylation on T cells. I. 5-Azacytidine induces T4 expression on T8+ T cells. *J. Immunol.*, **137**, 35–39.

Richardson, B., Powers, D., Hooper, F., Yung, R.L., and O'Rourke, K. (1994). Lymphocyte function-associated antigen 1 overexpression and T cell autoreactivity. *Arthritis Rheum.*, **37**, 1363–1372.

Richardson, B., Scheinbart, L., Strahler, J., Gross, L., Hanash, S., and Johnson, M. (1990). Evidence for impaired T cell DNA methylation in systemic lupus erythematosus and rheumatoid arthritis. *Arthritis Rheum.*, **33**, 1665–1673.

Richardson, B.C., Buckmaster, T., Keren, D.F., and Johnson, K.J. (1993). Evidence that macrophages are programmed to die after activating autologous, cloned, antigen-specific, CD4+ T cells. *E. J. Immunol.*, **23**, 1450–1455.

Richardson, B.C., Liebling, M.R., and Hudson, J.L. (1990). CD4+ cells treated with DNA methylation inhibitors induce autologous B cell differentiation. *Clin. Immunol. Immunopathol.*, **55**, 368–381.

Richardson, B.C., Strahler, J.R., Pivirotto, T.S., Quddus, J., Bayliss, G.E., Gross, L.A., O'Rourke, K.S., Powers, D., Hanash, S.M., and Johnson, M.A. (1992). Phenotypic and functional similarities between 5-azacytidine-treated T cells and a T cell subset in patients with active systemic lupus erythematosus. *Arthritis Rheum.*, **35**, 647–662.

Riggs, A.D. (1985). X-inactivation, DNA methylation, and differentiation revered. In A. Razin, H. Cedar, and A.D. Riggs (eds), *DNA Methylation: Biochemistry and Biological Significance*. Springer-Verlag, New York. pp. 269–278.

Scheinbart, L.S., Johnson, M.A., Gross, L.A., Edelstein, S.R., and Richardson, B.C. (1991). Procainamide inhibits DNA methyltransferase in a human T cell line. *J. Rheumatol.*, **18**, 530–534.

Stoger, R., Kubicka, P., Liu, C.G., Kafri, T., Razin, A., Cedar, H., and Barlow, D.P. (1993). Maternal-specific methylation of the imprinted mouse Igf2r locus identifies the expressed locus as carrying the imprinting signal. *Cell*, **73**, 61–71.

Tsao, B.P. and Grossman, J.M. (2001). Genetics and systemic lupus erythematosus. *Curr. Rheumatol. Rep.*, **3**, 183–190.

Walport, M.J. (2000). Lupus, DNase and defective disposal of cellular debris. *Nat. Genet.*, **25**, 135–136.

Yoder, J.A., Soman, N.S., Verdine, G.L., and Bestor, T.H. (1997). DNA (cytosine-5)-methyltransferases in mouse cells and tissues. Studies with a mechanism-based probe. *J. Mol. Biol.*, **270**, 385–395.

Yu, F., Thiesen, J., and Stratling, W.H. (2000). Histone deacetylase-independent transcriptional repression by methyl-CpG-binding protein 2. *Nucleic Acids Res.*, **28**, 2201–2206.

Yung, R., Chang, S., Hemati, N., Johnson, K., and Richardson, B. (1997). Mechanisms of drug-induced lupus. IV. Comparison of procainamide and hydralazine with analogs in vitro and in vivo. *Arthritis Rheum.*, **40**, 1436–1443.

Yung, R., Kaplan, M., Ray, D., Schneider, K., Mo, R.R., Johnson, K., and Richardson, B. (2001). Autoreactive murine Th1 and Th2 cells kill syngeneic macrophages and induce autoantibodies. *Lupus*, **10**, 539–546.

Yung, R., Powers, D., Johnson, K., Amento, E., Carr, D., Laing, T., Yang, J., Chang, S., Hemati, N., and Richardson, B. (1996). Mechanisms of drug-induced lupus. II. T cells overexpressing lymphocyte function-associated antigen 1 become autoreactive and cause a lupuslike disease in syngeneic mice. *J. Clin. Invest.*, **97**, 2866–2871.

Yung, R.L., Quddus, J., Chrisp, C.E., Johnson, K.J., and Richardson, B.C. (1995). Mechanism of drug-induced lupus. I. Cloned Th2 cells modified with DNA methylation inhibitors in vitro cause autoimmunity in vivo. *J. Immunol.*, **154**, 3025–3035.

Yung, R.L. and Richardson, B.C. (1994). Drug-induced lupus. *Rheum. Dis. Clin. N. Am.*, **20**, 61–86.

Yung, R.L. and Richardson, B.C. (2003). Drug-induced lupus. In M. Hochberg, A. Silman, J. Smolen, M. Weinblatt, and M. Weisman (eds) *Rheumatology*. Harcourt Health Sciences, London.

7

Complement Components C4A and C4B in Human Lupus

Yan Yang, Erwin K. Chung, Karl Lhotta, Yee Ling Wu, Gloria C. Higgins, Robert M. Rennebohm, Lee A. Hebert, Daniel J. Birmingham, Brad H. Rovin, and C. Yung Yu

1. Introduction

For over half a century low serum complement C4 concentrations have been recognized as a manifestation of systemic lupus erythematosus (SLE). However, the role of C4 in human SLE remains elusive. This is partly because of the unusually complex C4 genetics with frequent variations in gene number, gene size, protein isoforms, and expression levels. In this chapter we describe the strong association of complete C4A and C4B deficiencies and human SLE, and discuss the possible role of homozygous and partial deficiencies of C4A, which are present in 32–55% of SLE patients. Accumulated evidence provides support for the interpretation that C4A deficiency is a genetic risk factor for SLE. To explain the differences in SLE prevalence and disease severity among different ethnic groups, however, more elaborate analyses are needed to characterize the *C4A* and *C4B* gene dosages,

---

Yan Yang and Erwin K. Chung • Center for Molecular and Human Genetics, Columbus Children's Research Institute, 700 Children's Drive, Columbus, Ohio 43205 and Department of Molecular Virology, Immunology and Medical Genetics, The Ohio State University, Columbus, Ohio. **Karl Lhotta** • Division of Clinical Nephrology, Innsbruck University Hospital, Innsbruck, Austria. **Yee Ling Wu** • Center for Molecular and Human Genetics, Columbus Children's Research Institute, 700 Children's Drive, Columbus, Ohio 43205 and Integrated Biomedical Science Graduate Program, The Ohio State University, Columbus, Ohio. **Gloria C. Higgins and Robert M. Rennebohm** • Department of Pediatrics, The Ohio State University; Columbus, Ohio. **Lee A. Hebert** • Department of Internal Medicine, The Ohio State University; Columbus, Ohio. **Daniel J. Birmingham and Brad H. Rovin** • Integrated Biomedical Science Graduate Program, The Ohio State University, Columbus, Ohio and Department of Internal Medicine, The Ohio State University, Columbus, Ohio. **C. Yung Yu** • Center for Molecular and Human Genetics, Columbus Children's Research Institute, 700 Children's Drive, Columbus, Ohio 43205; Department of Molecular Virology, Immunology and Medical Genetics, The Ohio State University, Columbus, Ohio; Integrated Biomedical Science Graduate Program, The Ohio State University, Columbus, Ohio; Department of Pediatrics, The Ohio State University, Columbus, Ohio.

Molecular Autoimmunity: In commemoration of the 100th anniversary of the first description of human autoimmune disease, edited by Moncef Zouali. Springer Science+Business Media, Inc., New York, 2005.

RP-C4-CYP21-TNX (RCCX) modular variations, and quantitative and qualitative diversities of C4A and C4B proteins in SLE patients and controls.

2. Diversities of Complement Components C4A and C4B in Human Populations

The human complement component C4 represents an extraordinary paradigm of innate immune diversity (Yu *et al.*, 2003). There are two classes of proteins: the acidic C4A (hence the "A" designation) and the basic C4B (hence the "B" designation). In each class there are multiple polymorphic variants with different frequencies among human populations. The plasma or serum protein levels of total C4 vary widely (between about 100 and 800 mg/L) among different individuals (Porter, 1983).

The genetics of human C4 is complex. The *C4* gene is located at the class III region of the major histocompatibility complex (MHC). There may be one, two, three, or four copies of *C4* genes in an MHC and two to seven copies of *C4* genes with different combinations of long and short *C4A* and *C4B* genes have been found in diploid genomes from different subjects (Yang *et al.*, 1999; Blanchong *et al.*, 2000; Chung *et al.*, 2002a). In about half of the Caucasian populations, there are four copies of *C4* genes in an individual's diploid genome (gene dosage). Approximately 40% of normal Caucasians have a heterozygous deficiency of either C4A or C4B. In contrast, about one third of Caucasians have five or six *C4* genes. This increase in gene dosage results in increased plasma protein levels of C4A, C4B, or both (Yang *et al.*, 2003). Such genetic diversity of human *C4A* and *C4B* genes is one of the major determinants for the quantitative and qualitative variations of the C4A and C4B proteins that may be the result of the selection pressure imposed by the variety of microbes and parasites. A deficiency of C4A or C4B increases the likelihood or severity of viral and bacterial infections, and the susceptibility to autoimmune diseases.

2.1. Dichotomy in Gene Sizes, Polygenes, and RCCX Module Variants

There are two forms of *C4* genes: the long gene is 20.6 kb and the short gene is 14.2 kb in size. In Caucasians, 76–77% of the *C4* genes are long and 23–24% are short (Blanchong *et al.*, 2000; Yang *et al.*, 2003). Such gene size variation is attributable to the integration of a 6.36-kb endogenous retrovirus, HERV-K(C4), into intron 9 of the long gene (Dangel *et al.*, 1994; Mack *et al.*, 2004). Multiple mutations in HERV-K(C4) probably had knocked out most of its retroviral activities. The configuration of HERV-K(C4) is opposite to that of *C4*. Therefore, an antisense HERV-K(C4) sequence will be produced whenever a C4 transcript is synthesized from a long gene. The selection advantage for the coexistence of both long and short *C4* genes in human populations is uncertain. However, the long gene has become the predominant form among different human races.

The duplication of *C4* genes in the MHC is discretely modular, which includes Ser/Thr nuclear protein kinase gene *RP1* or *RP2* at the 5′ end, steroid

21-hydroxylase gene *CYP21A* or *CYP21B*, and extracellular matrix protein tenascin-X gene *TNXA* or *TNXB* at the 3' end. A duplicated RCCX module is either 32.7 kb or 26.2 kb in size, and usually contains a nonfunctional *CYP21A* with multiple point mutations, followed by gene fragments *TNXA* and *RP2*, and a functional *C4A* or *C4B* gene that is either long or short (Figure 7.1, panel A) (Shen *et al.*, 1994). The multiplication of RCCX modules can be clearly depicted by Southern blot analysis of *Pme*I-digested genomic DNA resolved by pulsed field gel electrophoresis, or by *Taq*I RFLP (Chung *et al.*, 2002b).

2.2. Diversity of Human C4A and C4B Proteins

About 40 polymorphic protein variants of human C4 have been detected, based on gross differences in electric charge and serologic variations. The most widely used method for C4 phenotyping is immunofixation of EDTA-plasma proteins resolved by high-voltage agarose gel electrophoresis (Figure 7.1C). The most common C4A and C4B allotypes are C4A3 and C4B1, respectively. Other common allotypes for C4A include A2, A4, and A6, and for C4B, B2, B3, and B5. These allotypes exhibit different frequencies among different races or ethnic groups. For example, in the Ohio population, C4B2 has a frequency of 9.4% in Caucasians, but 33.6% in Asian Chinese (Yang *et al.*, 2003; Yang, 2004).

The C4A and C4B isotypes are mainly defined by four specific amino acid residues at positions 1101, 1102, 1105 and 1106, located at the C4d region (Yu *et al.*, 1988). The C4B isotypic residues LSPVIH catalyze the formation of a covalent ester bond between the thioester carbonyl group of activated C4B (C4Bb) and a hydroxyl group from substrates. This transesterification reaction is rapid. However, due to hydrolysis, the half-life of the C4Bb thioester bond is relatively short, less than 1 s (Isenman and Young, 1986; Dodds *et al.*, 1996;). Hence, C4B is important for the propagation of the classical and the mannose-binding lectin (MBL) complement activation pathways, culminating in the rapid and focal formation of the membrane attack complex against microbes.

The C4A isotypic residues PCPVLD probably modulate the reactivity of the thioester bond from the activated C4A (C4Ab) molecule to efficiently form a covalent amide bond with substrates. While the reaction rate of activated C4A toward its targets is about four times slower than that of activated C4B, the isotypic residues also confer on C4Ab a relatively longer half-life against hydrolysis (~10 s) (Sepp *et al.*, 1993) and a higher affinity for complement receptor CR1 (Gatenby *et al.*, 1990; Gibb *et al.*, 1993; Reilly and Mold, 1997). Therefore, it is thought that C4A is important in the solubilization of immune aggregates, immunoclearance, and opsonization.

The 3-dimensional structure of a human C4d polypeptide from C4A has been solved by X-ray crystallography (van den Elsen *et al.*, 2002). The isotypic residues of C4A are found facing the thioester residues located on the convex side of a barrel-shaped structure. The major determinants of the Rodgers or Chido blood group antigens, which are VDLL for Rg1 and ADLR for Ch1, respectively, at positions 1188–1191 (Yu *et al.*, 1988), are located at the concave surface on the opposite side of the isotype thioester residues (van den Elsen *et al.*, 2002).

A. The MHC complement gene cluster and RCCX length variants

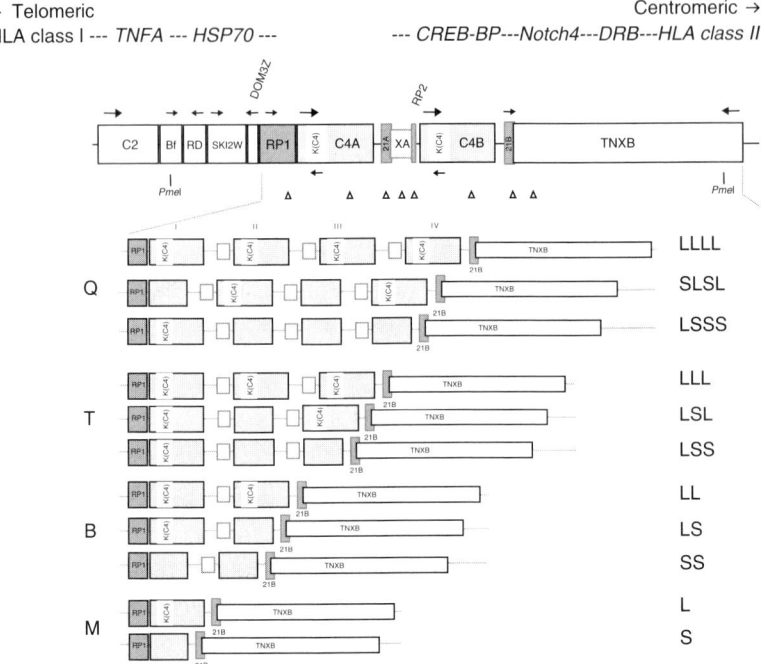

B. *Pme*I PFGE of RCCX modules

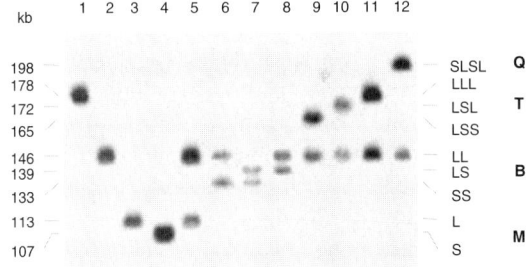

C. Immunofixation of EDTA plasma using goat antiserum against human C4

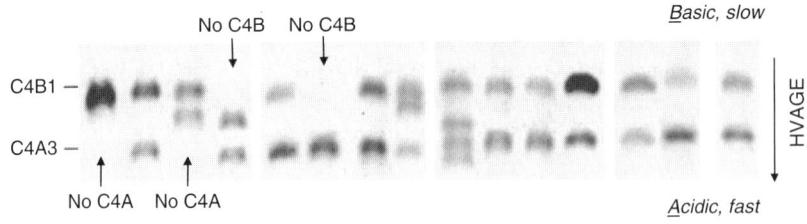

The deposition of activated C4 (C4b) or C3 (C3b) on host cell surfaces is probably a by-product of the complement activation, which is rendered nonharmful through the action of complement regulatory proteins together with factor I that degrades the activated C4b or C3b by proteolysis. The end products for such reactions are the covalent linkage of C4d or C3d on erythrocytes, lymphocytes, macrophages, dendritic cells, platelets, and endothelial cells of the vascular system (Isenman and Young, 1986; Atkinson *et al.*, 1988; Giles *et al.*, 1991; Taylor *et al.*, 2002). The surface deposition of C4d is a consistent marker for acute and chronic renal allograft rejections associated with the humoral immune response (Feucht, 2003). It was also observed that there was a substantial increase in C4d deposition on the surface of red cells from SLE patients (Giles *et al.*, 1991). Therefore, it appears that the quantitative variation of C4d deposition on cell surfaces is a relevant biomarker for allograft rejection or for SLE disease activity (Manzi *et al.*, 2004). A fundamental question to be addressed is whether the C4d deposition on host cell surfaces has any physiological or immunological consequences.

2.3. Genetic Determinants of C4 Plasma/Serum Protein Levels

A recent population study showed that in healthy Caucasians residing in Hungary, the serum C4A and C4B levels not only strongly associate with the corresponding gene dosage as expected, but also with the gene size of *C4A* and *C4B*. Short (S) *C4* genes tend to yield higher serum protein levels than long (L) *C4* genes. For example, among the individuals with four *C4* genes in a genome, those with homozygous LS/LS expressed serum C4 protein levels about 40% higher than those with homozygous LL/LL. Remarkably, African Americans have significantly higher frequency of short *C4* genes (42% versus 23.9% in Whites; $p < 10^{-11}$) (Blanchong *et al.*, 2000; Yang *et al.*, 2003; Yang, 2004), and the mean serum total C4 levels in Blacks is 40% higher than that of Whites. Most of the

---

Figure 7.1. Complex diversity of complement C4. (Panel A) The MHC complement gene cluster with quadrimodular (Q), trimodular (T), bimodular (B), and monomodular (M) RCCX length variants. The neighboring genes duplicated together with *C4* in modular fashion are serine/threonine nuclear kinase gene *RP* (or *STK19*), cytochrome P450 steroid 21-hydroxylase gene *CYP21*, and the extracellular matrix protein tenascin gene *TNX*. The size of an additional RCCX module with a long (L) *C4* gene is 32.7 kb, and for an additional duplicated RCCX module with a short (S) C4 gene is 26.3 kb. (Panel B) Pulsed field gel eletrophoresis (PFGE) of *Pme*I-digested genomic DNA from 12 individuals to elucidate the presence of RCCX length variants. *Lanes 1–4* are from subjects with homozygous LLL, LL, mono-L, and mono-S RCCX modules, respectively. *Lanes 5–12* are from subjects with heterozygous RCCX structures. (Panel C) Allotyping of human complement C4. EDTA-plasma samples from 15 human subjects were resolved by high-voltage agarose gel electrophoresis, processed by immunofixation using goat antiserum against human C4 and stained. The acidic C4A allotypes migrate faster than the basic C4B. A *vertical arrow* indicates the direction of electrophoretic migration. Notice the absence of C4A protein in *lanes* 1 and 3, the absence of C4B protein in *lanes* 4 and 6, and the differential band intensities of C4A and C4B in *lanes* 2, 5, and 8–15 (modified from Blanchong *et al.*, 2000; Chung *et al.*, 2002b; Yu *et al.*, 2003).

short *C4* genes code for C4B proteins (Yang *et al.*, 2003). Thus, not unexpectedly, in Blacks serum C4B protein levels are higher than C4A levels (Moulds *et al.*, 1991; Yang, 2004).

An unexpected finding is that the serum C4 levels also show a positive correlation with the body mass index, particularly in females (Gabrielsson *et al.*, 2003; Yang *et al.*, 2003). Many complement proteins including C4 are produced in adipose tissue. It is found that omental adipose tissue from both males and females synthesizes large quantities of C4 transcripts. Intriguingly, the serum C4 protein levels in females, but *not* in males, correlate with subcutaneous adipose tissue and total adipose depots, suggesting a sex difference in the extrahepatic biosynthesis of complement C4 (Gabrielsson *et al.*, 2003). It is therefore of interest to investigate whether the production of C4 in female adipose tissue would play a role on the gender differences of autoimmune diseases (Whitacre, 2001).

3. Complete Deficiencies of C4A and C4B in SLE and Immune Complex Diseases

To date, complete complement C4 deficiency has been firmly established in 26 members of 18 families (Yang *et al.*, 2004a). Fourteen of them (nine females and five males) were diagnosed with SLE according to the ACR criteria (Tan *et al.*, 1982). Of the remaining 12, seven had a lupus-like disorder such as photosensitive skin lesions and four had kidney diseases such as mesangioproliferative glomerulonephritis (GN), recurrent hematuria, membranous nephropathy, or Hencoh–Schölein purpura with end-stage kidney failure (Lhotta *et al.*, 2004a,b; Yang *et al.*, 2004b). Only one subject remained relatively healthy. The age of SLE disease onset/diagnosis among the C4-deficient subjects varied from 2 to 41 years. Four C4 deficient patients died between 2 and 25 years of age. Common manifestations of C4 deficiency are photosensitivity, severe skin lesions, digital erythema, Raynaud's phenomenon, arthritis, anti-Ro/SSA, rheumatoid factor, and recurrent bacterial or viral infections (Hauptmann *et al.*, 1986).

3.1. Molecular Basis of Complete C4 Deficiency

Fifteen different HLA haplotypes have been found in the C4-deficient patients. Homozygosities in HLA haplotypes were present in 73% of the complete C4 deficiency subjects. Altogether the molecular basis for 12 complete C4 deficiency patients have been elucidated. Three different RCCX structures are present. They are mono-L with a single *C4A* mutant gene, bimodular LS with mutant *C4A* and mutant *C4B* genes, and bimodular SS with two almost identical mutant *C4B* genes (Figure 7.2). There are two hotspots of deleterious mutations: one is located at exon 13 and the other is within a 2.6-kb genomic region spanning exons 20–29. The 2-bp insertion at codon 1213 from exon 29 has a frequency of 5–6% in Caucasian SLE patients, which is significantly higher than that in controls (1–2%) (Barta *et al.*, 1993; Sullivan *et al.*, 1999).

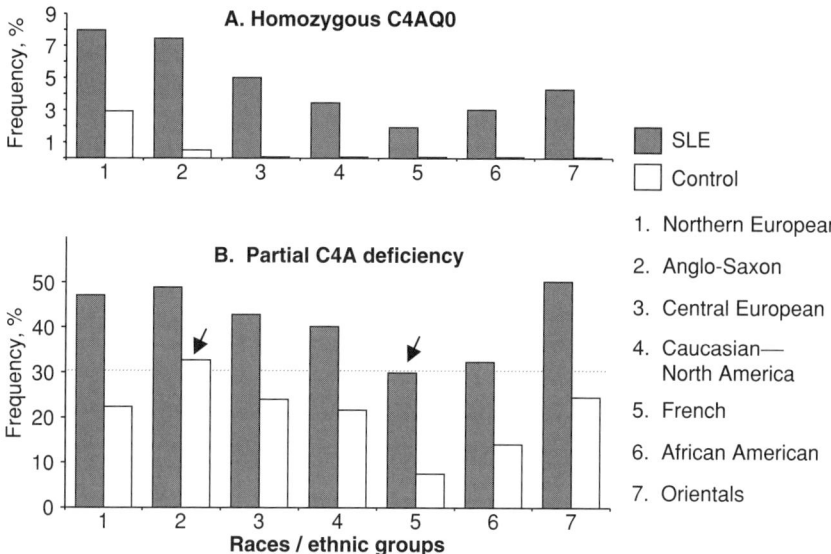

Figure 7.2. A comparison of frequencies of homozygous (panel A) and partial (panel B) C4A deficiency in SLE patients and matched controls in various ethnic groups/races. A *dotted* line marks the frequency of *partial* C4A deficiency in the French SLE patient population, which is even lower than that in the control (healthy) population in Anglo-Saxons (*arrows*). The data underscore the importance of using race- and ethnic-group matched-controls for data analysis. *Partial* C4A deficiency is the phenotype that showed a *lower* plasma or serum C4A than C4B protein levels. The higher C4B protein levels are a combination of higher gene copy number and/or more short genes for *C4B* than for *C4A* (modified from Yang *et al.*, 2004a).

3.2. Impairment of Immune Response in C4-Deficient Patients

An early study of the immunologic parameters in a Caucasian SLE patient with heterozygous HLA *A2 B12 Dw2/A2 Bw15 Dw8* and complete C4 deficiency revealed a depressed primary immune response after immunization with bacteriophage ϕX174, abnormal immunologic memory, and failure to switch from IgM to IgG during secondary immune response. Persistent lymphopenia and reduction in lymphocyte response to mitogen stimulation was also documented (Jackson *et al.*, 1979). In addition, the zymosan opsonization and the neutrophil chemotaxis activities by the C4-deficient serum were markedly impaired, when compared with those from healthy controls (Clark and Klebanoff, 1978). Failures in IgG or IgA class switching were also observed in two complete C4 deficiency patients recently, one with HLA *A24 Cw7 B38 DR13* and the other with HLA *A30 B18 DR7*. In addition, impairment in the immune response against hepatitis B surface antigen vaccine was observed in a patient with HLA *A30 B18 DR7* (Yang *et al.*, 2004b).

On the other hand, in two C4-deficient Moroccan siblings who had homozygous HLA haplotype *A11 B35 Cw4 DR1*, lupus-like disease, and persistent pulmonary infections were observed. Laboratory tests showed diminished opsonization and bactericidal activities by the C4-deficient sera. However, in other aspects the humoral and cellular immunity appeared normal (Mascart-Lemone et al., 1983). Such differences in the phenotypic manifestations underscore the relevance of genetic backgrounds and racial differences in immunity. It is worthwhile to point out that the autoimmune diseases associated with complete C4 deficiency often appear to be systemic in nature, characterized by the presence of circulating autoantibodies and a dysfunctional Th_2-type immune response. In keeping with this line of observation, none of the complete C4-deficient patients had been diagnosed to have organ-specific autoimmune diseases such as type 1 diabetes or multiple sclerosis.

4. Deficiencies of C4A or C4B in Human SLE

4.1. Low Complement Activity and C4 Protein Concentrations in SLE

It has been known for over half a century that SLE patients manifest reduced complement hemolytic activity (CH50) (Vaughan et al., 1951; Elliot and Mathieson, 1953). Reduced serum or plasma levels of complement C1q, C4, and C3 have been consistently observed in lupus patients, particularly those with lupus nephritis (Lewis et al., 1971; Cameron et al., 1976; Hebert et al., 1991). Serial analysis of serologic factors in SLE revealed that in many patients lower C4 levels occurred before the depression of other complement components. After the induction of remission, C4 had a tendency to return to normal levels more slowly than C3 (Gewurz et al., 1968; Kohler and Ten Bensel, 1969). The persistence of low C4 levels in many lupus nephritis patients before relapses and after remissions suggests the presence of a genetic factor as a cause for low C4 protein levels.

4.2. Homozygous or "Partial" Deficiency of C4A in SLE across Multiple Ethnic Groups

In the early 1980s, Bachelor's group in London (UK) and Dawkins' group in Perth (Australia) reported the association of C4 null alleles (C4Q0) or *partial* deficiency of C4A and/or C4B with human SLE (Christiansen et al., 1983; Fielder et al., 1983). Since then the associations of C4AQ0 or C4BQ0 with SLE have been examined extensively in many ethnic groups that are summarized as follows (Figure 7.2).

Historically, an *apparently* heterozygous or *partial* C4A deficiency was determined by the presence of lower C4A protein concentrations or band intensities than those of C4B proteins in phenotypic experiments such as C4 allotyping. The interpretation for such a phenomenon is the higher expression levels of C4B than C4A, which are the combined results of higher *C4B* gene dosages and short *C4B* genes. A genuine *heterozygous* C4A deficiency is defined as the presence of only one intact or functional *C4A* gene in a diploid genome. However, most published literature on *C4* genetics in the past 20 years did not distinguish an apparent or partial deficiency from a heterozygous deficiency. (Fielder et al., 1983; Christiansen et al., 1983; Dunckley et al., 1987; Zhao et al., 1989; Hartung et al., 1992; Petri et al., 1993; Fan et al., 1993; for complete references, please see Yang et al., 2004a).

Caucasians: The most intensively studied patient groups are Europeans and American Caucasians. In the healthy controls, the Nordics and Anglo-Saxons have C4AQ0 frequencies of 0.141 and 0.169, respectively. In the SLE populations, the C4AQ0 allelic frequencies are 0.315 for the Nordics and 0.319 for Anglo-Saxons. The frequencies of *homozygous* C4AQ0 were 7.5–8% in the patient groups, and 0.5–2.93% in the control groups. *Partial* deficiency of C4A had a frequency of 47% in northern European, British, or Australian SLE patients, and 22.4–32.7% in the control groups. With more than half of SLE patients possessing a homozygous or partial deficiency, C4A deficiency appears to be one of the most common genetic risk factors for SLE in these ethnic groups.

A slightly lower C4AQ0 allelic frequency, but a highly significant difference, was observed in Germans and Swiss, and in Caucasians residing in North America. The C4AQ0 allelic frequencies are between 0.236 and 0.264 for the patient groups, and 0.108 and 0.12 for the control groups. Homozygous and partial C4A deficiencies were present at 3.45–5.01% and 40.2–42.9%, respectively, in the western European SLE patient populations; and 0% and 21.7–24%, respectively, in the control populations (Figure 7.2, panels A and B).

The French probably have the lowest frequencies of C4AQ0 in both patient and control groups. However, the allelic frequency of C4AQ0 in SLE is still significantly different from that in the French healthy controls. It is of interest that the C4AQ0 allelic frequency is only 0.037 in the French control population, and 0.169 in the French SLE patients. The latter is very close to the C4AQ0 allelic frequency in the healthy controls of Anglo-Saxons (0.169) and Nordics (0.141). In other words, 31.8% of the French SLE patients have a homozygous or a *partial* C4A deficiency. Such phenotype frequency in the French lupus patients is even lower than that of healthy controls in Anglo-Saxons (33.2%). Such phenomenon underscores the importance of applying matched controls with the appropriate ethnic groups for data analyses (Figure 7.2, panel B).

African Americans: The C4AQ0 allelic frequencies in African American SLE patients and controls were slightly higher but still close to those in the French groups. The allelic frequency of C4AQ0 was 0.192 in the Black SLE patients and 0.07 in the matched controls ($\chi^2 = 43.8$, $p = 3.7 \times 10^{-11}$). About 35% of the African American SLE patients had a homozygous or a partial C4A deficiency, compared with 14.1% in matched controls.

Orientals: For Chinese, Korean, and Japanese, the SLE patients had C4AQ0 allelic frequencies of 0.293, compared with 0.123 in the corresponding control populations. In total, 54.5% of the Oriental SLE patients have partial or homozygous C4A deficiencies, which is >2 times higher than that of the matched controls (24.5%).

4.3. Deficiency of C4B in SLE Patients from Spanish, Mexican, and Australian Aborigines

While C4AQ0 is significantly associated with SLE, a difference in the C4BQ0 allelic frequencies between SLE patients and healthy controls was *not* observed in northern and western Europeans, African Americans, and most

Orientals. In contrast, the reverse situation is true for Spanish (De Juan *et al*., 1993; Naves *et al*., 1998), Mexican (Reveille *et al*., 1995; Reveille *et al*., 1998), and Australian Aborigine (Ranford *et al*., 1987; Christiansen *et al*., 1991) SLE patients. In these ethnic groups a significant increase in frequencies of C4B, but not C4A, deficiency was found in the SLE patient populations. Such a phenomenon would suggest a delicate shift of the physiologic roles of C4A and C4B among different ethnic groups or genetic backgrounds, or that a difference in the genetic milieu, such as racial backgrounds, could change the dependence on C4A or C4B in the emergence of autoimmunity.

4.4. Partial Deficiencies versus Polygenic Variations of C4A and C4B

In the past, higher plasma protein levels of C4B than of C4A was usually interpreted as a *partial* or apparent heterozygous deficiency of C4A caused by the presence of a "silent allele" for *C4A*, or by a *C4A* gene deletion. Such *apparent* or *partial* C4A or C4B deficiency actually reflects an unequal expression of *C4A* and *C4B* genes, which is mainly caused by polygenic and gene size variations. Unequal *C4A* and *C4B* gene number is likely in an individual whenever (a) a monomodular or trimodular RCCX structure is present, or (b) a bimodular structure has a homoexpression of either C4A or C4B proteins. Higher expression levels are observed for short *C4* genes, which more frequently code for C4B proteins. Therefore, it is essential to conduct a concurrent genotypic analysis to determine the number and size of *C4A* and *C4B* genes present in the patients, and phenotypic analysis to elucidate the C4A and C4B allotypes and protein levels.

5. Concluding Remarks and Perspectives

Complete deficiencies of C4A and C4B are among the strongest genetic risk factors associated with SLE or lupus-like diseases, across all HLA haplotypes and racial backgrounds. However, the age of disease onset and the disease severity vary substantially among the C4-deficient subjects, which underscores the importance of other genetic and environmental factors contributing to disease pathogenesis and progression. In contrast to the rarity of complete C4A and C4B deficiencies, partial and homozygous deficiencies of C4A are present in 32–55% of SLE patients from all races studied except Spanish, Mexican, and Australian Aborigines who had increased frequencies of C4B deficiency instead of C4A deficiency. This phenomenon underscores the relevance of C4A and C4B proteins in the fine control of autoimmunity. Different racial genetic backgrounds could change the thresholds and the requirement of C4A or C4B protein levels in immune tolerance and immune regulation. An important unanswered issue is the identification of a specific receptor or a chaperone that would link C4A and/or C4B proteins to the regulation of systemic autoimmunity.

In many lupus patients serum or plasma C4 levels fluctuate widely. Sometimes such fluctuations correlate with disease activity, especially in lupus nephritis. Consumption of C4 and activation of the complement pathways are involved in complement-mediated tissue injuries during the disease flares. Therefore, while a deficiency of C4A or C4B appears to be a genetic susceptibility factor for SLE (Atkinson and Schneider, 1999; Tsao, 2003; Yang et al., 2003), in about half of the lupus patients who have no apparent complement C4A or C4B deficiency, higher basal levels of C4A and/or C4B could instead increase their disease severity and vital organ involvement.

Another important aspect that has not been accurately addressed is the polygenic and gene size variations of *C4A* and *C4B* in the disease susceptibility and disease progression of SLE. A systematic analysis of *C4A* and *C4B* gene dosage and gene size, protein polymorphism and functional diversity, different basal plasma protein levels in various ethnic groups, and possible sex differences in extrahepatic tissue expression such as those in subcutaneous adipose tissue would refine our knowledge of the roles of C4A and C4B in the disease process.

After the establishment of the role(s) of heterozygous and/or homozygous deficiency of either C4A or C4B in SLE, a great challenge lies in how to use this information to help patients by developing more effective treatments or even a potential cure. In reality, the gap between knowledge and treatment is still great, and sustained multidisciplinary efforts will be needed to reach these ultimate goals. Perhaps innovative strategies for therapeutic interventions could include manipulation C4A and/or C4B protein expression in patients. Examples might include direct delivery of nonimmunogenic C4A or C4B proteins, creation of autologous or chimeric hepatocytes or adipocytes with high expression capability of C4A or C4B proteins, and prolongation of half-life of activated C4 proteins by site-directed mutagenesis of *C4* or by manipulating specific complement regulatory proteins through soluble competitors (Weisman et al., 1990; Morgan and Harris, 1999), inhibitory RNA or specific antibodies. Characterization of the immunologic and physiologic functions of C4A and C4B polymorphic variants will be fundamental to provide insights for development of therapeutic strategies. This research would be facilitated by the development of nonhuman primate and transgenic mouse models to validate research hypothesis and to allow pharmacologic testing of new therapies.

Acknowledgments

We would like to express our gratitude to our colleagues Dr. Joann Moulds (Drexel University, Philadelphia), Dr. Frank Arnett (University of Texas Medical School at Houston), Dr. Lokki Maisa (Finnish Red Cross, Helsinki, Finland), and Dr. Harvey R. Colten (Washington University, St. Louis), Dr. Geroge Fust (Semmelweis University, Budapest, Hungary) for collaborations. We would like to acknowledge contributions from Drs. Kristi L. Rupert, Zhenyu Yang, Carol A. Blanchong, Ms. Karla Jones, and Maddie Hebert. This work was supported by NIH grants 5R01 AR050078 from National Institute of Arthritis, Musculoskeletal and Skin Diseases, and 1P01 DK55546 from National Institute of Digestive, Diabetes and Kidney Diseases, Pittsburgh Supercomputing Center through NIH

Center for Research Resources Cooperative Agreement grant 1P41 RR06009, and the General Clinical Research Center of The Ohio State University.

References

Atkinson, J.P., Chan, A.C., Karp, D.R., Killion, C.C., Brown, R., Spinella, D., Shreffler, D.C., and Levine, R.P. (1988). Origins of the fourth component of complement related Chido and Rodgers blood-group antigens. *Complement*, **5**, 65–76.

Atkinson, J.P. and Schneider, P.M. (1999). Genetic susceptibility and class III complement genes. In R.G.Lahita (ed.), *Systemic Lupus Erythematosus*, 3rd edn. Academic Press, San Diego. pp. 91–104.

Barba, G., Rittner, C., and Schneider, P.M. (1993). Genetic basis of human complement C4A deficiency. Detection of a point mutation leading to nonexpression. *J. Clin. Invest.*, **91**, 1681–1686.

Blanchong, C.A., Zhou, B., Rupert, K.L., Chung, E.K., Jones, K.N., Sotos, J.F., Rennebohm, R.M., and Yu, C.Y. (2000). Deficiencies of human complement component C4A and C4B and heterozygosity in length variants of RP-C4-CYP21-TNX (RCCX) modules in Caucasians: The load of RCCX genetic diversity on MHC-associated disease. *J. Exp. Med.*, **191**, 2183–2196.

Cameron, J.S., Lessof, M.H., Ogg, C.S., Williams, B.D., and Williams, D.G. (1976). Disease activity in the nephritis of systemic lupus erythematosus in relation to serum complement concentrations. *Clin. Exp. Immunol.*, **25**, 418–427.

Christiansen, F.T., Dawkins, R.L., and Uko, G. (1983). Complement allotyping in SLE: Association with C4A null. *Aust. NZ J. Med.*, **13**, 483.

Christiansen, F.T., Zhang, W.J., Griffiths, M., Mallal, S.A., and Dawkins, R.L. (1991). Major histocompatibility complex (MHC) complement deficiency, ancestral haplotypes and systemic lupus erythematosus (SLE): C4 deficiency explains some but not all of the influences of the MHC. *J. Rheumatol.*, **18**, 1350–1358.

Chung, E.K., Yang, Y., Rennebohm, R.M., Lokki, M.L., Higgins, G.C., Jones, K.N., Zhou, B., Blanchong, C.A., and Yu, C.Y. (2002a). Genetic sophistication of human complement *C4A* and *C4B* and *RP-C4-CYP21-TNX* (RCCX) modules in the major histocompatibility complex (MHC). *Am. J. Hum. Genet.*, **71**, 823–837.

Chung, E.K., Yang, Y., Rupert, K.L., Jones, K.N., Rennebohm, R.M., Blanchong, C.A., and Yu, C.Y. (2002b). Determining the one, two, three or four long and short loci of human complement *C4* in a major histocompatibility complex haplotype encoding for C4A or C4B proteins. *Am. J. Hum. Genet.*, **71**, 810–822.

Clark, R. and Klebanoff, S. (1978). Role of the classical and alternative complement pathways in chemotaxis and opsonization: Studies of human serum deficient in C4. *J. Immunol.*, **120**, 1102–1108.

Dangel, A.W., Mendoza, A.R., Baker, B.J., Daniel, C.M., Carroll, M.C., Wu, L.-C., and Yu, C.Y. (1994). The dichotomous size variation of human complement C4 gene is mediated by a novel family of endogenous retroviruses which also establishes species-specific genomic patterns among Old World primates. *Immunogenetics*, **40**, 425–436.

De Juan, D., Martin-Villa, J.M., Gomez-Reino, J.J., Vicario, J.L., Corell, A., Martinez-Laso, J., Benmammar, D., and Arnaiz-Villena, A. (1993). Differential contribution of C4 and HLA-DQ genes to systemic lupus erythematosus susceptibility. *Hum. Genet.*, **91**, 579–584.

Dodds, A.W., Ren, X.-D., Willis, A.C., and Law, S.K.A. (1996). The reaction mechanism of the internal thioester in the human complement component C4. *Nature*, **379**, 177–179.

Dunckley, H., Gatenby, P. A., Hawkins, B., Naito. S., and Serjeanston, S.W. (1987). Deficiency of C4A is a genetic determinant of systemic lupus erythematosus in three ethnic groups. *J. Immunogenet.*, **14**, 209–218.

Elliot, J.A. and Mathieson, D.R. (1953). Complement in disseminated (systemic) lupus erythematosus. *AMA Arch. Dermatol. Syphilol.*, **68**, 119–128.

Fan, Q., Uring-Lambert, B., Weill, B., Gautreau, C., Menkes, C.J., and Delpech, M. (1993). Complement component C4 deficiencies and gene alterations in patients with systemic lupus erythematosus. *Eur. J. Immunogenet.*, **20**, 11–21.

Feucht, H.E. (2003). Complement C4d in graft capillaries – the missing link in the recognition of humoral alloreactivity. *Am. J. Transplant.*, **3**, 646–652.

Fielder, A.H.L., Walport, M.J., Batchelor, J.R., Rynes, R.I., Black, C.M., Dodi, I.A., and Hughes, G.R.V. (1983). Family study of the major histocompatibility complex in patients with systemic lupus erythematosus: Importance of null alleles of C4A and C4B in determining disease susceptibility. *Br. Med. J.*, **286**, 425–428.

Gabrielsson, B.G., Johansson, J.M., Lonn, M., Jernas, M., Olbers, T., Peltonen, M., Larsson, I., Lonn, L., Sjostrom, L., Carlsson, B., and Carlsson, L.M. (2003). High expression of complement components in omental adipose tissue in obese men. *Obes. Res.*, **11**, 699–708.

Gatenby, P.A., Barbosa, J.E., and Lachmann, P.J. (1990). Differences between C4A and C4B in the handling of immune complexes: The enhancement of CR1 binding is more important than the inhibition of immunoprecipitation. *Clin. Exp. Immunol.*, **79**, 158–163.

Gewurz, H., Pickering, R.J., Mergenhagen, S.E., and Good, R.A. (1968). The complement profile in acute glomerulonephritis systemic erythematosus and hypocomplementemic chronic glomerulonephritis. *Int. Arch. Allergy Appl. Immunol.*, **34**, 556–570.

Gibb, A.L., Freeman, A.M., Smith, R.A.G., Edmonds, S., and Sim, E. (1993). The interaction of soluble human complement receptor type 1 (sCR1, BR55730) with human complement component C4. *Biochim. Biophys. Acta*, **1180**, 313–320.

Giles, C.M., Davies, K.A., and Walport, M.J. (1991). *In vivo* and *in vitro* binding of C4 molecules on red cells: A correlation of numbers of molecules and agglutination. *Transfusion*, **31**, 222–228.

Hartung, K., Baur, M.P., Coldewey, R., Fricke, M., Kalden, J.R., Lakomek, H.J., Peter, H.H., Schendel, D., Schneider, P.M., Seuchter, S.A., Stangel, W., and Deicher, H.R.G. (1992). Major histocompatibility complex haplotypes and complement C4 alleles in systemic lupus erythematosus. *J. Clin. Invest.*, **90**, 1346–1351.

Hauptmann, G., Goetz, J., Uring-Lambert, B., and Grosshans, E. (1986). Component deficiencies. *Prog. Allergy*, **39**, 232–249.

Hebert, L.A., Cosio, F.G., and Neff, J.C. (1991). Diagnostic significance of hypocomplementemia. *Kidney Int.*, **39**, 811–821.

Isenman, D.E. and Young, J.R. (1986). Covalent binding properties of the CR1 and C4B isotypes of the fourth component of human complement on several C1-bearing cell surfaces. *J. Immunol.*, **136**, 2542–2550.

Jackson, C.G., Ochs, H.D., and Wedgwood, R.J. (1979). Immune response of a patient with deficiency of the fourth component of complement and systemic erythematosus. *N. Engl. J. Med.*, **300**, 1124–1129.

Kohler, P.F. and Ten Bensel, R. (1969). Serial complement component alterations in acute glomerulonephritis and systemic lupus erythematosus. *Clin. Exp. Immunol.*, **4**, 191–202.

Lewis, E.J., Carpenter, C.B., and Schur, P.H. (1971). Serum complement component levels in human glomerulonephritis. *Ann. Intern. Med.*, **75**, 555–560.

Lhotta, K., Wurzner, R., Rosenkranz, A.R., Beer, R., Rudisch, A., Neumair, F., and Mayer, G. (2004a). Cerebral vasculitis in a patient with hereditary complete C4 deficiency and systemic lupus erythematosus. *Lupus*, **13**, 139–141.

Lhotta, K., Wurzner, R., Rumpelt, H.J., Eder, P., and Mayer, G. (2004b). Membranous nephropathy in a patient with hereditary complete complement C4 deficiency. *Nephrol. Dial. Transplant.*, **19**, 990–993.

Mack, M., Bender, K., and Schneider, P.M. (2004). Detection of retroviral antisense transcripts and promoter activity of the HERV-K(C4) insertion in the MHC class III region. *Immunogenetics*, **56**, 321–332.

Manzi, S., Navratil, J.S., Ruffing, M.J., Liu, C.C., Danchenko, N., Nilson, S.E., Krishnaswami, S., King D.E., Kao A.H., and Ahearn J.M. (2004). Measurement of erythrocyte C4d and complement receptor 1 in systemic lupus erythematosus. *Arthitis Rheum.*, **50**, 3596–3604.

Mascart-Lemone, F., Hauptmann, G., Goetz, J., Duchateau, J., Delespesse, G., Vray, B., and Dab, I. (1983). Genetic deficiency of C4 presenting with recurrent infections and SLE-like disease. *Am. J. Med.*, **75**, 295–304.

Morgan, B.P. and Harris, C.L. (1999). *Complement Regulatory Proteins*. Academic Press, San Diego.

Moulds, J.M., Warner, N.B., and Arnett, F.C. (1991). Quantitative and antigenic differences in complement component C4 between American blacks and whites. *Complement Inflamm.*, **8**, 281–287.

Naves, M., Hajeer, A.H., Teh, L.S., vies, E.J., di-Ros, J., Perez-Pemen, P., Vilardell-Tarres, M., Thomson, W., Worthington, J., and Ollier, W.E. (1998). Complement C4B null alleles status confers risk for systematic lupus erythematosus in a Spanish population. *Eur. J. Immunogenet.*, **25**, 317–320.

Petri, M., Watson, R., Winkelstein, J.A., and McLean, R.H. (1993). Clinical expression of systemic lupus erythematosus in patients with C4A deficiency. *Medicine (Baltimore)*, **72**, 236–244.

Porter, R.R. (1983). Complement polymorphism, the major histocompatibility complex and associated diseases: A speculation. *Mol. Biol. Med.*, **1**, 161–168.

Ranford, P., Hay, J., Serjeantson, S.W., and Dunckley, H. (1987). A high frequency of inherited deficiency of complement component C4 in Darwin Aborigines. *Aust. NZ J. Med.*, **17**, 420–423.

Reilly, B.D. and Mold, C. (1997). Quantitative analysis of C4Ab and C4Bb binding to the C3b/C4b receptor (CR1, CD35). *Clin. Exp. Immunol.*, **110**, 310–316.

Reveille, J.D., Moulds, J.M., Ahn, C., Friedman, A.W., Baethge, B., Roseman, J., Straaton, K.V., and Alarcon, G.S. (1998). Systemic lupus erythematosus in three ethnic groups: I. The effects of HLA class II, C4, and CR1 alleles, socioeconomic factors, and ethnicity at disease onset. *Arthritis Rheum.*, **41**, 1161–1172.

Reveille, J.D., Moulds, J.M., and Arnett, F.C. (1995). Major histocompatibility complex class II and C4 alleles in Mexican Americans with systemic lupus erythematosus. *Tissue Antigens*, **45**, 91–97.

Sepp, A., Dodds, A.W., Anderson, M.J., Campbell, R.D., Willis, A.C., and Law, S.K.A. (1993). Covalent binding properties of the human complement protein C4 and hydrolysis rate of the internal thioester upon activation. *Protein Sci.*, **2**, 706–716.

Shen, L.M., Wu, L.C., Sanlioglu, S., Chen, R., Mendoza, A.R., Dangel, A., Carroll, M.C., Zipf, W., and Yu, C.Y. (1994). Structure and genetics of the partially duplicated gene RP located immediately upstream of the complement C4A and C4B genes in the HLA class III region: Molecular cloning, exon-intron structure, composite retroposon and breakpoint of gene duplication. *J. Biol. Chem.*, **269**, 8466–8476.

Sullivan, K.E., Kim, N.A., Goldman, D., and Petri, M.A. (1999). C4A deficiency due to a 2 bp insertion is increased in patients with systemic lupus erythematosus. *J. Rheumatol.*, **26**, 2144–2147.

Tan, E.M., Cohen, A.S., Fries, J.F., Masi, A.T., McShane, D.J., Rothfield, N.F., Schaller, J.G., Talal, N., and Winchester, R.J. (1982). The 1982 revised criteria for the classification of systemic lupus erythematosus. *Arthritis Rheum.*, **25**, 1271–1277.

Taylor, P.R., Pickering, M.C., Kosco-Vilbois, M., Walport, M.J., Botto, M., Gordon, S., and Martinex-Pomares, L. (2002). The follicular dendritic cell restricted epitope, FDC-M2, is complement C4; localization of immune complexes in mouse tissues. *Eur. J. Immunol.*, **32**, 1888–1896.

Tsao, B.P. (2003). The genetics of human systemic lupus erythematosus. *Trends Immunol.*, **24**, 595–602.

van den Elsen, J.M.H., Martin, A., Wong, V., Clemenza, L., Rose, D.R., and Isenman, D.E. (2002). X-ray crystal structure of the C4d fragment of human complement component C4. *J. Mol. Biol.*, **322**, 1103–1115.

Vaughan, J.H., Bayles, T.B., and Favour, C.B. (1951). The response of serum gammaglobulin level and complement titer to adrenocorticotrophic hormone therapy in lupus erythematosus disseminatus. *J. Lab. Clin. Med.*, **37**, 698–702.

Weisman, H.F., Bartow, T., Leppo, M.K., Marsh, H.C., Jr., Carson, G.R., Concino, M.F., Boyle, M.P., Roux, K.H., Weisfeldt, M.L., and Fearon, D.T. (1990). Soluble human complement receptor type 1: In vivo inhibitor of complement suppressing post-ischemic myocardial inflammation and necrosis. *Science*, **249**, 146–151.

Whitacre, C.C. (2001). Sex differences in autoimmune in autoimmune disease. *Nat. Immunol.*, **2**, 777–780.

Yang, Y. (2004). *The Genetic Complexity and Protein Polymorphism of Complement C4 in Health and Disease*. Ph.D. Thesis, The Ohio State University.

Yang, Y., Chung, E.K., Zhou, B., Blanchong, C.A., Yu, C.Y., Füst, G., Kovács, M., Vatay, A., Szalai, C., Karádi, I., and Varga, L. (2003). Diversity in intrinsic strengths of the human complement system: Serum C4 protein concentrations correlate with *C4* gene size and polygenic variations, hemolytic activities and body mass index. *J. Immunol.*, **171**, 2734–2745.

Yang, Y., Chung, E.K., Zhou, B., Lhotta, K., Hebert, L.A., Birmingham, D.J., Rovin, B.H., and Yu, C.Y. (2004a). The intricate role of complement C4 in human SLE. *Curr. Dir. Autoimmun.*, **7**, 98–132.

Yang, Y., Lhotta, K., Chung, E.K., Eder, P., Neumair, F., and Yu, C.Y. (2004b). Complete complement components C4A and C4B deficiencies in human kidney diseases and systemic lupus erythematosus. *J. Immunol.*, **173**, 2803–2814.

Yang, Z., Mendoza, A.R., Welch, T.R., Zipf, W.B., and Yu, C.Y. (1999). Modular variations of HLA class III genes for serine/threonine kinase RP, complement C4, steroid 21-hydroxylase CYP21 and tenascin TNX (RCCX): A mechanism for gene deletions and disease associations. *J. Biol. Chem.*, **274**, 12147–12156.

Yu, C.Y., Campbell, R.D., and Porter, R.R. (1988). A structural model for the location of the Rodgers and the Chido antigenic determinants and their correlation with the human complement C4A/C4B isotypes. *Immunogenetics*, **27**, 399–405.

Yu, C.Y., Chung, E.K., Yang, Y., Blanchong, C.A., Jacobsen, N., Saxena, K., Yang, Z., Miller, W., Varga, L., and Fust, G. (2003). Dancing with complement C4 and the RP-C4-CYP21-TNX (RCCX) modules of the major histocompatibility complex. *Prog. Nucleic Acid Res. Mol. Biol.*, **75**, 217–292.

Zhao, X.Z., Zhang, W.J., Tian, Y.W., Wu, F., Zhang, L., Jiang, X.D., Sun, Z.Z., Hu, C.F., Wang, W.Z., and Gan, L. (1989). Allotypic differences and frequencies of C4 null alleles (C4Q0) detected in patients with systemic lupus erythematosus (SLE). *Chinese Sci. Bullet.*, **34**, 237–240.

8

Non-MHC Genetic Polymorphisms with Functional Importance for Human Myasthenia Gravis

Ann Kari Lefvert

1. Introduction

Myasthenia gravis (MG) is commonly regarded as the prototype of an organ-specific antibody-mediated autoimmune disease. It is characterized by an immune response against the nicotinic acetylcholine receptor on the neuromuscular junction. The symptoms, weakness and increased fatigability, are caused by direct blockade and a reduction of the number of functional receptors on the neuromuscular junction by autoantibodies (Drachman, 1994). The disease can be transmitted from mother to the newborn child by antibodies that pass through the placenta (Lefvert and Osterman, 1983) and by serum antibodies to experimental animals. There is a statistical but not very impressive correlation between autoantibody concentration and disease severity. Moreover, acetylcholine receptor antibodies are found in several conditions not accompanied by neuromuscular symptoms, including some thymomas (Aarli et al., 1981), healthy first-degree relatives (Lefvert et al., 1985), monoclonal gammopathies (Eng et al., 1987), and primary biliary cirrhosis (Sundewall et al., 1985). It is intriguing that the healthy twins in two pairs of monozygotic twins discordant for MG had autoantibodies that were similar in concentration and subtype to those of their myasthenic sisters. IgG preparations from both the healthy and the myasthenic twins were equally effective in inducing experimental myasthenia in mice. The autoantigen-specific T cell reactivity was, however, greater in the myasthenia twins than in their healthy sisters (Kakoulidou et al., 2004). Another challenging observation was that treatment of patients with anti-CD4[+] antibodies resulted in long-lasting remission and abolished T cell autoreactivity, without decreasing autoantibody concentration (Åhlberg et al., 1994). Such findings clearly challenge the concept of a simple cause-and-effect relationship between autoantibodies and disease.

Ann Kari Lefvert • Immunological Research Laboratory, Center for Molecular Medicine and Department of Medicine, Karolinska Institutet, Karolinska Hospital, S-171 76 Stockholm, Sweden.

Molecular Autoimmunity: In commemoration of the 100th anniversary of the first description of human autoimmune disease, edited by Moncef Zouali. Springer Science+Business Media, Inc., New York, 2005.

Recent reports suggest that inflammatory mechanisms may play a more direct role in the development of the disease. First, transgenic mice that express IFÑ-γ at the neuromuscular junction develop a myasthenia-like disease (Danling et al., 1995). IL-1α and IL-1β are present together with postsynaptic components in innervated mature muscle fibers (Askanas et al., 1998). Second, there is an inflammatory reaction at the end plate and mononuclear cell infiltration in the skeletal muscles of some patients, especially those with thymomas (Maselli et al., 1991). In all, these findings suggest that mechanisms other than the autoantibodies and especially T cells and proinflammatory reactions might be of importance in MG.

The concordance rate for MG in identical twins is around 30%, suggesting that part of the pathogenesis of MG is determined by genetic factors. However, the reported positive associations with HLA class I and II are weak, with average odds ratios (ORs) of about 2:3. This suggests the contribution of other genetic factors and to elucidate this, several candidate genes have been examined (Garchon, 2003). The congenital myasthenic syndromes are clearly associated to mutations/variations in the genes for the acetylcholine receptor subunits on chromosome 2, and variations in the α-subunit gene is associated also with acquired autoimmune MG (Djabiri et al., 1997). IgCH Gm allotype genes located in the variable region of the Ig heavy-chain loci have been suggested to confer susceptibility for MG (Smith et al., 1984). There are a few studies of the genes of the T cell receptor (TCR) chains, showing overexpression of selected TCR Vβ gene families (Mantegazza et al., 1990; Gigliotti et al., 1996) and association to TCR Vα and Cα alleles (Oksenberg et al., 1989). The functional implications of the polymorphisms reported in these limited studies and their importance for the disease manifestations are unclear.

Our own studies concern the polymorphisms of the genes coding for proinflammatory and anti-inflammatory cytokines, costimulatory T cell factors and the β2-adrenergic receptor, and their functional effects on immune activation in MG. The reason for studying these candidate genes was to delineate additional mechanisms, beside the role of autoantibodies against the acetylcholine receptor, that contribute to disease manifestations. Proinflammatory and anti-inflammatory cytokines, such as IL-1, TNF, IL-4, IL-6, and IL-10, influence T cell and/or B cell activation. The T cell costimulatory factors CD28 and CTLA-4, together with their ligands CD80 and CD86, balance the activation of T cells and thus the immune reaction. The β2-adrenergic receptor represents a link between the sympathetic nervous system and the immune system. Since β2 receptors are present on cells of the immune system as well as on skeletal muscle cells, and regulate many of their functions, there is a possibility that these immune reactions might act to modulate the disease.

2. Proinflammatory and Anti-inflammatory Cytokines in MG

2.1. Association of MG to the High Secretory Alleles of TNF-α

TNF-α̃, the HLA-A1, B8, DR3, 8.1 ancestral haplotype contains the genes for TNF (including TNF-α and lymphotoxin α and β) that are tandemly arranged in the central region of the HLA class III region. This location and the potent proinflammatory and immunomodulatory functions of TNF-α have prompted speculations about its implication in the etiology of MHC-associated diseases. TNF-α is a true pleiotropic cytokine. Transgenesis indicates the biosynthesis of TNF-α in the thymus of normal mice, suggesting its role in the development/regulation of the immune response (Emilie et al., 1991). The TNF-α-308 A2 allele is a high secretory variant.

Our study of the Swedish Caucasian MG population showed that the frequency of TNF-α-308 A2/A2 genotype was increased (OR = 5.65) while the frequency of genotype A1/A1 was decreased (OR = 0.31) in patients compared to that of healthy individuals (Huang et al., 1999b). This association was determined by studying female patients with an early disease onset and thymic hyperplasia. The frequency of the TNF-α A2 allele in patients with thymic hyperplasia was increased as compared to patients with thymomas and healthy individuals. The association of TNF-α A2 and MG was stronger in patients with early onset than in those with late onset of the disease. It also was stronger in female than in male patients. There was no difference among patients with different levels of serum autoantibodies.

Since TNF-α-308 A2 is strongly linked with HLA-A1, B8, the similarity of the association of TNF-α-308 A2 with MG to that of HLA-B8 was investigated. The results suggest that associations of HLA-B8 and TNF-α-308 A2 are stronger than HLA-DR3 in MG. MG may be primarily associated to TNF-α, rather than to HLA-class I or II.

2.2. Functional Implications of the Association with the TNF-α-308 A2 Allele

TNF-α is necessary for the development of experimental autoimmune MG. Deficiency of the lymphotoxin gene, closely linked to the TNF-α gene, completely abolished experimental autoimmune MG in mice (Goluszko et al., 2001). Antibodies to TNF-α are now currently used for the treatment of rheumatoid arthritis. We have recently used this method to treat a severely ill patient with MG, who did not tolerate immunosuppressive agents. This patient showed a moderate clinical response and a rather pronounced decrease of the concentration of autoantibodies (Kakoulidou M, Bjelak S. Giscombe R, Pirskanen R, Lefvert AK, manuscript in preparation).

2.3. Association of MG to the high secretory Allele of IL-1β

IL-1β̃ IL-1β is an important proinflammatory cytokine. The IL-1 receptor antagonist (IL-1Ra) is a high-affinity antagonist that controls the activity exerted by IL-1 and the balance between IL-1 and IL-1Ra is important in both normal and disease states. *IL-1* and *IL-1Ra* genes have been mapped to the long arm of chromosome 2.

A biallelic polymorphism within the *IL-1β* gene, a TaqI restriction fragment length polymorphism (RFLP) in exon 5, influences the production of IL-1β protein. The 13.4-kb allele, usually referred to as A2, is a "high secretory" allele. A penta-allelic polymorphism in the *IL-1Ra* gene intron 2 is caused by variable numbers of an 86-bp tandem repeat (VNTR). Individuals with different copy numbers of this repeat sequence have corresponding numbers of protein-binding sites. Allele 2 of *IL-1Ra* is associated with autoimmune diseases such as multiple sclerosis, inflammatory bowel disease, and systemic lupus erythematosus and might be related to enhanced production of IL-1Ra (Dinarello, 2002).

Our study of a Swedish Caucasian MG population demonstrated an increased frequency of the genotype *IL-1β A2/A2* (OR = 4.63, p = 0.01) and the allele A2 and a concomitant decrease of allele A1 (Huang et al., 1998b).

Interesting results emerged when carriage of *IL-1β A2* was studied in relation to that of HLA-B8. The percentage of *IL-1β A2* carriage in patients not carrying HLA-B8 was significantly higher than that in healthy individuals (OR = 2.8, p = 0.008). The frequency of genotype A2/A2 was even higher in HLA-B8-negative MG patients (OR = 5.3, p = 0.03), while that of A1/A1 was lower (OR = 0.36, p = 0.008).

The *IL-1β A2* carriage was then studied in relation to *IL-1Ra* allele 2. MG patients who were noncarriers of *IL-1Ra A2* carried more frequently *IL-1β* allele 2 than healthy individuals (OR = 3.09, p = 0.003). Thus, the risk for development of MG was associated to *IL-1βA2*. For those homozygous for A2/A2, the risk was even higher, indicating an allele dosage effect on the initiation and development of MG. *IL-1β* TaqI RFLP on chromosome 2 provides a new genetic marker for MG patients. A recent study confirmed an association of MG to *IL-1α* variants in Italian patients (Sciacca et al., 2002).

2.4. Functional Implications of the Association with the IL-1β TaqI RFLP A2 Allele

IL-1$β^{-/-}$ mice do not develop experimental MG and our animal experiments support the important role of IL-1β. Mice lacking IL-1β have a much-reduced incidence and disease severity of experimental MG induced by the acetylcholine receptor. None of 16 IL-1$β^{-/-}$ mice died from disease, as compared to 4/17 of the wild-type mice. The number of mice with muscle weakness was 3/13 (23%), compared to 11/17 (65%) of the wild-type mice. In the knockout mice, both the T and the B cell responses to the acetylcholine receptor used for immunization were greatly reduced. (Huang et al., 2001). Finally, since a peptide with blocking effect on the IL-1 receptor I is now used for therapy of rheumatoid arthritis, this approach might be applied to human MG.

IL-1α and IL-1β are present together with postsynaptic components in innervated mature muscle fibers (Askanas et al., 1998). Cytokines such as IFN-γ and IL-2 directly influence the neuromuscular transmission, and patients with MG have mononuclear cell infiltration of skeletal muscle and of the neuromuscular junction (Maselli et al., 1991). A MG-like syndrome occurs in transgenic mice expressing IFN-γ in the neuromuscular junction (Gu et al., 1995). Taken together, we speculate that subtype(s) of MG associated with gene(s) located in the *IL-1* gene family might exhibit pathological mechanisms different from those active in subtype(s) associated with MHC genes. In IL-1 associated subtype(s) of MG, IL-1 might affect the neuromuscular junction directly. Alternatively, IL-1 has multiple biologic effects and can augment both cellular and humoral immunological responses, resulting in autoantibody production in MG patients.

2.5. Lack of Associations of MG to Genetic Variants of IL-4 and IL-6

Interleukin-4 (*IL-4*) is an important cofactor in the maturation of B cells. The human *IL-4* gene on chromosome 5 contains a dinucleotide repeat microsatellite (4R1) in the second intron and a variable number of tandem repeat (VNTR) polymorphisms located in the third intron of the *IL-4* gene. IL-6 has also been implicated as a B cell stimulatory factor. The gene contains a biallelic polymorphism at −174 in the promoter region and a polymorphism in the 3′ flanking AT-rich region. In the Swedish MG population there was no difference in these polymorphic sites between patients and healthy individuals (Huang *et al.*, 1999a,d).

2.6. IL-10 Is Associated to MG with High Autoantibody Levels

Interleukin-10 (*IL-10*) displays a broad spectrum of biological activities including immunosuppressive, anti-inflammatory, and B cell–stimulating properties (Lalani *et al.*, 1997). IL-10 downregulates the expression of MHC class II molecules and costimulatory factors such as intercellular adhesion molecules and CD80 and CD86. IL-10 exerts its anti-inflammatory function by inhibiting the synthesis of the proinflammatory cytokines TNF-α, IL-1α and IL-1β.

The IL-10 gene (*IL10*) is situated on chromosome 1. There are several polymorphisms within the *IL10*. At position −1082 a biallelic polymorphism is associated with IL-10 production by stimulated peripheral blood mononuclear cells (PBMCs). The presence of the G allele correlates with higher IL-10 secretion (Turner *et al.*, 1997). There are two CA repeats located in the promoter region of *IL10*.

The biallelic polymorphism and the two CA repeat microsatellites located in the promoter region of the *IL10* gene were analyzed in Swedish MG patients (Huang *et al.*, 1999c). The prevalence of a "high secretor" phenotype of IL-10 (IL10-1082 G/G) was higher in healthy individuals carrying the IL-1β TaqI polymorphism allele 2 ("high secretor" phenotype) than in healthy individuals lacking this allele. This balance between proinflammatory (*IL-1β*) and anti-inflammatory cytokine (*IL10*) genes was not present in MG patients. This emphasizes that the balances between IL-1 and IL-10 are important, and that genetic variations influencing these balances may deregulate the inflammatory reactions, thus providing the initiative event at the neuromuscular junction for triggering an autoantibody response.

The 12 alleles of *IL10.G* found in our populations were named after the length of PCR products. The frequency of allele 134 was increased in patients, mainly in those with the highest levels of serum autoantibodies (Huang *et al.*, 1999c). The same association is seen in patients with SLE and with Wegener's granulomatosis, indicating that this particular variant facilitates autoantibody production (Zhou *et al.*, 2002). The mechanism involved is not clear, but might, at least in part, depend on increased IL-10 secretion (Zheng *et al.*, 2001).

3. The β2-Adrenergic Receptor in MG

β2-adrenergic receptor (*β2-AR*) is a G protein–coupled receptor present on PBMCs, striated muscle cells, and cardiomyocytes. It represents a link between the immune and the sympathetic nervous systems. Patients with MG

have autoantibodies against both β1- and β2-AR (Xu et al., 1998). The density of β2-AR on peripheral muscle mononuclear cells is decreased. This decrease may be caused by autoantibodies against the β2̃-AR that have an agonist action and downregulate the expression of the receptor (Xu et al., 1997). Activation of the β2-AR by agonists suppresses IL-2 receptor expression, inhibits Th1 lymphocyte proliferation, and downregulates the production of Th1 cytokines (Feldman et al., 1987).

The gene encoding β2-AR is present on chromosome 5 at q31–32. Three single-nucleotide polymorphisms (SNPs) at amino acid positions 16, 27, and 164 are related to receptor functions. In addition, SNPs at the promoter region at positions −20, −47, −367, and −468 affect the expression of the receptor.

We found an increased prevalence of homozygosity for Arg16 in patients with generalized MG (Xu et al., 2000). This association was present in patients with early and late disease onset, and in patients with generalized disease, but not in patients with only ocular symptoms. It is of special interest that the two groups of patients with early and late onset had similar associations, but different MHC haplotypes (Hjelmstrom et al., 1995), suggesting that the SNP at amino acid position 16 constitutes an additional predisposition genetic factor. The frequencies of polymorphisms at amino acid positions 27 and 164 were similar in patients and healthy individuals. When analyzing subgroups of patients, homozygosity for Glu27 was negatively associated with the presence of antibodies against the β2-adrenergic receptor and to disease severity. One explanation might be that, in contrast to those carrying Gln27, individuals homozygous for Glu27 do not express the receptor in a fully mature form. The Glu27 receptor might be less immunogenic (Green et al., 1994).

The distribution of four SNPs in the promoter of the β2-AR gene at positions −20, −47, −367, and −468 was not directly associated with MG. After stratification, certain β2-adrenergic receptor SNPs were associated with thymomas. Four point mutations, −20C, −47C, −367C, −468G, significantly decrease reporter gene expression compared with −20T, −47T, −367T, and −468C, respectively (Scott et al., 1999). The frequencies of the alleles −20T, −47T, −367T, and −468C in patients with thymomas were increased (Zhao X, Gharizadeh B, Wang XB, Ghaderi M, Pirskanen R, Nyren P, Garcon HJ, Lefvert AK, manuscript in preparation). In contrast, there was no difference between patients with other thymic histopathological changes when compared to healthy individuals. The functional effects of Arg16 as well as the −20T, −47T, −367T, and −468C mutations remain obscure. The latter genotype is associated to a higher rate of transcription and should thus lead to adequate expression of CTLA-4.

4. The T Cell Receptor Cofactor CTLA-4 in MG

4.1. Association to MG with Thymoma and Increased Activation of the Immune System

CTLA-4 is an essential component of the immune system present on T cells, B cells, monocytes, and many other cells. It serves as a negative regulator for T cell activation. Animals deficient for the gene coding for CTLA-4 (*Ctla-4*) show a massive T cell lymphoproliferative disorder with increased numbers of activated T cells and elevated basal levels of serum immunoglobulins, resulting in autoimmune-like tissue destruction (Waterhouse et al., 1995). Defect in expression/function of CTLA-4 should thus result in an abnormal T cell activation and an exaggerated inflammatory/immune response. Ctla-4 variants that are associated to decreased expression/function have been described in many autoimmune disorders, including MG (Huang et al., 1998a; Kristiansen et al., 2000). The gene is located on human chromosome 2q33. Genetic variants of Ctla-4 includes an (AT)n microsatellite within the 3' untranslated region of exon 3 (3'UTR) at position +642, SNPs in the promoter region at position −318(C/T), position −1772(C/T) and position −1661(A/G), and a SNP(A/G) in the coding sequence 1 (CDS1) at position +49. These variants are in linkage disequilibrium and will be discussed together below.

Analysis of the length of the (AT)n microsatellite in the 3'UTR revealed the existence of a total of 16 alleles, with sizes ranging from 86 to 128 bp. (Huang et al., 1998a). There was no difference in the allelic distribution between healthy individuals and MG patients. The most common genotypes were 86/86 bp, followed by 104/104 bp, present in 56.1% and 14.6% of patients, respectively. When MG patients were subgrouped according to thymic histopathology, the frequency of the shortest allele, 86 bp, was decreased in patients with thymomas compared to those with hyperplasia and normal thymic histopathology (OR = 0.27, $p < 0.05$). The SNP at position +49 (A/G) in CDS1 showed aberrant distribution of the G allele in patients with thymomas. The G allele and the GG genotype were associated with thymomas when compared to patients with normal and hyperplastic thymic histopathology (OR = 3.52, $p < 0.01$ and OR = 8.44, $p < 0.01$, respectively). The distribution of the C/T promoter −318 variant was the same in patients and healthy individuals (Wang et al., 2002a). The frequency of the TC-1772 genotype was higher in MG patients with thymomas and the frequencies of the G-1661 allele and the GG-1661 genotype were lower in patients. There were linkage disequilibria between each SNP at position −1772, −1661, and −318 in the promoter, and at position +49 in Ctla-4 (Wang XB, Mao H, Kakoulidou M, Pirskanen R, Giscombe R, Lefvert AK, manuscript in preparation).

4.2. Functional Correlates to the Genetic Variants of Ctla-4

The (AT)n repeats in cytokine genes are generally related to mRNA stability, i.e. the longer alleles are unstable and, hence, the expression of Ctla-4 protein is reduced. Our investigation showed an association to the length of the (AT)n with both spontaneous and antigen-induced T cell activation (Huang et al., 2000), as measured by assessing telomerase activity, serum sIL-2Ra, and thymidine incorporation. There was a positive correlation between the length of the (AT)n and the CD28-mediated T cell activation. A follow-up study showed a decreased mRNA stability of the longer (AT)n as a probable cause of T cell activation, and it was evident that patients with MG had an aberrant expression of CTLA-4, leading to decreased expression on T cells and poor upregulation of the protein upon stimulation (Wang et al., 2002b). Sera from patients also contained high levels of soluble CTLA-4 (sCTLA-4). This soluble form is produced by alternative splicing and is a functional molecule with ligand-binding activities (Oaks *et al.*, 2000), but little is known concerning its significance and function in autoimmune diseases. In MG, the concentration of sCTLA-4 was generally higher than in healthy individuals. This was especially pronounced in patients with thymomas (Wang *et al.*, 2002b). Our preliminary investigations show that in the absence of inflammatory reaction, sCTLA-4 as well as other soluble T cell costimulatory molecules and their ligands are sensitive markers for immune activation.

4.3. The C/T SNP at −318

A SNP at −318 in the *Ctla-4* promoter region is associated with certain autoimmune diseases, but not with MG or MG subgroups. Since the −318 SNP occurs in a potential regulatory region, it is conceivable that the C/T transition may affect the expression of *Ctla-4*. We could indeed show that the −318 T allele is associated with a higher promoter activity than the −318 C allele (Wang *et al.*, 2002c).

4.4. The A/G SNP in CDS1

There were no differences in the +49 A/G SNP in CDS1 between patients and healthy individuals. The frequencies of allele G and genotype G/G were increased in patients with thymoma when compared to patients with normal and hyperplastic thymic histopathology. Patients with the G/G genotype also had signs of immune activation manifested as higher levels of serum IL-1β and higher percentage of $CD28^+$ T lymphocytes. There was a strong linkage between the 86-bp allele in the 3′-UTR and the +49 allele in CDS1. Our results suggest that the SNP at position +49 in CDS1 might be associated with MG manifestations. It is of interest that carriage of the G allele reduces the inhibitory function of CTLA-4 and, thus, contributes to the pathogenesis of Graves' disease (Kouki *et al.*, 2000).

4.5. Promoter SNPs −1772 (C/T) and −1661 (A/G)

Two SNPs in the promoter region of the *Ctla-4* gene at positions −1772 (C/T) and −1661 (A/G) were analyzed in patients with MG. The frequency of the CT-1772 genotype was higher in MG patients, especially in those with thymomas, compared to healthy individuals. The frequencies of the G-1661 allele and the GG-1661 genotype were lower in patients. There were linkage disequilibria between each SNP at position −1772, −1661, and −318 in the promoter, and at position +49 in CDS1. These two SNPs changed the transcription factor–binding site sequences. The T → C-1772 mutation deleted the NF-1-binding site while the A → G-1661 mutation created a c/EBPβ-binding site. Thus, these two SNPs in the promoter of *Ctla-4* result in inefficient transcription and may constitute a disease susceptibility factor (Wang XB, Mao H, Kakoulidou M, Pirskanen R, Giscombe R, Lefvert AK, manuscript in preparation).

4.6. CTLA-4 and Thymomas

Patients with thymomas constitute a distinct clinical entity in which the disease usually is severe, starts in middle age, and is as common in men as in women. Associations of MHC genes are different from those in patients with hyperplasia or normal thymic histology (Carlsson et al., 1990; Spurkland et al., 1991). The patients frequently have autoantibodies against muscle proteins and mononuclear cell infiltrations in the skeletal muscles. In this regard, it is of note that mice deficient for CTLA-4 have a massive lymphocyte infiltration in skeletal muscles. CD28 and CTLA-4 and their ligands are expressed in the thymus and may have a role in the induction of T cell tolerance (Degermann et al., 1994). The deficient mice exhibit abnormal composition of thymocytes, consisting of higher percentages of single positive T cells and lower percentages of double positive $CD4^+CD8^+$ T cells. Insufficient expression of CTLA-4 in the thymus due to certain genetic variants may decrease the avidity in the interactions between immature thymocytes and antigen-presenting cells, resulting in maturation of self-reactive T cells (Cibotti et al., 1997). Higher percentages of longer alleles of *Ctla-4* in persons who develop MG and thymoma could affect thymic selection, thus triggering an autoimmune response. For example, abnormal thymic selection exists in the thymus from MG patients (Truffault et al., 1997). This association between *Ctla-4* and MG with thymoma further confirms the hypothesis that the pathogenesis of MG in patients with thymoma appears to be different from that in other subgroups.

4.7. *Ctla-4* (AT)n Is Associated to ADCC

A possible additional mechanism for the pathogenic action of acetylcholine receptor antibodies might be antibody-dependent cell-mediated cytotoxicity (ADCC). Using a cell line expressing nicotinic acetylcholine receptor as target cells, we could demonstrate increased ADCC mediated by sera from MG patients (Xu et al., 1999). Sera from MG patients with thymomas induced a higher cytotoxic effect than those from other patients. Sera from thymoma patients who had extended (AT)n repeats in the *Ctla-4* gene mediated especially high cytotoxicity. ADCC mediated by acetylcholine receptor antibodies may thus be another possible pathogenic mechanism that could operate in MG patients, especially in those with thymomas.

5. Conclusions

MG is considered a prototype of autoantibody-mediated autoimmune disease. There are, however, several indications that this disorder also is influenced by other immune mechanisms. We have thus analyzed the contribution of proinflammatory and anti-inflammatory cytokines, the β2-adrenergic receptor, and the T cell costimulatory molecule CTLA-4, and have described genetic variants with possible functional implications for MG. One important result of this study is that MG can be divided into subgroups, according to the genetic associations. The functional implications of these associations are different, suggesting that the development/perpetuation of MG can be achieved by quite different immune mechanisms.

One subgroup includes patients with the high secretory allele A2 of TNF-α-308, which usually also carry HLA-B8, DR3. Our studies suggest that the association to TNF-α is indeed of primary importance, as supported by the resistance of mice lacking TNF-α to experimental MG. Recently, blockade of TNF-α has been shown to have beneficial effects on human MG. Another subgroup includes patients carrying the high secretory IL-1βTaqI RFLP A2 allele. Half of them were also carriers of allele A2 of TNF-α-308, showing that these subgroups overlap. Deficiency of IL-1β renders mice resistant to experimental MG. The therapeutic potential of IL-1 receptor–blocking agents has not yet been explored. Patients with thymomas clearly constitute a complete separate entity. They have no HLA-association, but rather strong associations to several variants of the *Ctla-4*, including the increased length of (AT)n in the 3′UTR, the G allele in CDS1 at position +49 and the C allele at −1772 in the promoter region. The common effect of all these variants is a reduced expression of CTLA-4 and, thus, a disturbed balance between activation and suppression of the T cell response. This is quite consistent with the clinical picture of severe disease and high autoantibody levels in thymoma patients, who also have associations to the β2-adrenergic receptor gene polymorphisms with unknown functional correlates. The associations to the variants of the *IL10* gene are more difficult to interpret, since the functional correlates are largely unknown. An important result might be that MG patients lack the naturally occurring balance between the high secretory variants of IL-1β and IL-10, which are present in healthy individuals but not in MG patients. This suggests that the balance between proinflammatory and anti-inflammatory cytokines is important and that genetic variants that influence these components may perturb the balance. To summarize, our findings support the notion that MG is a polygenic disease, with subgroups of patients having quite different genetic backgrounds. In future, this might lead to better design of treatments, aimed at influencing the activity and/or expression of certain cytokines, T cell costimulatory factors or the β2-adrenergic system.

Acknowledgments

The study was supported by the Swedish Research Council (05646), the foundations of the Karolinska Institutet, the European Commission Fifth Framework Programme (QLG1-CT-2001-01918), and the Palle Ferb foundation.

References

Aarli, J.A., Lefvert, A.K., and Tönder, O. (1981). Thymoma-specific antibodies in sera from patients with myasthenia gravis demonstrated by indirect haemagglutination. *J. Neuroimmunol.*, **1**, 421–427.

Åhlberg, R., Yi, Q., Pirskanen, R., Matell, G., Swerup, C., Rieber, P., Riethmuller, G., Holm, G., and Lefvert, A.K. (1994). Treatment of myasthenia gravis with anti-CD4 antibody: Improvement correlates to changes in T cell reactivity. *Neurology*, **44**, 1732–1737.

Askanas, V., Engel, W.K., and Alvarez, R.B. (1998). Fourteen newly recognized proteins at the human neuromuscular junction and their non-junctional accumulation in inclusion-body myositis. *Ann. NY Acad. Sci.*, **841**, 28–56.

Carlsson, B., Wallin, J., Pirskanen, R., Matell, G., and Smith, C.I.E. (1990). Different HLA DR-DQ associations in subgroups of idiopathic myasthenia gravis. *Immunogenetics*, **31**, 285–290.

Cibotti, R., Punt, J.A., Dash, K.S., Sharrow, S.O., and Singer, A. (1997). Surface molecules that drive T cell development in vitro in the absence of thymic epithelium and in the absence of lineage-specific signals. *Immunity*, **6**, 245–255.

Danling, G., Wogensen, L., Calcutt, N.A., Chunyao, X., Powell, H.C., and Sarvetnick, N. (1995). Myasthenia gravis-like syndrome induced by expression of interferon-γ in the neuromuscular junction. *J. Exp. Med.*, **181**, 547–557.

Degermann, S., Suhr, C.D., Glimcher, L.H., Sprent, J., and Lo, D. (1994). B7 expression on thymic medullary epithelium correlates with epithelium-mediated depletion of Vβ5+ thymocytes. *J. Immunol.*, **152**, 3254–3263.

Dinarello, C.A. (2002). The IL-1 family and inflammatory diseases. *Clin. Exp. Rheumatol.*, **20**, 1–13.

Djabiri, F., Caillat-Zucman, S., Gajdos, P., Jais, J.P., Gomez, L., Khalil, I., Charron, D., Bach, J.F., and Garchon, H.J. (1997). Association of the AChRalpha-subunit gene (CHRNA), DQA1*0101, and the DR3 haplotype in myasthenia gravis. Evidence for a three-gene disease model in a subgroup of patients. *J. Autoimmun.*, **10**, 407–413.

Drachman, D.B. (1994). Myasthenia gravis. *N. Engl. J. Med.*, **330**, 1797–1810.

Emilie, D., Crevon, M.C., Cohen-Kaminsky, S., Peuchnaur, M., Devergne, O., Berrih-Aknin, S., and Galanaud, P. (1991). In situ production of interleukins in hyperplastic thymus from myasthenia gravis patients. *Hum. Pathol.*, **22**, 461–468.

Eng, H., Lefvert, A.K., Mellstedt, H., and Österborg, A. (1987). Human monoclonal immunoglobulins that bind the human acetylcholine receptor. *Eur. J. Immunol.*, **17**, 1867–1869.

Feldman, R.D., Hunninghake, G.W., and McArdle, W.L. (1987). Beta-adrenergic-receptor-mediated suppression of interleukin 2 receptors in human lymphocytes. *J. Immunol.*, **139**, 3355–3359.

Garchon, H.J. (2003). Genetics of autoimmune myasthenia gravis, a model for antibody-mediated autoimmunity in man. *J. Autoimmun.*, **21**, 105–110.

Gigliotti, D., Lefvert, A.K., Jeddi-Tehrani, M., Esin, S., Hodara, V., Pirskanen, R,. Wigzell, H., and Andersson, R. (1996). Overexpression of selected TCR Vβ gene families within CD4+ and CD8+ T cell subsets of myasthenia gravis patients: A role for superantigens? *Mol. Med.*, **2**, 452–459.

Goluszko, E., Hjelmström, P., Deng, C., Poussin, M.A., Ruddle, N.H., and Christadoss, P.(2001). Lymphotoxin-alpha deficiency completely protects C57BL/6 mice from developing clinical experimental autoimmune myasthenia gravis. *J. Neuroimmunol.*, **113**, 109–118.

Green, S., Turki, J., Innis, M., and Liggett, S.B. (1994). Amino-terminal polymorphisms of the human β2-adrenergic receptor impart distinct agonist-promoted regulatory properties. *Biochemistry*, **33**, .9414–9419.

Gu, D., Wogensen, L., Calcutt, N.A., Xia, C., Zhu, S., Merlie, H.S., Fox, J.P., Lindstrom, J., Powell, H.C., and Sarvetnick, N. (1995). Myasthenia gravis-like syndrome induced by expression of interferon-γ in the neuromuscular junction. *J. Exp. Med.*, **181**, 547–557.

Hjelmstrom, P., Giscombe, R., Lefvert, A.K., Pirskanen, R., Kockum, I., Landin-Olsson, M., and Sanjeevi, C.B. (1995). Different HLA-DQ are positively and negatively associated in Swedish patients with myasthenia gravis. *Autoimmunity*, **22**, 59–65.

Huang, D.R., Giscombe, R., Matell, G., Pirskanen, R., and Lefvert, A.K. (1999a). Polymorphisms at −174 and in the 3′ flanking region of interleukin-6 (IL-6) gene in patients with myasthenia gravis. *J. Neuroimmunol.*, **101**, 197–200.

Huang D.R., Liu, L., Norén, K., Xia, S.Q., Trifunovic, J., Pirskanen, R., and Lefvert, A.K. (1998a). Genetic association of CTLA-4 to myasthenia gravis with thymoma. *J. Neuroimmunol.*, **88**, 192–198.

Huang, D.R., Pirskanen, R., Hjelmström, P., and Lefvert, A.K. (1998b). Polymorphisms in IL-1β and IL-1 receptor antagonist genes are associated with myasthenia gravis. *J. Neuroimmunol.*, **81**, 76–81.

Huang, D.R., Pirskanen, R., Matell, G., and Lefvert, A.K. (1999b). Tumour necrosis factor-α polymorphism and secretion in myasthenia gravis. *J. Neuroimmunol.*, **94**, 165–171.

Huang, D.R., Shi, F.D., Giscombe, R., Zhou, Y.H., Ljunggren, H.G., and Lefvert, A.K. (2001). Disruption of the Il-1β gene diminishes acetylcholine receptor-induced immune responses in a murine model of myasthenia gravis. *Eur. J. Immunol.*, **31**, 225–232.

Huang, D.R., Shi Qin, X., Zhou, Y.H., Pirskanen, R., Liu, L., and Lefvert, A.K. (1999c). Markers in the promotor region of Interleukin-10 (IL-10) gene in myasthenia gravis: Implications on diverse effects of IL-10 in the pathogenesis of the disease. *J. Neuroimmunol.*, **94**, 82–87.

Huang, D.R., Xia, S., Zhou, Y, Pirskanen, R., Liu, L., and Lefvert, A.K. (1999d). No evidence for Interleukin-4 gene conferring susceptibility to myasthenia gravis. *J. Neuroimmunol.*, **92**, 208–211.

Huang, D.R., Zhou, Y., Giscombe, R., Pirskanen, R., and Lefvert, A.K. (2000). Dinucleotide repeat expansion in CTLA-4 gene leads to T cell hyperreactivity via the CD28 pathway in myasthenia gravis. *J. Neuroimmunol.*, **105**, 69–77.

Kakoulidou, M., Åhlberg, R., Yi, Q., Giscombe, R., Pirskanen, R., and Lefvert, A.K. (2004). The autoimmune T and B cell repertoires in monozygotic twins discordant for myasthenia gravis. *J. Neuroimmunol.*, **148**, 183–191.

Kouki, T., Sawai, Y., Gardine, C.A., Fisfalen, M.E., Alegre, M.L., and DeGroot, L.J. (2000). CTLA-4 gene polymorphism at position 49 in exon 1 reduces the inhibitory function of CTLA-4 and contributes to the pathogenesis of Graves' disease. *J. Immunol.*, **165**, 6606–6611.

Kristiansen, O.P., Larsen, Z.M., and Pociot, F. (2000). CTLA-4 in autoimmune diseases—A general susceptibility gene to autoimmunity? *Genes Immun.*, **1**, 170–184.

Lalani, I., Bhol, K., and Ahmed, A.R. (1997). Interleukin-10: Biology, role in inflammation and autoimmunity. *Ann. Allergy Asthma Immunol.*, **79**, 469–483.

Lefvert, A.K. and Osterman, P.O. (1983). Newborn infants to myasthenic mothers: A clinical study and an investigation of acetylcholine receptor antibodies in 17 children. *Neurology*, **33**, 133–138.

Lefvert, A.K., Pirskanen, R., and Svanborg, E. (1985). Anti-idiotypic antibodies, acetylcholine receptor antibodies and disturbed neuromuscular function in healthy relatives to patients with myasthenia gravis. *J. Neuroimmunol.*, **9**, 41–53.

Mantegazza, R., Oksenberg, J.R., Baggi, F., Antozzi, C., Illeni, M.T., Pellegris, G., Cornelio, F., and Steinman, L. (1990). Increased incidence of certain TCR and HLA genes associated with myasthenia gravis in Italians. *J. Autoimmun.*, **3**, 431–440.

Maselli, R.A., Richman, D.P., and Wollmann, R.L. (1991). Inflammation at the neuromuscular junction in myasthenia gravis. *Neurology*, **41**, 1497–1504.

Oaks, M.K., Hallett, K.M., Penwell, R.T., Stauber, E.C., Warren, S.J., and Tector, A.J. (2000). A native soluble form of CTLA-4. *Cell. Immunol.*, **201**, 144–153.

Oksenberg, J.R., Sherritt, M., Begovich, A.B., Erlich, H.A., Bernard, C.C., Cavalli-Sforza, L.L., and Steinman, L. (1989). T-cell receptor V alpha and C alpha alleles associated with multiple sclerosis and myasthenia gravis. *Proc. Natl. Acad. Sci. USA*, **86**, 988–992.

Sciacca, F.L., Ferri, C., Veglia, F., Andreetta, F., Mantegazza, R., Cornelio, F., Franciotta, D., Piccolo, G., Cosi, V., Batocchi, A.P., Evoli, A., and Grimaldi, L.M.E. (2002). IL-1 genes in myasthenia gravis: IL-1A-889 polymorphism associated with sex and age of disease onset. *J. Neuroimmunol.*, **122**, 94–99.

Scott, M.G., Swan, C., Wheatley, A.P., and Hall, I.P. (1999). Identification of novel polymorphisms within the promoter region of the human beta2 adrenergic receptor gene. *Br. J. Pharmacol.*, **126**, 841–844.

Smith, C.I., Grubb, R., Hammarstrom, L., and Pirskanen, R. (1984). Gm allotypes in Finnish myasthenia gravis patients. *Neurology*, **34**, 1604–1605.

Spurkland, A., Gilhus, N.E., Ronningen, K.S., Aarli, J.A., and Vartdal, F. (1991). Myasthenia gravis patients with thymoma display different HLA associations. *Tissue Antigens*, **37**, 90–93.

Sundewall, A.C., Lefvert, A.K., and Olsson, R. (1985). Anti-acetylcholine receptor antibodies in primary biliary cirrhosis. *Acta Med. Scand.*, **217**, 519–525.

Truffault, F., Cohen-Kaminsky, S., Khalil, I., Levasseur, P., and Berrih-Aknin, S. (1997). Altered intrathymic T cell repertoire in human myasthenia gravis. *Ann. Neurol.*, **41**, 731–741.

Turner, D.M., William, D.M., Sankaran, D., Lazarus, M., Sinnott, P.J., and Hutchinson, I.V. (1997). An investigation of polymorphism in the interleukin-10 gene promoter. *Eur. J. Immunogenet.*, **24**, 1–8.

Non-MHC Genetic Polymorphisms

Wang, X.B., Kakoulidou, M., Qiu, Q., Giscombe, R., Huang, D.R., Pirskanen, R, and Lefvert, A.K. (2002a). CDS1 and promoter single nucleotide polymorphisms of the CTLA-4 gene in human myasthenia gravis. *Genes Immun.*, **3**, 46–49.

Wang, X.B., Kakoulidou, M., Giscombe, R., Qiu, Q., Huang, D.R., Pirskanen, R., and Lefvert, A.K. (2002b). Abnormal expression of CTLA-4 by T cells from patients with myasthenia gravis: Effect of an AT-rich gene sequence. *J. Neuroimmunol.*, **130**, 224–232.

Wang, X.B., Zhao, X., Giscombe, R., and Lefvert, A.K. (2002c). A CTLA-4 gene polymorphism at position −318 in the promoter region affects the expression of protein. *Genes Immun.*, **3**, 232–234.

Waterhouse, P., Penninger, J.M., Timms, E., Wakeham, A., Shahinian, A., Lee, K.P., Thompson, C.B., Griesser, H., and Mak, T.W. (1995). Lymphoproliferative disorders with early lethality in mice deficient in *Ctla-4*. *Science*, **270**, 985–988.

Xu, B.Y., Huang, D.R., Pirskanen, R., and Lefvert, A.K. (2000). β2-adrenergic receptor gene polymorphisms in myasthenia gravis. *Clin. Exp. Immunol.*, **119**, 156–160

Xu, B.Y., Pirskanen, R., and Lefvert, A.K. (1998). Antibodies against β1 and β2 adrenergic receptors in myasthenia gravis. *J. Neuroimmunol.*, **91**, 82–88.

Xu, B.Y., Pirskanen, R., and Lefvert, A.K.. (1999). Antibody-dependent cytotoxicity: An additional mechanism that might operate in human myasthenia gravis. *J. Neuroimmunol.*, **99**, 183–188.

Xu, B.Y., Yi, Q., Pirskanen, R., Matell, G., Eng, H., and Lefvert, A.K. (1997). Decreased β2 adrenergic receptor density on peripheral blood mononuclear cells in myasthenia gravis. *J. Autoimmun.*, **10**, 401–406.

Zheng, C., Huang, D., Liu, L., Wu, R., Bergenbrant-Glas, S., Österborg, A., Björkholm, M., Holm, G., Yi, Q., and Sundblad, A. (2001). Interleukin-10 gene promoter polymorphisms in multiple myeloma. *Int. J. Cancer*, **95**, 184–188.

Zhou, Y.H., Giscombe, R., Huang, D.R., and Lefvert, A.K.(2002). Novel genetic association of Wegener's granulomatosis to the interleukin-10 gene. *J. Rheumatol.*, **29**, 317–320.

9

Rheumatic Heart Disease: Molecular Basis of Autoimmune Reactions Leading to Valvular Lesions

Luiza Guilherme, Kellen C. Faé, and Jorge Kalil

1. Introduction

Rheumatic fever (RF) occurs as a delayed sequel of throat infection by *Streptococcus pyogenes*, affecting 3–4% of untreated children. The clinical signs and symptoms of RF are the same throughout the world. In the 1950s Jones established the major criteria for diagnosing initial attacks of RF, which comprised polyarthritis, carditis, and chorea. These criteria remain useful to date, despite small periodic changes. Arthritis is one of the earliest and most common features of the disease, present in 60–80% of patients. It usually affects the peripheral large joints; small joints and the axial skeleton are rarely involved. Knees, ankles, elbows, and wrists are most frequently affected. The arthritis is usually migratory and very painful. Carditis, the most serious manifestation of the disease, occurs 4–8 weeks (or later) after throat group A streptococcal (GAS) infection in 30–45% of individuals with RF, and usually presents as a pancarditis. Endocarditis is the most serious sequela and frequently leads to rheumatic heart disease (RHD). Valvular lesions and mitral and aortic regurgitation are the most common events caused by valvulitis leading to chronic RHD that still remains a major public health problem in developing countries.

Sydenham's chorea is characterized by involuntary movements, specially of the face and limbs, muscular weakness, and disturbances of speech, gait, and voluntary movements. Children usually exhibit concomitant psychological dysfunction, specially obsessive–compulsive disorder, increased emotional lability,

---

Luiza Guilherme and Kellen C. Faé • Heart Institute (InCor), School of Medicine, University of São Paulo, Brazil; and Institute for Immunology Investigation, Millenium Institute, Brazil. **Jorge Kalil** • Heart Institute (InCor), School of Medicine, University of São Paulo, Brazil; Institute for Immunology Investigation, Millenium Institute, Brazil; and Clinical Immunology and Allergy, Department of Clinical Medicine, University of São Paulo, School of Medicine, São Paulo, Brazil.

Molecular Autoimmunity: In commemoration of the 100th anniversary of the first description of human autoimmune disease, edited by Moncef Zouali. Springer Science+Business Media, Inc., New York, 2005.

hyperactivity, irritability, and age-regressed behavior. It is usually a delayed manifestation, and is often the sole manifestation of acute rheumatic fever (ARF). Other manifestations such as subcutaneous nodules and erythema marginatum can also occur during RF episodes and are characterized by nodules on the surface of joints and skin lesions, respectively.

In this chapter we first describe the etiopathogenic agent and then discuss the susceptibility genetic markers involved in the development of RF/RHD. The molecular basis of autoimmunity in RF/RHD is assessed by studies of molecular mimicry between streptococcal antigens and human tissue proteins mediated by B and T cell responses of human peripheral blood and T cell clones infiltrating heart lesions from RHD patients as well as in animal models. T cell receptor (TCR) analysis and cytokine production by mononuclear cells from peripheral blood and heart-infiltrating cells from RF/RHD patients are also presented.

2. The Etiopathogenic Agent: *Streptococcus pyogenes*

Studies conducted by Rebecca Lancefield (1941) classified streptococci groups by serology based on their cell wall polysaccharides (groups A, B, C, F, and G). *Streptococcus pyogenes* belongs to group A streptococci. Its cell wall is composed of N-acetyl β D-glucosamine linked to a polymeric rhamnose backbone. Group A streptococci contain M, T, and R surface proteins and lipoteichoic acid (LTA), involved in bacterial adherence to throat epithelial cells. The M protein, which extends from the cell wall, is composed of two polypeptide chains with approximately 450 amino acid residues, in an alpha-helical coiled-coil configuration. The amino-terminal portion is composed of two regions named A and B that contain repetitive amino acid residues (Figure 9.1). Antigenic variations on the 11 first amino acid residues located on the A repeat region define the 120 different serotypes identified to date. Some of them have been consistently found to be more frequently associated with RF, whereas others are more often associated with acute glomerulonephritis. The N-terminal B region has high homology between the serotypes described. The C-terminal half is conserved, and contains multiple repeat regions (Fishetti, 1991).

3. Genetic Susceptibility

RF and RHD are genetically controlled with universal frequencies of 3–4% of untreated children that present a throat infection by group A streptococci. Several genetic markers were studied but only associations with HLA class II antigens were consistently found. HLA class II genes are located in human chromosome 6, and are often associated with susceptibility to autoimmune diseases. Association with different HLA class II antigens and RF/RHD has been found in several populations. Figure 9.2 showed the distribution of the HLA alleles associated with the disease in different countries. The differences in the populations studied are probably due to the ability of the HLA class II molecules to present strain-specific streptococcal epitopes present in the more

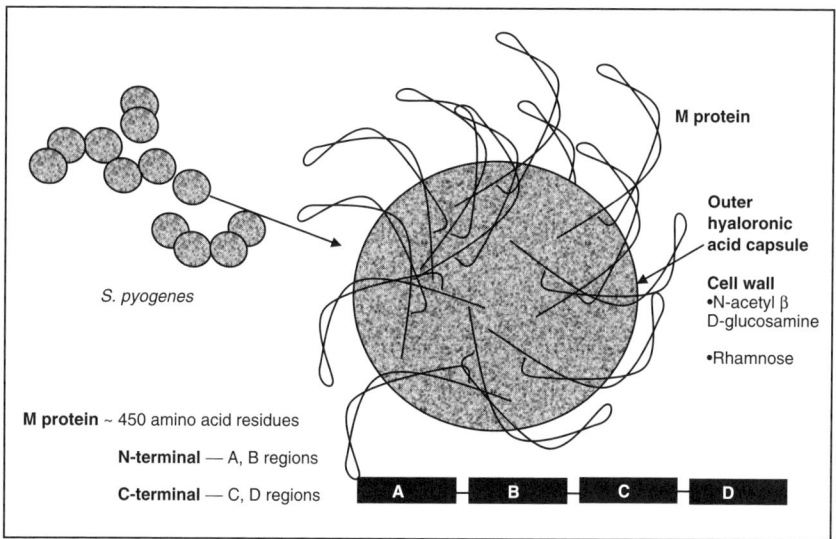

Figure 9.1. Representation of the major cell wall components of *Streptococcus pyogenes*. The structure of *S. pyogenes*, a Gram-positive bacterium, is formed by an outer hyaluronic acid capsule and the cell wall composed of N-acetyl β D-glucosamine linked to a polymeric rhamnose backbone. The external part of the cell wall contains surface adhesion proteins, such as M, T, and R proteins. Only M protein is shown here. This protein has a fibrillar α-helical coil-coiled structure with approximately 450 amino acids residues that are distributed in four repeat regions A and B in the N-terminal portion, and C and D in the C-terminal portion. The N-terminal region is polymorphic and the 11 first amino acids residues define the different serotypes of the bacteria. The A and B repeats share structural homologies with alpha-helical coiled-coil human proteins. The C-terminal region is conserved between the serotypes.

than 100 known streptococcal serotypes, some of which have peculiar geographic distribution.

Among the HLA class II alleles found, HLA-DR7 was the allele most consistently associated with the disease (Guilherme *et al.*, 1991; Ozkan *et al.*, 1993; Weidebach *et al.*, 1994; Guédez *et al.*, 1999; Visentainer *et al.*, 2000; Stanevicha *et al.*, 2003). The association of DR7 with DQB1*0302 and DQB*0401-2 alleles seems to be associated with the development of multiple valvular lesions (MVLs) and mitral valve regurgitation (MVR) respectively, in RHD patients from Latvia (Stanevicha *et al.*, 2003). The association of DR7 with different DQ-A alleles (DQA*0102 and DQA*0401) are also associated with MVR in RHD patients from Egypt (Guédez *et al.*, 1999) (Figure 9.1).

HLA-DR53, another HLA class II molecule, is in linkage disequilibrium with HLA-DR4, -DR7, and -DR9. In two studies the HLA-DR7 and -DR53 alleles were found to be in strong association with RF/RHD among mulatto Brazilian patients (Guilherme *et al.*, 1991; Weidebach *et al.*, 1994). HLA-DR4 and -DR9 were found to be associated with RF in American Caucasians, Arabian patients, and in Indians from Kashmir (Ayoub, 1984; Rajapakse *et al.*, 1987; Bhat *et al.*, 1997). Other class II antigens as DR1, DR2, DR3, and DR6 were also found to be associated with RF/RHD in other populations (Figure 9.2) (Anastasiou-Nana

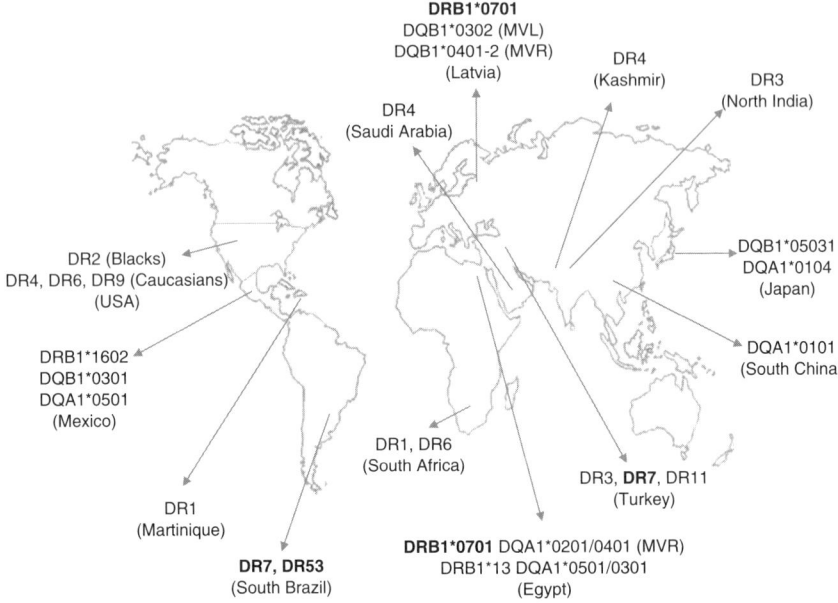

Figure 9.2. Distribution of HLA class II molecules associated with the development of RF/RHD in different populations. Several class II alleles were found in the populations studied and are indicated. HLADR7 was the most consistently allele associated with the disease (bold), and when associated with some DQB1 or DQA1 alleles, seems to be involved in the development of multiple valvular lesions (MVLs) or mitral valve regurgitation (MVR) in rheumatic heart disease (RHD) patients.

et al., 1986; Jhinghan et al., 1986; Monplaisir et al., 1986; Maharaj et al., 1987; Taneja et al., 1989; Reddy et al., 1990).

In Japanese RHD patients, susceptibility to the disease seems to be in part controlled by the HLA-DQA gene or by genes in linkage disequilibrium with HLA-DQA*0104, DQB1*05031 (Koyanagi et al., 1996). The alleles HLA-DQA*0501 and DQB*0301 in linkage disequilibrium with DRB1*1601 (DR2) were associated with RHD in a Mexican Mestizo population (Hernandez-Pacheco et al., 2003).

As mentioned above, the different HLA molecules associated with the disease are probably involved in the antigen presentation of streptococcal peptides as well as autoantigens to the T lymphocytes by molecular mimicry mechanism.

In addition, the description of genetic polymorphism in cytokines and other molecules directly involved in the control of immune responses will certainly further improve our current knowledge of the genes involved in RF/RHD susceptibility.

4. Molecular Mimicry and RF/RHD

RF/RHD is the most convincing example of molecular mimicry in human pathological autoimmunity, in light of the cross-reactions between streptococcal antigens and human tissue proteins. The M protein is the most important antigenic structure of the bacterium and shares structural homologies with alpha-helical coiled-coil human proteins such as cardiac myosin, tropomyosin, keratin, laminin, vimentin, and several valvular proteins. RF is mediated by both humoral and cellular immune responses. However, it is important to note that the data described in the last 20 years by us and others allow us to define RHD as a T cell–mediated disease.

4.1. The Humoral Immune Response

The presence of heart-reactive antibodies was known for more than 50 years. Several studies analyzed sera from both animals immunized with streptococcal antigens and RF/RHD patients and demonstrate the presence of cross-reactive antibodies to streptococcal antigens (membrane, cell wall, M protein, and soluble proteins) and human proteins. Monoclonal antibodies produced against streptococcal antigens confirmed these results. Among the human proteins, cardiac myosin is the most studied and seems to be the major cross-reactive antigen (reviewed by Cunningham, 2000).

4.2. The Cellular Immune Response

Focus on the cellular branch of immune response began in the early 1970s. In light of the important role played by T cells in RF, a number of studies have been performed using tonsils and human peripheral blood. Increased numbers of $CD4^+$ T cells have been observed (Bhatnagar et al., 1987). The cytotoxic activity of $CD8^+$ T cells from normal peripheral blood toward immortalized human heart cells was also described (Dale et al., 1981).

The predominance of $CD4^+$ T cells in rheumatic heart lesions was the first evidence of T cells involvement in RHD lesions (Raizada et al., 1983). The functional role of these infiltrating T cells was first described by us. We isolated these cells from heart lesions of severe RHD patients and established T cell clones. These T cell clones recognized M protein peptides and heart tissue–derived proteins. Our results indicated three M5 immunodominant regions (residues 1–25, 81–103, and 163–177) that cross-reacted with several heart protein fractions, mainly those derived from valvular tissue with molecular weights of 95–150, 43– and 30–43 kDa (Guilherme et al., 1995) (Figure 9.3). The three M5 immunodominant epitopes and the heart tissue proteins were also recognized by peripheral T cells from severe RHD patients (Guilherme et al., 2001). Through an analysis of the T cell repertoire, these results showed the significance of molecular mimicry between beta hemolytic streptococci and heart tissue, leading to local tissue damage in RHD.

The M protein immunodominant regions recognized by human heart–infiltrating T cells aligned with mice myosin/M5 protein cross-reactive T cell epitopes (NT4/5/6, B1B2, B2 epitopes) (Figure 9.3) (reviewed by Cunningham, 2000). The M5(81–96) epitope — included in the M5(81–103), an immunodominant region recognized by human intralesional T cells — was preferentially recognized by peripheral cells from HLA-DR7$^+$DR53$^+$ patients with severe RHD (Guilherme

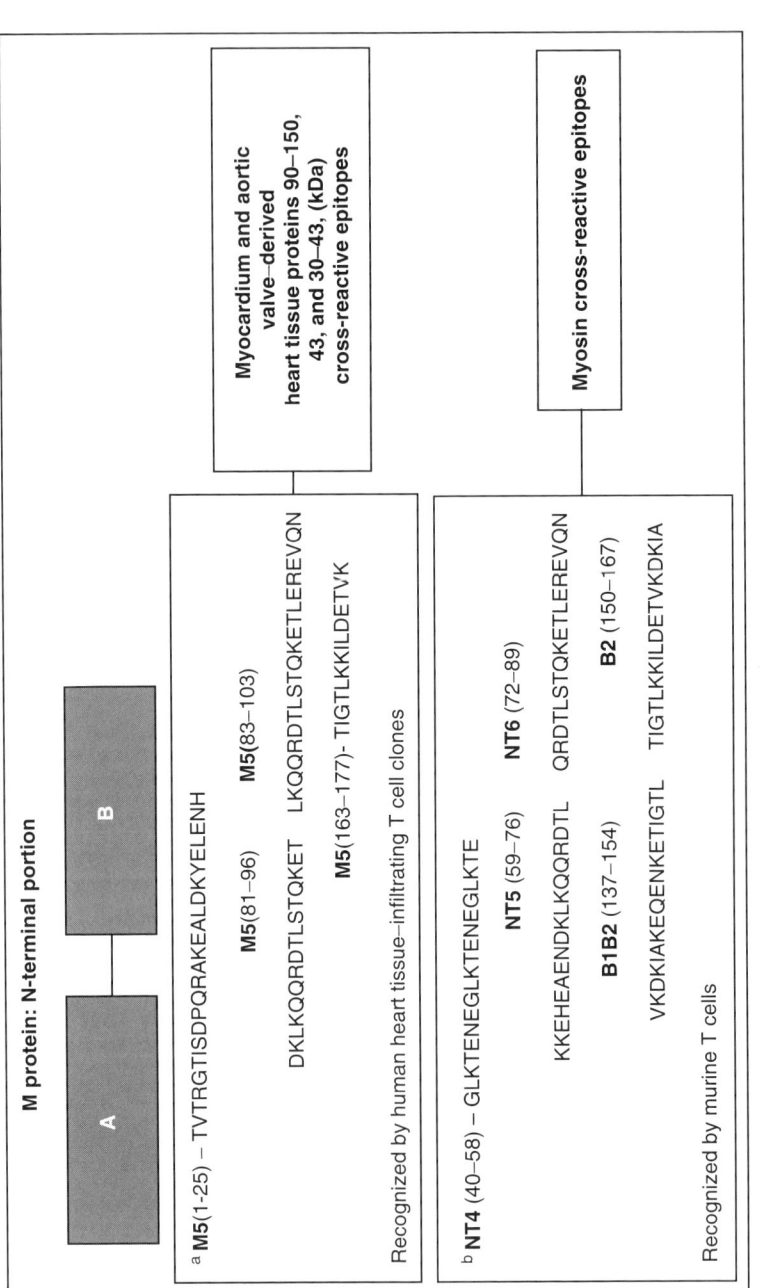

Figure 9.3. Immunodominant M5 protein epitopes that present cross-reaction with human heart proteins. Peptides from the M5 protein (regions 1–25, 81-103 (81–96, and 80–103 peptides), and 163–177) are recognized simultaneously with heart-derived proteins (95–150, 43–, and 30–43 kDa) by human heart-infiltrating T cell clones. NT4/5/6, B1B2, and B2 epitopes are recognized by murine T cells, cross-reacted with cardiac myosin, and align with M5 epitopes recognized by infiltrating human T cell clones. The peptide sequences M5(1–25), M5(81–96), M5(83–103), and M5(163–177) were based on the sequence of the M5 protein published by Manjula et al. (1984). The NT4/5/6, B1B2, and B2 peptides sequences were taken from Miller et al. (1988).

et al., 2001). These data suggest a role for HLA class II molecules DR7 and DR53 in presenting the streptococcal immunodominant peptide to the TCR and show the significance of the susceptibility of DR7+ RHD patients to developing severe RHD in Brazilian patients and probably in other populations in which this class II molecule is also associated with disease and with the development of valvular lesions (Figure 9.2).

4.3. Humoral and Cellular Immune Responses Interface in RF/RHD

The role of cross-reactive antibodies in the development of RHD was only recently established by several studies conducted by the Cunningham group. They showed that cross-reactive antibodies could bind to the endothelial surface, leading to inflammation, cellular infiltration, and valve scarring (reviewed by Cunningham, 2000). After binding of cross-reactive antibodies to the valvular endothelium, the ensued upregulation of the adhesion molecule VCAM-1 facilitates cellular infiltration (Roberts *et al.*, 2001). Once streptococcal primed CD4+ T cells infiltrate the heart, they react with heart tissue proteins by molecular mimicry and with the aid of specific cytokines a delayed-type hypersensitivity (DTH) reaction takes place and leads to heart lesions.

4.4. T Cell Receptor Usage

TCR is a component of the trimolecular complex (MHC, antigenic peptide, and TCR) involved in the activation of cellular immune responses. T cell activation and antigen recognition depend on T cell repertoire. There are 24 Vβ, 13 Jβ, 31 Vα, and 61 Jα known families. They vary in length and in amino acid sequence. The analysis of these variable regions can detect antigen-driven T cell expansions involved in autoimmune responses.

Since the M protein plays an important role in the host antistreptococcal immune response, a putative superantigenic property was investigated by some researchers in the 1990s. A superantigenic effect of streptococcal M5 protein preparations (pepsin-cleaved fragment, pepM5) for normal human T cells expressing some TCR-Vβ families was found (Kotb *et al.*, 1990; Tomai *et al.*, 1990; Watanabe-Ohnishi *et al.*, 1994). However, the superantigenic effect was in fact due to contamination with pyrogenic exotoxins that themselves had a potent superantigen effect (Li *et al.*, 1997).

To probe oligoclonal Vβ-chain expansions, we analyzed TCR-Vβ usage in the peripheral blood and heart-infiltrating T cell lines from severe RHD patients. Our results showed the expansion of several Vβ families with oligoclonal profiles in heart-infiltrating T cell lines, favoring the absence of superantigenicity of M proteins in RHD patients. When we compared the profile of T cells from the myocardium and valvular tissue we found only a few oligoclonal Vβ expansions shared by mitral valve and left atrium–derived T cell lines in the same individual. However, Jβ segments usage and the nucleotide sequences of complementarity-determining regions (CDR3) suggested that different antigenic peptides could be predominantly recognized in the mitral valve and the myocardium (Guilherme et al., 2000).

5. Cytokines

During throat infection by *S. pyogenes*, several streptococcal peptides are generated by antigen-presenting cells (APCs). These peptides, mainly from the M protein (Figure 9.3), are associated with HLA class II molecules (Figure 9.2) and, when presented to T cells, are able to trigger an inflammatory immune response. Since cytokines are likely to be important second signals following an infection triggering effective immune response in most individuals and probably a deleterious response in autoimmune disease, the cytokine production by mononuclear cells from RF/RHD patients was studied by several groups.

The increased production of proinflammatory cytokines (TNFα, IL-1, and IL-2) was observed in activated peripheral blood, tonsillar mononuclear cells, and also in the plasma of RF/RHD patients (Miller et al., 1989; Morris et al., 1993; Narin et al., 1995). During the acute phase of RHD, the production of IL-1, TNFα, and IL-2 in heart lesions was correlated with progression of the Aschoff nodule (Fraser et al., 1997).

Recently, we showed that intralesional mononuclear cells from heart lesions predominantly secrete IFN-γ and TNFα in both ARF and chronic RHD patients; however, only small numbers of IL-4-secreting cells in the valvular tissue were detected. *In vitro* analysis using heart tissue intralesional T cell lines stimulated with streptococcal M5 antigens confirmed these results. So, we showed the existence of predominant Th1 type cytokines produced mainly by CD4$^+$ T cells infiltrating valve tissues, suggesting that Th1 type cytokines could mediate the severe RHD valve lesions. The ability of myocardial infiltrating cells to produce regulatory cytokines such as IL-10 and IL-4 indicated that the presence of regulatory cells may have a role in the mildness of myocardial damage in RHD. In contrast, the small numbers of IL-4-secreting cells in the valvular tissue probably contribute to the progression and maintenance of valvular lesions (Guilherme et al., 2004).

6. Animal Models

Different protocols to induce rheumatic lesions in animal models have been attempted for over 60 years. M protein is the major antigen of streptococci and, as described above, the N-terminal portion seems to be directly related to the cross-reactions with heart tissue proteins that lead to RHD lesions by molecular mimicry. Although several animal models of myocarditis have been described in mice (Huber et al., 1996; reviewed by Cunningham, 2000), only recently was it possible to reproduce the valvulitis with Aschoff bodies in the Lewis rat. In this model the injection of M6 recombinant protein induced myocarditis and inflammatory valvular heart lesions. A lymph node CD4$^+$ T cell line obtained from an immunized rat recognized M6 recombinant protein and cardiac myosin (Quinn *et al.*, 2001). Beta chain cardiac myosin composed of light meromyosin fragment (LMM) and subfragment S2 also was capable of inducing myocarditis and/or valvulitis in the Lewis rat (Galvin *et al.*, 2002). In addition, the cardiac myosin S2 (amino acids 1052–1076) epitope was able to induce severe autoimmune myocarditis associated with an upregulation of inflammatory cytokine production in Lewis rats (Li *et al.*, 2004). The availability of these experimental models of myocarditis and valvulitis induced by streptococcal antigens and cardiac myosin will certainly contribute toward a better understanding of the pathogenesis of RHD and could prove useful for the development of vaccines.

7. Conclusions

The knowledge of the pathogenesis of RF/RHD makes it is possible to delineate a new picture of the disease that involves the remarkable points. First, molecular mimicry mechanism between streptococcal antigens and human tissues, mainly heart tissue, leads to rheumatic heart lesions in RHD patients. Second, $CD4^+$ T lymphocytes are the major effectors of heart lesions. Third, several streptococcal immunodominant peptides generate cross-recognition of several heart tissue–derived proteins. Fourth, several HLA class II molecules are associated with the disease worldwide, and HLA-DR7/DR53, combined with some HLA-DQ molecules, seem to be associated with the development of MVLs and/or MVR in RHD patients. Fifth, Th1 type cytokines seem to be predominant in heart lesions and the small numbers of mononuclear cells able to produce IL-4 (a regulatory cytokine) may account for the more severe tissue destruction in valves seen in RHD.

References

Anastasiou-Nana, M., Anderson, J.L., Carquist, J.F., and Nana, J.N. (1986). HLA DR typing and lymphocyte subset evaluation in rheumatic heart disease: A search for immune response factors. *Am. Heart J.*, **112**, 992–997.

Ayoub, E.M. (1984). The search for host determinants of susceptibility to rheumatic fever: The missing link. *Circulation*, **69**, 197–201.

Bhat, M.S., Wani, B.A., Koul, P.A., Bisati, S.D., Khan, M.A., and Shah, S.U. (1997). HLA antigen pattern of Kashmiri patients with rheumatic heart disease. *Indian J. Med. Res.*, **105**, 271–274.

Bhatnagar, P.K., Nijhawan, R., and Prakash, K. (1987). T cell subsets in acute rheumatic fever, rheumatic heart disease and acute glomerulonephritis cases. *Immunol. Lett.*, **15**, 217–219.

Cunningham, M.W. (2000). Pathogenesis of group A streptococcal infections. *Clin. Microbiol. Rev.*, **13**, 470–511.

Dale, J.B., Simpson, W.A., Ofek, I., and Beachey, E. (1981). Blastogenic responses of human lymphocytes to structurally defined polypeptide fragments of streptococcal M protein. *J. Immunol.*, **126**, 1499–1505.

Fishetti, V. (1991). Streptococcal M protein. *Sci. Am.*, **264**, 32–39.

Fraser, W.J., Haffejee, Z., Jankelow, D., Wadee, A., and Cooper, K. (1997). Rheumatic Aschoff nodules revisited. II. Cytokine expression corroborates recently proposed sequential stages. *Histopathology*, **31**, 460–464.

Galvin, J.E., Hemric, M.E., Kosanke, S.D., Factor, S.M., Quinn, A., and Cunningham, M.W. (2002). Induction of myocarditis and valvulitis in Lewis rats by different epitopes of cardiac myosin and its implications in rheumatic carditis. *Am. J. Pathol.*, **160**, 297–306.

Guédez, Y., Kotby, A., El-Demellaway, M., Galal, A., Thomson, G., Zaher, S., Kassem, S., and Kotb, M. (1999). HLA class II associations with rheumatic heart disease are more evident and consistent among clinically homogeneous patients. *Circulation*, **99**, 2784–2790.

Guilherme, L., Cunha-Neto, E., Coelho, V., Snitcowsky, R., Pomerantzeff, P.M.A., Assis, R.V., Pedra, F., Neumann, J., Goldberg, A., Patarroyo, M.E., Pillegi, F., and Kalil, J. (1995). Human heart-infiltrating T cell clones from rheumatic heart disease patients recognized both streptococcal and cardiac proteins. *Circulation*, **92**, 415–420.

Guilherme, L., Cury, P., Demarchi, L.M.F., Coelho, V., Abel, L., Lopez, A.P., Oshiro, S.E., Aliotti, S., Cunha-Neto, E., Pomerantzeff, P.M.A., Tanaka, A.C., and Kalil, J. (2004). Rheumatic heart disease: pro-inflammatory cytokines play a role in the progression and maintenance of valvular lesions. *Am. Pathol. J.*, **165**, 1583–1591.

Guilherme, L., Dulphy, N., Douay, C., Coelho, V., Cunha-Neto, E., Oshiro, S.E., Assis, R.V., Tanaka, A.C., Pomerantzeff, P.M., Charron, D., Toubert, A., and Kalil, J. (2000). Molecular evidences for antigen-driven immune responses in cardiac lesions of rheumatic heart disease patients. *Int. Immunol.*, **12**, 1063–1074.

Guilherme, L., Oshiro, S.E., Faé, K.C., Cunha-Neto, E., Renesto, G., Goldberg, A.C., Tanaka, A.C., Pomerantzeff, P., Kiss, M.H., Silva, C., Guzman, F., Patarroyo, M.E., Southwood, S., Sette, A., and Kalil, J. (2001). T cell reactivity against streptococcal antigens in the periphery mirrors reactivity of heart infiltrating T lymphocytes in rheumatic heart disease patients. *Infect. Immun.*, **69**, 5345–5351.

Guilherme, L., Weidebach, W., Kiss, M.H., Snitcowsky, R., and Kalil, J. (1991). Associations of human leukocyte class II antigens with rheumatic fever or rheumatic heart disease in a Brazilian population. *Circulation*, **83**, 1995–1998.

Hernandez-Pacheco, G., Aguilar-Garcia, J., Flores-Dominguez, C., Rodriguez-Perez, J.M., Perez-Hernandez, N., Alvarez-Leon, E., Reyes, P.A., and Vargas-Alarcon, G. (2003). MHC class II alleles in Mexican patients associated with rheumatic heart disease. *Int. J. Cardiol.*, **92**, 49–54.

Huber, S.A. and Cunnigham, M.W. (1996). Streptococcal M protein peptide with similarity to myosin includes CD4+ T cell-dependent myocarditis in MRL/++ mice and induces partial tolerance against coxsackie viral myocarditis. *J. Immunol.*, **156**: 3528–3534.

Jhinghan, B., Mehra, N.K., Reddy, K.S., Taneja, V., Vaidya, M.C., and Bhatia, M.L. (1986). HLA, blood groups and secretor status in patients with established rheumatic fever and rheumatic heart disease. *Tissue Antigens*, **27**, 172–178.

Kotb, M., Majumdar, G., Tomai, M., and Beachey, E.H. (1990). Accessory cell-independent stimulation of human T cells by streptococcal M protein superantigen. *J. Immunol.*, **145**, 1332–1336.

Koyanagi, T., Koga, Y., Nishi, H., Toshima, H., Sasazuki, T., Imaizumi, T., and Kimura, A. (1996). DNA typing of HLA class II genes in Japanese patients with rheumatic heart disease. *J. Mol. Cell. Cardiol.*, **28**, 1349–1353.

Lancefield, R.C. (1941). Specific relationship of cell composition to biological activity of hemolytic streptococci. *Harvey Lect.*, **36**: 251.

Li, P.L.L., Tiedemann, R.E., Moffat, L.S., and Fraser, J.D. (1997). The superantigen streptococcal pyrogenic exotoxin C (SPE-C) exhibits a novel mode of action. *J. Exp. Med.*, **186**, 375–391.

Li, Y., Heuser, J.S., Kosanke, S.D., Hemric, M., and Cunningham, M.W. (2004). Cryptic epitope identified in rat and human cardiac myosin S2 region induces myocarditis in the Lewis rat. *J. Immunol.*, **172**, 3225–3234.

Maharaj, B., Hammond, M.G., Appadoo, B., Leary, W.P., and Pudifin, D.J. (1987). HLA-A, B, DR and DQ antigens in black patients with severe chronic rheumatic heart disease. *Circulation*, **76**, 259–261.

Manjula, B.N., Acharya, A.S., Mische, S.M., Fairwell, T., and Fischetti, V.A. (1984). The complete amino acid sequence of a biologically active 197-residue fragment of M protein isolated from type 5 group A streptococci. *J. Biol. Chem.*, **259**, 3686–3693.

Miller, L., Gray, L., Beachey, E., and Kehoe, M. (1988). Antigenic variation among group A streptococcal M proteins. Nucleotide sequences of the serotype 5 M protein gene and its relationship with genes encoding types 6 and 24 M proteins. *J. Biol. Chem.*, **263**, 5668–5673.

Miller, L.C., Gray, E.D., Mansour, M., Abdin, Z.H., Kamel, R., Zaher, S., and Regelmann, W.E. (1989). Cytokines and immunoglobulin in rheumatic heart disease: Production by blood and tonsillar mononuclear cells. *J. Rheumatol.*, **16**, 1436–1442.

Monplaisir, N., Valette, I., and Bach, J.F. (1986). HLA antigens in 88 cases of rheumatic fever observed in Martinique. *Tissue Antigens*, **28**, 209–213.

Morris, K., Mohan, C., Wahi, P.L., Anand, I.S., and Ganguly, N.K. (1993). Enhancement of IL-1, IL-2 production and IL-2 receptor generation in patients with acute rheumatic fever and active rheumatic heart disease; a prospective study. *Clin. Exp. Immunol.*, **91**, 429–436.

Narin, N., Kütükçüler, N., Özyürek, R., Bakiler, A.R., Parlar, A., and Arcasoy, M. (1995). Lymphocyte subsets and plasma IL-1β, IL-2, and TNF-α concentrations in acute rheumatic fever and chronic rheumatic heart disease. *Clin. Immunol. Immunopathol.*, **77**, 172–176.

Ozkan, M., Carin, M., Sonmez, G., Senocak, M., Ozdemir, M., and Yakut, C. (1993). HLA antigens in Turkish race with rheumatic heart disease. *Circulation*, **87**, 1974–1978.

Quinn, A., Kosanke, S.D., Fischetti, V.A., Factor, S.M., and Cunningham, M.W. (2001). Induction of autoimmune valvular heart disease by recombinant streptococcal M protein. *Infect. Immun.*, **69**, 4072–4078.

Raizada, V., Williams, R.C., Jr., Chopra, P., Gopinath, N., Prakash, K., Sharma, K.B., Cherian, K.M., Panday, S., Arora, R., Nigam, M., Zabriskie, J.B., and Husby, G. (1983). Tissue distribution of lymphocytes in rheumatic heart valves as defined by monoclonal anti-T cells antibodies. *Am. J. Med.*, **74**, 225–237.

Rajapakse, N.A., Halim, K., Al-Orainey, L., Al-Nozha, M., and Al-Aska, A.K. (1987). A genetic marker for rheumatic heart disease. *Br. Heart J.*, **58**, 659–662.

Reddy, K.S., Narula, J., Bathia, R., Shailendri, K., Koicha, M., Taneja, V., Jhingan, B., Pothineni, R.B., Malaviya, A.N., and Mehra, N.K. (1990). Immunologic and immunogenetic studies in rheumatic fever and rheumatic heart disease. *Indian J. Pediatr.*, **57**, 93–97.

Roberts, S., Kosanke, S., Dunn, T.S., Jankelow, D., Duran, C.M.G., and Cunningham, M.W. (2001). Pathogenic mechanism in rheumatic carditis: Focus on valvular endothelium. *J. Infect. Dis.*, **183**, 507–511.

Stanevicha, V., Eglite, J., Sochevs, A., Gardovska, D., Zavadska, D., and Shantere, R. (2003). HLA class II associations with rheumatic heart disease among clinically homogeneous patients in children in Latvia. *Arthritis Res. Ther.*, **5**, 340–346.

Taneja, V., Mehra, N.K., Reddy, K.S., Narula, J., Tandon, R., Vaidva, M.C., and Bhatia, M.L. (1989). HLA-DR/DQ antigens and reactivity to B cell alloantigen D8/17 in Indian patients with rheumatic heart disease. *Circulation*, **80**, 335–340.

Tomai, M., Kotb, M., Majumdar, G., and Beachey, E.H. (1990). Superantigenicity of streptococcal M protein. *J. Exp. Med.*, **172**, 359–362.

Visentainer, J.E., Pereira, F.C., Dalalio, M.M., Tsuneto, L.T., Donadio, P.R., and Moliterno, R.A. (2000). Association of HLA-DR7 with rheumatic fever in the Brazilian population. *J. Rheumatol.*, **27**, 1518–1520.

Watanabe-Ohnishi, R., Aelion, J., Legros, L., Tomai, M.A., Sokurenko, E.V., Newton, D., Takahara, J., Irino, S., Rashed, S., and Kotb, M. (1994). Characterization of unique human TCR VB specificities for a family of streptococcal superantigens represented by rheumatogenic serotypes of M protein. *J. Immunol.*, **152**, 2066–2073.

Weidebach, W., Goldberg, A.C., Chiarella, J., Guilherme, L., Snitcowsky, R., Pileggi, F., and Kalil, J. (1994). HLA class II antigens in rheumatic fever: Analysis of the DR locus by RFLP and oligotyping. *Hum. Immunol.*, **40**, 253–258.

10

Autoimmunity against Desmogleins in Pemphigus Vulgaris

Christian Veldman and Michael Hertl

1. Introduction

Pemphigus vulgaris (PV) is an autoimmune disease of the skin caused by autoantibodies (autoAbs) against desmoglein 3 (Dsg3), a component of the intercellular adhesion structure of epidermal keratinocytes (Amagai *et al.*, 1991; Bedane *et al.*, 1996), leading to a loss of adhesion between keratinocytes. Several forms of pemphigus have been classified depending on the level of the intraepidermal split formation (Lever, 1953; Huilgol and Black, 1995; Hertl, 2000). In the PV group, the blisters are located just above the basal layer whereas in the pemphigus foliaceus (PF) group, the blisters occur within the upper layers of the epidermis (Bedane *et al.*, 1996). In many PV patients, the disease is disabling and often becomes devastating due to extensive blisters of skin and mucosa. The serum levels of anti-Dsg3 Abs correlate with the severity of disease. Current immunotherapy includes systemic administration of glucocorticoids and immunosuppressive agents such as azathioprine, cyclophospamide, mycophenolate mofetil, methotrexate and immunoadsorption or plasmapheresis. These treatments are not antigen (Ag)-specific and bear the risk of various side effects that may lead to complications during the course of therapy. In this chapter, the pathogenic role of autoAbs and the potential role of autoreactive T cells in the regulation of antibody production is discussed. Emphasis is put on the epitope recognition of autoaggressive T cells and the characterization of a subset of T cells that may be critical in the maintenance/restoration of tolerance against desmogleins. This information may be useful for the development of Ag-specific immunotherapies of PV.

Christian Veldman and Michael Hertl • Department of Dermatology, University of Erlangen-Nürnberg, 91052 Erlangen, Germany.

Molecular Autoimmunity: In commemoration of the 100th anniversary of the first description of human autoimmune disease, edited by Moncef Zouali. Springer Science+Business Media, Inc., New York, 2005.

2. Clinical Phenotype of Pemphigus Vulgaris

PV is a relatively rare disorder that primarily affects individuals in the 3rd to 5th decade without gender preference (Huilgol and Black, 1995; Nousari and Anhalt, 1999). Clinically, PV is characterized by flaccid blisters/erosions of the mucous membranes and the skin. In the majority of patients, the oral mucosa is primarily affected but other mucous membranes may be involved as well. Initial blisters rapidly rupture, leading to painful chronic lesions that may affect the larynx and the pharynx in addition to the oral mucosa. Once the disease progresses, skin lesions may occur at any site of the integument, but there is a preferential involvement of the trunk. Due to extensive blistering of skin and mucosa, the prognosis of PV used to be fatal before introduction of glucocorticoids as the major therapeutic strategy. The natural course of the disease is progressive, with death occurring within a few years of onset due to sepsis.

Neonatal pemphigus may occur due to the diaplacentar transfer of anti-Dsg3 IgG4 from mothers with PV to their unborn children. After birth, the newborns exhibit crusty erosions of the skin with the histopathological findings of PV. Once circulating autoAbs are degraded, these skin lesions disappear after a few months.

3. Epidemiology of Pemphigus and Association with HLA Class II Alleles

PV is the most common form of pemphigus, but is still a rare disease, the incidence varying from 0.1 to 0.5 per 100,000 and being higher among Jewish patients (Ahmed *et al.*, 1990). Several epidemiological studies in Jewish (Ahmed *et al.*, 1990), non-Jewish (Ahmed *et al.*, 1991), and Japanese PV patients demonstrated that the HLA-DRß1*0402 and HLA-DRß1*1401 alleles are highly prevalent in PV. Sinha *et al.* (1988) demonstrated that the DR14 susceptibility is strongly associated with a rare DQß allele (DQß1*0503), identifying this DQ allele as a major susceptibility factor for PV.

4. Pathogenesis of Pemphigus

The molecular basis for intraepithelial blister formation is the loss of adhesion between keratinocytes, called acantholysis, which is caused by autoAbs directed against intercellular adhesion structures of epidermal keratinocytes. AutoAb production in PV is polyclonal and most autoAbs are of the IgG4 subclass in PV patients with active disease (Bhol *et al.*, 1995; Tremeau-Martinage *et al.*, 1995; Späeth *et al.*, 2001). Patients in remission have mainly autoAbs of the IgG1 subtype, while healthy relatives of PV patients and healthy carriers of PV-prevalent HLA class II alleles have low levels of IgG1 autoAbs (Brandsen *et al.*, 1997; Kricheli *et al.*, 2000; Späeth *et al.*, 2001). Evidence for the pathogenicity of these circulating autoAb is provided by the observation that (a) the activity of PV correlates with autoAb titers, (b) newborns of mothers with active PV temporarily exhibit blisters due to the diaplacentar transfer of maternal autoAb, and (c) pemphigus-like lesions are induced in neonatal mice by transfer of IgG from PV patients (reviewed in Hertl, 2000) (Figure 10.1).

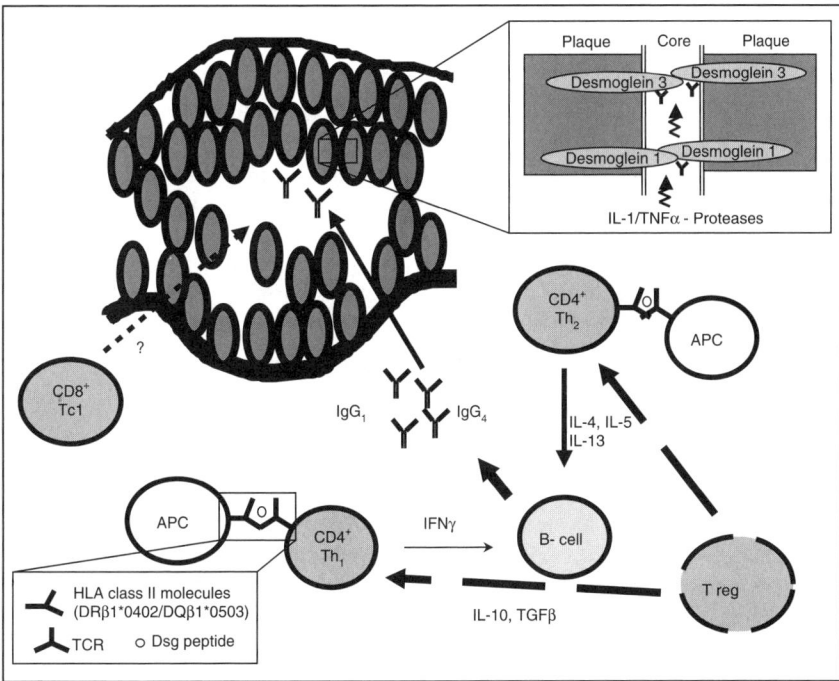

Figure 10.1. Hypothetical model for the immune pathogenesis of pemphigus. AutoAbs against demoglein 1 (Dsg1) and Dsg3 have been shown to induce loss of keratinocyte adhesion in pemphigus vulgaris (PV). Ab production by B cells depends on the help of Th1 cells for IgG1 and of Th2 cells for IgG4, IgA, and IgE autoAbs (solid lines). Autoreactive Th1 and Th2 cells recognize epitopes of the extracellular portion of Dsg1 and Dsg3 mainly in the context of the PV-associated HLA class II alleles HLA-DRβ1*0402 and DQβ1*0503. Regulatory T cells suppress autoreactive Th cells in a cytokine-dependent manner. CD8+ cytotoxic T (Tc1) cells have been identified in patients with PV; their function is unclear (dotted line). Upon binding of the autoAbs to demosomal target antigens, tumor necrosis factor (TNF) and interleukin-1 (IL-1) are released from epidermal keratinocytes, presumably enhancing the process of loss of desmosomal adhesion that results in blister formation.

5. Autoantibody Reactivity against Desmogleins

PV is caused by autoAbs against the extracellular domain (ECD) of Dsg3, a desmosomal adhesion molecule present on epidermal keratinocytes. Although Dsg3 is the major antigen targeted by autoAb in PV, recent studies showed that PV patients also have frequently autoAbs reacting against Dsg1 and against desmocollins, other transmembranous components of desmosomes (Emery *et al.*, 1995). Recently, a novel human desmosomal cadherin, Dsg4, sharing 41% identity with Dsg1 and 50% with Dsg3 was identified (Whittock and Bower, 2003). Moreover, serum Abs from a subset of patients with PV were shown to be also reactive with Dsg4 (Kljuic *et al.*, 2003). Although the pathogenic relevance of anti-Dsg4 Ab is not fully elucidated their occurrence seems to be associated with

Abs against Dsg1 in mucocutaneous PV and PF (Nishifuji et al., 2004). Two potential new target antigens belonging to the group of cholinergic receptors were identified. Pemphaxin is an annexin homolog that binds acetylcholine (Nguyen et al., 2000a). The second acetylcholine receptor targeted by serum IgG from PV patients is α9 acetylcholine receptor, which is also present on keratinocytes (Nguyen et al., 2000b). The precise role that autoAbs against acetylcholine receptors play in the pathogenesis of PV needs to be elucidated (Table 10.1).

Classical PV presents primarily with mucosal lesions and is associated with IgG against Dsg3 (Amagai et al., 1991). In contrast, sera of PV patients with mucocutaneous lesions contain higher levels of IgG4 than of IgG1 against Dsg3 *and* Dsg1, the autoantigen of PF. The epitope(s) of Dsg1 recognized by PV sera are located in the NH2-terminal region and are conformationally sensitive (Emery et al., 1995; Kowalczyk et al., 1995). An explanation for the association of characteristic autoAb profiles with distinct clinical variants of PV is provided by the differential expression pattern of Dsg1 and Dsg3 in cornified and non-cornified stratified epithelia (Shimizu et al., 1995). In the skin, Dsg1 is expressed in the upper epidermal layer, i.e., the granular layer, while Dsg3 is expressed predominantly in the suprabasilar epidermal layer (Amagai et al., 1996). In non-cornified stratified epithelia, such as the oral mucosa, Dsg3 is expressed throughout the epidermal layer while Dsg1 is poorly expressed. Dsg3 is thus a crucial target antigen for the development of oral and, to a lesser extent, of cutaneous lesions in PV. In contrast, autoAb against Dsg1 in PF do not cause mucosal blisters. The concert of anti-Dsg1 and -Dsg3 autoAbs leads to the formation of both mucosal and cutaneous blisters.

The transfer of IgG from PV patients with active PV into newborn mice causes acantholysis, while circulating IgG from PV patients in remission and HLA-matched normals does not (Anhalt, 1982; Amagai et al., 1991). Amagai et al. (1994) showed that the preabsorption of PV-IgG by recombinant Dsg3 protein removed all of the pathogenic autoAb, indicating that Dsg3 autoAbs are relevant for blister formation in PV. The specificity of the IgG response to Dsg3 has been extensively studied using peptides encompassing the entire ECD of Dsg3

Table 10.1. Autoantigens of Pemphigus Vulgaris

Target antigen	Reference
Desmoglein 3[a]	Amagai et al. (1991)
Desmoglein 1[a]	Emery et al. (1995)
	Kowalczyk et al. (1995)
	Ding et al. (1998)
Desmoglein 4[b]	Kljuic et al. (2003)
	Nishifuji et al. (2004)
Pemphaxin[c]	Nguyen et al. (2000a)
Acetylcholine receptor[c]	Nguyen et al. (2000b)

[a]Pathogenic antigen; [b]potentially pathogenic antigen; [c]pathogenic role is unknown.

(Bhol *et al.*, 1995; Sekiguchi *et al.*, 2000). IgG affinity-purified from PV sera on the N-terminal ECD1–2 of Dsg3 causes suprabasilar acantholysis, the typical histologic finding of PV. In contrast, IgG affinity-purified on a recombinant protein representing the ECD3–5 of Dsg3 did not induce acantholysis upon injection into neonatal mice (Amagai *et al.*, 1992). IgG1 and IgG4 from patients with active PV recognize epitopes in the ECD1 and ECD2. Additional *in vitro* data indicate that IgG4 directed against the ECD2, and to a lesser extent, against the ECD1 causes acantholysis (Bhol *et al.*, 1995) (Figure 10.2). In summary, the observations strongly suggest that IgG4 against the ECD2 of Dsg3 is presumably the main acantholytic Ab while IgG4 against the ECD1 may act as a facilitator or enhancer of this process. A recent study demonstrated that PV sera also recognize intracellular epitopes of Dsg3 (Ohata *et al.*, 2001). The significance of this finding is yet unclear.

Several studies suggest that binding of PV IgG to epidermal keratinocytes induces a rapid and transient [Ca^{++}] flux that may result in an altered Ab-transmitted signaling, leading to loss of cell adhesion (Seishima *et al.*, 1995). Acantholysis may be induced by proteases such as plasminogen activator upon binding of autoAbs to keratinocytes (Hashimoto *et al.*, 1983). Recent studies suggest that phospholipase C plays an important role in transmembrane signaling, leading to cell–cell detachment exerted by pemphigus IgG binding to the cell surface (Esaki *et al.*, 1995).

A recent study by Amagai *et al.* (2000b) has shed additional light on the association of anti-Dsg1/Dsg3 autoAb with the clinical phenotype. They showed that the toxin of bullous impetigo is a protease that selectively degrades Dsg1 leading to subcorneal split formation, which is also characteristic for PF. This finding supports the idea that anti-Dsg1 are exclusively responsible for the pathology of PF lesions.

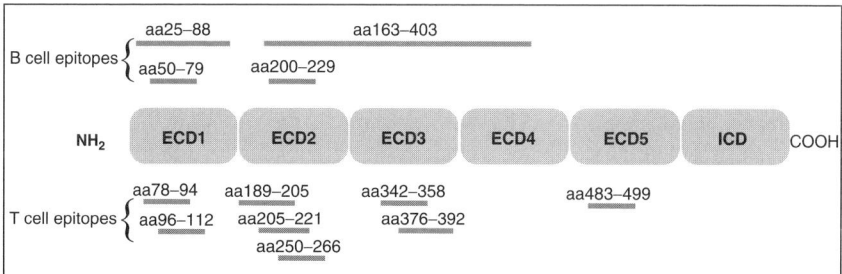

Figure 10.2. T and B cell epitopes of desmoglein 3 (Dsg3), the autoantigen of pemphigus vulgaris (PV). Shown is a diagram of the extracellular portion of Dsg3, which consists of five domains (ECD1–ECD5) and an intracellular domain. B cell epitopes (top) have been identified in the ECD1 and ECD2 domains (aa50–79 and aa200–229, Bhol *et al.*, 1995; aa25–88, Sekiguchi *et al.*, 2001) and in the middle region (Sekiguchi *et al.*, 2001). Recent evidence suggests that B cell epitopes are also located on the intracellular portion of Dsg3. T cell epitopes of Dsg3 (bottom) have been identified all over the extracellular domain. Similar to the B cell epitopes, the major T cell epitopes are clustered around the NH_2 terminus (ECD1, ECD2) of the Dsg3 ectodomain (Veldman *et al.*, 2004a).

6. Autoreactive T Lymphocytes in Pemphigus

Current concepts strongly suggest that autoreactive T cells play a crucial role in the initiation and perpetuation of both Ab- and cell-mediated autoimmune diseases. Autoreactive T cells may provide critical help for B cells to continuously produce pathogenic autoAbs in PV. Involvement of $CD4^+$ T lymphocytes in the pathogenesis of PV has been suggested by the aforementioned strong association of PV with HLA-DRß1*0402 and HLA-DQß1*0503 (Sinha *et al.*, 1988; Ahmed *et al.*, 1990; Hertl and Riechers, 1999). Using an ELISPOT assay, anti-Dsg3 autoAbs secreted by autoreactve B cells were detected upon *in vitro* stimulation of peripheral lymphocytes from PV patients with Dsg3 (Nishifuji *et al.*, 2000). In contrast, activation of autoreactive B cells was virtually absent upon depletion of the peripheral lymphocytes from $CD4^+$ T cells (Nishifuji *et al.*, 2000). In addition, adoptive *in vivo* transfer of splenocytes from $Dsg3^{-/-}$ mice immunized with Dsg3 into $Dsg3^{+/+}$ $Rag2^{-/-}$ mice led to the induction of Dsg3-specific autoAbs accompanied by mucosal erosions characteristic for PV. Transfer of either Dsg3-reactive T or B cells into $Dsg3^{+/+}$ $Rag2^{-/-}$ mice did not lead to autoAb production and a PV phenotype, demonstrating that the interaction of both Dsg3-reactive T and B cells were required to induce autoAb production (Tsunoda *et al.*, 2002).

The majority of peripheral T cell lines and clones generated from several patients with PV expressed a $CD4^+$ memory phenotype (Hertl *et al.*, 1998). Peripheral $CD4^+$ T cell responses to the extracellular domain of Dsg3 were identified in PV patients by several independent investigations (Wucherpfennig *et al.*, 1995; Lin *et al.*, 1997; Hertl *et al.*, 1998). Their phenotype, cytokine profile, immunogenetic restriction, and epitope specificity varied. Both Dsg3-reactive Th1 (Hertl *et al.*, 1998) and Th2 (Lin *et al.*, 1997) cells were identified that recognized portions of the extracellular domain of Dsg3 in the context of PV-associated HLA class II alleles. By ELISPOT assay, autoreactive Th1/Th2 cells were detectable at similar frequencies in acute onset PV (Eming *et al.*, 2000). To clarify this issue, a recent study sought to quantitate peripheral Dsg3-responsive Th1 and Th2 cells in PV patients and healthy controls by the MACS cytokine secretion assay (Veldman *et al.*, 2003). Both Dsg3-autoreactive Th1 and Th2 cells were isolated from patients with acute onset, chronic active, and remittent PV. The appearance of Dsg3-reactive Th2 was constant at the different disease stages, while Dsg3-reactive Th1 cells were detected at a significantly higher frequency in chronic active PV. A synergistic interplay of autoreactive Th1 and Th2 cells seems to be critical for promoting IgG1 and IgG4 secretion by Dsg3-reactive B cells in the pathogenesis of PV. This is supported by the finding that neither the frequency of Dsg3-reactive Th2 cells in PV patients nor that of Th1 cells in Dsg3-reactive healthy donors was directly related to the titers or the presence of Dsg3-specific IgG autoAbs, respectively. Noteworthy, the titers of serum autoAbs against Dsg3 correlated best with the ratio of autoreactive Th1/Th2 cells, suggesting that both Th1 and Th2 cells may be critically involved in the regulation of autoAb production. Thus, both autoreactive Th1 and Th2 cells may be involved in the regulation of the production of pathogenic autoAbs by B cells in PV since sera of patients with PV contain Th1-regulated IgG1 and Th2-regulated IgG4, IgA, and IgE autoAbs directed against Dsg3 (Bhol *et al.*, 1995; Späeth *et al.*, 2001). Ongoing *in vitro* coculture studies with autoreactive Th cells and B cells will hopefully help to clarify the role that these Th subsets play in the maintenance of autoimmunity versus tolerance against Dsg3 (Figure 10.1).

CD8$^+$ T cells that are responsive to Dsg3 were occasionally detected in patients with active PV (unpublished observation). They uniformly secreted IL-2 and IFN-γ but not IL-4 or IL-5 upon *in vitro* stimulation with Dsg3. Their function has not yet been thoroughly characterized.

A recent study demonstrated that T cell recognition of Dsg3 was restricted by the PV-associated HLA class II alleles DRB1*0402 and DQB1*0503. We found that only antigen-presenting cells (APCs) expressing HLA-DRB1*0402 and HLA-DQB1*0503 were capable of presenting Dsg3 to autoreactive Th1 and Th2 clones, and that their proliferative response was blocked by anti-DR and anti-DQ Abs, respectively (Veldman *et al.*, 2003). Therefore, HLA-DRB1*0402 and HLA-DQB1*0503 appear to be the major HLA class II alleles involved in restriction of the T cell responses against Dsg3 peptides, a finding in line with previous studies that identified DRB1*0402 as a restriction element for the presentation of Dsg3-derived peptides (Lin *et al.*, 1997; Hertl *et al.*, 1998b; Riechers *et al.*, 1999). Noteworthy, selected Dsg3-reactive T cell clones (TCCs) also were restricted by non-PV-associated HLA class II alleles; however, all of these alleles were homologous to DRß1*0402 with regard to peptide-binding motifs (Hertl *et al.*, 1998b; Riechers *et al.*, 1999) and shared a negative charge at the critical peptide-binding site position 70 of the variable ß1-chain. The importance of these distinct peptide-binding motifs of the aforementioned HLA class II alleles is supported by the identification of Dsg3 peptides that carry a positive charge at position 4 that may be critical for binding to the negatively charged P4 pockets of DRB1*0402 (DRß70 and 71) and DQB1*0503 (DQß57) (Wucherpfennig *et al.*, 1995).

Of particular interest was the detection of Dsg3-reactive Th1 cells in healthy individuals (Hertl *et al.*, 1998), specifically those who carried the PV-associated HLA class II alleles DRB1*0402 and HLA-DQB1*0503 (Veldman *et al.*, 2003). In addition, Dsg3-reactive Th1 clones derived from these Dsg3-reactive healthy donors were indeed restricted by HLA-DRB1*0402 and -DQB1*0503, respectively. This finding strongly suggests that T cell recognition of Dsg3 in PV patients and healthy individuals depends on the presentation of Dsg3 peptides by distinct HLA class II alleles independent from the development of PV.

Epitopes of Dsg3 that are recognized by autoreactive CD4$^+$ T cells have been identified utilizing long-term TCCs (reviewed in Riechers *et al.*, 1999). Wucherpfennig *et al.* (1995) proposed several candidate peptides of Dsg3 based on their potential binding motifs with DRß1*0402. In fact, three peptides of the extracellular domain of Dsg3 induced a proliferative *in vitro* response of peripheral lymphocytes from PV patients. Our group identified additional Dsg3 peptides that were homologous to these peptides (Hertl *et al.*, 1998; Riechers *et al.*, 1999; Veldman *et al.*, 2004a). The finding that autoreactive T cells from PV patients and normals recognize identical epitopes of Dsg3 supports the hypothesis that PV is the consequence of a loss of tolerance at the B cell level, rather than at the T cell level (Veldman et al., 2004a).

7. Regulatory T Lymphocytes in Pemphigus

There is now compelling evidence that CD4$^+$ T cells, specialized in suppressing immune responses, play a critical role in immune regulation. Three major populations of T regulatory (Treg) cells have been identified based on their distinct phenotype (CD4$^+$CD25$^+$) or cytokine secretion patterns (Tr1 and Th3 cells). While the CD4$^+$CD25$^+$ subset mediates suppression in a non-Ag-specific manner, the later Tr cell types may act in an Ag-specific way. Tr1 cells can be distinguished from Th3 cells because the former preferentially exert their regulatory effects via production of IL-10 (Roncarolo et al., 2001), while Th3 cells preferentially secrete the immunosuppressive cytokine TGF-β. Tr1 cells exist naturally in the human mucosa and maintain intestinal homeostasis against bacterial pathogens (Khoo et al., 1997) and parasites (Satoguina et al., 2002) via the production of IL-10 and TGF-β. Similarly, MHC-autoreactive Tr1-like TCCs isolated from the peripheral blood of healthy donors suppressed antigen-specific T cell responses by the secreting IL-10 and TGF-β (Kitani et al., 2000).

In a recent study, we assessed whether the presence or absence of Dsg3-specific Tr1 cells in Dsg3-responsive healthy donors and PV patients, respectively, may account for the development of tolerance versus autoimmunity against Dsg3. In fact, Dsg3-reactive IL-10-secreting Tr1 cells were identified in 80% of healthy carriers of PV-associated HLA class II alleles and in only 17% of PV patients whose cells suppressed the proliferative response of Dsg3-reactive Th cells. The Dsg3-specific Tr1 cells secreted IL-10, TGF-β, and IL-5 upon Ag stimulation, proliferated in response to IL-2, but not to Dsg3 or mitogenic stimuli, and inhibited the proliferative response of Dsg3- and TT-responsive Th clones in both Ag-specific (Dsg3) and cell number–dependent manners. Moreover, their inhibitory effect was blocked by Ab against IL-10, TGF-β, and by paraformaldehyde fixation. These observations strongly suggest that (a) Dsg3-responsive Tr1 cells predominate in healthy individuals, (b) their growth requires the presence of IL-2, and (c) they exert Dsg3-dependent inhibitory function by the secretion of IL-10 and TGF-β. These findings suggest that Dsg3-specific Tr may be involved in the maintenance of peripheral tolerance to Dsg3 in healthy individuals and in the restoration of tolerance against Dsg3 in PV patients.

There is evidence that Tr1 cells may indeed act in an Ag-specific manner. In nickel (Ni) allergy, nonallergic subjects carry Ni-specific T cells that fulfill the criteria of Tr1 cells based on their cytokine profile (higher IL-10, IL-5, IFN-γ, and low IL-4 levels) and their ability to suppress the proliferative response of Ni-activated Th1 cells (Cavani et al., 2000) and may thus be critically involved in the downregulation of Ni-specific Th cell responses *in vivo*. IL-10$^+$ Tr cells also were detected in patients allergic to bee venom upon specific immunotherapy with phospholipase A, which suppressed the proliferative response of allergen-specific Th cells (Akdis et al., 1998). Moreover, the expression of IL-10 increased during specific immunotherapy with phospholipase A, suggesting that the protective effect of this regimen was directly correlated to the presence of IL-10$^+$ allergen–specific Tr cells.

8. Passive Animal Models of Pemphigus Vulgaris

The most impressive evidence for the central role of Dsg3 in intraepidermal adhesion was provided by Koch *et al.* (1997). They genetically engineered mice with a targeted disruption of the Dsg3 gene. These mice were normal at birth, but developed a runting phenotype later on. They presented with oral erosions/blisters, leading to the observed weight loss due to the inhibited food uptake, and developed cutaneous blisters only when the skin was traumatized. Noteworthy, the Dsg3$^{-/-}$ mice developed telogen hair loss. This finding provided strong support to the idea that anti-Dsg3 autoAb induce mucosal, but not cutaneous lesions, in PV.

Using the Dsg3$^{-/-}$ mouse model, Mahoney *et al.* (1999) dissected the relationship between the epidermal distribution of Dsg3 and Dsg1, and the pathogenic role of circulating autoAbs targeting these structures. The role of Dsg3 in limiting blister formation in PF was demonstrated by injecting Dsg1-reactive IgG into Dsg3$^{+/+}$ and Dsg3$^{-/-}$ mice. Upon transfer of PF IgG, Dsg3$^{+/+}$ mice developed small cutaneous blisters, while the Dsg3$^{-/-}$ mice developed gross blisters on the skin and mucous membranes strongly expressing Dsg3, but little Dsg1 (Shirakata *et al.*, 1998). These data also account for the observation that PV patients with anti-Dsg3 autoAbs only have exclusively oral lesions. Since blocking of both Dsg1 and Dsg3 is necessary to inhibit desmosomal adhesion in the skin, once anti-Dsg1 Ab are present, skin lesions occur (Ding *et al.*, 1997).

9. Active Animal Model of Pemphigus Vulgaris

Amagai *et al.* (2000a) have taken advantage of the availability of Dsg3$^{-/-}$ mice to establish an active *in vivo* model of PV. Dsg3$^{-/-}$ mice immunized with recombinant mouse Dsg3 produced anti-Dsg3 Abs. Their splenocytes were then transferred into immunodeficient Rag$^{-/-}$ mice that expressed Dsg3. The recipient mice produced anti-Dsg3 autoAbs and developed erosions of the mucous membranes with typical histological findings of PV. In addition, the mice showed telogen hair loss, as seen in Dsg3$^{-/-}$ mice. This first active *in vivo* model of pemphigus will be useful for understanding how autoimmunity develops in PV and for evaluating therapeutic strategies aimed at specifically interfering with the T cell–dependent autoAb production by autoreactive B cells.

10. Conclusions

Pemphigus encompasses a group of life-threatening autoimmune blistering disorders characterized by intraepithelial blister formation. The molecular basis for intraepithelial blister formation is the loss of adhesion between keratinocytes, called acantholysis, caused by autoAbs against intercellular adhesion structures of epidermal keratinocytes. Clinically, PV is characterized by extensive bullae and erosions of the mucous membranes and also of the skin (if anti-Dsg1 autoAbs are present). Evidence for the pathogenicity of these circulating autoAbs is provided by several observations, including the finding that pemphigus-like lesions are induced in neonatal mice by transfer of Dsg3-specific IgG. Involvement of Th cells in the pathogenesis of PV has been suggested by several epidemiological studies showing that HLA-DRB1*0402 is associated with PV in Jewish and HLA-DQB1*0503 in non-Jewish populations. Both Dsg3-reactive Th1 and Th2 cells were identified in PV patients, and appear to recognize epitopes of the ECD of Dsg3 in association with HLA-DRB1*0402 and HLA-DQB1*0503. Noteworthy, autoreactive Th cells recognizing identical epitopes of the Dsg3 ectodomain were also identified in healthy individuals expressing the PV-associated HLA class II alleles. These findings suggest that PV is the consequence of a loss of self-tolerance against Dsg3 at the B cell level and that active immune regulation may be operative in Dsg3-responsive healthy individuals. Since autoreactive Th cells specific for identical Dsg3 epitopes were detected both in patients and in healthy donors, this tolerance loss probably is not the consequence of deletion or immune deviation of autoreactive T cells. In addition, our observations strongly suggest that immunological tolerance against Dsg3 may be mediated by Dsg3-specific type 1 Tr cells that mediate suppression by the release of the immunosuppressive cytokines IL-10 and TGF-β. These findings provide a sound explanation for the existence of B cell tolerance to Dsg3 in healthy individuals carrying autoaggressive T cells reactive to Dsg3 epitopes identical to those recognized by T cells from the PV patients. Thus, Dsg3-responsive Tr1 cells may represent an ideal tool to therapeutically restore Dsg3-specific immune tolerance in PV.

References

Ahmed, A.R., Yunis, E.J., and Khatri, K. (1990). Major histocompatibility complex haplotype studies in Ashkenazi Jewish patients with pemphigus vulgaris. *Proc. Natl. Acad. Sci. USA*, **87**, 7658–7662.

Ahmed, A.R., Wagner, R., and Khatri, K. (1991). Major histocompatibility complex haplotypes and class II genes in non-Jewish patients with pemphigus vulgaris. *Proc. Natl. Acad. Sci. USA*, **88**, 5056–5061.

Akdis, C.A., Blesken, T., Akdis, M., Wuthrich, B., and Blaser, K. (1998). Role of interleukin 10 in specific immunotherapy. *J. Clin. Invest.*, **102**, 98–106.

Amagai, M., Klaus-Kovtun, V., and Stanley, J.R. (1991). Auto-Ab against a novel epithelial cadherin in pemphigus vulgaris, a disease of cell adhesion. *Cell*, **67**, 869–877.

Amagai, M., Karpati, S., Prussick, R., Klaus-Kovtun, V., and Stanley, J.R. (1992). Autoantibodies against the amino-terminal cadherin-like binding domain of pemphigus vulgaris antigen are pathogenic. *J. Clin. Invest.*, **90**, 919–926.

Amagai, M., Hashimoto, T., Shimizu, N., and Nishikawa, T. (1994). Absorption of pathogenic autoantibodies by the extracellular domain of pemphigus vulgaris antigen (Dsg 3) produced by baculovirus. *J. Clin. Invest.*, **94**, 59–67.

Amagai, M., Koch, P.J., Nishikawa, T., and Stanley, J.R. (1996). Pemphigus vulgaris antigen (desmoglein 3) is localized in the lower epidermis, the site of blister formation in patients. *J. Invest. Dermatol.*, **106**, 351–355.

Amagai, M., Tsunoda, K., Suzuki, H., Nishifuji, K., Koyasu, S., and Nishikawa, T. (2000a). Use of autoantigen-knockout mice in developing an active autoimmune disease model for pemphigus. *J. Clin. Invest.*, **105**, 625–631.

Amagai, M., Matsuyoshi, N., Wang, Z.H., Andl, C., and Stanley, J.R. (2000b). Toxin in bullous impetigo and staphylococcal scalded skin syndrome targets desmoglein 1. *Nat. Med.*, **6**, 1275–1277.

Anhalt, G.J., Labib, K.S., Vorhees, J.S., Beals, T.F., and Diaz, L.A. (1982). Induction of pemphigus in neonatal mice by passive transfer of IgG from patients with the disease. *N. Engl. J. Med.*, **306**, 1189–1192.

Bedane, C., Prost, C., Thomine, E., Intrator, L., Joly, P., Caux, F., Blecker, M., Bernard, P., Leboutet, M.J., Tron, F., Lauret, P., Bonnetblanc, J.M., and Dubertret, L. (1996). Binding of autoantibodies is not restricted to desmosomes in pemphigus vulgaris: Comparison of 14 cases of pemphigus vulgaris and 10 cases of pemphigus foliaceus studied by western immunoblot and immunoelectron microscopy. *Arch. Dermatol. Res.*, **288**, 343–352.

Bhol, K., Ahmed, A.R., Aoki, V., Mohimen, A., Nagarwalla, N., and Natarajan, K. (1995). Correlation of peptide specificity and IgG subclass with pathogenic and nonpathogenic autoantibodies in pemphigus vulgaris: A model for autoimmunity. *Proc. Natl. Acad. Sci. USA*, **92**, 5239–5243.

Brandsen, R., Frusic-Zlotkin, M., Lyubimov, H., Yunes, F., Michel, B., Tamir, A., Milner, Y., and Brenner, S. (1997). Circulating pemphigus IgG in families of patients with pemphigus: Comparison of indirect immunofluorescence, direct immunofluorescence, and immunoblotting. *J. Am. Acad. Dermatol.*, **36**, 44–52.

Cavani, A., Nasorri, F., Prezzi, C., Sebastiani, S., Albanesi, C., and Girolomoni, G. (2000). Human CD4+ T lymphocytes with remarkable regulatory functions on dendritic cells and nickel-specific Th1 immune responses. *J. Invest. Dermatol.*, **114**, 295–302.

Ding, X., Fairley, J.A., Diaz, L.A., Lopez-Swiderski, A., Mascaro, J.M., Jr., and Aoki, V. (1997). Mucosal and mucocutaneous (generalized) pemphigus vulgaris show distinct autoantibody profiles. *J. Invest. Dermatol.*, **109**, 592–596.

Emery, D.J., Diaz, L.A., Fairly, J.A., Lopez, A., Taylor, A.F., and Giudice, G.J. (1995). Pemphigus foliaceus and pemphigus vulgaris autoantibodies react with the extracellular domain of desmoglein 1. *J. Invest. Dermatol.*, **104**, 323–328.

Eming, R., Büdinger, L., Riechers, R., Christensen, O., Bohlen, H., Kalish, R., and Hertl, M. (2000). Frequency analysis of autoreactive T helper 1 and 2 cells in bullous pemphigoid and pemphigus vulgaris by ELISPOT analysis. *Br. J. Dermatol.*, **143**, 1279–1282.

Esaki, C., Kitajima, Y., Osada, K., Yamada, T., and Seishima, M. (1995). Pharmacologic evidence for involvement of phospholipase C in pemphigus IgG-induced inositol 1,4,5-trisphosphate generation, intracellular calcium increase, and plasminogen activator secretion in DJM-1 cells, a squamous cell carcinoma line. *J. Invest. Dermatol.*, **105**, 329–333.

Hashimoto, T., Shafran, M., Webber, P.A., Lazarus, G.S., and Singer, K.H. (1983). Anti-cell surface pemphigus autoantibody stimulates plasminogen activator activity of human epidermal cells. A mechanism for the loss of epidermal cohesion and blister formation. *J. Exp. Med.*, **157**, 259–272.

Hertl, M., Amagai, M., Ishii, K., Stanley, J., Sundaram, H., and Katz, S.I. (1998a). Recognition of desmoglein 3 by autoreactive T cells in pemphigus vulgaris patients and normals. *J. Invest. Dermatol.*, **110**, 62–66.

Hertl, M. (1998b). Molecular characterization of autoantibodies in pemphigus. In Conrad, K., Humbel, R.L., Meurer, M., Shoenfeld, Y., and Tan, E.M. (eds) *Pathogenic and Diagnostic Relevance of Autoantibodies*. Pabst Science Publishers, Berlin.

Hertl, M. and Riechers, R. (1999). Analysis of the T cells that are potentially involved in autoantibody production in pemphigus vulgaris. *J. Dermatol.*, **26**, 748–752.

Hertl, M. (2000). Humoral and cellular autoimmunity in autoimmune bullous skin disorders. *Int. Arch. Allergy Immunol.*, **122**, 91–100.

Huilgol, S.C., and Black, M.M. (1995). Management of the immunobullous disorders. II. Pemphigus. *Clin. Exp. Dermatol.*, **20**, 283–293.

Khoo, U.Y., Proctor, I.E., and Macpherson, A.J. (1997). CD4+ T cell down-regulation in human intestinal mucosa: Evidence for intestinal tolerance to luminal bacterial antigens. *J. Immunol.*, **158**, 3626–3634.

Kitani, A., Chua, K., Nakamura, K., and Strober, W. (2000). Activated self-MHC-reactive T cells have the cytokine phenotype of Th3/T regulatory cell 1 T cells. *J. Immunol.*, **165**, 691–702.

Kljuic, A., Bazzi, H., Sundberg, J.P., Martinez-Mir, A., O'Shaughnessy, R., Mahoney, M.G., Levy, M., Montagutelli, X., Ahmad, W., Aita, V.M., Gordon, D., Uitto, J., Whiting, D., Ott, J., Fischer, S., Gilliam, T.C., Jahoda, C.A., Morris, R.J., Panteleyev, A.A., Nguyen, V.T., and Christiano, A.M. (2003). Desmoglein 4 in hair follicle differentiation and epidermal adhesion: Evidence from inherited hypotrichosis and acquired pemphigus vulgaris. *Cell*, **113**, 249–260.

Koch, P.J., Mahoney, M.G., Ishikawa, H., Pulkkinen, L., Uitto, J., Shultz, L., Murphy, G.F., Whitacker-Menezes, D., and Stanley, J.R. (1997). Targeted disruption of the pemphigus vulgaris antigen (desmoglein 3) gene in mice causes loss of keratinocyte cell adhesion with a phenotype similar to pemphigus vulgaris. *J. Cell Biol.*, **137**, 1091–1102.

Kowalczyk, A.P., Green, K.J., Stanley, J.R., Hashimoto, T., Borgwardt, J.E., and Anderson, J.E. (1995). Pemphigus sera recognize conformationally sensitive epitopes in the amino-terminal region of desmoglein-1. *J. Invest. Dermatol.*, **105**, 147–152.

Kricheli, D., David, M., and Frusic-Zlotkin, M. (2000). The distribution of pemphigus vulgaris IgG subclasses and reactivity with desmoglein 3 and 1 in pemphigus patients and first-degree relatives. *Br. J. Dermatol.*, **143**, 337–342.

Lever, W.F. (1953). Pemphigus. *Medicine*, **32**, 1–123.

Lin, M.S., Swartz, S.L., Lopez, A., Ding, X., Fernandez-Vina, M.A., Stastny, P., Fairley, J.A., and Diaz, L. (1997). Development and characterization of desmoglein 3-specific T cells from patients with pemphigus vulgaris. *J. Clin. Invest.*, **99**, 31–40.

Mahoney, M.G., Wang, Z., Rothenberger, K.L., Koch, P.J., Amagai, M., and Stanley, J.R. (1999). Explanation for localization of blisters in pemphigus patients. *J. Clin. Invest.*, **103**, 461–468.

Nguyen, V.T., Ndoye, A., and Grando, S.A. (2000a). Pemphigus vulgaris antibody identifies pemphaxin. A novel keratinocyte annexin-like molecule binding acetylcholine. *J. Biol. Chem.*, **275**, 29466–29476.

Nguyen, V.T., Ndoye, A., and Grando, S.A. (2000b). Novel human alpha? acetylcholine receptor regulating keratinocyte adhesion is targeted by pemphigus vulgaris autoimmunity. *Am. J. Pathol.*, **157**, 1377–1391.

Nishifuji, K., Amagai, M., Kuwana, M., Iwasaki, T., and Nishikawa, T. (2000). Detection of antigen-specific B cells in patients with pemphigus vulgaris by enzyme-linked immunospot assay: Requirement of T cell collaboration for autoantibody production. *J. Invest. Dermatol.*, **114**, 88–94.

Nishifuji, K., Nagasaka, T., Ota, T., Whittock, N.V., and Amagai, M. (2004). Is desmoglein 4 involved in blister formation in pemphigus or impetigo? *J. Invest. Dermatol.*, **122**, A33.

Nousari, H.C. and Anhalt, G.J. (1999). Pemphigus and bullous pemphigoid. *Lancet*, **354**, 667–672.

Ochsendorf, F.R., Schofer, H., and Milbradt, R. (1987). Pemphigus erythematosus – detection of anti-DNA antibodies. *Hautarzt*, **38**, 400–403.

Ohata, Y., Amagai, M., Ishii, K., and Hashimoto, T. (2001). Immunoreactivity against intracellular domains of desmogleins in pemphigus. *J. Dermatol. Sci.*, **25**, 64–71.

Riechers, R., Grötzinger, A., and Hertl, M. (1999). HLA class II restriction of autoreactive T cell responses in pemphigus vulgaris and potential application for a specific immunotherapy. *Autoimmunity*, **30**, 183–196.

Roncarolo, M.G., Bacchetta, R., Bordignon, C., Narula, S., and Levings, M.K. (2001). Type 1 T regulatory cells. *Immunol. Rev.*, **182**, 68–79.

Satoguina, J., Mempel, M., Larbi, J., Badusche, M., Loliger, C., Adjei, O., Gachelin, G., Fleischer, B., and Hoerauf, A. (2002). Antigen-specific T regulatory-1 cells are associated with immunosuppression in a chronic helminth infection (onchocerciasis). *Microbes Infect.*, **4**, 1291–1300.

Seishima, M., Kitajima, Y., Hashimoto, T., Mori, S., Osada, K., and Esaki, C. (1995). Pemphigus IgG, but not bullous pemphigoid IgG, causes a transient increase in intracellular calcium and inositol 1,4,5-triphosphate in DJM-1 cells, a squamous cell carcinoma line. *J. Invest. Dermatol.*, **104**, 33–37.

Sekiguchi, M., Futei, Y., Fujii, Y., Iwasaki, T., Nishikawa, T., and Amagai, M. (2001). Dominant autoimmune epitopes recognized by pemphigus antibodies map to the N-terminal adhesive region of desmogleins. *J. Immunol.*, **167**, 5439–5448.

Shimizu, H., Nishikawa, T., Hashimoto, T., Kikuchi, A., Ishiko, A., and Masunaga, T. (1995). Pemphigus vulgaris and pemphigus foliaceus sera show an inversely graded binding pattern to extracellular regions of desmosomes in different layers of human epidermis. *J. Invest. Dermatol.*, **105**, 153–159.

Shirakata, Y., Hashimoto, K., Nishikawa, T., Hanakawa, Y., and Amagai, M. (1998). Lack of mucosal involvement in pemphigus foliaceus may be due to low expression of desmoglein 1. *J. Invest. Dermatol.*, **110**, 76–78.

Sinha, A.A., Brautbar, C., Szafer, F., Friedmann, A., Tzfoni, E., Todd, J.A., Steinman, L., and McDevitt, H.O. (1988). A newly characterized HLA-DQß allele associated with pemphigus vulgaris. *Science*, **239**, 1026–1029.

Späth, S., Riechers, R., Borradori, L., Zillikens, D., Büdinger, L., and Hertl, M. (2001). Detection of autoantibodies of various subclasses (IgG1, IgG4, IgA, IgE) against desmoglein 3 in patients with acute onset and chronic pemphigus vulgaris. *Br. J. Dermatol.*, **144**, 1183–1188.

Tremeau-Martinage, C., Bazex, J., and Oksman, F. (1995). Immunoglobulin G subclass distribution of anti-intercellular substance antibodies in pemphigus. *Annu. Dermatol. Venereol.*, **122**, 409–411.

Tsunoda, K., Ota, T., Suzuki, H., Ohyama, M., Nagai, T., Nishikawa, T., Amagai, M., and Koyasu, S. (2002). Pathogenic autoantibody production requires loss of tolerance against desmoglein 3 in both T and B cells in experimental pemphigus vulgaris. *Eur. J. Immunol.*, **32**, 627–633.

Veldman, C. and Hertl, M. (2001). Pemphigus—Paradigm of autoantibody-mediated autoimmunity. *Skin Pharmacol. Appl. Skin Physiol.*, **14**, 408–418.

Veldman, C., Stauber, A., Wassmuth, R., Uter, W., Schuler, G., and Hertl, M. (2003). Dichotomy of autoreactive Th1 and Th2 cell responses to desmoglein 3 in patients with pemphigus vulgaris (PV) and healthy carriers of PV-associated HLA class II alleles. *J. Immunol.*, **170**, 635–642.

Veldman, C.M., Gebhard, K.L., Uter, W., Wassmuth, R., Grotzinger, J., Schultz, E., and Hertl, M. (2004a). T cell recognition of desmoglein 3 peptides in patients with pemphigus vulgaris and healthy individuals. *J. Immunol.*, **172**, 3883–3892.

Veldman, C., Hohne, A., Dieckmann, D., Schuler, G., and Hertl, M. (2004b). Type I regulatory T cells specific for desmoglein 3 are more frequently detected in healthy individuals than in patients with pemphigus vulgaris. *J. Immunol.*, **172**, 6468–6475.

Whittock, N.V. and Bower, C. (2003). Genetic evidence for a novel human desmosomal cadherin, desmoglein 4. *J. Invest. Dermatol.*, **120**, 523–530.

Wucherpfennig, K.W., Yu, W.B., Bhol, K., Monos, D.S., Argyris, E., Karr, R.W., Ahmed, A.R., and Strominger, J.L. (1995). Structural basis for major histocompatibility complex (MHC)-linked susceptibility to autoimmunity: Charged residues of a single MHC binding pocket confer selective presentation of self-peptides in pemphigus vulgaris. *Proc. Natl. Acad. Sci. USA*, **92**, 11935–11939.

11

The Molecular Basis of Celiac Disease

Liesbeth Spaenij-Dekking and Frits Koning

1. Introduction

Celiac disease (CD) is an intestinal disorder caused by an inflammatory T cell response to gluten peptides bound to histocompatibility leucocyte antigen (HLA)-DQ2 or HLA-DQ8. Both these HLA molecules have a preference for peptides that contain negatively charged amino acids. Gluten is a heterogeneous mixture of proteins comprising alpha-, gamma-, and omega-gliadins and low molecular weight (LWM)- and high molecular weight (HMW)-glutenins. None of these molecules contains significant amounts of negatively charged amino acids, raising the question of how gluten-derived peptides can bind to HLA-DQ2 and/or -DQ8. This paradox was solved when it became clear that, in the small intestine, glutamine residues in gluten can be modified by the enzyme tissue transglutaminase (tTG). The conversion of glutamine to glutamic acid introduces negative charges in gluten peptides at sites that are critical for binding to HLA-DQ2 or -DQ8. It is now clear that gluten contains a multitude of peptides that can be modified by tTG, bind to HLA-DQ2/8, and induce T cell responses. Similar peptides are found in barley, rye, and oats, raising the question of why so few HLA-DQ2-and/or HLA-DQ8-positive individuals develop CD. Yet, CD is presently the best-understood HLA-associated disease. The observation that posttranslational modification of gluten is critical for the generation of a repertoire of T cell stimulatory peptides may be relevant for other HLA-associated diseases as well. Moreover, unraveling the mechanisms that underlie the development of CD may have important implications for our understanding of these other diseases. In this chapter, we describe a number of key observations that have been made in recent years and discuss the potential for the development of improved diagnosis, safer foods, and novel therapies for patients.

Liesbeth Spaenij-Dekking and Frits Koning • Department of Immunohematology and Blood Transfusion, E3-Q, Leiden University Medical Center, PO Box 9600, 2300 RC Leiden, The Netherlands.

Molecular Autoimmunity: In commemoration of the 100th anniversary of the first description of human autoimmune disease, edited by Moncef Zouali. Springer Science+Business Media, Inc., New York, 2005.

2. T Cell Recognition of Gluten Peptides

It is well established that CD is almost exclusively associated with HLA-DQ2 and -DQ8. While approximately 95% of the patients are HLA-DQ2-positive, the remainder are usually HLA-DQ8-positive (Marsh, 1992). Moreover, Lundin et al. (1993) found that HLA-DQ2- and HLA-DQ8-restricted, gluten-specific T cells are present in the small intestine of CD patients. These findings indicated that these DQ molecules have unique peptide-binding properties that set them apart from other HLA class II molecules. Several studies have investigated the peptide-binding properties of HLA-DQ2 (Table 11.1) (Kwok et al., 1996; van de Wal et al., 1996; Vartdal et al., 1996; Godkin et al., 1997). HLA-DQ2 was found to selectively bind peptides with large hydrophobic residues at positions p1 and p9, a negatively charged amino acid at positions p4 and/or p7, and a proline residue or a negatively charged amino acid at position p6 (Table 11.1). Similarly, HLA-DQ8 was found to have a preference for anchor residues with a negative charge. These results were striking because negatively charged amino acids are very rare in gluten molecules.

Gluten is widely used in the food industry. It is an important source of nitrogen and amino acids; it is cheap and possesses unique elastic properties that facilitate the preparation of dough with good baking properties. Gluten consists of a mixture of proteins, including the alpha-, gamma-, and omega-gliadins and LMW- and HMW-glutenins. Since all these gluten gene families contain many alleles, up to 100 different gluten molecules can be found in a single wheat variety. Gluten molecules are very rich in glutamine (~30%), proline (~20%), and the bulky hydrophobic amino acids leucine, phenylalanine, and tyrosine, a composition that is tightly linked to both the unique baking properties of gluten and its harmful properties (see below). Similar molecules are found in barley, rye, and to a lesser extent, in oats.

This extensive heterogeneity and complexity of gluten has complicated approaches taken to identify T cell stimulatory gluten peptides. In one approach, T cell stimulatory gluten peptides were purified from a crude mixture of pepsin/(chymo)trypsin-treated gluten by repeated rpHPLC separations (Sjostrom et al., 1998; van de Wal et al., 1998a; Vader et al., 2002a). Gluten-specific T cells were used to identify HPLC fractions that contained T cell stimulatory capacity, and the T cell stimulatory peptides were finally characterized by tandem mass

Table 11.1. Peptide-Binding Motifs for HLA-DQ2

HLA-haplotype	Position in the binding groove								
	1	2	3	4	5	6	7	8	9
DR3DQ2	FWYILV[a]	—	—	DEVLI	—	PAE	DE	—	FYLWI
DR7DQ2	FWYILV	—	no P	DEVLI	—	PAE	DE	—	FYLWI

[a] Represented are the amino acids that are favored at the positions indicated. Amino acids are indicated by the single letter code.

The Molecular Basis of Celiac Disease

spectrometry. In another approach, recombinant gliadin molecules were generated that were found to stimulate gluten-specific T cells (Arentz-Hansen et al., 2000). As the amino acid sequence of these recombinant gliadins was known, this allowed the rapid identification of the T cell stimulatory sequences. In a third approach Vader et al. (2002a) screened gluten-specific T cells against a panel of 250 randomly chosen synthetic gluten peptides and found reactivity against three of them. Finally, algorithms have been designed to identify sequences in gluten molecules with T cell stimulatory properties, as discussed below (Vader et al., 2002b). Together, these studies have revealed that distinct T cell stimulatory sequences can be found in many gluten molecules (Table 11.2).

Early in these studies it became clear that gluten modification by the enzyme tTG plays an important role in the generation of T cell stimulatory gluten peptides. It was already well established that CD patients make autoantibodies that can be used as disease markers. In 1997 it was found that these autoantibodies are directed to the enzyme tTG (Dieterich et al., 1997), a ubiquitous enzyme found in all organs and known to be released upon cellular damage and to cross-link proteins in order to control tissue damage. This cross-linking occurs by forming a covalent bond between a glutamine in one protein and the amino group of a lysine in another. In the absence of a lysine residue, however, the enzyme reactivity can result in deamidation, the conversion of a glutamine residue into the negatively charged glutamic acid. Given the preference of HLA-DQ2 and HLA-DQ8 for negative charges and the abundance of glutamine residues in gluten it was logical to determine if tissue tranglutaminase would be involved in the generation of T cell stimulatory gluten peptides. We tested this possibility with an HLA-DQ8-restricted gliadin peptide recognized by DQ8-restricted T cells (van de Wal et al., 1998a, Table 11.3). In the nine amino acid core of this gliadin peptide, no negatively charged amino acids were present. However, the peptide contained four glutamine residues, including two at positions 1 and 9, sites where

Table 11.2. Characteristics of a Selected Group of T Cell Stimulatory Gluten Peptides

Gluten peptide	Sequence	Protein source	Binding to[a]:		
			DR3 DQ2	DR7 DQ2	DQ8
Glia-α2	PQPQLPYPQ	α-Gliadin	+	+/–	–
Glia-α9	PFPQPQLPY	α-Gliadin	+	–	–
Glia-α20	FRPQQPYPQ	α-Gliadin	+	–	–
Glu-5	QLPQQPQQF	Unknown	+/–	–	–
Glia-γ2	FPQQPQQPF	γ-Gliadin	+	–	–
Glia-γ1	PQQSFPQQQ	γ-Gliadin	+	+	–
Glia-γ30	IIQPQQPAQ	γ-Gliadin	+	+	–
Glt-156	FSQQQQSPF	LMW-glutenin	+	+	–
Glia-α	QGSFQPSQQ	α-Gliadin	–	–	+
Glt	QGYYPTSPQ	HMW-glutenin	–	–	+

[a] + = high binding affinity; +/– = intermediate binding affinity; – = no binding affinity.

Table 11.3. T Cell Stimulatory Gluten Peptides and Homologs in the Hordeins of Barley, the Secalins of Rye, and the Avenins of Oats

Gluten peptides[a]	Homologous peptides	Sequence	T cell stimulatory capacity
Glia-α2		PQPQLPYPQ	+
	Hor-α2	PQPQQPFPQ	+
	Sec-α2	PQPQQPFPQ	+
Glia-α9		PFPQPQLPY	+
	Hor-α9	PFPQPQQPF	+
	Sec-α9	PFPQPQQPF	+
	Av-α9[A]	PYPEQQEPF	+
	Av-α9[B]	PYPEQQQPF	+
Glia-α20		FRPQQPYPQ	+
	Hor-α20	FPPQQPFPQ	−
Glia-γ1		PQQSFPQQQ	+
	Hor-γ1	PQQAFPQQP	−
	Sec-γ1	PQQSFPQQP	+
Glia-γ2		FPQQPQQPF	+
	Av-γ2[A]	FVQQQQQPF	−
	Av-γ2[B]	FVQQQQPFV	+

[a] Glia = gliadin-derived; Hor = hordein-derived; Sec = secalin-derived; Av = avenin-derived.

HLA-DQ8 prefers negatively charged residues. Strikingly, tTG was found to selectively modify this peptide through conversion of the glutamine residues at the p1 and p9 positions into glutamic acid (van de Wal *et al.*, 1998b). As the result of this modification, a 100-fold less of the peptide was required for optimal T cell stimulation. Similarly, others identified an HLA-DQ2-restricted gliadin peptide and found that this peptide was only recognized after deamidation of particular Q residues, either as a result of chemical modification or treatment with tTG (Molberg *et al.*, 1998; Sjostrom *et al.*, 1998). In subsequent studies, a large number of gluten peptides have been identified that can stimulate T cells of CD patients (Arentz-Hansen *et al.*, 2000, 2002; van de Wal *et al.*, 2000; Vader *et al.*, 2002a,b). It was also observed that some of those peptides cluster in regions that are resistant to degradation by enzymes in the gastrointestinal tract, resulting in highly immunogenic multivalent gluten fragments (Shan *et al.*, 2002). Importantly, it also became clear that T cell responses to more than one gluten peptide are found in all patients. All of them give rise to the secretion of inflammatory cytokines, and are likely involved in the disease process (Vader *et al.*, 2002a).

Most of the peptides identified to date require modification by tTG before they can bind to HLA-DQ2 and stimulate T cells. Thus, modification of gluten by tTG plays an important role in the generation of an extensive repertoire of T cell stimulatory gluten peptides. It should be noted, however, that several gluten peptides do not require tTG modification for T cell recognition (van de Wal., 2000; Vader *et al.*, 2002a), indicating that native gluten can also induce T cell reactivity.

It should also be pointed out that gluten might also act through the innate arm of the immune system. Recent studies indicate that activation of the innate immune system by gliadin strongly enhances the gluten-specific T cell response (Maiuri *et al.*, 2003). The actual mechanism through which gluten exerts this effect, however, remains to be determined.

3. The Specificity of tTG Is Linked to Gluten Toxicity

A striking feature of gluten modification by tTG is that only particular glutamine residues are modified. In the HLA-DQ8-restricted gliadin peptide, for example, four glutamine residues are present, but tTG only modifies those at the p1 and p9 positions, but not those at the p5 and p8 positions (van de Wal *et al.*, 1998b). Replacement of these latter glutamine residues by glutamic acid, however, completely abrogates the T cell stimulatory properties of this peptide (van de Wal *et al.*, 1998b). Similarly, in other gluten peptides only those glutamine residues are modified that are important for the generation of potent T cell stimulatory peptides (Arentz-Hansen *et al.*, 2000; Vader *et al.*, 2002a,b). The selective modification of gluten peptides by tTG is thus tightly linked to gluten toxicity.

An analysis of the modification of large series of gluten peptides by tTG indicated that the spacing between glutamine and proline residues in gluten peptides has a major influence on the modification pattern (Vader *et al.*, 2002b). In the sequences QP and QXXP, Q is not a target for tTG. In contrast, in the sequences QXP, QXXF, and QXPF, Q is deamidated by tTG (Vader *et al.*, 2002b). These observations are highly significant since glutamine and proline are the two most abundant amino acids in gluten and the above mentioned sequences are very frequently found in gluten molecules. Consequently, this observation was used for the design of an algorithm that predicts peptides that are recognized by gluten-specific T cells of patients (Vader *et al.*, 2002b). In this algorithm, two factors were taken into consideration: the specificity of tTG and the previously defined HLA-DQ2-specific peptide-binding motif that favors peptides with a negative charge at positions 4 and 7 in the bound peptide. As in the sequences QXP and QXPY the Q residues are the target of tTG, an algorithm incorporating these motifs was constructed (XXXQXPQXPY) and used to search the gluten database in which it selected 14 peptides. Seven of these peptides were indeed found to have potent T cell stimulatory capacity. As the gluten database contains thousands of potential epitopes, this algorithm displayed a very high predictive value.

4. Additional T Cell Stimulatory Peptides in Barley, Rye, and Oats

To date, T cell stimulatory peptides have been identified in the alpha- and gamma-gliadins as well as in the LMW- and HMW-glutenins (Molberg *et al.*, 1998, 2003; van de Wal *et al.*, 1998a,b; Arentz-Hansen *et al.*, 2000, 2002; Vader *et al.*, 2002a,b, 2003a). With the exception of the omega-gliadins, all types of gluten molecules thus appear harmful for consumption by CD patients. Other cereals can also be harmful for CD patients, in particular barley and rye, while oats is often considered safe. We have therefore investigated the presence of potential T cell stimulatory peptides in these cereals. Database searches with the predictive algorithm XXXQXPQXPY readily identified peptides in the hordeins of barley and in the secalins of rye, including the T cell stimulatory peptide QQPFQQPQQPFPQ (underlined residues are modified by tTG) that is also present in gluten (Vader et al., 2002b). In contrast, this algorithm failed to score hits in the avenins of oats (Vader et al., 2002b). Subsequently, we performed an extensive search in hordein, secalin, and avenin databases to identify additional gluten homologs. This resulted in the identification of alpha- and gamma-gliadin homologs in hordeins, secalins, *and* avenins (Table 11.3; Vader et al., 2003a). Upon synthesis, several of these peptides, although not identical to their counterparts in gluten, did stimulate polyclonal and monoclonal T cells from CD patients (Vader et al., 2003a). Reactivity was found to hordein-, secalin- and avenin-derived peptides (Table 11.3; Vader et al., 2003a). These results indicate that gluten-specific T cells can also be stimulated by peptides found in other cereals. Consequently, the toxicity of these cereals can, to a large extent, be explained by cross-reactivity toward homologous peptides found in the gluten-like molecules in cereals (Vader et al., 2003a). Finally, although some T cell stimulatory peptides evidently are present in the avenins of oats, many more of such peptides are found in the gluten and gluten-like molecules in wheat, barley, and rye. The relative safety of oats for CD patients may thus directly correlate with the relatively low abundance of T cell stimulatory peptides in oats. Nevertheless, patients that are intolerant to oats have recently been identified, indicating that oats is harmful for a small subset of patients (Lundin et al., 2003).

5. The HLA Gene Dose Effect Is Linked to the Level of Gluten Presentation

Two types of HLA-DQ2 molecules are known. While the DQ2 molecule associated with the DR3 haplotype predisposes to CD, that associated with DR7 does not (Sollid et al., 1989). Yet, both molecules have almost identical peptide-binding properties and would thus be expected to present gluten peptides to T cells (van de Wal et al., 1997). An extensive analysis of the peptide-binding properties of DR7DQ2, however, revealed that this molecule can only bind and present a small subset of gluten peptides (Vader et al., 2003b). This is due to a subtle difference in the binding groove of DR7DQ2 that prohibits the presence of a proline at position 3 in the bound peptides (Table 11.1; van de Wal et al., 1997). In contrast, DR3DQ2 does tolerate a proline at position 3, and since this residue is invariably found in T cell stimulatory gluten peptides, DR3DQ2 can present a far larger repertoire of such peptides compared to DR7DQ2 (Table 11.1; Vader et al., 2003b). This result implies the existence of a threshold: a low level of gluten presentation does not lead to disease development while the presentation of a greater number of gluten peptides does.

This result converges with the previous observation of the existence of a strong HLA gene dose effect (Mearin et al., 1983). Individuals with a double dose of HLA-DQ2 have an approximately 5-fold greater chance of developing CD compared to individuals with a single dose. This gene dose effect is also reflected in the magnitude of the gluten-specific T cells response, further indicating that the level of gluten presentation plays a decisive role in disease development (Vader et al., 2003b).

6. Generation of Safer Foods for Patients

The identification of gluten peptides with T cell stimulatory properties allows the development of tests that can screen food for the presence of such peptides. For this purpose we have generated monoclonal antibodies to T cell stimulatory peptides from alpha- and gamma-gliadins (Spaenij-Dekking et al., 2004). Moreover, we are in the process of generating monoclonal antibodies to T cell stimulatory peptides from LMW- and HMW-glutenins. The alpha- and gamma-gliadin-directed antibodies are highly specific for sequences in gluten molecules. A competition assay has been developed with these antibodies that can simultaneously measure the presence of multiple gluten components (Spaenij-Dekking et al., 2004). As this assay is the first to detect T cell stimulatory peptides and as it can measure both intact gluten proteins and fragments thereof, it offers significant advantages over assays that merely measure gluten content.

We have also tested the possibility to detoxify gluten by introducing amino acid substitutions at positions critical for HLA-DQ2 binding or T cell recognition. First, we analyzed natural variants of a LMW-glutenin peptide. While some of these variants possess T cell stimulatory properties, others do not. As the only difference between these peptides was a proline to leucine substitution at position 8 of the peptide this indicated that this substitution was sufficient to abrogate the T cell stimulatory properties of the glutenin peptide (Vader et al., 2003a). Similarly, a proline to glutamine substitution in an alpha-gliadin peptide had a severe impact on the T cell stimulatory properties (Vader et al., 2003a). A subsequent analysis of codon usage in gluten revealed that single base-pair substitutions in gluten genes would suffice to introduce the proline to leucine and proline to glutamine substitutions in gluten proteins (Vader et al., 2003a). These results indicate that genetic modification of gluten may aid in the development of safer foods for CD patients.

7. A Hypothesis for Disease Development

In an oversimplified view CD results from interactions between a large number of potentially harmful peptides in gluten with HLA-DQ molecules. The more HLA-DQ molecules are available, the more gluten peptides can be bound, the larger the chance that a gluten-specific T cell response is initiated. After initiation, HLA-DQ expression is upregulated, an inflammatory response and tissue damage occurs, leading to the release and activation of tTG, which, in turn,

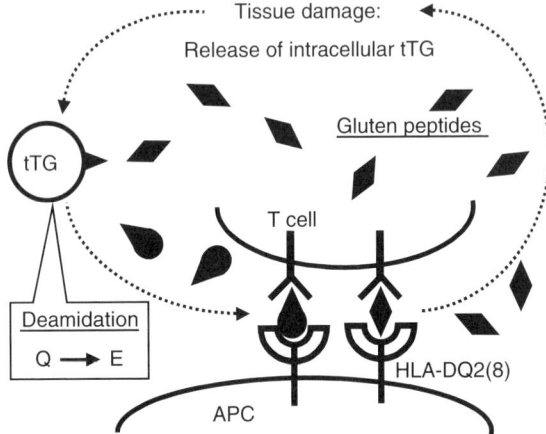

Figure 11.1. Release of intracellular tissue transglutaminase results in an amplification of the gluten-specific T cell response through the generation of gluten peptides that have higher HLA-DQ binding affinity. APC = antigen-presenting cell; tTG = tissue transglutaminase; HLA-DQ2(8) = human leucocyte antigen-DQ2 or -DQ8.

enhances gluten modification. Subsequently, increased formation of HLA-DQ2 gluten complexes will trigger further T cell responses, more inflammation, more tissue damage, setting in motion a vicious circle that can no longer be stopped, except by gluten withdrawal (Figure 11.1).

8. Future Research and Perspectives

We now have detailed insight into the nature of the gluten-specific T cell response in overt disease. It is unclear, however, what triggers the disease. Nor is it known why the large majority of HLA-DQ2-and/or HLA-DQ8-positive individuals do not develop the disease in spite of the fact that they continuously generate immunogenic HLA-gluten complexes in their intestine. It seems likely that this will occasionally trigger T cell responses. Apparently, whereas most individuals can control these responses, other progress to disease development. There is now strong evidence that, in addition to HLA-DQ, other genetic factors also play a role. Identification of those genes and unraveling of their mode of action represent important research goals for the coming years. Additionally, the influence of environmental factors cannot be ruled out. In this respect, it is important to point out that inflammation in the small intestine will lead to an upregulation of HLA-DQ molecules and this may lower the threshold for the development of a gluten-specific response.

As pointed out above, the current knowledge may lead to the development of safer food products for CD patients. Additionally, efforts aiming to develop drugs to prevent peptide binding to HLA-DQ and/or to inhibit the activity of tTG may lead to alternative treatment protocols.

Acknowledgments

This work was supported by grants from the EC (BHM4-CT98-3087 and QLK1-2000-00657), the "Stimuleringsfonds Voedingsonderzoek LUMC", the Dutch Organization for Scientific Research (ZonMW grant 912-02-028), and the Celiac Disease Consortium, an Innovative Cluster approved by the Netherlands Genomics Initiative and partially funded by the Dutch Government (BSIK03009).

References

Arentz-Hansen, H., Körner, R., Molberg, Ø., Quarsten, H., Vader, W., Kooy, Y.M.C., Lundin, K.E.A., Koning, F., Roepstorff, P., Sollid, L.M., and McAdam, S. (2000). The intestinal T cell response to (α-gliadin in adult celiac disease is focused on a single deamidated glutamine targeted by tissue transglutaminase. *J. Exp. Med.*, **191**, 603–612.

Arentz-Hansen, H., McAdam, S.N., Molberg, Ø., Fleckenstein, B., Lundin, K.E.A., Jorgensen, T.J., Jung, G., Roepstorff, P., and Sollid, L.M. (2002). Celiac lesion T cells recognize epitopes that cluster in regions of gliadins rich in proline residues. *Gastroenterology*, **123**, 803–809.

Dieterich, W., Ehnis, T., Bauer, M., Donner, P., Volta, U., Riecken, E.O., and Schuppan, D. (1997). Identification of tissue transglutaminase as the autoantigen of celiac disease. *Nat. Med.*, **3**, 797–801.

Godkin, A., Friede, T., Davenport, M., Stevanovic, S., Willis, A., Jewell, D., Hill, A., and Rammensee, H.G. (1997). Use of eluted peptide sequence data to identify the binding characteristics of peptides to the insulin-dependent diabetes susceptibility allele HLA-DQ8 (DQ3.2). *Int. Immunol.*, **9**, 905–911.

Kwok, W.W., Domeier, M.E., Raymond, F.C., Byers, P., and Nepom, G.T. (1996). Allele-specific motifs characterize HLA-DQ interactions with a diabetes-associated peptide derived from glutamic acid decarboxylase. *J. Immunol.*, **156**, 2171–2177.

Lundin, K.E.A., Nilsen, E.M., Scott, H.G., Løberg, E.M., Gjøen, A., Bratlie, J., Skar, V., Mendez, E., Lovik, A., and Kett, K. (2003). Oats induced villous atrophy in coeliac disease. *Gut*, **52**, 1149–1152.

Lundin, K.E., Scott, H., Hansen, T., Paulsen, G., Halstensen, T.S., Fausa, O., Thorsby, E., and Sollid, L.M. (1993). Gliadin-specific, HLA-DQ(α1*0501,β1*0201) restricted T cells isolated from the small intestinal mucosa of celiac disease patients. *J. Exp. Med.*, **178**, 187–196.

Marsh, M.N. (1992). Gluten, major histocompatibility complex, and the small intestine. *Gastroenterology*, **102**, 330–354.

Maiuri, L., Ciacci, C., Ricciardelli, I., Vacca, L., Raia, V., Auricchio, S., Picard, J., Osman, M., Quaratino, S., and Londei, M. (2003). Association between innate response to gliadin and activation of pathogenic T cells in coeliac disease. *Lancet*, **362**, 30–37.

Mearin, M.L., Biemond, I., Pena, A.S., Polanco, I., Vazquez, C., Schreuder, G.T., de Vries, R.R., and van Rood, J.J. (1983). HLA-DR phenotypes in Spanish coeliac children: Their contribution to the understanding of the genetics of the disease. *Gut*, **24**, 532–537.

Molberg, Ø., Flaete, N.S., Jensen, T., Lundin, K.E.A., Arentz-Hansen, H., Anderson, O.D., Kjersti Uhlen, A., and Sollid, L.M. (2003). Intestinal T-cell responses to high-molecular-weight glutenins in celiac disease. *Gastroenterology*, **125**, 337–344.

Molberg, Ø., McAdam, S., Körner, R., Quarsten, H., Kristiansen, C., Madsen, L., Fugger, L., Scott, H., Norén, O., Roepstorff, P., Lundin, K.E.A., Sjöström, H., and Sollid, L.M. (1998). Tissue transglutaminase selectively modifies gliadin peptides that are recognized by gut derived T cells in celiac disease. *Nat. Med.*, **4**, 713–717.

Shan, L., Molberg, Ø., Parrot, I., Hausch, F., Filiz, F., Gray, G.M., Sollid, L.M., and Khosla, C. (2002). Structural basis for gluten intolerance in celiac sprue. *Science*, **297**, 2275–2279.

Sjostrom, H., Lundin, K.E.A., Molberg, Ø., Korner, R., McAdam, S., Anthonsen, D., Quarsten, H., Noren, O., Roepstorff, P., Thorsby, E., and Sollid, L.M. (1998). Identification of a gliadin T cell epitope in coeliac disease: General importance of gliadin deamidation for intestinal T cell recognition. *Scand. J. Immunol.*, **48**, 111–115.

Sollid, L.M., Markussen, G., Ek, J., Gjerde, H., Vartdal, F., and Thorsby, E. (1989). Evidence for a primary association of coeliac disease to a particular HLA-DQ alpha/beta heterodimer. *J. Exp. Med.*, **169**, 345–350.

Spaenij-Dekking, E.H.A., Kooy-Winkelaar, E.M.C., Drijfhout, J.W., and Koning, F. (2004). A novel and sensitive method for the detection of T cell stimulatory epitopes of α/β- and γ-gliadin. *Gut*, **53**, 1267–1273.

Vader, W., de Ru, A., van de Wal, Y., Kooy, Y., Benckhuijsen, W., Mearin, L., Drijfhout, J.W., van Veelen, P., and Koning, F. (2002b). Specificity of tissue transglutaminase explains cereal toxicity in celiac disease. *J. Exp. Med.*, **195**, 643–649.

Vader, W., Kooy, Y., van Veelen, P., de Ru, A., Harris, D., Benckhuijsen, W., Pena, S., Mearin, L., Drijfhout, J.W., and Koning, F. (2002a). The gluten response in children with recent onset celiac disease. A highly diverse response towards multiple gliadin and glutenin derived peptides. *Gastroenterology*, **122**, 1729–1737.

Vader, W., Stepniak, D., Bunnik, E.M., Kooy, Y., de Haan, W., Drijfhout, J.W., van Veelen, P.A., and Koning, F. (2003a). Characterization of cereal toxicity for celiac disease patients based on protein homology in grains. *Gastroenterology*, **125**, 1105–1113.

Vader, W., Stepniak, D., Kooy, Y., Mearin, M.L., Thompson, A., Spaenij, L., and Koning, F. (2003b). The HLA-DQ2 gene dose effect in celiac disease is directly related to the magnitude and breadth of gluten-specific T-cell responses. *Proc. Natl. Acad. Sci. USA*, **100**, 12390–12395.

van de Wal, Y., Kooy, Y.M.C., Drijfhout, J.W., Amons, R., and Koning, F. (1996). Peptide binding characteristics of the coeliac disease-associated DQ(α1*0501,β1*0201) molecule. *Immunogenetics*, **44**, 246–253.

van de Wal, Y., Kooy, Y.M.C., Drijfhout, J.W., Amons, R., Papadopoulos, G.K., and Koning, F. (1997). Unique peptide-binding characteristics of the disease-associated DQ(a1*0501, ß1*0201) versus the non-disease associated DQ(a1*0201, a1*0202) molecule. *Immunogenetics*, **46**, 484–492.

van de Wal, Y., Kooy, Y.M.C., van Veelen, P., August, S.A., Drijfhout, J.W., and Koning, F. (2000). Glutenin is involved in the gluten-driven mucosal T cell response. *Eur. J. Immunol.*, **29**, 3133–3139.

van de Wal, Y., Kooy, Y., van Veelen, P., Pena, S., Mearin, L., Molberg, Ø., Lundin, L., Mutis, T., Benckhuijsen, W., Drijfhout, J.W., and Koning, F. (1998a). Small intestinal cells of celiac disease patients recognize a natural pepsin fragment of gliadin. *Proc. Natl. Acad. Sci. USA*, **95**, 10050–10054.

van de Wal, Y., Kooy, Y.M.C., van Veelen, P., Peña, A.S., Mearin, M.L., Papadopoulos, G.K., and Koning, F. (1998b). Selective deamidation by tissue transglutaminase strongly enhances gliadin-specific T cell reactivity. *J. Immunol.*, **161**, 1585–1588.

Vartdal, F., Johansen, B.H., Friede, T., Thorpe, C.J., Stevanovic, S., Eriksen, J.E., Sletten, K., Thorsby, E., Rammensee, H.G., and Sollid, L.M. (1996). The peptide binding motif of the disease-associated HLA-DQ(a1*0501,ß1*0201) molecule. *Eur. J. Immunol.*, **26**, 2764–2774.

12

Intrathymic Expression of Neuromuscular Acetylcholine Receptors and the Immunopathogenesis of Myasthenia Gravis

Arnold I. Levinson, Yi Zheng, Glen Gaulton, and Decheng Song

1. Introduction

Myasthenia gravis (MG) is a disease characterized by weakness of striated muscles. The weakness is due to impaired neuromuscular transmission resulting from a reduction in the number of receptors for the neurotransmitter, acetylcholine (ACh), at the postsynaptic myoneural junction. This reduction is mediated by the action of anti-acetylcholine receptor (anti-AChR) antibodies (reviewed in Levinson *et al.*, 1987). MG is a prototypic autoimmune disease; the immune effector mechanisms and autoantigenic target have been delineated (Patrick and Lindstrom, 1973; Almon *et al.*, 1974; Toyka *et al.*, 1975; Engel *et al.*, 1977; Drachman *et al.*, 1978; Levinson *et al.*, 1987). However, the events that lead to the abrogation of self-tolerance to the neuromuscular acetylcholine receptors (nAChRs) remain a mystery. The thymus gland has long been considered to hold the key to solving this mystery, although the nature of its involvement remains to be elucidated (Wekerle *et al.*, 1978). Several lines of evidence support this view (reviewed in Levinson and Wheatley, 1995). The studies described herein relate our

---

Arnold I. Levinson and Yi Zheng • Allergy and Immunology Section, University of Pennsylvania School of Medicine, 421Curie Boulevard, Philadelphia, Pennsylvania. **Glen Gaulton** • Department of Laboratory Medicine and Pathology, University of Pennsylvania School of Medicine, 421Curie Boulevard, Philadelphia, Pennsylvania. **Decheng Song** • Allergy and Immunology Section, University of Pennsylvania School of Medicine, 421Curie Boulevard, Philadelphia, Pennsylvania.

Molecular Autoimmunity: In commemoration of the 100th anniversary of the first description of human autoimmune disease, edited by Moncef Zouali. Springer Science+Business Media, Inc., New York, 2005.

efforts to determine how intrathymic expression of the AChRs is involved in the immunopathogenesis of MG. We review our work characterizing the expression of nAChRs in the thymus and advance a new hypothesis that accounts for the intrathymic expression of this autoantigen in disease pathogenesis.

2. Evidence Supporting the Role of the Thymus in MG Pathogenesis

Attention was originally focused on a potential pathogenic role of the thymus following the recognition of striking pathologic changes in this organ at the time of autopsy. A pattern of germinal center (GC) hyperplasia is observed in the thymus of 70% of patients with early onset of MG (before the age of 40 years) (reviewed in Shiono et al., 2003). Another 10% of MG patients' thymi display thymomas, i.e., tumors of epithelial cell origin (reviewed in Marx et al., 2003). The architecture of the hyperplastic thymi is generally preserved with well-demarcated cortical and medullary regions. However, the medulla is crowded by numerous GCs that display the architectural features and cellular constituents of GCs seen in the secondary follicles of peripheral lymph nodes from healthy subjects. The GCs extend into thymic medullary perivascular spaces. By contrast, the thymic architecture is severely altered in patients with thymoma. Normal-appearing thymocytes are admixed with neoplastic epithelial cells with the loss of a distinct corticomedullary demarcation. Thymoma lymphocytes show the immunophenotypic properties of normal immature cortical thymocytes.

Clinical evidence pertaining to the pathogenic role for the thymus in MG is derived from the apparent beneficial role of thymectomy on this disease (Olanow et al., 1987; Genkins et al., 1993). Although the first controlled trial is still in the planning stages, thymectomy continues to be a first-line therapy, particularly in young patients with thymic hyperplasia. Thymectomy is typically followed by a gradual fall in anti-AChR antibodies. However, it is not known if the fall in autoantibody titer is linked to the clinical improvement or what additional mechanisms may be in play.

Functional studies of thymus cell suspensions obtained from MG patients have also provided clues to the nature of thymic involvement in the pathogenesis of MG. Whereas B cells and plasma cells are rare intramedullary inhabitants of normal thymi, the numbers of B cells and spontaneous immunoglobulin-secreting cells are increased in cell suspensions of hyperplastic MG thymus versus control thymus (subjects undergoing elective cardiothoracic surgery). This finding has been interpreted to mean that such B cells have undergone *in vivo* activation (Levinson et al., 1981; Zweiman et al., 1989; Levinson et al., 1990). Single-cell suspensions prepared from MG thymi with GC hyperplasia secrete anti-AChR antibodies (Newsom-Davis et al., 1981; Fuji et al., 1986; Lisak et al., 1986). However, the B cell repertoire in the hyperplastic MG thymus may reflect systemic as well as local immune events since thymic B cells specific for influenza and tetanus toxoid have also been detected (Newsom-Davis et al., 1981; Lisak et al., 1986). The latter have only been found when patients are booster-immunized to tetanus toxoid 3–4 weeks before thymectomy (Lisak et al., 1986). Such anti-

AChR antibodies are not produced by cells recovered from the thymomas of MG patients (Shiono et al., 2003). By contrast, these tumors appear to be sources of anti-IL-12 and anti-interferon-α antibodies. These autoantibodies are found in the serum of a majority of MG patients with thymomas and in a lesser number of patients with late onset MG without thymomas (Shiono et al., 2003). Their role in the pathogenesis of these subsets of MG patients has not been determined.

nAChR-reactive human T cells bearing a helper/inducer phenotype have also been detected in and propagated as long-term lines and clones from MG but not from normal thymus (Melms et al., 1988; Sommer et al., 1990). These autoreactive T cells are found in greater numbers in the thymus than in the blood of the same MG patient (Sommer et al., 1990). However, it is not clear if AChR-reactive T cells isolated from MG thymus are sensitized *in situ* or in the periphery with subsequent intrathymic localization, since migration of immunocompetent T cells into the thymus is known to occur (Naparstek et al., 1982, 1993; Michie et al., 1988; Hirokawa et al., 1989; Agus et al., 1991; Gossman et al., 1991; Jamieson et al., 1991; King et al., 1992; Westermann et al., 1996).

3. Expression of Neuromuscular AChRs by Thymic Cells

The focus of our more recent studies has been on expression of the autoantigen AChR in the thymus and the potential role it plays in MG pathogenesis. Reports that a neuromuscular type of AChR was expressed on cells resident in the thymus spawned the hypothesis more than 25 years ago that this organ represents a potentially important site for initiating or perpetuating the autoimmune response in MG (Wekerle et al., 1978). This idea runs counter to current dogma, which defines the pivotal role played by thymic self-proteins, particularly those expressed on epithelial cells, in the induction of self-tolerance (Klein and Kyewski, 2000).

nAChRs are expressed in two major forms (reviewed in Levinson, 2001). The so-called mature or junctional form is expressed on innervated muscles and the immature or fetal form is expressed on noninnervated tissue. At the mature (innervated) myoneural junction, nAChRs comprise four subunits labeled α, β, δ, and ε. Two alpha subunits and one each of the other subunits are assembled, like the whalebone in a corset, to form an asymmetric hourglass channel spanning the membrane. Two alternatively spliced alpha subunit isoforms have been characterized: $P3A^-$ and $P3A^+$. The larger $P3A^+$ isoform, which includes an additional sequence of 25 amino acids between exons 3 and 4, is found only in humans and other primates. In fetal muscle, as in adult denervated muscle or nonjunctional membrane, a γ subunit replaces the ε subunit found on nAChRα at mature, innervated muscle endplates.

Most of the attention on thymic nAChRs was formerly focused on the alpha subunit (nAChRα) since it contains the disease-relevant B and T cell epitopes (Hohlfeld et al., 1987; Fuji and Lindstrom, 1988; Oshima et al., 1990; Zhang et al., 1990). Expression of this subunit was reported on an array of thymic cells including epithelial cells (Engel et al., 1977), thymocytes (Fuchs et al., 1980), and myoid cells (Kao and Drachman, 1978; Wekerle et al., 1978; Schluep et al.,

1987). Until recently, myoid cells were viewed as the principal AChR-expressing cells in the thymus (Schluep et al., 1987). These cells phenotypically resemble skeletal muscle cells. They are found in the medulla of both normal and MG thymus. They are found adjacent to GCs in patients with thymic hyperplasia and are not present in MG thymomas (reviewed in Marx et al., 2003).

Given the uncertainty with which resident thymus cell types express this autoantigen, others, and we, took a molecular approach to study this question. We were particularly intrigued by the possibility that neuromuscular AChRs were expressed on thymic epithelial cells (TECs). Using reverse transcription–PCR (RT–PCR), we initially reported that AChRα mRNA was expressed in normal mouse (Wheatley et al., 1992, 1993) and normal human and MG thymus (Wheatley et al., 1993). We also reported that AChRα mRNA was expressed on transformed murine thymic cortical and medullary epithelial cell lines and thymic dendritic cell lines (Wheatley et al., 1992, 1993). We provided evidence for the first time that mRNAs encoding both the P3A$^+$ and P3A$^-$ nAChRα isoforms were expressed in normal and MG thymus, and in normal human TECs (Wheatley et al., 1993; Zheng et al., 1998, 1999). We also reported that the nucleotide sequences of these isoforms were identical to their counterparts expressed at the myoneural junction and provided evidence for the expression of a third, albeit minor, AChRα isoform. Others subsequently reported that AChRα protein as well as mRNA was expressed on human TECs (Wakkach et al., 1996).

The RT–PCR studies have engendered considerable debate about the expression of other nAChR subunits on thymic cells and whether they are expressed as components of intact receptors. Some of the reported discrepancies may reflect differences in the ages of the thymus donors and differences in the design of the RT–PCRs. A distillation of these studies suggests that the ε and β mRNAs are expressed in most normal and MG thymus specimens with variable expression of δ and γ subunits (Naveneetham et al., 2001; Bruno et al., 2004). Myoid cells appear to be the principal cell type expressing the γ subunit. Expression of the AChR subunits appears to be concentrated in the thymic medullary compartment. At the protein level, intact receptors, particularly of the fetal form, have been identified on myoid cells, but not on TECs. Likewise, based on *in vitro* studies functional receptors appear to be expressed on thymic myoid cells but not on TECs (Naveneetham et al., 2001; Shiono et al., 2003).

We expanded our studies to address additional features of AChR expression in the thymus in an effort to gain a better understanding of how intrathymic expression of this autoantigen might be linked to the development of disease. We found that the smaller P3A$^-$ isoform was present in 5-fold excess in both control and MG thymus, and a 2.5-fold excess in a normal human TEC line, relative to the larger P3A$^+$ isoform (Figure 12.1) (Zheng et al., 1998, 1999). This represents a different relationship between the expression of these isoforms than previously described in both healthy and MG muscle, where they were found to be expressed equivalently at the mRNA level (Beeson et al., 1990). These results indicate that the expression of the P3A$^-$ and P3A$^+$ isoforms is differentially regulated in thymus and muscle compartments. The mechanisms accounting for this discrepancy remain to be determined.

Itrathymic Expression of Neuromuscular Acetylcholine Receptors

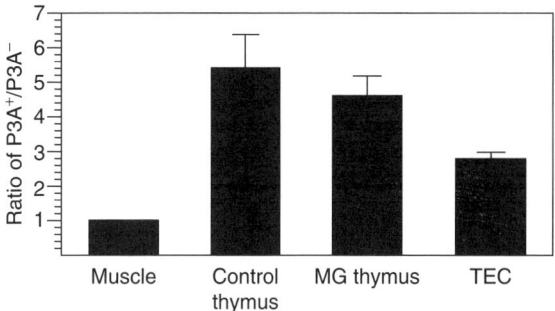

Figure 12.1. Relative expression of AChRα P3A⁻ and P3A⁺ isoforms in thymus and a nontransfered human thymus epithelial cell (TEC) line. Compilation of data from experiments using 14 myasthenia gravis (MG) thymi, 7 control thymi, and 4 separate TECs. The signals for the P3A⁻ and P3A⁺ bands on Southern blots were quantitated on a phosphorimager. The P3A⁻/P3A⁺ ratios are shown. Expression of P3A⁻ exceeded that of P3A⁺ by a factor of 5.5 ± 0.9 (mean $\pm$ SEM) in control thymus, 4.7 ± 0.05 in MG thymus, and 2.8 ± 0.2 in TEC. (Copyright, *Clin. Immunol.*, 91:1999.)

We also observed that P3A⁻ mRNA expression was 2.5-fold greater in MG thymus than in control thymus (Figure 12.2A). Similarly, expression of the larger P3A⁺ isoform was 2.8-fold greater in MG thymus than in control thymus (Zheng *et al.*, 1998, 1999). These results parallel findings in skeletal muscle where mRNA expression was found to be greater in MG muscle than in control muscle (Guyon *et al.*, 1993). The mechanisms accounting for the increased expression of AChRα in MG versus control thymus are unknown. This difference may reflect

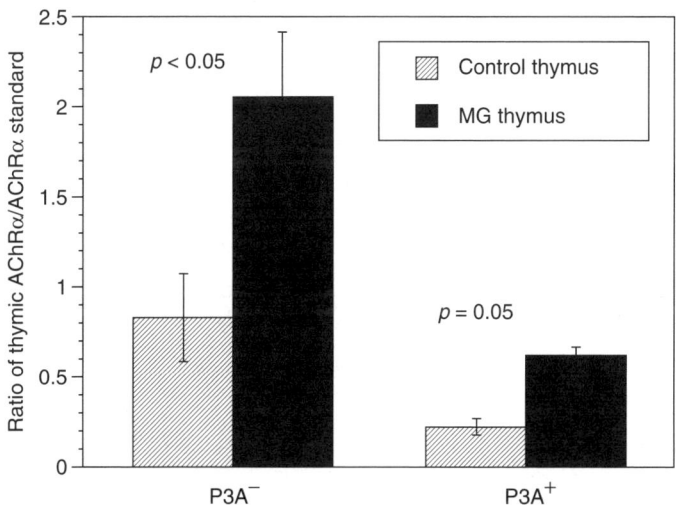

Figure 12.2A. Semiquantitative reverse transcription–PCR (RT–PCR). Compilation of results from thymus specimens of 7 control subjects and 14 myasthenia gravis (MG) patients. The signal intensity of the AChRα bands is normalized to that of the standard by calculating the ratio of thymic AChRα/AChRα standard. The expression of P3A⁻ and P3A⁺ isoforms in MG thymi is 2.5- and 2.8-fold greater, respectively, than that in control thymi. (Copyright, *Clin. Immunol.*, 91:1999.)

the antecedent action of local environmental factors including anti-ACHRα antibodies and cytokines.

Indeed, the hyperplastic thymus in MG is characterized by increased epithelial cell production of IL-1 and IL-6 (Cohen-Kaminsky et al., 1991; Emilie et al., 1991). Since there is evidence that cytokines elaborated by normal TEC have autocrine function (Galy and Spits, 1991), it seemed plausible that these cytokines or perhaps others produced by cells in the thymus might regulate TEC expression of AChR. Therefore, we sought to determine if IL-1, IL-4, IL-6, or interferon-γ (IFN-γ) altered the expression of nAChRα mRNA by a nontransformed human TEC line. The cytokines had no effect whereas IFN-γ increased expression of P3A$^-$ and P3$^+$ isoforms by factors of 2.7 and 2.8, respectively (Figure 12.2B) (Zheng et al., 1998, 1999).

Thymic myoid cells and medullary epithelial cells constitutively express MHC class I antigens. By contrast, MHC class II antigens are not expressed by myoid ells in situ and appear to be expressed in sparse amounts by rare thymic medullary epithelial cells. These observations have raised questions about the potential of these cell types to prime AChR-specific helper T cells in situ (Shiono et al., 2003). Of note, IFN-γ upregulates the in vitro expression of MHC class II antigens on TECs (Berrih-Aknin et al., 1985; Galy and Spits, 1991). The dual effect of IFN-γ on AChRα and MHC II antigens raises the possibility that this cytokine, and/or perhaps others, alters expression of thymic AChRα in vivo in a manner that leads to

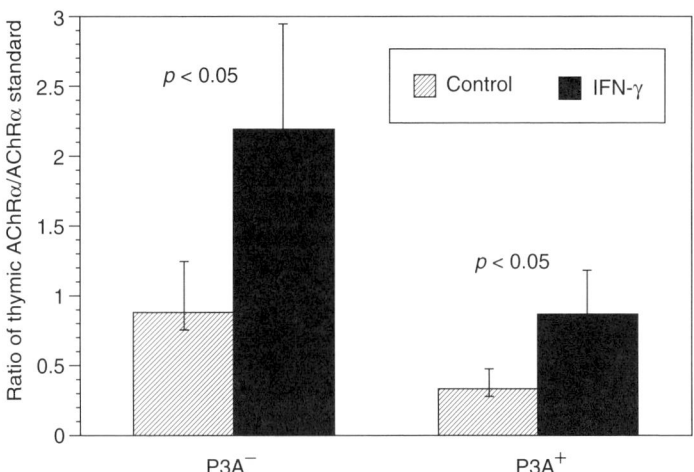

Figure 12.2B. Semiquantitative reverse transcription–PCR (RT–PCR) compilation of results from six thymus epithelial cell (TEC) experiments depicting the effect of IFÑ-γ on expression of AChRα isoforms in TEC. To determine the effect of IFN-γ on the expression of AChRα P3A$^-$ and P3A$^+$ isoforms, we compared the normalized signal intensities of the isoforms (ratio of thymic AChRα/AChRα standard) detected in untreated and IFN-γ-treated TEC cultures. Expression of P3A$^-$ mRNA was significantly greater in IFN-γ-treated (2.19 ± 0.75, mean ± SEM) than in untreated cultures (0.89 ± 0.36, $p < 0.05$, Student's t-test). Likewise, expression of P3A$^+$ mRNA was significantly greater in IFN-γ-treated (0.9 ± 0.31) than in untreated cultures (0.36 ± 0.15, $p < 0.05$). (Copyright, Clin. Immunol., 91:1999.)

the development or perpetuation of MG. Understanding how this might happen requires a brief review of the thymus' role in the development of T cell tolerance and a consideration of how the thymus could serve as a site of immune activation.

4. The Thymus and Central Immune Tolerance

The thymus plays a fundamental role in the generation of the peripheral T cell repertoire (Sprent *et al.*, 1988; Blackman *et al.*, 1990; Kisielow and Boehmer, 1990; Adelstein *et al.*, 1991; Iwabuchi *et al.*, 1992; Bonomo and Matzinger, 1993; Hugo *et al.*, 1994; Oehen *et al.*, 1994; Hoffman *et al.*, 1995; Alam *et al.*, 1996; Anderson *et al.*, 1996; Klein and Kyewski, 2000). It is generally believed that thymocytes with low-affinity receptors for self are positively selected for export to the peripheral lymphoid tissues where they comprise the T cell repertoire that recognizes exogenous antigens (Kisielow and Boehmer, 1990; Anderson *et al.*, 1996). In contrast, T cell tolerance to self derives largely from the process of central deletion/inactivation, wherein developing thymocytes with high-affinity receptors for self-peptide undergo apoptosis or anergy. Although there is widespread agreement that presentation of self-peptides by cortical epithelial cells is necessary for positive selection (Kisielow and Boehmer, 1990; Alam *et al.*, 1996; Anderson *et al.*, 1996) there is still controversy over the identity of the thymic antigen-presenting cells (APCs) involved in negative selection. Indeed, there is evidence supporting roles for thymic medullary epithelial cells and bone marrow–derived macrophages and dendritic cells in this process (Blackman *et al.*, 1990; Kisielow and Boehmer, 1990; Bonomo and Matzinger, 1993; Hugo *et al.*, 1994; Hoffman *et al.*, 1995). However, central deletion is not complete even though a broad array of self-peptides is "promiscuously" expressed on medullary TECs (Klein and Kyewski, 2000). Self-reactive T cells appear to escape from the thymus in small numbers, perhaps due to the fact that limiting levels of self-antigens limit the efficiency of tolerance induction (Adelstein *et al.*, 1991; Iwabuchi *et al.*, 1992; Oehen *et al.*, 1994). However, such self-reactive T cells are thought to be rendered unresponsive because of their anergic or ignorant status, i.e., they never encounter self-antigens in the periphery in a manner that leads to immune activation, or they are suppressed by regulatory T cells (Shevach, 2000).

5. The Thymus and T Cell Trafficking

The thymus is generally not considered to be a sight of immune activation. Based on the classic studies of Gowans, traffic of lymphocytes was unidirectional, i.e., out of the thymus into the blood and peripheral lymphoid organs (Gowans and Knight, 1964). However, there is evidence that small numbers of peripheral immunocompetent T cells migrate to the thymus, entering via the medulla (Naparstek *et al.*, 1982, 1993; Michie *et al.*, 1988; Agus *et al.*, 1991). Most of the thymic immigrants are T cells activated in the peripheral immune system with an even smaller contribution made by resting T cells (Michie *et al.*, 1988; Agus *et al.*, 1991; Naparstek *et al.*, 1993). It is not known if the rate or number of thymic immigrants is increased by an inflammatory reaction in the thymus. Likewise, it is not known if self-reactive T cell immigrants are activated if they encounter their specific antigens in the thymus. Thymus T cell immigrants specific for the lymphocytic choriomeningitis virus (LCMV) clear infectious foci from the thymus (Hirokawa *et al.*, 1989; Gossmann *et al.*, 1991; King *et al.*, 1992). Therefore, peripheral T cells can be activated when they engage specific foreign antigens in the thymus. When self-reactive T cells encounter their antigens in other compartments in the presence of required costimulatory signals, they can be activated to express their differentiation program (Mondino *et al.*, 1996). One mechanism that leads to a milieu that promotes the abrogation of tolerance peripherally is infection. Local infection can lead to the upregulation of MHC antigens and costimulatory molecules on cells that express low levels of self-antigens and, thereby, to activation of autoreactive T cells (Mondino *et al.*, 1996).

6. Development of an Experimental Model to Examine Peripheral T Cell Entry and Activation in the Thymus

Delineation of the molecular events, particularly in the thymus, that trigger MG has been hampered by the lack of a model system. Thymic pathology is not a feature of experimental models of MG in rodents (Meinl *et al.*, 1991). Although such models have provided insight into the pathogenesis of MG (reviewed in Christadoss *et al.*, 2000), they have not served to elucidate the role played by the thymus. In recent studies, we have addressed this problem with the aid of a new experimental model that investigates potential intrathymic events that initiate human MG (Levinson *et al.*, 2003). The salient feature of the experimental model is the establishment of an inflammatory reaction in the thymic medulla, the thymic entry site of immigrant peripheral T cells. We generated molecular variants of the well-characterized thymotrophic Gross murine leukemia virus (G-MLV), GD17, which had previously been shown to exclusively infect medullary thymic epithelium following intrathymic injection in naive mice. The variants were constructed to allow for acceptance of a broad array of genes of interest. Thymotropic MLV vectors were created by ligating a 425-bp fragment containing the U3 region of GD-17 into the LTR backbone of the well-defined M-MLV vector LXSH.

The vectors used in our studies are presented in linear form in Figure 12.4. The parental LXSH vector includes a 5′ M-MSV LTR, the psi packaging site and 5′ gag region, the hygromycin resistance gene under control of the SV40 promoter, and the 3′ LTR of M-MLV. For our experimental protocol, we modified this vector by insertion of the Lac z gene (LBSHG). We utilized LBSHG and LXSHG as our experimental and control vectors, respectively. As was true for GD17, we found that these vectors also target expression of encoded genes to the thymic medullary epithelium.

BALB/c mice were immunized to β-galactosidase (β-gal) and then injected intrathymically (i.t.) with the β-gal-encoding vector LBSHG or the control vector LXSHG. Hematoxylin- and eosin-stained sections of thymus obtained 4 days after i.t. injection of LBSHG, but not LXSHG, showed obliteration of the cortical/medullary architecture with marked cellular expansion of the medulla. To determine whether this local inflammatory reaction nonspecifically augmented

Itrathymic Expression of Neuromuscular Acetylcholine Receptors

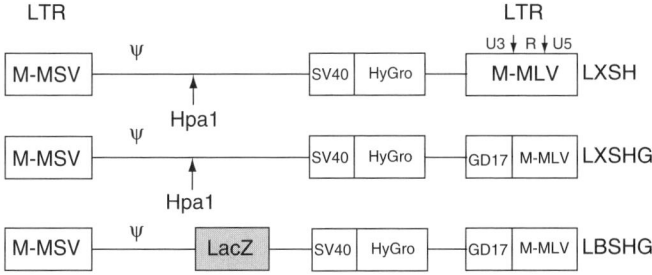

Figure 12.3. Schematic diagram of MLV-based vectors. The vectors used in these studies are presented in linear form. The parental LXSH vector includes the 5′ M-MCV LTR, the psi packaging site and the 5′ ga region, the hygromycin resistance gene under the control of the SV40 promoter, and the 3′ LTR of M-MLV. Vectors are modified by insertion of either GD-17 U3 and/or LacZ. (Copyright, *Ann. NY Acad. Sci.*, 998:2003).

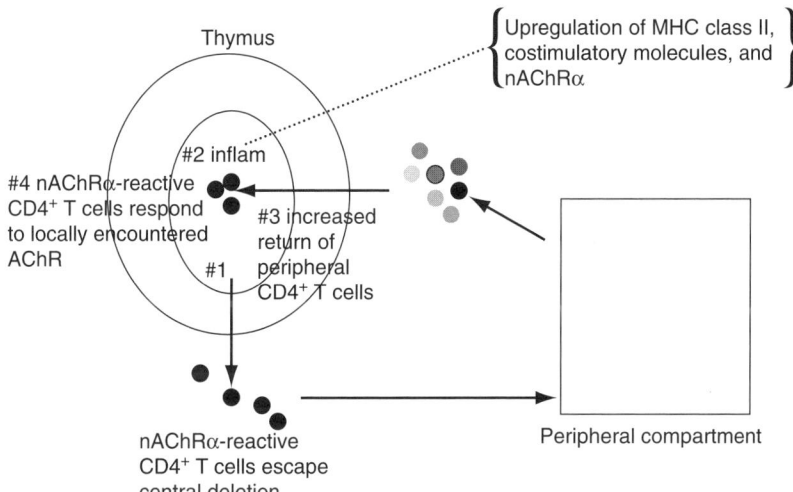

Figure 12.4. A new hypothesis bearing on the intrathymic pathogenesis of myasthenia gravis (MG). The hypothesis (see text for details) describes events leading to the activation of peripheral AChRα-specific CD4+ T following their entry into the thymic medulla. (Copyright. *Ann. NY Acad. Sci.*, 998:2003).

the entry of peripheral T cells into the thymus, β-gal-immunized mice were injected i.v. with a population of CFSE-labeled CD4+ T cells specific for an unrelated antigen 4 days after i.t. injection of LBSHG or LXSHG. The CD4+ T cells were derived from a transgenic mouse bearing a T cell receptor that recognized an influenza hemagglutinin peptide (provided by Dr. Andrew Caton, Wistar Institute, Philadelphia, Pennsylvania). Animals that received LBSHG had 4.2-fold more CFSE-labeled CD4+ thymic immigrants than animals that received the control vector.

This model provides us with an opportunity to test a new hypothesis bearing on the intrathymic pathogenesis of MG (Figure 12.4). The hypothesis states that an inflammatory reaction to an unrelated antigen within the medulla of the

Table 12.1. Rationale for Intrathymic Pathogenesis Hypothesis

---
nAChRα-reactive CD4+ T cells can be found in the blood of healthy donors as well as MG patients
nAChRα-reactive T and B cells are recovered from MG thymus but not "control" thymus
nAChRα is constitutively expressed on thymic myoid cells and thymic epithelial cells
nAChRα mRNA and MHC class II protein expression on human thymic epithelial cells is upregulated by interferon-γ
Peripheral T cells traffic to thymus where they enter the medulla

---

thymus promotes the entry of peripheral AChRα-reactive CD4+ T cells that had escaped central deletion. These cells enter the thymus in the medullary compartment where they encounter AChRα peptides expressed by APCs. The concomitant intrathymic inflammatory reaction creates a milieu that favors activation of these autoreactive T cells, i.e., upregulation of MHC class II antigens and costimulatory molecules on APCs, and perhaps upregulation of AChR expression on TECs. In this scenario, AChR-expressing thymic medullary epithelial cells in addition to local professional APCs might serve as the agents of T cell priming. Presentation of AChRα epitopes to the CD4+ thymic immigrants would lead to their activation, help for locally stimulated αAChR-reactive B cells, the production of anti-AChR antibodies, and GCs, and the development of MG. The rationale for this hypothesis is outlined in Table 12.1.

7. Conclusions

There is a large body of circumstantial evidence highlighting a primary role of the thymus in the pathogenesis of MG. Nevertheless, the etiologic link remains to be forged. We are reexamining the hypothesis that AChR expressed in the thymus drives the pathogenic autoimmune response. To this end, we have established a model of intrathymic inflammation that is localized to the thymic medulla and demonstrated that this inflammatory process promotes the nonspecific entry of peripheral CD4+ T cells into the thymus. Using this model, we are in the process of determining whether (a) AChR-reactive CD4+ T cell homing to the thymus is augmented by a concurrent intrathymic inflammatory response to an unrelated antigen, and (b) AChR-reactive T cell immigrants undergo activation following their engagement of autoantigen in this inflammatory milieu, provide help for the production of anti-AChR antibodies by immigrant autoreactive B cells, and thereby promote the development of a myasthenic syndrome.

Acknowledgments

Studies described in this chapter were supported by a grant from the Muscular Dystrophy Association and National Institutes of Health grant AI 50058. The authors thank Cecelia Willitt for assistance with preparation of the manuscript.

References

Adelstein, S., Pritchard-Briscoe, H., Anderson, T.A., Crosbie, J., Gammon, G, Loblay, R.H., Basten, A., and Goodnow, C.C. (1991). Induction of self-tolerance in T cells but not B cells of transgenic mice expressing little self-antigen. *Science*, **251**, 1223–1225.

Agus, D., Surh, C.D., and Sprent, J. (1991). Reentry of T cells to the ault thymus is restricted to activated cells. *J. Exp. Med.*, **173**, 1039–1046.

Alam, S.M., Travers, P.J., Wung, J.L., Nasholds, W., Redpath, S., Jameson, S.C., and Gascoigne, N.R. (1996). T-cell-receptor affinity and thymocyte positive selection. *Nature*, **381**, 616–620.

Almon, R.R., Andrew, C.G., and Appel, S.H. (1974). Serum globulin in myasthenia gravis: Inhibition of bungarotoxin binding to acetylcholine receptors. *Science*, **186**, 55–57.

Anderson, G., Moore, N.C., Owen, J.J., and Jenkinson, E.J. (1996). Cellular interactions in thymocyte development. *Annu. Rev. Immunol.*, **14**, 73–99.

Beeson, D., Morris, A., Vincent, A., and Newsom-Davis, J. (1990). The human muscle nicotinic acetylcholine receptor α-subunit exists as two isoforms: A novel exon. *EMBO J.*, **9**, 2101–2106.

Berrih-Aknin, S., Arenzana-Seisdedos, F., Cohen, S., Devos, R., Charron, D., and Virelizier, J. 1985). Interferon-gamma modulates HLA class II antigen expression on cultured human thymic epithelial cells. *J. Immunol.*, **35**, 1165–1171.

Blackman, M., Kappler, J., and Marrack, P. (1990). The role of the TCR in positive and negative selection of developing cells. *Science*, **248**, 1335–1341.

Bonomo, A. and Matzinger, P. (1993). Thymus epithelium induces tissue-specific tolerance. *J. Exp. Med.*, **177**, 1153–1164.

Bruno, R., Sabater, L., Tolosa, E., Sospedra, M., Ferrer-Francesch, X., Coll, J., Foz, M., Melms, A., and Pujol-Borrell, R. (2004). Different patterns of nicotinic acetylcholine receptor subunit transcription in human thymus. *J. Neuroimmunol.*, 149, 147–159.

Christadoss, P., Poussin, M., and Deng, C. (2000). Animal models of myasthenia gravis. *Clin. Immunol.*, **94**, 75–87.

Cohen-Kaminsky, S., Delattre, R., Devergne, O., Rouet, P., Gimond, D, Berrih-Aknin, S., and Galanaud, P. (1991). Synergistic induction of interleukin-6 production and gene expression in human thymic epithelial cells by LPS and cytokines. *Cell Immunol.*, **138**, 79–93.

Drachman, D.B., Angus, C.W., Adams, R.N., Michelson, J.D. and Hoffman, G.J. (1978). Myasthenic antibodies cross-link acetylcholine receptors to accelerate degradation. *N. Engl. J. Med.*, **298**, 1116–1122.

Emilie, D., Creven, M.C., Cohen-Kaminsky, S., Peuchmaur, M., Devergne, O., Berrih-Aknin, S., and Galanaud, P. (1991). In situ production of interleukins in hyperplastic thymus from myasthenia gravis patients. *Hum. Pathol.*, **22**, 461–468.

Engel, A.G., Lambert, E.H., and Howard, F.M. (1977). Immune complexes (IgG and C3) at the motor endplate in myasthenia gravis. Ultrastructural and light microscopic localization and electrophysiologic correlations. *Mayo Clin. Proc.*, **52**, 267–280.

Engel, W.K., Trotter, J.L., MacFarlin, D.E., and McIntosh, C.L. (1977). Thymic epithelial cells contain acetylcholine receptor. *Lancet*, **1**, 1310.

Fuchs, S., Schmidt-Hopfeldd, I., and Tridente, G. (1980). Thymic lymphocytes bear a surface antigen which cross-reacts with acetylcholine receptor. *Nature*, **287**, 162–164.

Fuji, Y., Hashimoto, J., Monden, Y., Ito, T., Nakahara, K., and Kawashima, Y. (1986). Specific activation of lymphocytes against acetylcholine receptor in myasthenia gravis. *J. Immunol.*, **136**, 887–891.

Fuji, Y. and Lindstrom, J. (1988). Specificity of the T cell immune response to acetylcholine receptor in experimental autoimmune myasthenia gravis. *J. Immunol.*, **140**, 1830–1837.

Galy, A H.M. and Spits, H. (1991). IL-1, IL-4, and IFN-γ differentially regulate cytokine production and cell surface molecule expression in cultured human thymic epithelial cells. *J. Immunol.*, **147**, 3823–3830.

Genkins, G., Sivar, M., and Tartter, P.I. (1993). Treatment strategies in myasthenia gravis. *Ann. NY Acad. Sci.*, **681**, 603–608.

Gossmann, J., Lohler, J., and Lehmann-Grube, F. (1991). Entry of antivirally active T lymphocytes into the thymus of virus-infected mice. *J. Immunol.*, **146**, 293–297.

Gowans, J.L. and Knight, E. (1964). The route of re-circulation of lymphocytes in the rat. *Proc. R. Soc. London B. Biol. Sci.*, **159**, 257–282.

Guyon, T., Levasseur, P., Truffault, F., Cottin, C., Ohta, K., Itoh, N., and Ohta, M. (1993). Nicotinic acetylcholine receptor α subunit variants in human myasthenia gravis: Quantification of steady-state levels of messenger RNA in muscle biopsy using the polymerase chain reaction. *J. Clin. Invest.*, **94**, 16–24.

Hirokawa, K., Utsuyama, M., and Sado, T. (1989). Immunohistological analysis of immigration of thymocyte-precursors into the thymus: Evidence for immigration of peripheral T cells into the thymic medulla. *Cell Immunol.*, **119**, 160–170.

Hoffman, M.W., Heath, W.R., Ruschmeyer, D., and Miller, J.F. (1995). Deletion of high-avidity T cells by thymic epithelium. *Proc. Natl. Acad. Sci. USA*, **92**, 9851–9855.

Hohlfeld, R., Toyka, K.V., Tzartos, S.J., Carson, W., and Conti-Tronconi, B. (1987). Human helper T lymphocytes in myasthenia gravis recognize the nicotinic receptor a subunit. *Proc. Natl. Acad. Sci.*, **84**, 5379–5383.

Hugo, P., Kappler, J.W., Godfrey, D.I., and Marrack, P.C. (1994). Thymic epithelial cell lines that mediate positive selection can also induce thymocyte clonal deletion. *J. Immunol.*, **152**, 1022–1031.

Iwabuchi, K., Nakayama, K., McCoy, R.L., Wang, F., Nishimura, T., Habu, S., Murphy, K.M., and Loh, D.Y. (1992). Cellular and peptide requirements for in vitro clonal deletion of immature thymocytes. *Proc. Natl. Acad. Sci.*, **89**, 9000–9004.

Jamieson, B.D., Somasundaram, T., and Ahmed, R. (1991). Abrogation of tolerance to a chronic viral infection. *J. Immunol.*, **147**, 3521–3529.

Kao, I. and Drachman, D.B. (1978). Thymic muscle cells bear acetylcholine receptors: Possible relation to myasthenia gravis. *Science*, **195**, 74–75.

King, C., Jamieson, B.D., Reddy, K., Bali, N., Concepcion, R.J., and Ahmed, R. (1992). Viral infection of the thymus. *J. Virol.*, **66**, 3155–3160.

Kisielow, P. and Boehmer, M. (1990). Negative and positive selection of immature thymocytes: Timing and the role of the ligand for T cell receptor. *Semin. Immunol.*, **2**, 35–44.

Klein, L. and Kyewski, B. (2000). "Promiscuous" expression of tissue antigens in the thymus: A key to T-cell tolerance and autoimmunity? *J. Mol. Med.*, **78**, 483–494.

Levinson, A.I. (2001). Myasthenia gravis. In R. Rich, R.R., Fleisher, T.A., Shearer, W.T., Kotzin, B.L., and Schroeder, H.W. Jr. (eds) *Principles and Practice of Clinical Immunology*, Vol. 2. Mosby, St. Louis.

Levinson, A.I., Dziarski, A., Lisak, R.P., Zweiman, B., Moskovitz, A., Brenner, T., Abramsky, O. (1981). Comparative immunoglobulin synthesis by blood lymphocytes of myasthenics and normals. *Ann. NY Acad. Sci.*, **377**, 385–392.

Levinson, A.I. and Wheatley, L. (1995). The thymus and the pathogenesis of myasthenia gravis. *Clin. Immunol. Immunopathol.*, **78**, 1–5.

Levinson, A.I., Zheng, Y., Gaulton, G., Moore, J., Pletcher, C.H., Song, D., and Wheatley, L.M. (2003). A new model linking intrathymic acetylcholine receptor expression and the pathogenesis of myasthenia gravis. *Ann. NY Acad. Sci.*, **998**, 257–265.

Levinson, A.I., Zweiman, B., and Lisak, R.P. (1987). Immunopathogenesis and treatment of myasthenia gravis. *J. Clin. Immunol.*, **7**, 187–197.

Levinson, A.I., Zweiman, B., and Lisak, R.P. (1990). Pokeweed mitogen-induced immunoglobulin secretory responses of thymic B cells in myasthenia gravis: Selective secretion of IgG vs. IgM cannot be explained by helper functions of thymic T cells. *Clin. Immunol. Immunopathol.*, **57**, 211–217.

Lisak, R.P., Levinson, A.I., Zweiman, B., and Kornstein, M.J. (1986). Antibodies to acetylcholine receptor and tetanus toxoid: In vitro synthesis by thymic lymphocytes. *J. Immunol.*, **137**, 1221–1225.

Marx, A., Muller-Hermelink, H.K., and Strobel, P. (2003). The role of thymomas in the development of myasthenia gravis. *Ann. NY Acad. Sci.*, **998**, 223–236.

Meinl, E., Klinkert, W.E., Wekerle, H. (1991). The thymus in myasthenia gravis. Changes typical for the human disease are absent in experimental autoimmune myasthenia gravis of the Lewis rat. *Am. J. Pathol.*, **139**, 999–1008.

Melms, A., Schalke, B.C.G., Kirchner, T., Muller-Hermelink, H.K., Albert, E., and Werkele, H. (1988). Thymus in myasthenia gravis. Isolation of T-lymphocyte lines specific for acetylcholine receptor from thymuses of myasthenic patients. *J. Clin. Invest.*, **81**, 902–908.

Michie, S.A., Kirkpatrick, E.A., and Rouse, R.V. (1988). Rare peripheral T cells migrate to and persist in normal mouse thymus. *J. Exp. Med.*, **168**, 1929–1934.

Mondino, A., Khourts, A., and Jenkins, M.K. (1996). The anatomy of T-cell activation and tolerance. *Proc. Natl. Acad. Sci.*, **93**, 2245–2252.

Naparstek, Y., Ben-Nun, A., Holoshitz, J., Reshef, T., Frenkel, A., Rosenberg, M., Cohen, I.R. (1993). T lymphocyte lines producing or vaccinating against autoimmune encephalomyelitis (EAE). Functional activation induces peanut agglutinin receptors and accumulation in the brain and thymus of line cells. *Eur. J. Immunol.*, **13**, 418–423.

Naparstek, Y., Holoshitz, J., Eissenstein, S., Reshef, T., Rappaport, S., Chemke, J., Ben-Nun, A., and Cohen, I.R. (1982). Effector T lymphocyte line cells migrate to the thymus and persist there. *Nature*, **300**, 262–264.

Naveneetham, D., Penn, A.S., Howard, J.F., and Conti-Fine, B.M. (2001). Human thymuses express incomplete sets of muscle acetylcholine receptor subunit transcripts that seldom include the δ subunit. *Muscle Nerve*, **24**, 203–210.

Newsom-Davis, J., Willcox, N., and Calder, L. (1981). Thymus cells in myasthenia gravis selectively enhance production of anti-acetylcholine-receptor antibody by autologous blood lymphocytes. *N. Engl. J. Med.*, **305**, 1313–1318.

Oehen, S.U., Ohashi, P.S., Burki, K., Hengartner, H., Zinkernagel, R.M., and Aichele, P.(1994). Escape of thymocytes and mature T cells from clonal deletion due to limiting tolerogen expression levels. *Cell. Immunol.*, **58**, 342–352.

Olanow, C.W., Wechsler, A.S., Sirotkin-Roses, M., Stajich, J., and Roses, A.D. (1987). Thymectomy as primary therapy in myasthenia gravis. *Ann. NY Acad. Sci.*, **505**, 595–606.

Oshima, M., Ashizawa, T., Pollack, M.S., and Atassi, M.Z. (1990). Autoimmune T cell recognition of human acetylcholine receptor: The sites of T cell recognition in myasthenia gravis on the extracellular part of the α-subunit. *Eur. J. Immunol.*, **20**, 2563–2569.

Patrick, J. and Lindstrom, J.M. (1973). Autoimmune response to acetylcholine receptor. *Science*, **180**, 871–872.

Schluep, M.N., Wilcox, N., Vincent, A., Dhoot, G.K., and Newsom-Davis, J. (1987). Acetylcholine in human thymic myoid cells in situ: An immunologic study. *Ann. Neurol.*, **22**, 212–222.

Shevach, E. (2000). Regulatory T cells in autoimmunity. *Adv. Rev. Immunol.*, **18**, 423–449.

Shiono, H., Roxanis, I., Zhang, W., Sims, G.P., Meager, A., Jacobson, L.W., Liu, J.L., Matthews, I., Wong, Y.L., Bonifati, M., Micklem, K., Stott, D.I., Todd, J.A., Beeson, D., Vincent, A., and Willcox, N. (2003). Scenarios for autommunization of T and B cells in myasthenia gravis. *Ann. NY Acad. Sci.*, **998**, 237–256.

Sommer, N., Willcox, N., Harcourt, G.C., and Newsom-Davis, J. (1990). Myasthenic thymus and thymoma are selectively enriched in acetylcholine receptor-reactive T cells. *Ann. Neurol.*, **28**(3), 312–319.

Sprent, J., Lo, D., Gao, E.-K., and Ron, Y. (1988). T cell selection in the thymus. *Immunol. Rev.*, **101**, 173–190.

Toyka, K.V., Drachman, D.B., Pestronk, A., and Kao, J. (1975). Myasthenia gravis: Passive transfer from man to mouse. *Science*, **190**, 397–399.

Wakkach, A., Guyon, T., Bruand, C., Tzartos, S., Cohen-Kaminsky, S., Berrih-Aknin, S. (1996). Expression of acetylcholine receptor genes in human thymic epithelial cells. Implications for myasthenia gravis. *J. Immunol.*, **157**, 3752–3760.

Wekerle, H., Ketelson, U.P., Zurn, A.D., and Fulpius, B.W. (1978). Intrathymic pathogenesis of myasthenia gravis: Transient expression of acetylcholine receptors on thymus-derived myogenic cells. *Eur. J. Immunol.*, **8**, 579–582.

Westermann, J., Smith, T., Peters, U., Tschernig, T., Pabst, R., Steinhoff, G., Sparshott, S.M., and Bell, E.B. (1996). Both activated and nonactivated leukocytes from the periphery continuously enter the thymic medulla of adult rats: Phenotypes, sources and magnitude of traffic. *Eur. J. Immunol.*, **26**, 1866–1874.

Wheatley, L., Urso, D., Tumas, K., Maltzman, J., Loh, E., and Levinson, A.I. (1992). Molecular characterization of the nicotinic acetylcholine receptor alpha chain in mouse thymus. *J. Immunol.*, **148**, 3105–3109.

Wheatley, L.M., Urso, D., Zheng, Y., Loh, E., and Levinson, A.I. (1993). Molecular analysis of intrathymic nicotinic acetylcholine receptor. *Ann. NY Acad. Sci.*, **681**, 74–82.

Zhang, Y., Schluep, M., Frutiger, S., Hughes, G.J., Jeannet, M., Steck, A., and Barkas, T. (1990). Immunologic heterogeneity of autoreactive T lymphocytes against the nicotinic acetylcholine receptor in myasthenic patients. *Eur. J. Immunol.*, **20**, 2577–2583.

Zheng, Y., Wheatley, L.M., Liu, T., and Levinson, A.I. (1998). Regulation of nicotinic acetylcholine receptor alpha subunit mRNA expression in myasthenic thymus. *Ann. NY Acad. Sci.*, **841**, 393–396.

Zheng, Y., Wheatley, L.M., Liu, T., and Levinson, A.I. (1999). Acetylcholine receptor alpha subunit mRNA expression in human thymus: Augmented expression in myasthenia gravis and upregulation by interferon-α. *Clin. Immunol.* **91**, 170–177

Zweiman, B., Levinson A.I., and Lisak R.P. (1989). Phenotypic characterization of thymic B lymphocytes in myasthenia gravis. *J. Clin. Immunol.*, **9**, 242–247.

13

Autoantibodies and Nephritis: Different Roads May Lead to Rome

Paola Migliorini, Consuelo Anzilotti, Laura Caponi, and Federico Pratesi

1. Introduction

Two major hypotheses may explain the development of autoantibody-mediated nephritis: the deposition of circulating immune complexes (CICs) or the *in situ* formation of immune complexes (ICs) (Figure 13.1). In the first, CICs form whenever an antibody encounters and binds its target antigen in the circulation. Under normal conditions CICs can be rapidly cleared: erythrocytes bind CICs through a mechanism involving complement receptor 1 (CR1) on the surface of erythrocytes and the complement proteins C1q, C4b, C3b, and C3b1, all of which are found on the ICs that activate the complement system. This mechanism allows the erythrocytes to shuttle ICs through the circulation until they reach the liver or the spleen, at which point the ICs are transferred to the monocyte phagocytic system and removed from circulation. When CICs are produced in excessive amounts or have certain characteristics (small size, slight antigen excess), clearance mechanisms may become saturated or inefficient (Davies *et al.*, 2002; Birmingham *et al.*, 2003). In such cases, CICs may deposit in certain tissues where the anatomical and hydrodynamic conditions are favorable; for example, in the kidney, CICs tend to form deposits in the mesangial and subendothelial areas. In the second hypothesis, *in situ* ICs form when circulating autoantibodies derived from autoreactive B cell clones react with a tissue self-antigen or with a circulating self-antigen bound to the kidney (planted antigens).

Paola Migliorini and Consuelo Anzilotti • Clinical Immunology Unit, Department of Internal Medicine, Via Roma, 67-56126, Pisa, Italy. **Laura Caponi** • Department of Experimental Pathology, University of Pisa, Via Roma 55, Pisa, Italy. **Federico Pratesi** • Clinical Immunology Unit, Department of Internal Medicine, Via Roma, 67-56126, Pisa, Italy.

Molecular Autoimmunity: In commemoration of the 100th anniversary of the first description of human autoimmune disease, edited by Moncef Zouali. Springer Science+Business Media, Inc., New York, 2005.

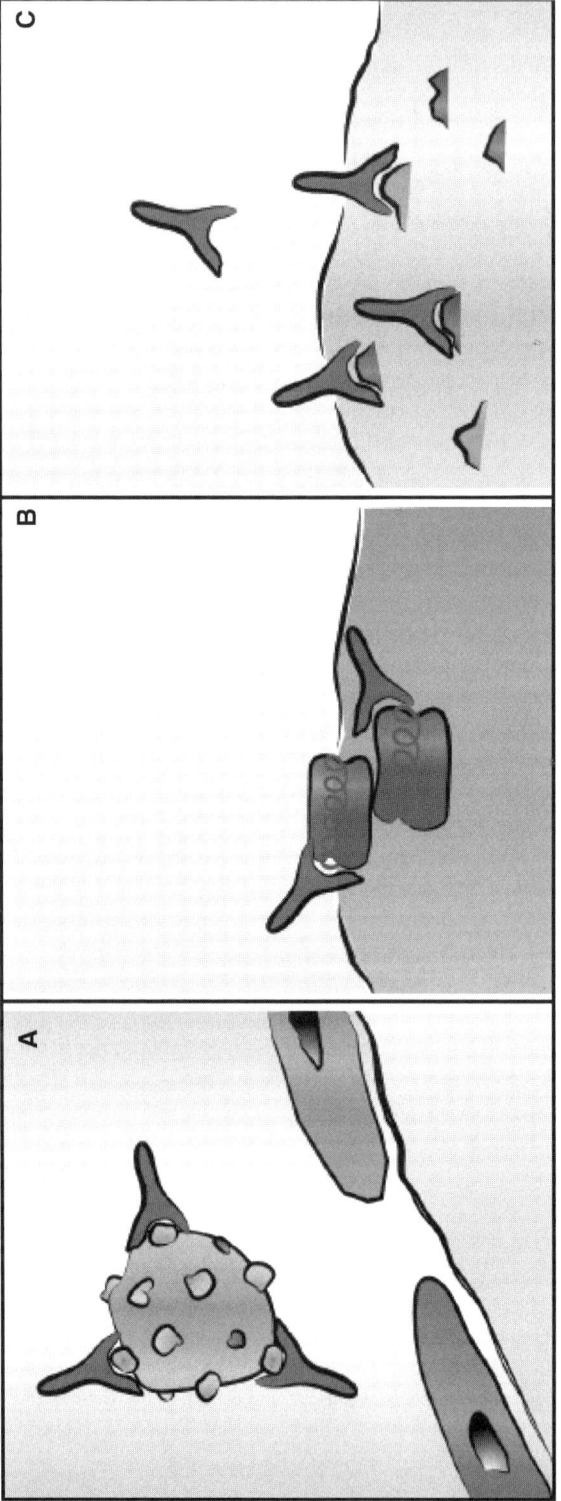

Figure 13.1. Possible mechanisms of formation of immune deposits in the kidney. (A) Circulating immune complexes (CICs). The circulating antigen is bound by specific antibodies. The resulting CIC may be trapped within the glomeruli. (B) Binding of antibodies to planted antigens. Antigens able to bind glomerular structures (e.g., nucleosomes) may be subsequently targeted by antibodies. The resulting IC is formed locally. (C) Binding of antibodies to renal intrinsic antigens. Antibodies are able to bind antigens located in glomerular structures. The IC is formed *in situ*.

The identification of nephritogenic antibodies is difficult because sera from patients with autoimmune disorders and nephritis contain multiple types of autoantibodies, but only some of them are potentially pathogenic. Moreover, autoantibodies are often able to react with several antigens, but not all of them may be relevant to kidney injury. In many disorders, however, the formation of renal immune deposits is not entirely explained, and new data are continuously emerging to challenge interpretations hitherto considered firm and conclusive.

If deposition of CICs and *in situ* formation of ICs represent the ends of a spectrum of two possible mechanisms for the development of nephritis, different classes/types of glomerulonephritis can be categorized along this spectrum. In this chapter we describe poststreptococcal nephritis as a representative example of CIC-mediated nephritis, and Goodpasture's syndrome as a disease mediated by antibodies specific for a renal antigen. Systemic lupus nephritis is described as an example of nephritis where different damage mechanisms probably coexist and play a role in the different phases of the disease.

2. Acute Poststreptococcal Glomerulonephritis

Acute poststreptococcal glomerulonephritis (APSGN) is the most common postinfectious renal disease following group A streptococci (GAS) infection, and the first form of glomerular disease in which immunological mechanisms were suspected to play a role. In fact, its evolution is characterized by a serum sickness–like latent period followed by hypocomplementemia and nephritis (Nordstrand *et al.*, 1999). Researchers originally believed that the pathogenic mechanism underlying APSGN was the renal deposition of CICs. This theory was consistent with the clinical picture: the elevated serum levels of IgG and IgM in a high percentage of patients; the pattern of CIC levels, which are high during the acute phase and usually return to normal within 6–9 months after the attack, but linger in patients with persisting hematuria and proteinuria (Lin, 1982); and the finding of extracellular streptococcal antigens typical of nephritogenic strains in the patients' CICs (Friedman *et al.*, 1984). However, various findings in recent decades have shed doubt on this hypothesis. For example, increased CIC titers have also been found in the sera of patients with other types of streptococcal disease. In APSGN patients, streptococcal antigens are present in CICs, but not in ICs eluted from kidney biopsies; the latter are endogenous complexes containing IgG that have probably been modified enzymatically by a streptococcal product (McIntosh *et al.*, 1978; Rodriguez-Iturbe *et al.*, 1980). The glomerular deposition of preformed ICs would cause complement activation by the classical pathway, but the presence of properdin in the kidney and the fact that C3 deposition precedes that of IgG indicate that complement activation occurs before IgG is deposited, possibly by the alternative pathway.

At present, the prevailing theory is that deposition in the glomeruli of a streptococcal antigen common to nephritogenic strains induces an inflammatory process by activating complement *in situ*, followed by C3 deposition, immune response activation, tissue destruction, IgG deposition, and amplification of the inflammatory reaction (Nordstrand et al., 1999). IgG deposition may be caused by autoantibodies to glomerular epitopes (e.g., cryptic epitopes exposed after tissue damage or epitopes that cross-react with streptococcal antigens, i.e., molecular mimicry), CICs, and/or antibodies to already planted streptococcal antigens. Accordingly, several streptococcal products have been suggested as possible nephritogenic agents, with different mechanisms of action. (i) M proteins (surface molecules that confer resistance to phagocytosis) may be antigenically cross-reactive with the glomerular basement membrane (GBM) and directly bind its constituents (Glurich et al., 1991); some M-like proteins can act like Fc receptors (FcR) and strains carrying these FcR-like M proteins can induce circulating anti-IgG antibodies, thus contributing to IC formation in APSGN (Burova et al., 2003). (ii) Preabsorbing antigen (PA-Ag), a streptococcal antigen able to activate the complement cascade via the alternative pathway, can induce the deposition of C3 (without IgG) in the glomeruli of rabbits (Yoshizawa et al., 1997). (iii) Streptococcal pyrogenic esotoxin B (Spe-B), an extracellular plasmin-binding protein secreted by nephritis-associated GAS, has been found only in kidney biopsies from APSGN patients; it could contribute to the disease pathogenesis either behaving like a planted antigen or blocking the inactivation of plasmin, as SpeB-bound plasmin is not inactivated by α_2-antiplasmin (Cu et al., 1998). (iv) Streptokinase is a 46-kDa extracellular protein that plays a role in the streptococcal invasion of tissues due to its ability to convert plasminogen to plasmin. In a mouse model its presence was necessary for APSGN induction by a GAS nephritis isolate and it was detected in the glomeruli as early as 4 days after infection. Its ability to target the glomeruli seems to be restricted to nephritogenic strains, and streptokinase from nephritis isolates binds more tightly to human glomeruli than non-nephritis-associated streptokinases. Here again the deposited protein could play a role in the initiation of the disease process, and ICs might be involved in the later stages of the disease (Nordstrand et al., 1998). (v) Nephritis-associated plasmin receptor (NAPlr) with glycolytic activity (streptococcal glyceraldehyde-3-phosphate dehydrogenase) has been found in glomeruli from kidney biopsies of APSGN patients, but not in controls. It could contribute to the pathogenesis of the disease by maintaining activated plasmin in the glomeruli and playing a role in complement activation (Yoshizawa et al., 2004). (vi) During nephritogenic GAS infection, neuraminidase is produced and neuraminidase-treated leukocytes accumulate preferentially in the kidney. Furthermore, neuraminidase could induce changes in IgG molecules, and render them immunogenic (McIntosh et al., 1978; Marin et al., 1997). (vii) Streptococcal enolase is a recently discovered cross-reactive antigen located on the surface of GAS. It may play an important role in the initiation of autoimmune diseases linked to streptococcal infection. There is a cross-reactivity between streptococcal enolase and human enolase, and patients with acute rheumatic fever have higher levels of antibodies that react with human and bacterial enolase than those with streptococcal pharyngitis or healthy subjects (Fontan et al., 2000), suggesting that streptococcal enolase may induce autoantibodies via molecular mimicry. In fact, α-enolase is a target of autoantibodies in several autoimmune disorders with renal involvement, such as systemic lupus erythematosus (SLE) and mixed cryoglobulinemia (MC) (Pratesi et al., 2000).

Thus, after having been considered a prototypic CIC-mediated form of glomerulonephritis, the autoimmune nature of APSGN has been challenged by several findings; rather than the pathogenetic mechanism underlying APSGN, CICs could represent an epiphenomenon in GAS infection.

3. Goodpasture's Syndrome

Goodpasture's syndrome, a prototypical autoantibody-mediated disease, is associated with antibodies specific for constituents of the GBM. The term Goodpasture's syndrome is applied to hemorrhagic lung involvement in the presence of anti-GBM antibodies. The term "anti-GBM disease" is now used to refer to the nephritic syndromes associated with the presence of these autoantibodies. Anti-GBM antibodies deposit linearly along the GBM and tubular basement membrane, but complement components are detected in only one third of the cases. The antibodies that form linear deposits on the glomerular and tubular basement membranes are essentially specific for the noncollagenous domains (NC1s) of collagen type IV. Ubiquitary components of collagen IV are alpha 1 and 2 chains; in the kidney, lung, eye, ear, and choroids plexus alpha 3, 4, 5, and 6 chains are also expressed. Anti-GBM antibodies react with the alpha 3 NC1 domain, specifically with the conformational epitopes formed by adjacent molecules, which are cryptic in intact GBM (Borza and Hudson, 2003). Some sera also contain antibodies that are reactive with alpha 2, 4, and 6 chains (Dehan et al., 1996) or with the amino-terminal portion of alpha 3 (Ryan et al., 1998). Sera from patients with anti-GBM disease react with purified collagen IV or with recombinant alpha 3 chains, and a higher titer of antibodies is more frequent in patients with lung involvement. The severity of renal involvement, however, is not always proportional to the amount of circulating antibodies (Yamamoto and Wilson, 1987). Anti-GBM antibodies are characterized by a very high affinity for antigens, a high association rate, and a low dissociation rate. Therefore, these antibodies bind rapidly and remain bound to GBM, a property that probably explains the fulminant nature of the disease and its resistance to therapy (Rutgers et al., 2000).

The experimental model for Goodpasture's disease is experimental autoimmune glomerulonephritis (EAG), which can be induced in animals by immunization with GBM extracts or collagen type IV. EAG is characterized by circulating and deposited anti-GBM antibodies, accompanied by focal necrotizing glomerulonephritis with crescent formation. The observation that complement components are detectable in glomerular lesions and that the experimentally induced disease is milder in mice deficient in C3 or C4 (Sheerin et al., 1997) suggests that complement can contribute to tissue damage. However, higher amounts of anti-GBM antibodies induce renal damage even in the absence of complement. In contrast, mice are completely protected from disease in the absence of the γ chain of the FcRs. After administration of anti-GBM antibodies, linear deposits of complement and immunoglobulins are observed in mice knockout for the γ-chain of the FcRs, but no hypercellularity and inflammatory infiltrates can be detected (Park et al., 1998). On the contrary, mice lacking FcγRIIb on the nonpermissive H-2^b haplotype develop pulmonary hemorrhage and crescentic glomerulonephritis with a "ribbon deposition" pattern of ICs in their glomeruli in response to immunization with bovine collagen type IV (Nakamura et al., 2000).

The role of T cells in the induction of anti-GBM disease is less clear. Initially, the therapeutic effects of cyclosporin A on anti-GBM antibody production and proteinuria in BN rats with EAG suggested that T cell help is required (Reynolds et al., 1991). Such a role has then been demonstrated in bursectomized chicks that, upon immunization with heterologous GBM, developed nephritis in the absence of autoantibodies. Oral tolerization of Wistar Kyoto rats (Reynolds and Pusey, 2001) and mice (Kalluri et al., 1997) leads to a significant reduction in circulating IgG2a but not IgG1, suggesting a downregulation of the Th1 response. Moreover, in rats that have been orally tolerized there is a dose-dependent reduction in the proliferative response of the splenic T cells to GBM antigens *in vitro*. Its *in vivo* counterpart probably is a significant reduction in the severity of the disease. In addition to indirect effects, a direct role of T cells in the induction of glomerular injury has also been investigated. In Wistar rats, rapidly progressive crescentic glomerulonephritis was elicited by a single immunization with denatured mouse collagen type IV alpha 3 chain NC1 (Wu et al., 2001) or with a peptide encompassing the amino acids 28–40 (Wu et al., 2004). Anti-GBM antibodies eluted from nephritic kidneys do not bind this epitope and the typical linear deposits appear days after proteinuria and the histological signs of crescentic glomerulonephritis. These data, combined with the observation that T cells from immunized animals transfer the disease (Wu et al., 2002), suggest that anti-GBM antibodies are not the cause of the disease in this model, but are rather the consequence of T cell–mediated glomerular injury.

4. Lupus Nephritis

Systemic lupus (SLE) is an autoimmune disease characterized by several autoantibodies directed against intracellular antigens. Kidney involvement is frequent and indeed constitutes one of the primary causes of morbidity and mortality. It is generally agreed that antibodies are the principal agents at work in lupus nephritis, forming immune deposits by different mechanisms, more than one of which may be involved. Their central role in causing renal damage has been convincingly demonstrated in a genetically manipulated autoimmune mouse strain. MRL/*lpr-lpr* mice lacking Ig heavy chain *Jh* genes, and therefore lacking B cells and autoantibodies, do not develop glomerular, tubular, or interstitial damage, even in the presence of the *lpr* mutation (Shlomchik et al., 1994). While not excluding the contribution of T cells, soluble mediators, and B cell functions other than antibody formation (Chan et al., 1999; Tipping and Holdsworth, 2003), antibodies have been demonstrated to be a key factor in the pathogenesis of nephritis. Three mechanisms may lead to their targeting to the kidney: deposition of CICs, binding of antibodies to antigens previously complexed to glomerular structures, and binding of antibodies to intrinsic glomerular antigens.

Given the close relationship between lupus and anti-DNA antibodies, the latter immunoglobulins have been proposed to cause renal damage by the three mechanisms described above. As far as CIC deposition is concerned, experimental studies have yielded conflicting results: circulating DNA seemed to be rapidly cleared from the circulation (Emlen and Mannik, 1984), and, intravenous administration of preformed DNA–anti-DNA ICs to normal mice led only to a transient deposition in glomeruli without inflammation. Moreover, convincing evidence for the presence of DNA–anti-DNA complexes in the circulation of SLE patients is still lacking. Therefore, other pathogenetic factors in addition to CIC deposition are presumably at work. In fact, the DNA that circulates in the plasma of SLE patients is contained in nucleosomes (Rumore and Steinman, 1990), particles composed of DNA linked to histones and nonhistonic proteins. Since nucleosomes are exposed in clusters together with other internal antigens on the cell surface during apoptosis, they may act as antigenic stimuli. The histone components have a cationic charge, which allows them to bind to the glomerulus by interacting with its negatively charged components, mainly heparan sulphate in the basement membrane. Thus, nucleosomes can mediate the binding of circulating anti-DNA antibodies to glomeruli (Berden *et al*., 1999). The results of a number of studies in both murine and human lupus support this scenario. Nucleosomes have been found in immune deposits of nephritic kidneys in human SLE (van Bruggen *et al*., 1997a), and there is corroborative evidence for a link between antinucleosome antibodies and lupus nephritis. Antinucleosome antibodies have been eluted from nephritic kidneys of lupus-prone mice, where they are produced in the early disease stages, even before anti-DNA antibodies are produced (Amoura *et al*., 1994). When hybridomas synthesizing antinuclear antibodies of different specificities were inoculated to normal mice, anti-DNA and antinucleosomal antibodies, but not antihistone antibodies, deposited in the GBM (van Bruggen *et al*., 1997b).

In addition to DNA, anti-DNA antibodies can cross-react with a large number of intracellular and extracellular molecules, and this polyreactivity could play an important role in conferring a pathogenetic potential, allowing them to bind intrinsic renal antigens (Madaio and Shlomchik, 1996). This was demonstrated by the presence of anti-DNA antibodies in the eluates from nephritic kidneys in human SLE and in murine models. The immunoglobulins deposited in the kidney consist mainly of IgG, often able to bind several targets such as DNA, polynucleotides, ribonucleoproteins, and gp70 (Pankewycz *et al*., 1987). Similarly, eluates obtained from nephritic SLE kidneys were found to contain polyreactive antibodies. In addition to anti-DNA antibodies, some of the specificities highly enriched in the kidneys were anti-C1q and anti-Sm antibodies (Mannik *et al*., 2003).

The ability of anti-DNA antibodies to bind intrinsic renal antigens has been extensively investigated. In early studies, polyclonal or monoclonal anti-DNA antibodies were infused into isolated kidneys or injected into normal mice

(Madaio et al., 1987; Raz et al., 1989), resulting in glomerular deposition, and, in some cases, in renal damage (Ehrenstein et al., 1995). In fact, not all SLE patients with high anti-DNA antibody titers have clinically significant nephritis, while SLE patients low titers of anti-DNA antibodies may exhibit overt nephritis. Therefore it appears likely that some but not all anti-DNA antibodies have a nephritogenic potential.

More recently, a number of glomerular antigens targeted by anti-DNA antibodies have been isolated and characterized. Laminin, a constituent of the GBM, is an extracellular protein composed of several chains that regulates the kidney's filtering efficiency. Anti-laminin antibodies also display anti-DNA binding activity (Foster et al., 1993) and anti-DNA antibodies cross-react with laminin (Sabbaga et al., 1989). Spontaneously produced autoantibodies directed toward laminin were found in MRL/*lpr-lpr* mice (Foster et al., 1993) and in the Graft versus Host Disease (GvHD), another experimental model of lupus (Peutz-Kootstra et al., 2000). In GvHD mice, the glomerular expression of laminin chains alters as the disease progresses, and concomitant modifications of the fine specificity of antilaminin antibodies have been observed. Other proteins located in the kidney may constitute antigenic targets of nephritogenic antibodies in lupus nephritis. Murine monoclonal anti-DNA antibodies with nephritic potential (Mostoslavsky et al., 2001) demonstrate binding ability against α-actinin, a structural protein expressed in the cytoplasm but also on the cell membrane of various glomerular cells, especially podocytes (Deocharan et al., 2002). Furthermore, it has been suggested that cross-reactivity with α-actinin is a property of anti-DNA antibodies isolated from SLE patients with active nephritis (Mason et al., 2004). Since mice lacking α-actinin 4 (the isoform most frequently expressed in the kidney) develop severe nephritis (Kos et al., 2003) and because high titers of anti-α-actinin antibodies are present in the sera and in renal eluates of mice with nephritis (Deocharan et al., 2002), it is conceivable that alterations of this protein may lead to nephritis. In addition to α-actinin and laminin, other intrinsic glomerular antigens may be recognized by anti-DNA antibodies, but further studies are needed to ascertain the role of these antibodies in the pathogenesis of lupus nephritis.

Once the immune deposits are formed in the glomeruli, autoantibodies, for example, anti-C1q antibodies, may enhance the IC renal disease (Trouw et al., 2004). It is known that anti-C1q antibodies are associated with active SLE and in particular with active nephritis (Siegert et al., 1991). However, administration of anti-C1q antibodies is followed by glomerular deposition, but neither significant albuminuria nor proliferative lesions are detected in the injected animals (Trouw et al., 2003b). Experiments in $Rag2^{-/-}$ mice confirm that autologous immunoglobulins are needed for the renal deposition of anti-C1q antibodies, indicating that they can form *in situ* ICs reacting with preexisting immune deposits (Trouw et al., 2003a). Taken together, these data suggest that while anti-dsDNA antibodies, which are polyreactive and able to bind renal or planted antigens, may directly initiate the formation of renal immune deposits, anti-C1q antibodies may amplify a preexisting damage. It should be stressed that anti-C1q antibodies may also contribute to the development of nephritis by lowering the levels of C1q and thus impairing its protective role.

Pathogenic antibodies may cause tissue damage by activating complement, by recruiting effector cells via FcRs or, more directly, by penetrating into living cells. First, the complement activation induced by antigen–antibody complexes may impair glomerular structures through the membrane attack complex, and by means of mediators recruited by chemotactic components produced during activation of the cascade. While complement is a mediator of inflammation, it also plays a role in IC clearance by binding to CR1 receptors, as was demonstrated in a study using preformed DNA–anti-DNA ICs (Craig et al., 2000). Therefore, any impairment of this clearance ability—whether acquired or genetically determined—can amplify the pathogenetic potential of ICs. Likewise, a deficiency in one of the components of the classical complement pathway constitutes an important susceptibility factor for SLE. Recently, a more extended protective role has been postulated for the complement system, because C1q plays a fundamental role in the clearance of apoptotic cells, a mechanism that may explain the occurrence of glomerulonephritis in C1q-deficient mice (Botto and Walport, 2002).

Second, the binding of antibodies to FcRs located on various cell types may have a wide range of consequences, resulting in either an effector function for human FcγRI, FcγRIIa, and FcγRIII or in a downmodulation of the immune response, when FcγRIIb is recruited. The effector functions include antibody-dependent cell cytotoxicity, phagocytosis and release of inflammatory mediators, and probably also internalization of antigens for processing (Takai, 2002; Reefman et al., 2003). The ability of preformed ICs to elicit inflammation in C3-and C4-deficient mice supports the theory that FcRs are key mediators of the inflammatory reaction (Sylvestre et al., 1996). The importance of FcRs in nephritis has been clearly demonstrated in (NZB × NZW) F_1 lupus-prone mice lacking the Fcγ chain (Clynes et al., 1998). These mice lack FcγRI and FcγRIII, which are necessary for the clearance of nonsoluble ICs, but nevertheless express the inhibitory FcγRIIb. Like the parental strain, they develop autoantibodies, but their nephritis is delayed and milder. ICs and complement are deposited in the kidney without any sign of inflammation, indicating that Fc binding of IC, rather than complement fixation, contributes to the development of nephritis. Moreover, the ratio of inhibitor versus activating Fcγ receptors in the kidney conditions the response elicited by the ICs that are deposited. This ratio is regulated by several factors, including the cytokines present in the microenvironment (Ravetch, 2002). Among the activating Fcγ receptors, FcγRIIa is the most frequent isoform in inflammatory cells and it is only present in humans. Mice transgenic for this receptor seem to be particularly sensitive to antibody-mediated inflammation, and one of the clinical manifestations is glomerulonephritis with IC deposition (Tan Sardjono et al., 2003). Conversely, FcγRIIb$^{-/-}$ mice develop inflammation with crescent glomerulonephritis and other characteristics of Goodpasture's syndrome (Nakamura et al., 2000). The fact that double-deficient FcγRI/III mice are protected from nephrotoxic nephritis suggests that both high-affinity FcγRI and low-affinity FcγRIII are significant mediators of nephritis (Tarzi et al., 2003). In the search for a polymorphism that confers disease susceptibility, genes encoding FcRs were investigated in SLE patients and epidemiological studies were performed in different populations. Two alleles of FcγIIRa that encode receptors with different affinities have been found. A higher frequency of the allele encoding the lower affinity receptor (possibly responsible for a delayed clearance of ICs) has been reported in various studies (Salmon et al., 1996; Michel et al., 2000), but it did not confer a higher risk for nephritis (Karassa et al., 2002).

Third, autoantibodies, and in particular anti-DNA antibodies, may also penetrate living cells. It is known that they can bind different antigens, including cell surface molecules (Raz et al., 1993; Puccetti et al., 1995). For example, ribosomal P proteins, one of the antigens bound by some anti-DNA antibodies (Caponi et al., 2002), have been found on the surface of various cell types, including endothelial cells, fibroblasts, hepatocytes, and astrocytes (Reichlin, 1996; Sun et al., 1996). Similarly, a subset of anti-DNA antibodies bind α-enolase, another protein that is expressed also on the cell surface (Pratesi et al., 2000). The binding to particular membrane proteins may mediate antibody internalization, and, when internalized, the antibody may still bind to intracellular targets.

The observation that antibodies directed against intracellular antigens may be able to penetrate into the cell has recently found confirmation and attracted renewed attention, because it could explain the perturbation seen in the physiological functions of the cells (Madaio and Yanase, 1998; Putterman, 2004). One of the cell surface receptors that mediate binding and internalization of anti-DNA antibodies is brush border myosin 1 (Yanase et al., 1997), a protein widely expressed in tissues. Its binding seems to induce a vesicle-mediated transport of the antibody into the cell. Once internalized, the antibody may interact with other potential targets. For example, antibodies with double specificities for DNA and the enzyme DNase may interfere with its activity, making the cells resistant to apoptotic stimuli (Madaio et al., 1996).

It is generally held that the loss of tolerance to nuclear antigens such as dsDNA and nucleosomes can be a central step in the pathogenesis of lupus nephritis. However, severe glomerulonephritis with proteinuria can be present even in the absence of antinuclear antibodies, anti-dsDNA, and antinucleosome antibodies, suggesting that breaking tolerance to chromatin and dsDNA is not required for the pathogenesis of lupus nephritis (Waters et al., 2004). This raises the possibility that other autoantibodies are able to cause glomerulonephritis in SLE.

5. Other Nephritogenic Autoantibodies

Among the non-anti-DNA autoantibodies involved in kidney damage, those directed to α-enolase have recently been shown to be associated with nephritis. They are detectable in 27% of SLE patients (70% of them have active nephritis) (Pratesi et al., 2000), in 30% of mixed cryoglobulinemia patients with renal involvement (Sabbatini et al., 1997), in 40% of ANCA-positive vasculitis patients (Moodie et al., 1993), in 69% of patients with the primary form of the membranous nephritis, and in 58% of patients with secondary membranous nephritis (Wakui et al., 1999). Taken together, these data suggest that antibodies specific for α-enolase may play a role in nephritis.

Alpha-enolase is a ubiquitous glycolytic enzyme. Although it is expressed in virtually all tissues, the kidney and the thymus contain the highest amount of the enzyme. It is present in the cytoplasm, but in a variety of cells it is also expressed on the membrane, where it acts as a member of the plasminogen receptors family (Miles et al., 1991; Moscato et al., 2000). In normal kidney, α-enolase is highly expressed in tubuli and almost undetectable in glomeruli. In SLE and MC nephritic kidneys, the enzyme is overexpressed in tubuli and present in active inflammatory lesions. In SLE, enolase is also detectable in different sites of the glomeruli (mesangium, glomerular, and parietal epithelium, and especially in the crescents) (Migliorini et al., 2002). It is not clear how enolase expression is regulated in the kidney and how inflammatory stimuli lead to its overexpression. As α-enolase is ubiquitous, it is possible that anti-enolase antibodies form CICs, but the high amounts of enzyme in the kidney also allow *in situ* formation of complexes. In this regard, membrane expression of enolase is of utmost importance, because antienolase antibodies could directly cause glomerular and tubular injury by complement fixation. Autoantibodies may also interfere with enolase function as plasminogen receptor.

Plasminogen receptors are a heterogeneous group of proteins with carboxy-terminal lysines, characterized by a low affinity for plasminogen, a high density, and a ubiquitous distribution on different cell types. Plasminogen activation to plasmin takes place on fibrin surfaces or on cell membranes, where plasmin has increased fibrinolytic activity and is protected from inactivation. Fibrin deposits are abundant in nephritis and represent a marker of poor prognosis. The role of the fibrinolytic pathway in renal disorders has been analyzed in mice deficient in plasminogen, or in tissue plasminogen activator, or in urokinase-like plasminogen activator (Kitching *et al.*, 1997). The proliferative glomerulonephritis due to *in situ* formation of ICs is more severe in mice deficient in fibrinolysis, as compared to wild-type mice. Thus, a downregulation of the fibrinolytic activity that impairs fibrin clearance may worsen glomerular inflammatory lesions. Recently, an anti-enolase monoclonal antibody was shown to be able to inhibit plasmin generation on the surface of cells (Lopez-Alemany *et al.*, 2003). Moreover, anti-enolase antibodies able to inhibit plasminogen binding to α-enolase have been detected in patients with SLE or MC and active nephritis (Moscato *et al.*, 2000). The role of autoantibodies endowed with such a property in the induction and perpetuation of renal damage remains under investigation. It could help unraveling the pathogenic mechanisms underlying tissue damage.

6. Conclusions

Our understanding of antibody-mediated renal damage has not changed drastically in recent years. However, within this framework a lot of new data emerged to change our way of looking at "old" diseases. For example, APSGN, which in the past has been regarded as a typical CIC-mediated disease, may also show ICs formed *in situ*. In Goodpasture's syndrome, once considered to be the classical model of autoantibody-mediated disease, it is now realized that nephritogenic T cells play an important role. Complement, formerly thought to be a pathogenic factor in nephritis, is now believed to be crucial for the protection of immune-mediated renal injury. At the same time, components of the fibrinolytic system play a role as autoantibody targets in renal damage, and a disregulation of this pathway may be a mechanism in different renal diseases.

Furthermore, new hypotheses are emerging to explain how the immune system interacts with the external world. One of them suggests that a great deal more is involved than the mere discrimination between self and nonself: the immune system is able to handle all elements, even endogenous ones, that pose a potential danger to the host (Matzinger, 1994). Thus, tissues themselves play a central role: when healthy, they may induce tolerance and when distressed, they may stimulate immunity.

In this view, autoimmunity can be regarded as a defect in the "cross-talk" between the immune system and the body tissues, a perspective that opens up fascinating new avenues for research on the mechanisms that regulate the expression of renal antigens in health and disease.

References

Amoura, Z., Chabre, H., Koutouzov, S., Lotton, C., Cabrespines, A., Bach, J.F., and Jacob, L. (1994). Nucleosome-restricted antibodies are detected before anti-dsDNA and/or antihistone antibodies in serum of MRL-Mp lpr/lpr and +/+ mice, and are present in kidney eluates of lupus mice with proteinuria. *Arthritis Rheum.*, **37**, 1684–1688.

Berden, J.H., Licht, R., van Bruggen, M.C., and Tax, W.J. (1999). Role of nucleosomes for induction and glomerular binding of autoantibodies in lupus nephritis. *Curr. Opin. Nephrol. Hypertens.*, **8**, 299–306.

Birmingham, D.J., Chen, W., Liang, G., Schmitt, H.C., Gavit, K., and Nagaraja, H.N. (2003). A CR1 polymorphism associated with constitutive erythrocyte CR1 levels affects binding to C4b but not C3b. *Immunology*, **108**, 531–538.

Borza, D.B. and Hudson, B.G. (2003). Molecular characterization of the target antigens of anti-glomerular basement membrane antibody disease. *Springer Semin. Immunopathol.*, **24**, 345–361.

Botto, M. and Walport, M.J. (2002). C1q, autoimmunity and apoptosis. *Immunobiology*, **205**, 395–406.

Burova, L., Thern, A., Pigarevsky, P., Gladilina, M., Seliverstova, V., Gavrilova, E., Nagornev, V., Schalen, C., and Totolian, A. (2003). Role of group A streptococcal IgG binding proteins in triggering experimental glomerulonephritis in the rabbit. *Acta Pathol. Microbiol. Immunol. Scand.*, **111**, 955–962.

Caponi, L., Chimenti, D., Pratesi, F., and Migliorini, P. (2002). Anti-ribosomal antibodies from lupus patients bind DNA. *Clin. Exp. Immunol.*, **130**, 541–547.

Chan, O.T., Hannum, L.G., Haberman, A.M., Madaio, M.P., and Shlomchik, M.J. (1999). A novel mouse with B cells but lacking serum antibody reveals an antibody-independent role for B cells in murine lupus. *J. Exp. Med.*, **189**, 1639–1648.

Clynes, R., Dumitru, C., and Ravetch, J.V. (1998). Uncoupling of immune complex formation and kidney damage in autoimmune glomerulonephritis. *Science*, **279**, 1052–1054.

Craig, M.L., Bankovich, A.J., McElhenny, J.L., and Taylor, R.P. (2000). Clearance of anti-double-stranded DNA antibodies: The natural immune complex clearance mechanism. *Arthritis Rheum.*, **43**, 2265–2275.

Cu, G.A., Mezzano, S., Bannan, J.D., and Zabriskie, J.B. (1998). Immunohistochemical and serological evidence for the role of streptococcal proteinase in acute post-streptococcal glomerulonephritis. *Kidney Int.*, **54**, 819–826.

Davies, K.A., Robson, M.G., Peters, A.M., Norsworthy, P., Nash, J.T., and Walport, M.J. (2002). Defective Fc-dependent processing of immune complexes in patients with systemic lupus erythematosus. *Arthritis Rheum.*, **46**, 1028–1038.

Dehan, P., Weber, M., Zhang, X., Reeders, S.T., Foidart, J.M., and Tryggvason, K. (1996). Sera from patients with anti-GBM nephritis including Goodpasture syndrome show heterogenous reactivity to recombinant NC1 domain of type IV collagen alpha chains. *Nephrol. Dial. Transplant.*, **11**, 2215–2222.

Deocharan, B., Qing, X., Lichauco, J., and Putterman, C. (2002). Alpha-actinin is a cross-reactive renal target for pathogenic anti-DNA antibodies. *J. Immunol.*, **168**, 3072–3078.

Ehrenstein, M.R., Katz, D.R., Griffiths, M.H., Papadaki, L., Winkler, T.H., Kalden, J.R., and Isenberg, D.A. (1995). Human IgG anti-DNA antibodies deposit in kidneys and induce proteinuria in SCID mice. *Kidney Int.*, **48**, 705–711.

Emlen, W. and Mannik, M. (1984). Effect of DNA size and strandedness on the *in vivo* clearance and organ localization of DNA. *Clin. Exp. Immunol.*, **56**, 185–192.

Fontan, P.A., Pancholi, V., Nociari, M.M., and Fischietti, V.A. (2000). Antibodies to streptococcal surface enolase react with human alpha-enolase: Implications in poststreptococcal sequelae. *J. Infect. Dis.*, **182**, 1712–1721.

Foster, M.H., Sabbaga, J., Line, S.R., Thompson, K.S., Barrett, K.J., and Madaio, M.P. (1993). Molecular analysis of spontaneous nephrotropic anti-laminin antibodies in an autoimmune MRL-lpr/lpr mouse. *J. Immunol.*, **151**, 814–824.

Friedman, J., van de Rijn, I., Ohkuni, H., Fischetti, V.A., and Zabriskie, J.B. (1984). Immunological studies of post-streptococcal sequelae. Evidence for presence of streptococcal antigens in circulating immune complexes. *J. Clin. Invest.*, **74**, 1027–1034.

Glurich, I., Winters, B., Albini, B., and Stinson, M. (1991). Identification of *Streptococcus pyogenes* proteins that bind to rabbit kidney *in vitro* and *in vivo*. *Microb. Pathog.*, **10**, 209–220.

Kalluri, R., Danoff, T.M., Okada, H., and Neilson, E.G. (1997). Susceptibility to anti-glomerular basement membrane disease and Goodpasture syndrome is linked to MHC class II genes and the emergence of T cell-mediated immunity in mice. *J. Clin. Invest.*, **100**, 2263–2275.

Karassa, F.B., Trikalinos, T.A., Ioannidis, J.P., and FcgammaRIIa-SLE Meta-Analysis Investigators (2002). Role of the Fc receptor IIa polymorphism in susceptibility to systemic lupus erythematosus and lupus nephritis: A meta-analysis. *Arthritis Rheum.*, **46**, 1563–1571.

Kitching, A.R., Holdsworth, S.R., Ploplis, V.A., Plow, E.F., Collen, D., Carmeliet, P., and Tipping, P.G. (1997). Plasminogen and plasminogen activators protect against renal injury in crescentic glomerulonephritis. *J. Exp. Med.*, **185**, 963–968.

Kos, C.H., Le, T.C., Sinha, S., Henderson, J.M., Kim, S.H., Sugimoto, H., Kalluri, R., Gerszten, R.E., and Pollak, M.R. (2003). Mice deficient in alpha-actinin-4 have severe glomerular disease. *J. Clin. Invest.*, **111**, 1683–1690.

Lin, C.Y. (1982). Serial studies of circulating immune complexes in poststeptococcal glomerulonephritis. *Pediatrics*, **70**, 725–727.

Lopez-Alemany, R., Longstaff, C., Hawley, S., Mirshahi, M., Fabregas, P., Jardi, M., Merton, E., Miles, L.A., and Felez, J. (2003). Inhibition of cell surface mediated plasminogen activation by a monoclonal antibody against alpha-enolase. *Am. J. Hematol.*, **72**, 234–242.

Madaio, M.P., Carlson, J., Cataldo, J., Ucci, A., Migliorini, P., and Pankewycz, O. (1987). Murine monoclonal anti-DNA antibodies bind directly to glomerular antigens and form immune deposits. *J. Immunol.*, **138**, 2883–2889.

Madaio, M.P., Fabbi, M., Tiso, M., Daga, A., and Puccetti, A. (1996). Spontaneously produced anti-DNA/DNase I autoantibodies modulate nuclear apoptosis in living cells. *Eur. J. Immunol.*, **26**, 3035–3041.

Madaio, M.P. and Shlomchik, M.J. (1996). Emerging concepts regarding B cells and autoantibodies in murine lupus nephritis. B cells have multiple roles; all autoantibodies are not equal. *J. Am. Soc. Nephrol.*, **7**, 387–396.

Madaio, M.P. and Yanase, K. (1998). Cellular penetration and nuclear localization of anti-DNA antibodies: Mechanisms, consequences, implications and applications. *J. Autoimmun.*, **11**, 535–538.

Mannik, M., Merrill, C.E., Stamps, L.D., and Wener, M.H. (2003). Multiple autoantibodies form the glomerular immune deposits in patients with systemic lupus erythematosus. *J. Rheumatol.*, **30**, 1495–1504.

Marin, C., Mosquera, J., and Rodirguez-Iturbe, B. (1997). Histological evidence of neuraminidase involvement in acute nephritis: Desialized leukocytes infiltrate the kidney in acute post-streptococcal glomerulonephritis. *Clin. Nephrol.*, **47**, 217–221.

Mason, L.J., Ravirajan, C.T., Rahman, A., Putterman, C., and Isenberg, D.A. (2004). Is alpha-actinin a target for pathogenic anti-DNA antibodies in lupus nephritis? *Arthritis Rheum.*, **50**, 866–870.
Matzinger, P. (1994). Tolerance, danger, and the extended family. *Annu. Rev. Immunol.*, **12**, 991–1045.
McIntosh, R.M., Garcia, R., Rubio, L., Rabideau, D., Allen, J.E., Carr, R.I., and Rodriguez-Iturbe, B. (1978). Evidence of an autologous immune complex pathogenic mechanism in acute poststreptococcal glomerulonephritis. *Kidney Int.*, **14**, 501–510.
Michel, M., Piette, J.C., Roullet, E., Duron, F., Frances, C., Nahum, L., Pelletier, N., Crassard, I., Nunez, S., Michel, C., Bach, J., and Tournier-Lasserve, E. (2000). The R131 low-affinity allele of the Fc gamma RIIA receptor is associated with systemic lupus erythematosus but not with other autoimmune diseases in French Caucasians. *Am. J. Med.*, **108**, 580–583.
Migliorini, P., Pratesi, F., Bongiorni, F., Moscato, S., Scavuzzo, M., and Bombardieri, S. (2002). The targets of nephritogenic antibodies in systemic autoimmune disorders. *Autoimmun. Rev.*, **1**, 168–173.
Miles, L.A., Dahlberg, C.M., Plescia, J., Felez, J., Kato, K., and Plow, E.J., (1991). Role of cell-surface lysines in plasminogen binding to cells: Identification of alpha-enolase as a candidate plasminogen receptor. *Biochemistry*, **30**, 1682–1691.
Moodie, F.D.L., Leaker, B., Cambridge, G., Totty, N.F., and Segal, A.W. (1993). Alpha-enolase: A novel cytosolic autoantigen in ANCA positive vasculitis. *Kidney Int.*, **43**, 675–681.
Moscato, S., Pratesi, F., Sabbatini, A., Chimenti, D., Scavuzzo, M., Passatino, R., Bombardieri, S., Giallongo, A., and Migliorini, P. (2000) Surface expression of a glycolytic enzyme, alpha-enolase, recognized by autoantibodies in connective tissue disorders. *Eur. J. Immunol.*, **30**, 3575–3584.
Mostoslavsky, G., Fischel, R., Yachimovich, N., Yarkoni, Y., Rosenmann, E., Monestier, M., Baniyash, M., and Eilat, D. (2001). Lupus anti-DNA autoantibodies cross-react with a glomerular structural protein: A case for tissue injury by molecular mimicry. *Eur. J. Immunol.*, **31**, 1221–1227.
Nakamura, A., Yuasa, T., Ujike, A., Ono, M., Nukiwa, T., Ravetch, J.V., and Takai, T. (2000). Fcgamma receptor IIB-deficient mice develop Goodpasture's syndrome upon immunization with type IV collagen: A novel murine model for autoimmune glomerular basement membrane disease. *J. Exp. Med.*, **191**, 899–906.
Nordstrand, A., Norgren, M., Ferretti, J.J., and Holm S.E. (1998). Streptokinase as a mediator of acute post-streptococcal glomerulonephritis in an experimental mouse model. *Infect. Immun.*, **66**, 315–321.
Nordstrand, A., Norgren, M., and Holm, S.E. (1999). Pathogenic mechanism of acute post-streptococcal glomerulonephritis. *Scand. J. Infect. Dis.*, **31**, 523–537.
Pankewycz, O.G., Migliorini, P., and Madaio, M.P. (1987). Polyreactive autoantibodies are nephritogenic in murine lupus nephritis. *J. Immunol.*, **139**, 3287–3294.
Park, S.Y., Ueda, S., Ohno, H., Hamano, Y., Tanaka, M., Shiratori, T., Yamazaki, T., Arase, H., Arase, N., Karasawa, A., Sato, S., Ledermann, B., Kondo, Y., Okumura, K., Ra, C., and Saito, T. (1998). Resistance of Fc receptor-deficient mice to fatal glomerulonephritis. *J. Clin. Invest.*, **102**, 1229–1238.
Peutz-Kootstra, C.J., Hansen, K., De Heer, E, Abrass, C.K., and Bruijn, J.A. (2000). Differential expression of laminin chains and anti-laminin autoantibodies in experimental lupus nephritis. *J. Pathol.*, **192**, 404–412.
Pratesi, F., Moscato, S., Sabbatini, A., Chimenti, D., Bombardieri, S. and Migliorini, P. (2000). Autoantibodies specific for alpha-enolase in systemic autoimmune disorders. *J. Rheumatol.*, **27**, 109–115.
Puccetti, A., Madaio, M.P., Bellese, G., and Migliorini, P. (1995). Anti-DNA antibodies bind to DNase I. *J. Exp. Med.*, **181**, 1797–1804.
Putterman, C. (2004). New approaches to the renal pathogenicity of anti-DNA antibodies in systemic lupus erythematosus. *Autoimmun. Rev.*, **3**, 7–11.
Ravetch, J.V. (2002). A full complement of receptors in immune complex diseases. *J. Clin. Invest.*, **110**, 1759–1761.
Raz, E., Ben-Bassat, H., Davidi, T., Shlomai, Z., and Eilat, D. (1993). Cross-reactions of anti-DNA autoantibodies with cell surface proteins. *Eur. J. Immunol.*, **23**, 383–390.
Raz, E., Brezis, M., Rosenmann, E., and Eilat, D. (1989). Anti-DNA antibodies bind directly to renal antigens and induce kidney dysfunction in the isolated perfused rat kidney. *J. Immunol.*, **142**, 3076–3082.

Reefman, E., Dijstelbloem, H.M., Limburg, P.C., Kallenberg, C.G., and Bijl, M. (2003). Fcgamma receptors in the initiation and progression of systemic lupus erythematosus. *Immunol. Cell Biol.*, **81**, 382–389.

Reichlin, M. (1996). Presence of ribosomal P protein on the surface of human umbilical vein endothelial cells. *J. Rheumatol.*, **23**, 1123–1125.

Reynolds, J., Cashman, S.J., Evans, D.J., and Pusey, C.D. (1991). Cyclosporin A in the prevention and treatment of experimental autoimmune glomerulonephritis in the brown Norway rat. *Clin. Exp. Immunol.*, **85**, 28–32.

Reynolds, J. and Pusey, C.D. (2001). Oral administration of glomerular basement membrane prevents the development of experimental autoimmune glomerulonephritis in the WKY rat. *J. Am. Soc. Nephrol.*, **12**, 61–70.

Rodriguez-Iturbe, B., Rabideau, D., Garcia, R., Rubio, L., and McIntosh, R.M. (1980). Characterization of the glomerular antibody in acute poststreptococcal glomerulonephritis. *Ann. Intern. Med.*, **92**, 478–481.

Rumore, P.M. and Steinman, C.R. (1990). Endogenous circulating DNA in systemic lupus erythematosus. Occurrence as multimeric complexes bound to histone. *J. Clin. Invest.*, **86**, 69–74.

Rutgers A., Meyers K.E., Canziani, G., Kalluri, R, Lin, J., and Madaio, M.P. (2000). High affinity of anti-GBM antibodies from Goodpasture and transplanted Alport patients to alpha3(IV)NC1 collagen. *Kidney Int.*, **58**, 115–122.

Ryan, J.J., Mason, P.J., Pusey, C.D., and Turner, N. (1998). Recombinant alpha-chains of type IV collagen demonstrate that the amino terminal of the Goodpasture autoantigen is crucial for antibody recognition. *Clin. Exp. Immunol.*, **113**, 17–27.

Sabbaga, J., Line, S.R., Potocnjak, P., and Madaio, M.P. (1989). A murine nephritogenic monoclonal anti-DNA autoantibody binds directly to mouse laminin, the major non-collagenous protein component of the glomerular basement membrane. *Eur. J. Immunol.*, **19**, 137–143.

Sabbatini, A., Dolcher, M.P., Marchini, B., Chimenti, D., Moscato, S., Pratesi, F., Bombardieri, S., and Migliorini, P. (1997). Alpha-enolase is a renal-specific antigen associated with kidney involvement in mixed cryoglobulinemia. *Clin. Exp. Rheumatol.*, **15**, 655–658.

Salmon, J.E., Millard, S., Schachter, L.A., Arnett, F.C., Ginzler, E.M., Gourley, M.F., Ramsey-Goldman, R., Peterson, M.G.E., and Kimberly, R.P. (1996). Fc-gamma-RIIA alleles are heritable risk factors for lupus nephritis in African Americans. *J. Clin. Invest.*, **97**, 1348–1354.

Sheerin, N.S., Springall, T., Carroll, M.C., Hartley, B., and Sacks, S.H. (1997). Protection against anti-glomerular basement membrane (GBM)-mediated nephritis in C3- and C4-deficient mice. *Clin. Exp. Immunol.*, **110**, 403–409.

Shlomchik, M.J., Madaio, M.P., Ni, D., Trounstein, M., and Huszar, D. (1994). The role of B cells in lpr/lpr-induced autoimmunity. *J. Exp. Med.*, **180**, 1295–1306.

Siegert, C., Daha, M., Westedt, M.L., van der Voort, E., and Breedveld, F. (1991). IgG autoantibodies against C1q are correlated with nephritis, hypocomplementemia, and dsDNA antibodies in systemic lupus erythematosus. *J. Rheumatol.*, **18**, 230–234.

Sun, K.H., Liu, W.T., Tang, S.J., Tsai, C.Y., Hsieh, S.C., Wu, T.H., Han, S.H., and Yu, C.L. (1996). The expression of acidic ribosomal phosphoproteins on the surface membrane of different tissues in autoimmune and normal mice which are the target molecules for anti-double-stranded DNA antibodies. *Immunology*, **87**, 362–371.

Sylvestre, D., Clynes, R., Ma, M., Warren, H., Carroll, M.C., and Ravetch, J.V. (1996). Immunoglobulin G-mediated inflammatory responses develop normally in complement-deficient mice. *J. Exp. Med.*, **184**, 2385–2392.

Takai, T. (2002). Roles of Fc receptors in autoimmunity. *Nat. Rev. Immunol.*, **2**, 580–592.

Tan Sardjono, C., Mottram, P.L., and Hogarth, P.M. (2003). The role of FcgammaRIIa as an inflammatory mediator in rheumatoid arthritis and systemic lupus erythematosus. *Immunol. Cell Biol.*, **81**, 374–381.

Tarzi, R.M., Davies, K.A., Claassens, J.W., Verbeek, J.S., Walport, M.J., and Cook, H.T. (2003). Both Fcgamma receptor I and Fcgamma receptor III mediate disease in accelerated nephrotoxic nephritis. *Am. J. Pathol.*, **162**, 1677–1683.

Tipping, P.G. and Holdsworth, S.R. (2003). T cells in glomerulonephritis. *Springer Semin. Immunopathol.*, **24**, 377–393.

Trouw, L.A., Duijs, J.M., van Kooten, C., and Daha, M.R. (2003a). Immune deposition of C1q and anti-C1q antibodies in the kidney is dependent on the presence of glomerular IgG. *Mol. Immunol.*, **40**, 595–602.

Trouw, L.A., Groeneveld, T.W., Seelen, M.A., Duijs, J.M., Bajema, I.M., Prins, F.A., Kishore, U., Salant, D.J., Verbeek, J.S., van Kooten, C., and Daha, M.R. (2004). Anti-C1q autoantibodies deposit in glomeruli but are only pathogenic in combination with glomerular C1q-containing immune complexes. *J. Clin. Invest.*, **114**, 679–688.

Trouw, L.A., Seelen, M.A., Duijs, J.M., Benediktsson, H., Van Kooten, C., and Daha, M.R. (2003b). Glomerular deposition of C1q and anti-C1q antibodies in mice following injection of antimouse C1q antibodies. *Clin. Exp. Immunol.*, **132**, 32–39.

van Bruggen, M.C., Kramers, C., Walgreen, B., Elema, J.D., Kallenberg, C.G., van den Born, J., Smeenk, R.J., Assmann, K.J., Muller, S., Monestier, M., and Berden, J.H. (1997a). Nucleosomes and histones are present in glomerular deposits in human lupus nephritis. *Nephrol. Dial. Transplant.*, **12**, 57–66.

van Bruggen, M.C., Walgreen, B., Rijke, T.P., Tamboer, W., Kramers, K., Smeenk, R.J., Monestier, M., Fournie, G.J., and Berden, J.H. (1997b). Antigen specificity of anti-nuclear antibodies complexed to nucleosomes determines glomerular basement membrane binding *in vivo*. *Eur. J. Immunol.*, **27**, 1564–1569.

Wakui, H., Imai, H., Komatsuda, A., and Miura, A.B. (1999). Circulating antibodies against alpha-enolase in patients with primary membranous nephropathy (MN). *Clin. Exp. Immunol.*, **118**, 445–450.

Waters, S.T., McDuffie, M., Bagavant, H., Deshmukh, U.S., Gaskin, F., Jiang, C., Tung, K.S., and Fu, S.M. (2004). Breaking tolerance to double stranded DNA, nucleosome, and other nuclear antigens is not required for the pathogenesis of lupus glomerulonephritis. *J. Exp. Med.*, **199**, 255–264.

Wu, J., Arends, J., Borillo, J., Zhou, C., Merszei, J., McMahon, J., and Lou, Y.H. (2004). A self T cell epitope induces autoantibody response: Mechanism for production of antibodies to diverse glomerular basement membrane antigens. *J. Immunol.*, **172**, 4567–4574.

Wu, J., Hicks, J., Borillo, J., Glass, W.F., 2nd, and Lou Y.H. (2002). CD4(+) T cells specific to a glomerular basement membrane antigen mediate glomerulonephritis. *J. Clin. Invest.*, **109**, 517–524.

Wu, J., Hicks, J., Ou, C., Singleton, D., Borillo, J., and Lou, Y.H. (2001). Glomerulonephritis induced by recombinant collagen IV alpha 3 chain noncollagen domain 1 is not associated with glomerular basement membrane antibody: A potential T cell-mediated mechanism. *J. Immunol.*, **167**, 2388–2395.

Yamamoto, T. and Wilson, C.B. (1987). Binding of anti-basement membrane antibody to alveolar basement membrane after intratracheal gasoline instillation in rabbits. *Am. J. Pathol.*, **126**, 497–505.

Yanase, K., Smith, R.M., Puccetti, A., Jarett, L., and Madaio, M.P. (1997). Receptor-mediated cellular entry of nuclear localizing anti-DNA antibodies via myosin 1. *J. Clin. Invest.*, **100**, 25–31.

Yoshizawa, N., Oshima, S., Takeuchi, A., Kondo, S., Oda, T., Shimizu, J., Nishiyama, J., Ishida, A., Nakabayashi, I., Tazawa, K., and Sakurai, Y. (1997). Experimental acute glomerulonephritis induced in the rabbit with a specific streptococcal antigen. *Clin. Exp. Immunol.*, **107**, 61–67.

Yoshizawa, N., Yamakami, K., Fujino, M., Oda, T., Tamura, K., Matsumoto, K., Sugisaki, T., and Boyle, M.D. (2004). Nephritis-associated plasmin receptor and acute poststreptococcal glomerulonephritis: Characterization of the antigen and associated immune response. *J. Am. Soc. Nephrol.*, **15**, 1785–1793.

14

Estrogen, Interferon-gamma, and Lupus

S. Ansar Ahmed and Ebru Karpuzoglu-Sahin

1. Introduction

To maintain immune homeostasis, the immune system must actively and rigorously impose restraints on the induction of damaging immune responses to self-antigens. Under normal circumstances, effective mechanisms exist to either eliminate or silence dangerous autoreactive cells. Failure to do so will likely result in the genesis of autoimmune diseases. To date, the cause(s) of autoimmune diseases is not known, but several key factors have been identified that promote the onset and/or affect the course of autoimmune diseases. These include genetic, infectious, environmental, and hormonal factors. Autoimmune diseases can afflict almost any tissue type, and it is likely that there may be differing or disparate triggering and/or pathologic mechanisms in the induction of autoimmune lesions in diverse tissues. This situation may be comparable to cancer, where oncopathogenic events differ among various types of cancers. Each autoimmune disease ought to be treated independently and generalizations must be treated with caution. Given the diversity of autoimmune diseases, it is not surprising that even though genetic factors are vital for the induction of autoimmune disease, there has been no identification of a genetic defect that is common to all autoimmune diseases. Interestingly, an overwhelmingly common feature of most autoimmune diseases is that women are more susceptible to these disorders compared to men (Ansar Ahmed et al., 1999; Ansar Ahmed, 2000). It is noteworthy that the degree of susceptibility of women to different autoimmune diseases varies. Diseases such as systemic lupus erythematosus (SLE), Sjögren's syndrome, thyroiditis, and scleroderma principally occur in women and demonstrate a very high female-to-male susceptibility ratio (Whitacre, 2001). Diseases such

---

S. Ansar Ahmed and Ebru Karpuzoglu-Sahin • Center for Molecular Medicine and Infectious Diseases, Department of Biomedical Sciences and Pathobiology, Virginia–Maryland Regional College of Veterinary Medicine, Virginia Tech (Virginia Polytechnic Institute and State University), Blacksburg, Virginia 24061-0342.

Molecular Autoimmunity: In commemoration of the 100th anniversary of the first description of human autoimmune disease, edited by Moncef Zouali. Springer Science+Business Media, Inc., New York, 2005.

as rheumatoid arthritis, myasthenia gravis, and multiple sclerosis (MS) also occur predominantly in women, albeit the female-to-male susceptibility ratio is lower when compared to that of SLE and thyroiditis. In a minority of autoimmune diseases (such as diabetes and ulcerative colitis), both women and men appear to be equally susceptible. An epidemiological study that sought to determine the prevalence of 24 established autoimmune diseases estimated that 1 in 31 Americans are afflicted with 1 of these 24 autoimmune diseases (Jacobson *et al.*, 1997). The prevalence of autoimmune diseases likely would have been much higher had several more diseases with possible autoimmune components been included. Significantly, most diseases in this survey were more common in women, leading to an estimated risk 2.7 times higher than that in men to acquire an autoimmune disease.

Most autoimmune diseases occur predominantly in female animals compared to males. Female (NZB × NZW)F_1 mice (B/W) manifest lupus disease months earlier than males. These classic findings have been established for many experimental autoimmune diseases and have been extensively reviewed (Ansar Ahmed *et al.*, 1999; Ansar Ahmed, 2000).

The precise reason(s) underlying the marked gender differences in susceptibility to autoimmune diseases is a subject of intense investigation. Possible explanations for the gender differences include: the effects of sex hormones, regulation by sex chromosomal genes, and microchimerism. Much work has been done with regard to the effects of sex hormones in autoimmune diseases. Studies in a number of animal models of autoimmune diseases make a strong case for the role of sex hormones in autoimmune diseases. This chapter highlights the importance of sex factors, especially estrogens, in autoimmune diseases. Studies on the effects of estrogens on immunity and autoimmunity are of increasing biomedical importance, especially considering that human exposure to estrogens occurs through various sources. These include: natural estrogens (17β-estradiol) that vary during the lifetimes of women, from low levels in prepubertal life to high levels during estrus and during pregnancy; exogenous estrogens given for medical reasons [e.g., diethylstilbestrol (DES), hormone replacement therapy, estrogen-based oral contraceptives]; and environmental estrogens (or endocrine-disrupting chemicals) that are omnipresent in the environment (pesticides, plastic products, detergents, industrial by-products, municipal sewage–contaminated water that contains metabolites of estrogen-based contraceptive drugs). It is possible that an individual may be exposed to all of these main sources of estrogenic compounds. It is conceivable that these combined estrogens may more profoundly modulate the immune system of individuals.

2. Estrogen and Lupus: Human and Animal Studies

It is beyond the scope of this chapter to discuss the role of estrogens in all autoimmune diseases. Therefore, we focus primarily on estrogen effects on lupus, since the effects of this hormone have been noted in humans and in murine models of lupus. SLE patients have autoantibodies against many self-antigens including double-stranded DNA (dsDNA), red blood cells, platelets, leukocytes, and clotting factors, which leads to the formation of immune complexes. The deposition of these immune complexes triggers inflammation, culminating in widespread tissue damage (Abdou *et al.*, 1981). Gender is a strong risk factor for SLE since this disease primarily affects women in the reproductive years and the female-to-male susceptibility ratio can be as high as 13:1 (Rider and Abdou, 2001). SLE has been associated with situations where levels of gonadal hormones are changing such as during pregnancy, postpartum period, menopause, and during estrogen administration. The first onset of the disease is unlikely to occur before puberty or after menopause. Pregnancy has been associated with flares of lupus (Wilder, 1998). SLE disease activity fluctuates with the menstrual cycle (Bruce and Laskin, 1997) and lessens after menopause (Mok *et al.*, 1999). The flares of lupus have been reported to increase during *in vitro* fertilization when levels of female hormones, particularly estrogen, are clinically manipulated (Guballa *et al.*, 2000). Further, although not unequivocal, exogenous estrogen administration such as oral contraceptives and the use of HRT (Petri, 2001) have been reported to affect the disease course. Female lupus patients tend to have increased 16α-hydroxyesterone and estriol, as well as increased oxidation at C-17 position compared to controls (Lahita, 1999). The precise contribution of altered sex hormone metabolism to lupus, although provocative, is not clear.

Estrogen has been associated with B cell activation and T cell dysregulation. For example, direct exposure of peripheral blood mononuclear cells (PBMCs) from lupus patients to 17β-estradiol induced the secretion of anti-dsDNA antibodies and enhanced the secretion of immunoglobulins (Kanda *et al.*, 1999). In similar cultures of PBMCs from healthy donors, estrogen enhanced immunoglobulin levels but did not induce anti-dsDNA autoantibodies, thereby suggesting that estrogen has differential effects in SLE and normal individuals. This estrogen-induced increase in autoantibodies to dsDNA and secretion of immunoglobulins may be related to IL-10, since estrogen was found to increase IL-10 secretion by monocytes, and anti-IL-10 partially blocked the increase in B cell secretion of autoantibodies and immunoglobulins (Kanda *et al.*, 1999). Interestingly, aberrant T cell activation was evident in lupus T cells when cultured in the presence of estrogen. T cells from female lupus patients had a dose-dependent increase in calcineurin steady-state mRNA levels and an increase in phosphatase activity. In contrast to the estrogen effects on T cells from female SLE patients, estrogen had no effect on calcineurin in T cells from normal females, normal males, or lupus males (Rider and Abdou, 2001). This suggests that estrogen has differential effects depending upon the source of target T cells. These effects of estrogen are mediated through the estrogen receptor (ER), since culturing of female SLE T cells in the presence of an ER antagonist blocks the estrogen-induced increase in calcineurin and phosphatase activity. Additionally, differences in the response of lupus T cells to estrogen have been noted. Estrogen increased CD40L mRNA and the amount of CD40L expression on T cells from SLE patients, but not on T cells from normal individuals. It is conceivable that the estrogen-dependent increases in CD40L expression could hyperstimulate SLE T cells and may contribute to the pathogenesis of SLE (Rider *et al.*, 2001).

Much evidence supporting the role of sex hormones in lupus has come from studies in B/W mice, where mere alterations in the levels of sex hormones can have profound effects on the disease. For example, relatively resistant male

B/W mice (that have a delayed onset of the disease) can be made susceptible by the administration of estrogens or depletion of male hormones (by orchiectomy or administration of anti-androgens) (Ansar Ahmed et al., 1999). Conversely, susceptible female B/W mice can be made relatively resistant to the disease by the administration of androgens or the estrogen antagonist, tamoxifen (Ansar Ahmed et al., 1999). In yet another model of lupus, MRL/lpr mice, which develop an aggressive disease together with lymphadenopathy, estrogen was shown to be a potent accelerator of lupus (Carlsten et al., 1990). Estrogen treatment of MRL/lpr mice resulted in the appearance of forbidden autoreactive clones in the liver (Okuyama et al., 1992). These cells include $\alpha\beta TCR^{Intermediate}$, $V\beta3^+$, or $V\beta8^+$ T cells that are often deleted in the thymus.

Studies in our laboratory have shown that 17β-estradiol treatment of non-autoimmune mice (e.g., C57BL/6) induce autoantibodies against dsDNA and phospholipids, which are common in lupus (Ansar Ahmed et al., 1999; Ansar Ahmed, 2000). This is an important conceptual finding that implies that estrogen can override B cell tolerance to induce autoimmunity even in normal mice. Similarly, others have also shown that 17β-estradiol can break B cell tolerance in non-autoimmune BALB/c mice transgenic for the heavy-chain of pathogenic anti-DNA antibodies. 17β-estradiol can induce high-affinity autoantibodies against DNA as well as immune complex glomerulonephritis (Grimaldi et al., 2001). Hybridomas generated from estrogen-treated mice express high-affinity, unmutated anti-DNA antibodies. This indicates that naive B cells that are normally deleted or anergized are rescued from tolerance induction (Bynoe et al., 2000). Further, like 17β-estradiol, the synthetic estrogen DES has also been shown to induce autoantibodies in both normal and lupus-prone mice (Forsberg, 2000; Yurino et al., 2004).

Recently, a new class of estrogens called environmental estrogens (or endocrine-disrupting chemicals) has been identified in the environment (Sonnenschein and Soto, 1998). These compounds are generally considered to be weak estrogens, yet they can disrupt the endocrine system through many mechanisms including hormone mimicry, blocking or altering hormonal binding to receptors, binding to ERs (or conceivably other receptors) to alter gene regulation, and/or altering the metabolism of natural estrogen. A current concern is whether environmental estrogens can affect autoimmune diseases. Experimental evidence suggests that this may be likely. For example, studies using the well-documented environmental estrogen, Bisphenol-A (BPA), which is present in resins, plastics, dental sealants, adhesives, flame retardants, and optical lens materials suggest that it can promote autoimmunity. Administration of BPA to lupus-prone B/W mice resulted in increased autoantibody secretion by B-1 cells ($CD5^+$, IgM^{hi}, $B220^{Int}$, $IL-5R^+$) (Yurino et al., 2004). Direct in vitro exposure of B-1 cells to BPA induced IgM autoantibodies to a level that was comparable to that noticed in cultures exposed to DES or estradiol. These effects were more pronounced in B-1 cells from aged mice (8–12 months of age) compared to young mice (1 month of age). This supports the argument that environmental estrogens may affect autoimmune diseases.

3. Mechanisms of Estrogen Effects on the Immune System

3.1. Estrogen Exerts Its Biological Effects on Cells by Both Estrogen Receptor–Dependent and –Independent Mechanisms

A concise description of ERs and their interactions with DNA is helpful to appreciate the molecular mode of action of estrogens and to understand the diversity of potential effects of estrogens on the immune system. Estrogen exerts its biological functions on target tissues by both ER-dependent and ER-independent mechanisms. Estrogen binds to two specific, but distinct receptors, ERα and ERβ, which belong to the nuclear hormone receptor family. ERs are ligand-activated transcription factors and are located both intracellularly and presumably on the cell surface. Heat shock proteins, such as hsp90, bind to unliganded ERs to maintain the receptors in an inactive but functionally prepared state for ligand binding (Pratt and Toft, 1997). Each type of ER appears to be differentially expressed in various tissues. It is conceivable that the differences in the relative expression of ERs in different tissues may result in the selective actions of estrogen in the immune system. ERα and ERβ comprise several domains: A/B domain (located near the NH_2 terminus), C domain (DNA-binding domain), D domain (hinge region), and E/F domain (ligand-binding domain, COOH terminal) (Figure 14.1A). ERα and ERβ share an amino acid homology of 97% in the DNA-binding domain, while they have a homology of only 60% in the ligand-binding domain. This suggests that two ERs can interact with the same genes but bind differentially to ligands (Kuiper et al., 1997). ERα has two transactivation functional regions: activation factor-1 (AF-1) in the DNA-binding domain and AF-2 in the ligand-binding domain. These regions synergize with each other and provide the response to estrogen. ERβ also has an AF-1 region but most of its activity comes from AF-2 (McInerney et al., 1998). The hinge region contains a nuclear localization signal and links the C domain to the multifunctional carboxyl-terminus E/F domain. The hormone-dependent AF-2 region in the E/F domain is important in ligand-dependent transcriptional activity and in interaction with coactivators. Further, this region is responsible for ligand-dependent activation by nuclear receptors, interactions with heat shock protein (hsp), nuclear translocation, and transactivation of target genes (Meier, 1997). Differential binding of ERs to transcriptional cofactors (coactivators, corepressors, and coregulators) will likely have a profound impact on transcription of estrogen-responsive genes. The coactivator proteins steroid receptor coactivator-1 (SRC-1) and transcriptional intermediary factor-2 (GRIP1/TIF2) enhance gene expression by remodeling chromatin and allowing interactions with the basal transcription machinery (Cavailles et al., 1995; Henttu et al., 1997; Johansson et al., 1999). In contrast, corepressors and coregulators inhibit gene activation or turn off the activated genes. For example, coregulatory factors RIP140 and SHP compete with SRC-1 coactivator proteins such as TIF2 for AF-2 regions and may recruit deacetylases to estrogen target genes.

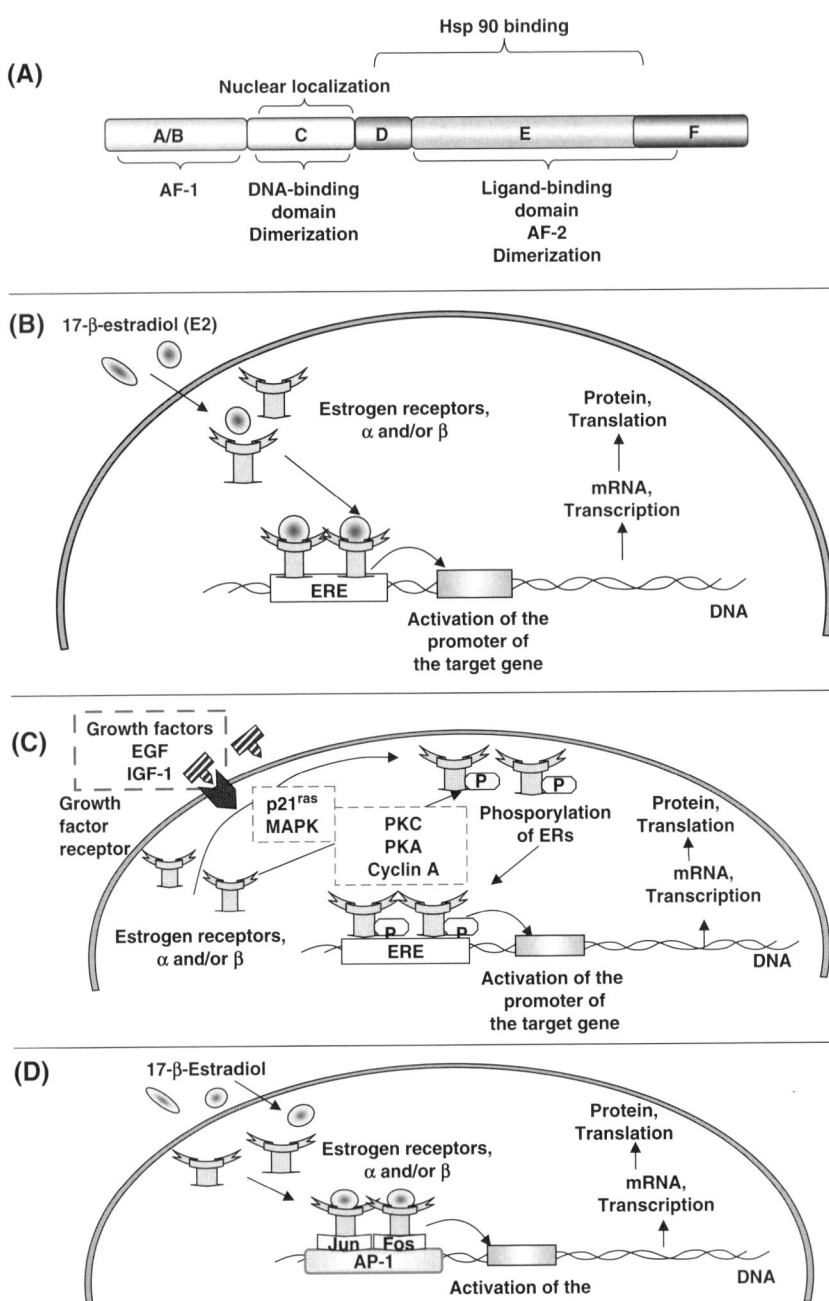

Figure 14.1A: The Structure of the Estrogen Receptor. The estrogen receptor is composed of several domains, including A/B domain (located near the amino terminus), C domain (the DNA binding domain, D (the hinge region), and E/F domain (the ligand-binding domain near the COOH terminus. Activation Factors are depicted as AF-1 and AF-2. **Figure 14.1B:** *The ligand-dependent*

Continues

Lymphoid and nonlymphoid cells in both mice and humans possess ERs (Ansar Ahmed, 2000; Weihua *et al.*, 2003). ERα is expressed on nearly all lymphocyte subsets from the thymus and thymic stromal cells. The double-positive CD4$^+$CD8$^+$ thymocytes and thymic stromal cells express ERβ (Mor *et al.*, 2001). Studies in many laboratories are under way to identify the type of ERs in distinct cells of the immune system. ERα is obligatory for functional development of the thymus, since ERα-deficient mice have hypoplasia of the thymus and T cell developmental impediments such as a decrease in numbers of mature CD4$^+$CD8$^-$ thymocytes (Erlandsson *et al.*, 2003). The expression of ERβ is essential for estrogen-mediated thymic cortex atrophy. In SLE patients, monocytes, T cells, and B cells express transcripts of ER (Suenaga *et al.*, 2001). Reverse transcription and polymerase chain amplification indicate that ERα and ERβ mRNA are expressed in human T cells (Rider and Abdou, 2001).

ERβ and ERα are essential for regulating B cell responses, splenic size, and cytokine responses (Erlandsson *et al.*, 2003). Much of the data supporting this are derived from mice lacking ERα and/or ERβ genes. ERα-deficient mice have decreased splenic size, while ERβ-deficient mice exhibit splenomegaly, thus suggesting that ERs are important in maintaining normal spleen size (Erlandsson *et al.*, 2003). Complete B lymphopoesis in the bone marrow and spleen was only observed in wild-type mice with intact ERα and ERβ, when compared to ERα and ERβ-deficient mice given estrogen. The ERα-deficient mice given estrogen had fewer B cell subpopulations in the bone marrow, whereas wild-type mice given estrogen had decreased early hematopoietic B cell progenitors and a shift toward a mature B cell subpopulation. ERα is mainly responsible for regulating estrogen-induced B cell changes and immunoglobulin secretion. The loss of ERβ resulted in hyperplasia in the bone marrow, increased B cell numbers in the blood, and lymphadenopathy (Shim *et al.*, 2003). The expression of ERs is also important for T helper cell activities. ERα, but not ERβ, is essential for enhanced estrogen-induced Th1 cell responses such as increased IFNγ secretion (Maret *et al.*, 2003).

Recent investigations have shown that estrogen-responsive genes can be activated by a variety of mechanisms (Nilsson and Gustafsson, 2002), which are depicted in Figure 14.1B4-2. First, in the classical pathway (Figure 14.1), the binding of the ER to estrogen results in a conformational change involving dimerization

---

Figure 14.1. Continued *Classical Estrogen Receptor Pathway.* The binding of estrogen receptor to estrogen results in a conformational change in the receptor. This estrogen-estrogen receptor complex is then translocated to the nucleus. This complex will specifically bind to the estrogen response elements (ERE) in the promoter of target genes and affect their transcription. **Figure 14.1C:** *Ligand-Independent Pathway.* Estrogen receptors can be activated even in the absence of their usual ligand, estrogen. For example, estrogen receptors can be phosphorylated and activated following signaling mediated by growth factors, Epidermal Growth Factor (EGF) and other molecules. The activated estrogen receptor can bind to ERE to affect the transcription of target genes. **Figure 14.1D:** *Ligand-dependent, but Estrogen Receptor Element (ERE)-independent Pathway.* Estrogen receptors can "cross-talk" with other transcriptional molecules. For example, estrogen-estrogen receptor complex can bind to Jun and Fos, that are in turn bound to AP-1 site. The coactivators recruited by Jun/Fos bind to the estrogen receptors and affect the transcription of genes.

and release of the ER from heat shock proteins. This conformational change of the ER involves exposing a hydrophobic surface in the AF-2 region of the ERs, allowing the coactivator proteins to bind. Transcription of the target gene is mediated by binding of the DNA-binding domain of the ER to an estrogen response element (ERE) that is present in the promoter of estrogen-responsive genes. Once bound to an ERE, the ER dimer interacts with coactivators to promote chromatin remodeling and recruits other transcription factors. This results in the upregulation or downregulation of target gene expression. Interestingly, both ERα and ERβ can exist intracellularly as homodimers or heterodimers (Levin, 2002). The different transcriptional activities of homodimers or heterodimers formed by ERα and ERβ may explain the selective actions of estrogen on different cell types and genes (Katzenellenbogen, 1996). Although these ERs are both expressed in tissues and form functional heterodimers, when coexpressed, ERβ inhibits the transcriptional activity of ERα at saturating hormone levels (Matthews and Gustafsson, 2003). Thus, overall estrogen responsiveness may be determined by the ERα versus ERβ ratio in cells where both receptors are expressed.

Second, in a ligand-independent pathway, ERs can be activated in the absence of their usual ligand, estrogen (Figure 14.1C). ERs can be activated by many nonestrogenic physiological molecules, such as growth factors (EGF, IGF-1), cell cycle proteins, and protein kinases. EGF-mediated signaling involving the MAP kinase pathway can activate ERs by phosphorylation. The activated ER can then bind to EREs to affect gene transcription (Bunone *et al.*, 1996). ERs can also be phosphorylated by cyclins (Rogatsky *et al.*, 1999), general regulators such as protein kinase C (PKC) (Lahooti *et al.*, 1998), or protein kinase A (PKA) (Bunone *et al.*, 1996).

Third, in the ERE-independent pathway, ERs can "cross-talk" with other transcription molecules (Figure 14.1D). For example, estrogen-activated ER complexes can physically interact with key transcription molecules including fos-jun complexes, NF-κB, and GC box-bound SP-1 to modulate the transcription of target genes (Kushner *et al.*, 2000). Both ERα and ERβ when complexed with estrogen bind to Jun and Fos located on the AP-1 site. The coactivator (p160) recruited by Jun/Fos binds to the ERs and triggers increased transcription of the target gene. In addition, the interaction between ERα and c-Rel prevents NF-κB from binding to the IL-6 promoter and inhibits protein expression of IL-6 (Ray *et al.*, 1997). ERα also interacts with the transcription factor Sp-1, regardless of the presence or absence of estrogen (Porter *et al.*, 1997).

Fourth, in the nongenomic pathway, also called the cell surface ER signaling pathway, estrogen can induce rapid signaling by binding to ERs on the cell surface that reside in cell membrane domains named caveolae (Levin, 2002). Estrogen treatment stimulates the protein synthesis of caveolae structural coat protein, caveolin-1, which facilitates ER translocation to the cell membrane (Razandi *et al.*, 2002). The estrogen–ER complex in the caveolae may participate in signaling by activating G proteins. The activation of G proteins leads to rapid and specific signaling through activation of phosphoinositol 3-kinase/Akt (PI3K) and MAPK pathways resulting in rapid gene transcription and biological effects implying that this pathway does not involve ER-dependent genomic alterations

(Pedram et al., 2002). Estrogen–ER signaling can also occur via other signaling molecules such as p21ras (Migliaccio et al., 1996), B-Raf (Singh et al., 1999), and Src (Migliaccio et al., 2000). The existence of this pathway in cells of the immune system has also been demonstrated (Guo et al., 2002).

3.2. Estrogen Alterations of B cells

One important mechanism by which estrogen promotes lupus-like features in mice may be via inducing alterations in B cell differentiation and maturation. We have shown that estrogen treatment of non-autoimmune C57BL/6 mice enhanced B cell differentiation into plasma cells that secreted autoantibodies against dsDNA and phospholipids (Ansar Ahmed et al., 1999; Ansar Ahmed, 2000). Further, in a lupus model (BALB/c mice transgenic for the heavy-chain of a pathogenic anti-DNA antibody), estrogen decreased immature transitional B cells and increased mature anti-dsDNA autoantibody-secreting marginal zone (MZ) B cells, implicating this B cell subset in autoimmunity (Grimaldi et al., 2001).

Defective apoptosis is well known to be involved in autoimmunity. In BALB/c mice transgenic for the heavy-chain of a pathogenic anti-DNA antibody, the expression of the anti-apoptotic protein, Bcl-2, in splenic B cells of estrogen-treated mice was increased (Bynoe et al., 2000). Further, estrogen treatment protected isolated primary B cells from B cell receptor–mediated apoptosis. Estrogen upregulated several genes that are involved in B cell activation and survival, including *cd22*, *shp-1*, *bcl-2*, and *vcam-1*. These effects of estrogen are mediated through the ERs (Grimaldi et al., 2002). In our model of estrogen treatment of non-autoimmune mice, we also noticed that splenic B cells demonstrated a remarkable ability to survive in culture. This survival of splenic B cells from estrogen-treated mice is even more pronounced with anti-CD40 antibody treatment with concomitant upregulation of antiapoptotic proteins, Bcl-2 and Bcl-x_L. Thus, estrogen can alter B cell development, leading to the survival, expansion, and activation of a population of autoreactive B cells (Ansar Ahmed et al., 1999; Grimaldi et al., 2001). In humans, direct exposure of PBMC from SLE patients to 17β-estradiol resulted in decreased apoptosis of blood monocytes and decreased secretion of TNFα, which may allow survival of autoreactive cells in SLE patients (Evans et al., 1997).

It is possible that estrogens can regulate the levels of B cell activation factor (BAFF), an aspect that is currently being examined in our laboratory. This factor is crucial for the survival of MZ and transitional-2 B cells and is potentially implicated in a number of autoimmune diseases (Mackay and Kalled, 2002).

3.3. Estrogen Effects on Cytokines

The immune system uses cytokines as molecular messengers to coordinate the functioning of diverse cells of lymphoid and nonlymphoid organs. It is therefore not surprising that in an immune-dysregulated state such as autoimmune disease, significant alterations in the profile and levels of cytokines or response to these cytokines are evident. One mechanism by which estrogens influence autoimmunity is by regulating the secretion of cytokines and signaling responses to these cytokines. 17β-estradiol affects several cytokines and chemokines including IFNγ, IL-1α, IL-4, IL-6, TNFα, IL-10, and IL-12 (Deshpande *et al.*, 1997; Carruba *et al.*, 2003). It is beyond the scope of this chapter to discuss at length the various cytokines that sex hormones regulate. Instead, we focus here on IFNγ, a proinflammatory cytokine that plays an important role in autoimmune diseases, including lupus, and is highly responsive to estrogens.

4. IFNγ in SLE and Other Autoimmune Diseases

Several lines of evidence implicate IFNγ in inflammatory autoimmune disease (Billiau, 1996; Schwarting *et al.*, 1998). First, signaling through the IFNγ receptor is essential for the initiation and progression of lupus nephritis in lupus-prone, MRL-lpr^{fas} mice (Schwarting *et al.*, 1998). Second, the IFNγ-related transcription factor T-bet has been shown to regulate IgG2$_a$ class switching and induction of pathogenic autoantibodies in murine lupus (Peng *et al.*, 2002). Third, during the course of SLE, the increased level and the expression of IFNγ was accompanied by an increase in IgG2$_a$ and IgG3 levels (Reininger *et al.*, 1996). Fourth, in the absence of IFNγ gene, the levels of anti-dsDNA and ss-DNA autoantibodies, and immune complex–mediated glomerulonephritis are decreased in murine models of SLE (Carvalho-Pinto *et al.*, 2002). Fifth, blocking IFNγ by the administration of plasmids with cDNA encoding the IFNγ-receptor Fc into MRL-lpr mice (at preclinical or advanced stages of lupus) resulted in decreased IFNγ levels in the sera and decreased hallmarks of lupus, such as autoantibodies, hyperplasia, glomerulonephritis, and mortality (Lawson *et al.*, 2000). Sixth, addition of estrogen to CD4$^+$ T cell lines from MS patients results in increased production of IFNγ and IL-10 (Gilmore *et al.*, 1997). Finally, in the absence of the IL-12p40 gene, the IFNγ levels are decreased in lupus patients and the survival is increased (Kikawada *et al.*, 2003).

In several strains of mice including non-autoimmune (C57BL/6J, CBA/Ca) and autoimmune (B/W) mice, females secrete more IFNγ compared to males (Huygen and Palfliet, 1983; Haas *et al.*, 1998; Karpuzoglu-Sahin *et al.*, 2001). Transcription of the IFNγ gene was also increased in unstimulated T cells from females of the B/W strain of mice (Sato *et al.*, 1995). Studies from our laboratory have shown that several estrogenic compounds (17β-estradiol, DES, genistein, α-zeralanone) influence the transcription of the IFNγ gene or the levels of IFNγ protein (Karpuzoglu-Sahin *et al.*, 2001; Calemine *et al.*, 2003). In the case of 17β-estradiol, splenocytes from estrogen-treated C57BL/6 mice, when activated with T cell mitogens, had increased IFNγ gene expression and higher levels of IFNγ. In this strain, estrogen had no marked effect on IL-4 mRNA or protein levels. Estrogen treatment of CD-1 mice also increased IFNγ mRNA expression in Concanavalin-A-stimulated splenocytes (Fox *et al.*, 1991). Culturing of PBMC from nonpregnant women with estrogen increased IFNγ secretion compared to similar cultures from men (Grasso and Muscettola, 1990). Administration of physiological doses of 17β estradiol to castrated female mice results in the selective development of IFNγ-producing cells from the lymph nodes (Maret *et al.*, 2003). ERα, but not ERβ, was necessary for enhanced estrogen-driven Th1 cell responsiveness (Maret *et al.*, 2003).

5. Conclusions

It is unquestionable that a majority of autoimmune diseases tend to occur more frequently in females. The effects of sex hormones, such as estrogens, are also clearly evident in animal models of autoimmune diseases. However, caution must be exercised when generalizing the effects of estrogens to all autoimmune diseases. Estrogen effects on autoimmunity can vary from increased disease severity (e.g., lupus) to amelioration (e.g., rheumatoid arthritis, experimental allergic encephalomyelitis) of the disease. While the precise mechanisms are not readily apparent, multiple factors can influence the outcome of estrogen-induced immune responses. It is noteworthy that despite decades of intense research in endocrinology, the complexity of estrogen action on reproductive tissues is only now being appreciated. Our understanding of estrogenic effects on the immune system is limited by comparison. Extrapolating the data from reproductive to immune tissues, estrogen effects on target genes are critically influenced by factors such as ERα, ERβ (ERαα, ERββ, or ERαβ ratios), cofactors, coactivators, and coregulators. Further, estrogen-responsive genes can be activated by both ligand- or ER-dependent and -independent mechanisms. The ERs in turn can cross-talk with other transcriptional molecules such as NF-κB and AP-1 that are crucial in immune regulation and influence gene transcription. It is therefore conceivable that differences in expression of ERα or ERβ, responses of ER to estrogen, ER-mediated signaling, and coactivators and coregulators can explain the differing responses to estrogens in various cell types of the immune system. It should not be surprising that estrogen effects on autoimmune diseases can vary in different experimental settings and in different animal models. Further, estrogen can have different immunological effects in autoimmune compared to healthy individuals. For example, estrogen can have varied effects in cells from lupus patients compared to healthy donors in relation to cytokine induction, calcineurin and CD40L expression on T cells (Rider and Abdou, 2001; Rider *et al.*, 2001).

It is clear from many experimental studies that the outcome of estrogen-induced immunomodulation, including cytokine induction, is dependent upon many factors, such as the dose of estrogen exposure, type of cells, activation status of cells, stage of estrus cycle, and genetic background. With regard to the dose, at physiological concentrations ($\sim 10^{-8}$ M), estrogen significantly decreased the spontaneous secretion of IL-6, IL-1rα, and IL-1β in whole blood cultures from post-menopausal women. At concentrations higher than physiological levels or at pharmacological doses, estrogen enhanced the production of IL-1 and IL-6 (Rogers and Eastell, 2001). With increasing concentrations of estrogen, the cytokine profile from the supernatants of whole human blood cells shifted from Th1 to Th2 (Matalka, 2003). The IFNγ, IL-12, and IL-10 levels were increased in cultures that were incubated with preovulatory doses of estrogen (Matalka, 2003). The PHA- and

LPS-stimulated whole blood cells cultured with estrogen (at concentrations similar to those experienced during pregnancy) had suppressed IFNγ IL-12 and the IFNγ/IL-10 ratio and increased IL-10 secretion (Matalka, 2003). The increase in IL-10 switches the Th1 profile to Th2 cytokines, which is important for successful pregnancy. The levels of IL-1α, IL-1β, and TNFα were highest during proestrus and/or estrus (De *et al*., 1992; Willis *et al*., 2003). Estrogen may have differential effects depending upon the tissue: it decreases TNFα in LPS-stimulated murine splenocytes and bone marrow–derived macrophages (Deshpande *et al*., 1997), but increases TNFα in peritoneal rat macrophages (Chao *et al*., 2000).

Estrogen can promote autoimmunity by both prolactin-dependent and -independent mechanisms. Further, it is well accepted that vulnerability to lupus depends on the presence of an appropriate genetic susceptibility component coupled with an unknown environmental trigger. The degree to which autoreactive B cells are eliminated from the naive B cell repertoire is genetically regulated and may determine whether a nonspontaneously autoimmune host will develop autoimmunity following an environmental trigger. Recent studies have shown that lupus-like serology was evident in BALB/c but not in DBA/2 mice, following immunization with a peptide mimotope of DNA (Wang *et al*., 2003). This was largely due to differences in the B cell compartment, since DBA/2 mice have fewer B cells specific for the DNA mimotope compared to susceptible BALB/c. Further, BALB/c mice possess more autoreactive cells in the native repertoire, and demonstrate less antigen-induced B cell apoptosis. In contrast, DBA/2 mice have a stronger B cell receptor signal and more stringent central tolerance, which correlate with resistance to lupus induction. It is likely that response to estrogen's ability to influence autoimmunity may similarly be different among various genetic backgrounds. In this context, in the anti-DNA heavy-chain transgenic model, estrogen readily promoted lupus in BALB/c mice when compared to C57BL/6 mice. We have similarly shown that genes in the MHC and non-MHC regions influence immunomodulation by testosterone (Ansar Ahmed *et al*., 1999).

In conclusion, the effects of estrogen on autoimmune diseases have been vigorously demonstrated in many experimental settings and to a lesser extent in clinical situations. However, estrogen effects may vary in individual autoimmune diseases. This should not be perplexing, since individual autoimmune diseases tend to differ. Moreover, estrogen effects on target cells are complex. Generalization ought to be avoided and more attention should be paid to effects of estrogens on individual diseases. This will hopefully enhance our mechanistic understanding of autoimmune diseases and allow the design of more effective strategies for therapy.

Acknowledgments

This study was supported by grants from NIH-1RO1 AI051880 and 1RO1-ES08043, and USDA/HATCH program. We gratefully acknowledge the editorial assistance of Mrs. Andrea Lengi and Mrs. Rebecca Phillips.

References

Abdou, N.I., Wall, H., Lindsley, H.B., Halsey, J.F., and Suzuki, T. (1981). Network theory in autoimmunity. In vitro suppression of serum anti-DNA antibody binding to DNA by anti-idiotypic antibody in systemic lupus erythematosus. *J. Clin. Invest.*, **67**, 1297–1304.

Ansar Ahmed, S. (2000). The immune system as a potential target for environmental estrogens (endocrine disrupters): A new emerging field. *Toxicology*, **150**, 191–206.

Ansar Ahmed, S., Hissong, B.D., Verthelyi, D., Donner, K., Becker, K., and Karpuzoglu-Sahin, E. (1999). Gender and risk of autoimmune diseases: Possible role of estrogenic compounds. *Environ. Health Perspect.*, **107** (Suppl. 5), 681–686.

Billiau, A. (1996). Interferon-gamma in autoimmunity. *Cytokine Growth Factor Rev.*, **7**, 25–34.

Bruce, I.N. and Laskin, C.A. (1997). Sex hormones in systemic lupus erythematosus: A controversy for modern times. *J. Rheumatol.*, **24**, 1461–1463.

Bunone, G., Briand, P.A., Miksicek, R.J., and Picard, D. (1996). Activation of the unliganded estrogen receptor by EGF involves the MAP kinase pathway and direct phosphorylation. *EMBO J.*, **15**, 2174–2183.

Bynoe, M.S., Grimaldi, C.M., and Diamond, B. (2000). Estrogen up-regulates Bcl-2 and blocks tolerance induction of naive B cells. *Proc. Natl. Acad. Sci. USA*, **97**, 2703–2708.

Calemine, J., Zalenka, J., Karpuzoglu-Sahin, E., Ward, D.L., Lengi, A., and Ahmed, S.A. (2003). The immune system of geriatric mice is modulated by estrogenic endocrine disruptors (diethylstilbestrol, alpha-zearalanol, and genistein): Effects on interferon-gamma. *Toxicology*, **194**, 115–128.

Carlsten, H., Tarkowski, A., Holmdahl, R., and Nilsson, L.A. (1990). Oestrogen is a potent disease accelerator in SLE-prone MRL lpr/lpr mice. *Clin. Exp. Immunol.*, **80**, 467–473.

Carruba, G., D'Agostino, P., Miele, M., Calabro, M., Barbera, C., Bella, G.D., Milano, S., Ferlazzo, V., Caruso, R., Rosa, M.L., Cocciadiferro, L., Campisi, I., Castagnetta, L., and Cillari, E. (2003). Estrogen regulates cytokine production and apoptosis in PMA-differentiated, macrophage-like U937 cells. *J. Cell. Biochem.*, **90**, 187–196.

Carvalho-Pinto, C.E., Garcia, M.I., Mellado, M., Rodriguez-Frade, J.M., Martin-Caballero, J., Flores, J., Martinez, A.C., and Balomenos, D. (2002). Autocrine production of IFN-gamma by macrophages controls their recruitment to kidney and the development of glomerulonephritis in MRL/lpr mice. *J. Immunol.*, **169**, 1058–1067.

Cavailles, V., Dauvois, S., L'Horset, F., Lopez, G., Hoare, S., Kushner, P.J., and Parker, M.G. (1995). Nuclear factor RIP140 modulates transcriptional activation by the estrogen receptor. *EMBO J.*, **14**, 3741–3751.

Chao, T.C., Chao, H.H., Chen, M.F., Greager, J.A., and Walter, R.J. (2000). Female sex hormones modulate the function of LPS-treated macrophages. *Am. J. Reprod. Immunol.*, **44**, 310–318.

De, M., Sanford, T.R., and Wood, G.W. (1992). Interleukin-1, interleukin-6, and tumor necrosis factor alpha are produced in the mouse uterus during the estrous cycle and are induced by estrogen and progesterone. *Dev. Biol.*, **151**, 297–305.

Deshpande, R., Khalili, H., Pergolizzi, R.G., Michael, S.D., and Chang, M.D. (1997). Estradiol down-regulates LPS-induced cytokine production and NFkB activation in murine macrophages. *Am. J. Reprod. Immunol.*, **38**, 46–54.

Erlandsson, M.C., Jonsson, C.A., Islander, U., Ohlsson, C., and Carlsten, H. (2003). Oestrogen receptor specificity in oestradiol-mediated effects on B lymphopoiesis and immunoglobulin production in male mice. *Immunology*, **108**, 346–351.

Evans, M.J., MacLaughlin, S., Marvin, R.D., and Abdou, N.I. (1997). Estrogen decreases in vitro apoptosis of peripheral blood mononuclear cells from women with normal menstrual cycles and decreases TNF-alpha production in SLE but not in normal cultures. *Clin. Immunol. Immunopathol.*, **82**, 258–262.

Forsberg, J.G. (2000). Neonatal estrogen treatment and its consequences for thymus development, serum level of autoantibodies to cardiolipin, and the delayed-type hypersensitivity response. *J. Toxicol. Environ. Health. Part A*, **60**, 185–213.

Fox, H.S., Bond, B.L., and Parslow, T.G. (1991). Estrogen regulates the IFN-gamma promoter. *J. Immunol.*, **146**, 4362–4367.

Gilmore, W., Weiner, L.P., and Correale, J. (1997). Effect of estradiol on cytokine secretion by proteolipid protein-specific T cell clones isolated from multiple sclerosis patients and normal control subjects. *J. Immunol.*, **158**, 446–451.

Grasso, G. and Muscettola, M. (1990). The influence of beta-estradiol and progesterone on interferon gamma production in vitro. *Int. J. Neurosci.*, **51**, 315–317.

Grimaldi, C.M., Cleary, J., Dagtas, A.S., Moussai, D., and Diamond, B. (2002). Estrogen alters thresholds for B cell apoptosis and activation. *J. Clin. Invest.*, **109**, 1625–1633.

Grimaldi, C.M., Michael, D.J., and Diamond, B. (2001). Cutting edge: Expansion and activation of a population of autoreactive marginal zone B cells in a model of estrogen-induced lupus. *J. Immunol.*, **167**, 1886–1890.

Guballa, N., Sammaritano, L., Schwartzman, S., Buyon, J., and Lockshin, M.D. (2000). Ovulation induction and in vitro fertilization in systemic lupus erythematosus and antiphospholipid syndrome. *Arthritis Rheum.*, **43**, 550–556.

Guo, Z., Krucken, J., Benten, W.P., and Wunderlich, F. (2002). Estradiol-induced nongenomic calcium signaling regulates genotropic signaling in macrophages. *J. Biol. Chem.*, **277**, 7044–7050.

Haas, C., Ryffel, B., and Le Hir, M. (1998). IFN-gamma receptor deletion prevents autoantibody production and glomerulonephritis in lupus-prone (NZB × NZW)F1 mice. *J. Immunol.*, **160**, 3713–3718.

Henttu, P.M., Kalkhoven, E., and Parker, M.G. (1997). AF-2 activity and recruitment of steroid receptor coactivator 1 to the estrogen receptor depend on a lysine residue conserved in nuclear receptors. *Mol. Cell. Biol.*, **17**, 1832–1839.

Huygen, K. and Palfliet, K. (1983). Strain variation in interferon gamma production of BCG-sensitized mice challenged with PPD. I. CBA/Ca mice are low producers in vivo, but high producers in vitro. *Cell. Immunol.*, **80**, 329–334.

Jacobson, D.L., Gange, S.J., Rose, N.R., and Graham, N.M. (1997). Epidemiology and estimated population burden of selected autoimmune diseases in the United States. *Clin. Immunol. Immunopathol.*, **84**, 223–243.

Johansson, L., Thomsen, J.S., Damdimopoulos, A.E., Spyrou, G., Gustafsson, J.A., and Treuter, E. (1999). The orphan nuclear receptor SHP inhibits agonist-dependent transcriptional activity of estrogen receptors ERalpha and ERbeta. *J. Biol. Chem.*, **274**, 345–353.

Kanda, N., Tsuchida, T., and Tamaki, K. (1999). Estrogen enhancement of anti-double-stranded DNA antibody and immunoglobulin G production in peripheral blood mononuclear cells from patients with systemic lupus erythematosus. *Arthritis Rheum.*, **42**, 328–337.

Karpuzoglu-Sahin, E., Hissong, B.D., and Ansar Ahmed, S. (2001). Interferon-gamma levels are upregulated by 17-beta-estradiol and diethylstilbestrol. *J. Reprod. Immunol.*, **52**, 113–127.

Katzenellenbogen, B.S. (1996). Estrogen receptors: Bioactivities and interactions with cell signaling pathways. *Biol. Reprod.*, **54**, 287–293.

Kikawada, E., Lenda, D.M., and Kelley, V.R. (2003). IL-12 deficiency in MRL-Fas(lpr) mice delays nephritis and intrarenal IFN-gamma expression, and diminishes systemic pathology. *J. Immunol.*, **170**, 3915–3925.

Kuiper, G.G., Carlsson, B., Grandien, K., Enmark, E., Haggblad, J., Nilsson, S., and Gustafsson, J.A. (1997). Comparison of the ligand binding specificity and transcript tissue distribution of estrogen receptors alpha and beta. *Endocrinology*, **138**, 863–870.

Kushner, P.J., Agard, D.A., Greene, G.L., Scanlan, T.S., Shiau, A.K., Uht, R.M., and Webb, P. (2000). Estrogen receptor pathways to AP-1. *J. Steroid Biochem. Mol. Biol.*, **74**, 311–317.

Lahita, R.G. (1999). Gender and age in lupus. In R.G. Lahita (ed.), *Systemic Lupus Erythematosus*. Academic Press, New York. pp. 129–143.

Lahooti, H., Thorsen, T., and Aakvaag, A. (1998). Modulation of mouse estrogen receptor transcription activity by protein kinase C delta. *J. Mol. Endocrinol.*, **20**, 245–259.

Lawson, B.R., Prud'homme, G.J., Chang, Y., Gardner, H.A., Kuan, J., Kono, D.H., and Theofilopoulos, A.N. (2000). Treatment of murine lupus with cDNA encoding IFN-gammaR/Fc. *J. Clin. Invest.*, **106**, 207–215.

Levin, E.R. (2002). Cellular functions of plasma membrane estrogen receptors. *Steroids*, **67**, 471–475.

Mackay, F. and Kalled, S.L. (2002). TNF ligands and receptors in autoimmunity: An update. *Curr. Opin. Immunol.*, **14**, 783–790.

Maret, A., Coudert, J.D., Garidou, L., Foucras, G., Gourdy, P., Krust, A., Dupont, S., Chambon, P., Druet, P., Bayard, F., and Guery, J.C. (2003). Estradiol enhances primary antigen-specific CD4 T cell responses and Th1 development in vivo. Essential role of estrogen receptor alpha expression in hematopoietic cells. *Eur. J. Immunol.*, **33**, 512–521.

Matalka, K.Z. (2003). The effect of estradiol, but not progesterone, on the production of cytokines in stimulated whole blood, is concentration-dependent. *Neuroendocrinol. Lett.*, **24**, 185–191.

Matthews, J. and Gustafsson, J.A. (2003). Estrogen signaling: A subtle balance between ER alpha and ER beta. *Mol. Interv.*, **3**, 281–292.

McInerney, E.M., Weis, K.E., Sun, J., Mosselman, S., and Katzenellenbogen, B.S. (1998). Transcription activation by the human estrogen receptor subtype beta (ER beta) studied with ER beta and ER alpha receptor chimeras. *Endocrinology*, **139**, 4513–4522.

Meier, C.A. (1997). Regulation of gene expression by nuclear hormone receptors. *J. Recept. Signal Transduct. Res.*, **17**, 319–335.

Migliaccio, A., Castoria, G., Di Domenico, M., de Falco, A., Bilancio, A., Lombardi, M., Barone, M.V., Ametrano, D., Zannini, M.S., Abbondanza, C., and Auricchio, F. (2000). Steroid-induced androgen receptor-oestradiol receptor beta-Src complex triggers prostate cancer cell proliferation. *EMBO J.*, **19**, 5406–5417.

Migliaccio, A., Di Domenico, M., Castoria, G., de Falco, A., Bontempo, P., Nola, E., and Auricchio, F. (1996). Tyrosine kinase/p21ras/MAP-kinase pathway activation by estradiol-receptor complex in MCF-7 cells. *EMBO J.*, **15**, 1292–1300.

Mok, C.C., Lau, C.S., Ho, C.T., and Wong, R.W. (1999). Do flares of systemic lupus erythematosus decline after menopause? *Scand. J. Rheumatol.*, **28**, 357–362.

Mor, G., Munoz, A., Redlinger, R., Jr., Silva, I., Song, J., Lim, C., and Kohen, F. (2001). The role of the Fas/Fas ligand system in estrogen-induced thymic alteration. *Am. J. Reprod. Immunol.*, **46**, 298–307.

Nilsson, S. and Gustafsson, J.A. (2002). Biological role of estrogen and estrogen receptors. *Crit. Rev. Biochem. Mol. Biol.*, **37**, 1–28.

Okuyama, R., Abo, T., Seki, S., Ohteki, T., Sugiura, K., Kusumi, A., and Kumagai, K. (1992). Estrogen administration activates extrathymic T cell differentiation in the liver. *J. Exp. Med.*, **175**, 661–669.

Pedram, A., Razandi, M., Aitkenhead, M., Hughes, C.C., and Levin, E.R. (2002). Integration of the non-genomic and genomic actions of estrogen. Membrane-initiated signaling by steroid to transcription and cell biology. *J. Biol. Chem.*, **277**, 50768–50775.

Peng, S.L., Szabo, S.J., and Glimcher, L.H. (2002). T-bet regulates IgG class switching and pathogenic autoantibody production. *Proc. Natl. Acad. Sci. USA*, **99**, 5545–5550.

Petri, M. (2001). Exogenous estrogen in systemic lupus erythematosus: Oral contraceptives and hormone replacement therapy. *Lupus*, **10**, 222–226.

Porter, W., Saville, B., Hoivik, D., and Safe, S. (1997). Functional synergy between the transcription factor Sp1 and the estrogen receptor. *Mol. Endocrinol.*, **11**, 1569–1580.

Pratt, W. B. and Toft, D.O. (1997). Steroid receptor interactions with heat shock protein and immunophilin chaperones. *Endocr. Rev.*, **18**, 306–360.

Ray, P., Ghosh, S.K., Zhang, D.H., and Ray, A. (1997). Repression of interleukin-6 gene expression by 17 beta-estradiol: Inhibition of the DNA-binding activity of the transcription factors NF-IL6 and NF-kappa B by the estrogen receptor. *FEBS Lett.*, **409**, 79–85.

Razandi, M., Oh, P., Pedram, A., Schnitzer, J., and Levin, E.R. (2002). ERs associate with and regulate the production of caveolin: Implications for signaling and cellular actions. *Mol. Endocrinol.*, **16**, 100–115.

Reininger, L., Santiago, M.L., Takahashi, S., Fossati, L., and Izui, S. (1996). T helper cell subsets in the pathogenesis of systemic lupus erythematosus. *Ann. Med. Interne (Paris)*, **147**, 467–471.

Rider, V. and Abdou, N.I. (2001). Gender differences in autoimmunity: Molecular basis for estrogen effects in systemic lupus erythematosus. *Int. Immunopharmacol.*, **1**, 1009–1024.

Rider, V., Jones, S., Evans, M., Bassiri, H., Afsar, Z., and Abdou, N.I. (2001). Estrogen increases CD40 ligand expression in T cells from women with systemic lupus erythematosus. *J. Rheumatol.*, **28**, 2644–2649.

Rogatsky, I., Trowbridge, J.M., and Garabedian, M.J. (1999). Potentiation of human estrogen receptor alpha transcriptional activation through phosphorylation of serines 104 and 106 by the cyclin A-CDK2 complex. *J. Biol. Chem.*, **274**, 22296–22302.

Rogers, A. and Eastell, R. (2001). The effect of 17beta-estradiol on production of cytokines in cultures of peripheral blood. *Bone*, **29**, 30–34.
Sato, M.N., Minoprio, P., Avrameas, S., and Ternynck, T. (1995). Defects in the regulation of anti-DNA antibody production in aged lupus-prone (NZB × NZW)F1 mice: Analysis of T-cell lymphokine synthesis. *Immunology*, **85**, 26–32.
Schwarting, A., Wada, T., Kinoshita, K., Tesch, G., and Kelley, V.R. (1998). IFN-gamma receptor signaling is essential for the initiation, acceleration, and destruction of autoimmune kidney disease in MRL-Fas(lpr) mice. *J. Immunol.*, **161**, 494–503.
Shim, G.J., Wang, L., Andersson, S., Nagy, N., Kis, L.L., Zhang, Q., Makela, S., Warner, M., and Gustafsson, J.A. (2003). Disruption of the estrogen receptor beta gene in mice causes myeloproliferative disease resembling chronic myeloid leukemia with lymphoid blast crisis. *Proc. Natl. Acad. Sci. USA*, **100**, 6694–6699.
Singh, M., Setalo, G., Jr., Guan, X., Warren, M., and Toran-Allerand, C.D. (1999). Estrogen-induced activation of mitogen-activated protein kinase in cerebral cortical explants: Convergence of estrogen and neurotrophin signaling pathways. *J. Neurosci.*, **19**, 1179–1188.
Sonnenschein, C. and Soto, A.M. (1998). An updated review of environmental estrogen and androgen mimics and antagonists. *J. Steroid Biochem. Mol. Biol.*, **65**, 143–150.
Suenaga, R., Rider, V., Evans, M.J., and Abdou, N.I. (2001). In vitro-activated human lupus T cells express normal estrogen receptor proteins which bind to the estrogen response element. *Lupus*, **10**, 116–122.
Wang, C., Khalil, M., Ravetch, J., and Diamond, B. (2003). The naive B cell repertoire predisposes to antigen-induced systemic lupus erythematosus. *J. Immunol.*, **170**, 4826–4832.
Weihua, Z., Andersson, S., Cheng, G., Simpson, E.R., Warner, M., and Gustafsson, J.A. (2003). Update on estrogen signaling. *FEBS Lett.*, **546**, 17–24.
Whitacre, C.C. (2001). Sex differences in autoimmune disease. *Nat. Immunol.*, **2**, 777–780.
Wilder, R.L. (1998). Hormones, pregnancy, and autoimmune diseases. *Ann. NY Acad. Sci.*, **840**, 45–50.
Willis, C., Morris, J.M., Danis, V., and Gallery, E.D. (2003). Cytokine production by peripheral blood monocytes during the normal human ovulatory menstrual cycle. *Hum. Reprod.*, **18**, 1173–1178.
Yurino, H., Ishikawa, S., Sato, T., Akadegawa, K., Ito, T., Ueha, S., Inadera, H., and Matsushima, K. (2004). Endocrine disruptors (environmental estrogens) enhance autoantibody production by b1 cells. *Toxicol. Sci.*, **81**, 139–147.

15

Extent of Regulatory T Cell Influence on Major Histocompatibility Complex Class II Gene Control of Susceptibility in Murine Autoimmune Thyroiditis

Yi-chi M. Kong, Gerald P. Morris, and Chella S. David

1. Introduction

Murine experimental autoimmune thyroiditis (EAT) has served as a model in studies of autoimmune diseases for nearly 3 decades, due to the early recognition of the association of major histocompatibility complex (MHC) with susceptibility, and the identification of thyroglobulin (Tg) as an autoantigen. In 1981, we demonstrated the presence of autoreactive T cells responding to syngeneic mouse (m) Tg immunization in a susceptible strain (ElRehewy et al., 1981) and in 1982, we pinpointed EAT susceptibility to *H2A* class II genes (Beisel et al., 1982). Also, in 1982, we reported the induction of resistance with mTg, either from exogenous source or from endogenous release by thyroid-stimulating hormone (TSH) (Kong et al., 1982). This resistance was specific and protected mice from EAT induction. The cells with suppressive function were thymus-derived. We hypothesized that, in normal susceptible individuals, there is a clonal balance of suppressor T cells, keeping the autoreactive T cells in check. This balance toward suppression or autoimmunity can shift according to tolerogenic and immunogenic signals. Subsequent availability of CD4 and CD8 monoclonal antibodies (mAbs) enabled the allocation of suppressor T cells to the CD4$^+$ subset in 1989 (Kong et al., 1989). The recent availability of CD25-depleting mAb further characterized

---

Yi-chi M. Kong and Gerald P. Morris • Department of Immunology and Microbiology, Wayne State University School of Medicine, Detroit, Michigan 48201. **Chella S. David** • Department of Immunology, Mayo Clinic, Rochester, Minnesota 55905.

Molecular Autoimmunity: In commemoration of the 100th anniversary of the first description of human autoimmune disease, edited by Moncef Zouali. Springer Science+Business Media, Inc., New York, 2005.

them as CD4+CD25+ (Morris et al., 2003). Clearly, the early mTg-specific suppressor T cell is one and the same as the current CD4+CD25+ regulatory T cell.

This chapter briefly reviews the progressive characterization of CD4+CD25+ regulatory T cells in EAT. The known class II association has enabled a comparison of their contribution to susceptibility and resistance. We have also examined their role in a recently uncovered *H2E(H2A⁻)* transgenic mouse model, which responds to EAT induction with human (h) Tg, but not mTg. Depletion of CD4+CD25+ regulatory T cells did not alter this stringent H2E restriction. Thus, while CD4+CD25+ T cell regulation can modulate EAT development, it does not supersede MHC restriction for certain Tg epitopes.

2. Major Histocompatibility Complex Class II Gene Control of Susceptibility

The recognition of genetic linkage of susceptibility in EAT to the MHC *H2* gene complex has pioneered more than 3 decades of research into many autoimmune diseases in humans and in other species. Induction of EAT with the self-antigen, mTg, was compared in multiple inbred mouse strains of different *H2* haplotypes, and susceptibility was distinguished from resistance by the marked infiltration of mononuclear cells (Vladutiu and Rose, 1971). Subsequent availability of intra-*H2*-recombinant strains enabled the identification of class II *I-A* gene as the locus encoding susceptibility or resistance alleles (Beisel et al., 1982).

A strong *HLA* class II association with autoimmune thyroid disease, including Hashimoto's thyroiditis and Graves' disease, has been more difficult to gain widespread acceptance, despite implications from patient studies, due to the polygenic nature of the human class II genes as well as ethnic differences. Firm association of autoimmunity with specific *HLA* class II genes was further made difficult by the now-recognized class II gene influences within a single host, which could jointly affect the extent of susceptibility. Several years ago, we initiated transgenic studies, introducing either *H2* or *HLA* transgenes into mice, in order to determine the influence of a single class II transgene independent of other *HLA* genes, and in the absence of endogenous H2 class II molecules (reviewed in Kong and David, 2000; Kong et al., 2003). Briefly, for the purpose of this review, we first ascertained that our transgenic techniques were applicable to *HLA* transgenic studies by using a known susceptibility allele for mice. *H2* class II gene from the susceptible *k* haplotype ($Aa^k Ab^k$) was introduced into B10.M, a resistant *f* haplotype, and the mice became susceptible to EAT induction with mTg (Kong et al., 1997). We then introduced *HLA-DRB1*0301* (DR3) into the resistant B10.M mice, as well as into *Ab*-knockout mice that express neither the A nor E class II molecules (Kong et al., 1996). Both DR3 transgenic groups became susceptible to EAT induction with either mTg or hTg. In contrast, introduction of *HLA-DRB1*1502* (DR2) or *HLA-DRB1*0401* (DR4) transgene into the class II–deficient mice did not render them EAT-susceptible, demonstrating the efficacy of single transgenic studies in revealing *HLA-DRB1* polymorphism as a determinant in EAT susceptibility (Wan et al., 2002).

Clearly, the MHC class II gene control of susceptibility to autoimmune thyroiditis is strong and invariant, as demonstrable by the use of *H2*-disparate inbred strains, intra-*H2*-recombinant strains, as well as *H2* and *HLA* class II transgenic strains. Although other factors, such as the presence of other interacting MHC class II genes (reviewed in Kong *et al.*, 2003), MHC class I and non-MHC background genes, environmental and dietary differences, etc. (reviewed in Kong, 1999), have been shown to play a contributory role, their participation tends to be secondary or minor in that these factors are not revealed unless the host also possesses the appropriate class II susceptibility haplotype. However, we reported earlier an important contributory factor of T cell regulation in susceptible mice; these CD4$^+$ regulatory (suppressor) T cells mediate mTg-induced tolerance to withstand EAT induction (Kong *et al.*, 1982, 1989). These findings have recently been applied to examine the extent of regulatory T cell modulation on EAT susceptibility.

3. Establishment of CD4$^+$ T Cells as Mediators of Induced Resistance

3.1. Protection from EAT Induction by Elevating the Circulatory Thyroglobulin Level

Susceptible mice do not develop thyroiditis spontaneously, despite the continuous presence of low levels of circulatory Tg and the presence of autoreactive T cells. Indeed, thyroiditogenic, autoreactive T cells capable of responding to syngeneic mTg administration and mediating thyroid pathology were demonstrated in susceptible, but not resistant, mice (ElRehewy *et al.*, 1981). Clearly, suppressor mechanisms exist in susceptible mice, which can be overcome upon immunization with mTg. We subsequently observed that pretreatment of susceptible mice with mTg rendered the mice tolerant to immunization with mTg and adjuvant (Kong *et al.*, 1982). Three parameters are used to measure this resistance: much reduced to absent infiltration of mononuclear cells into the thyroid, very low mTg antibody level, and minimal T cell proliferative response *in vitro* to mTg. Raising the circulatory mTg level 3- to 5-fold above baseline for 2–3 days is key to successful tolerance induction (Kong *et al.*, 1982; Lewis *et al.*, 1987). The increase can be achieved with three different protocols: (i) two intravenous doses of 100 μg deaggregated (d) mTg 7 days apart, or daily doses of 10 μg mTg for 10 days (Lewis *et al.*, 1987), (ii) intravenous administration of 20 μg bacterial lipopolysaccharide (LPS) to reduce reticuloendothelial clearance 24 hr before two subtolerogenic, 20-μg doses of dmTg (Lewis *et al.*, 1992), and (iii) TSH injections (Kong *et al.*, 1982) or infusion with a peritoneal osmotic pump for 3 days to stimulate endogenous mTg release (Lewis *et al.*, 1987). As circulatory Tg was a by-product of thyroid hormone release and had no apparent function, we hypothesized that there was a clonal balance between regulatory T cells and autoreactive T cells in the normal, susceptible host and that one major function of circulatory Tg was to activate low levels of suppressor T cells to keep autoreactive T cells in check (Kong *et al.*, 1982). But this low level was unable to withstand insults from autoantigens and various polyclonal activators from infectious agents and other chronic inflammation. When these regulatory T cells were further expanded by mTg for 2–3 days, they altered the balance to favor protection against strong antigenic challenge.

3.2. CD4$^+$ Regulatory T Cells as Mediators of Induced Resistance

Because mTg-activated regulatory T cells were few and no known markers existed at the time, unresponsiveness was relegated to the phenomenon of anergy by skeptics. However, anergy in autoreactive T cells induced by pretreatment with dmTg was insufficient to explain the finding that infusion of one spleen equivalent of normal T cells did not reconstitute the response of tolerant mice to challenge, indicating an active inhibition of autoreactive T cells (Kong et al., 1989). The hypothesis of active suppression was also tested by the transfer of cells from tolerant mice. This transfer of suppression was possible to demonstrate, but with some difficulty, because the use of adjuvant to induce autoimmune disease could overwhelm the regulatory T cells (Fuller et al., 1993). Since EAT tolerance can be established within 3 days after the two tolerogenic doses and lasts for at least 73 days (Fuller et al., 1993), this window was applied to examine suppression directly in the tolerized host. The availability of CD4 and CD8 mAbs for depletion studies *in vivo* enabled us to demonstrate that only depletion of CD4$^+$ T cells abrogated tolerance, again supporting the hypothesis of active suppression. Thus, regulatory T cells in mice with established tolerance belonged to the CD4$^+$ subset (Kong et al., 1989).

3.3. Effect of Cytokines on CD4$^+$ Regulatory T Cell Induction and Function

3.3.1. Noninvolvement of IL-4 and IL-10

To determine the mechanisms of CD4$^+$ T cells regulating induced resistance to EAT induction, several cytokines with inhibitory or proinflammatory activity were examined. The possible role of Th2 cells in regulatory T cell function was suggested by the involvement of interleukins IL-4 and IL-10. Repeated doses of IL-10 into animals prevented the onset of diabetes (Pennline et al., 1994), experimental autoimmune encephalomyelitis (EAE) (Rott et al., 1994), and EAT (Mignon-Godefroy et al., 1995). Also, IL-4 was shown to protect animals from EAE (Racke et al., 1994) and diabetes (Rapoport et al., 1993). Accordingly, we examined the role of IL-4 and IL-10 in tolerance to EAT induction. Antibodies to IL-4 and IL-10 were initially given separately in conjunction with tolerance induction, and the animals were then challenged with mTg and LPS. No apparent effect of either cytokine was observed, and the mice became tolerant to EAT (Zhang and Kong, 1998). In the event that CD4$^+$ regulatory T cells in tolerant mice released IL-10 to inhibit the response of autoreactive T cells during challenge, IL-10 mAb was injected in conjunction with immunization. No effect on the already established tolerance was seen and the animals remained unresponsive. We then combined IL-4 and IL-10 mAb administration at the time of tolerance induction, and, alternatively, administered IL-10 mAb to IL-4-knockout mice. Again, no interference with tolerance establishment was observed. The response of control IL-4-deficient mice to EAT induction was comparable to that of the wild type, indicating that the absence of IL-4 does not increase thyroiditis severity. These data demonstrate the noninvolvement of the two Th2 cytokines in EAT tolerance, but do not rule out any joint involvement with other cytokines, such as TGF-ß.

3.3.2. Interference with Tolerance Induction by IL-1 and IL-12

EAT is a cell-mediated autoimmune disease involving Th1 cells and proinflammatory cytokines, but the two major Th2 cytokines apparently are not involved in regulating EAT tolerance. We selected IL-1 and IL-12 as representatives of proinflammatory cytokines to test their capacity to interfere with tolerance induction. IL-1 has multiple biologic effects (Dinarello, 1989), one of which is its adjuvant effect for non-self-protein antigens (Staruch and Wood, 1983). Because of the short $t_{1/2}$, IL-1ß was found to be most effective when given 3 hr after, but not before, dmTg and less effective when given 24 hr after (Nabozny and Kong, 1992). IL-12 exerts its action primarily on Th1 cell differentiation and has been shown to accelerate the onset of and increase the incidence of cell-mediated autoimmune diseases, such as diabetes (Trembleau *et al.*, 1995), collagen-induced arthritis (Germann *et al.*, 1995), EAE (Leonard *et al.*, 1997), and experimental autoimmune uveitis (Caspi, 1998). On interference with induction of EAT tolerance, IL-12, with a longer $t_{1/2}$ than IL-1ß, was effective when given twice, on the day of and the day after dmTg (Zhang *et al.*, 2001). The treated animals were not protected against challenge, as tolerance had not been established. The interference of tolerance induction by IL-12 was not neutralized by the presence of anti-IFN-γ. On the other hand, T cell costimulatory molecules may be involved through IL-12 activity. Either CD40L or CD28 mAb ameliorated the priming effect of dmTg and IL-12, compared to the IgG control group. Thus, both cytokines can convert a tolerogenic signal to a priming signal, thereby increasing thyroiditis severity after challenge. However, similar to the stronger adjuvants, complete Freund's adjuvant or LPS, neither cytokine can overcome established tolerance when given at the time of challenge. These data suggest that proinflammatory signals derived from ongoing immune responses or from cytokine immunotherapy could potentially interfere with the continuing induction of peripheral tolerance to self-antigens in susceptible individuals.

4. CD25 Expression on CD4+ Regulatory T Cells in Induced Resistance

4.1. Abrogation of Established Tolerance by CD4+CD25+ T Cell Depletion

CD25 (IL-2Rα) has been increasingly used as a marker for CD4+ regulatory T cells important for maintaining peripheral tolerance to self-antigens. For example, the cotransfer of CD4+CD25+ T cells into nude mice (on EAT-resistant

BALB/c background) blocks the development of several autoimmune diseases including thyroiditis, which can arise by the transfer of CD4$^+$CD25$^-$ T cells alone (Sakaguchi et al., 1995). In vivo depletion of CD4$^+$CD25$^+$ T cells enabled the induction of autoimmune gastritis in otherwise resistant mice (McHugh and Shevach, 2002). We examined the participation of these cells in dmTg-induced resistance to EAT induction (Morris et al., 2003). After the establishment of tolerance, mice were depleted of CD4$^+$CD25$^+$ T cells with CD25 mAb, their peripheral blood leukocytes verified for depletion, and challenged with mTg and LPS. Depletion of CD4$^+$CD25$^+$ T cells led to the loss of tolerance in most animals, whereas dmTg-pretreated mice given rat IgG remained tolerant. Thus, the CD4$^+$ regulatory T cells mediating EAT tolerance also carry the CD25 marker.

CD4$^+$CD25$^+$ regulatory T cells isolated from tolerized mice also inhibited the *in vitro* proliferative response of mTg-primed cells to mTg in a dose-dependent manner, as reported recently (Morris et al., 2003). Figure 15.1 shows that, whereas CD4$^+$CD25$^+$ regulatory T cells were unresponsive to mTg stimulus as expected, they did suppress the response of mTg-primed cells at a ratio of 0.5:1 to 0.25:1, losing their effect at 0.125:1. The development of an *in vitro* coculture model of regulatory T cell suppression of an autoantigen-specific T cell response should enable a closer examination of potential regulatory mechanisms of autoimmune response, and how they correlate with EAT tolerance *in vivo*. As indicated above, we have earlier ruled out IL-4 and IL-10 playing an inhibitory role in induced resistance (Zhang and Kong, 1998), and indeed anti-IL10 had no effect in moderating *in vitro* suppression of mTg-primed T cell responses

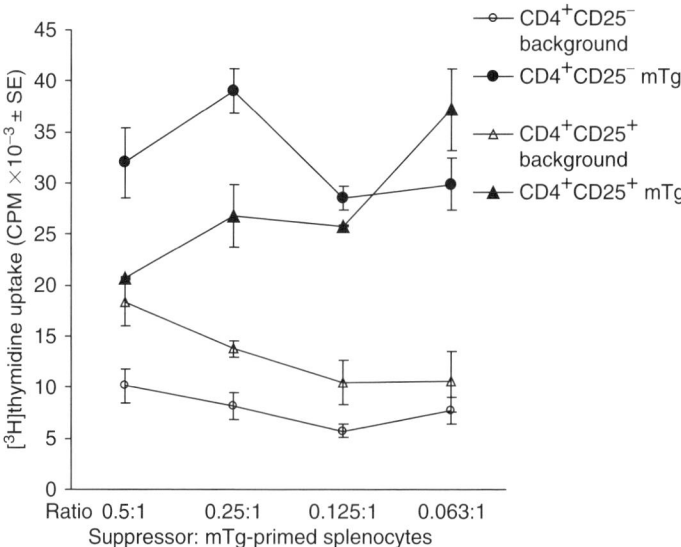

Figure 15.1. CD4$^+$CD25$^+$ T cells from mTg-tolerized mice suppress mTg-primed T cell proliferative response to antigen *in vitro*. Enriched CD4$^+$CD25$^+$ or CD4$^+$CD25$^-$ T cells from mTg-tolerized mice were cultured at the indicated ratios of suppressor to 4×10^5 mTg-primed splenocytes for 5 days. Response to mTg (40 μg/ml) was assessed by ^{3}[H]thymidine uptake.

(G. P. Morris and Y. M. Kong, unpublished data). It remains to be determined if other cytokines are involved.

4.2. Interference with CD4$^+$CD25$^+$ Regulatory T Cell Function by Cross-Linking TNFR Family Molecules

The exact mechanisms of suppression by CD4$^+$CD25$^+$ regulatory T cells remain to be elucidated, and may vary with the model systems. Until a soluble cytokine can be definitively characterized, the contact-dependent manner for efficient suppression (Thornton and Shevach, 2000) implicates the participation of costimulatory molecules. One prime candidate for a costimulatory role on T cells is CD137 (4-1BB), a member of the tumor necrosis factor receptor (TNFR) superfamily, which has been shown to influence the development of inflammatory immune responses (Cannons et al., 2001). CD137 mAb given in vivo not only promoted antitumor immunity (Melero et al., 1998), but also ameliorated EAE (Sun et al., 2002), indicating potential roles in either promoting or inhibiting EAT. We evaluated the role of CD137 in induced resistance to EAT. First, we observed that the administration of CD137 mAb in conjunction with dmTg interfered with tolerance induction. Similar to IL-12, the tolerogenic signal was converted to a priming signal by cross-linking CD137, as shown by enhanced thyroiditis severity after challenge (Morris et al., 2003).

The effect of anti-CD137 treatment on established tolerance was then tested in vivo at the time of immunization; protection by CD4$^+$CD25$^+$ regulatory T cells was severely hampered. This reversal of expected suppression can be duplicated in vitro; the presence of anti-CD137 inhibited the suppression of CD4$^+$CD25$^-$ T cell proliferation by CD4$^+$CD25$^+$ regulatory T cells (Morris et al., 2003). To differentiate if the effect was on tolerized cells or on mTg-primed cells, CD137 mAb was added to separate cultures of isolated CD4$^+$CD25$^+$ regulatory T cells and CD4$^+$CD25$^-$ T cells. No direct effect on tolerized cells by CD137 mAb plus mTg was observed, but it enhanced the response of mTg-primed cells. We concluded that signaling through CD137 likely circumvented the suppressive action of CD4$^+$CD25$^+$ regulatory T cells.

We also examined the role of another T cell costimulatory molecule GITR, a glucocorticoid-induced TNF family member (Tone et al., 2003), in EAT tolerance. Stimulation by mAb to GITR has been shown to break self-tolerance (Shimizu et al., 2002). Similar to CD137 mAb, administration of GITR mAb along with dmTg interfered with tolerance induction and, in mice with established tolerance, injection of GITR mAb at challenge abrogated tolerance. In both instances, the mice subsequently developed EAT (G. P. Morris and Y. M. Kong, unpublished data).

The use of mAbs to T cell costimulatory molecules of the TNFR family, CD137 and GITR, in EAT tolerance, strengthened by pretreatment with mTg, supports that there exist mechanisms in the host to circumvent the tolerant status and tilt the balance back toward an autoimmune response, possibly with undesirable sequelae. Efforts to manipulate T cell regulation and costimulation in the form of immunotherapy, such as for tumor, must take these possibilities into account (Wei et al., 2004).

5. Naturally Existing CD4⁺CD25⁺ T Cells as Peripheral Barrier to Autoimmune Thyroiditis

As we have hypothesized earlier and discussed above, there is a clonal balance of regulatory T cells and autoreactive T cells, with the former keeping the latter in check in normal, susceptible individuals (Kong et al., 1982). This hypothesis has certainly been borne out by transfer experiments in which CD4⁺CD25⁻ T cells, transferred without CD4⁺CD25⁺ T cells, mediated the development of several autoimmune diseases including thyroiditis (Sakaguchi et al., 1995). The depletion of CD4⁺CD25⁺ T cells also enabled resistant mice to respond to induction of autoimmune gastritis (McHugh and Shevach, 2002). We have postulated that such a natural barrier to autoimmune thyroiditis development represents the first level of regulatory influence, while mTg-induced resistance to withstand EAT induction represents the second level of regulation (Kong et al., 1989). With the identification of CD4⁺CD25⁺ regulatory T cells as mediators of induced resistance, we examined if naturally occurring regulatory T cells were indeed CD4⁺CD25⁺ (Morris and Kong, 2004). We used the protocol of inducing EAT with repeated injections of mTg without adjuvant, which generally results in EAT development in about 50% of the animals (ElRehewy et al., 1981). Table 15.1 shows that depletion of CD4⁺CD25⁺ T cells before repeated doses of mTg 16 times over 4 weeks greatly increased the incidence of thyroiditis, compared to prior treatment with control IgG. The difference in thyroid infiltration was also accompanied by higher mTg Ab levels and greater *in vitro* proliferative response to mTg. It thus appears that CD4⁺CD25⁺ regulatory T cells naturally exist and maintain peripheral tolerance to self-antigens in EAT.

Table 15.1. *In Vivo* Depletion of CD4⁺CD25⁺ T Cells Lowers the Threshold for EAT Induction without Adjuvant

	Thyroiditis				
	Number of mice with % thyroid involvement			Incidence	
	0	> 0–10%	> 10–20%	Pos./total	%
Rat IgG	8	2	2	4/12	33*
Rat CD25 antibody	2	4	5	9/11	82*

Mice were preinjected i.v. with rat antibody to CD25 or control antibody at days −14 and −10. EAT was induced in susceptible CBA mice by administration of aggregated mouse thyroglobulin (40 µg) i.v. beginning at day 0, on 4 successive days for 4 weeks. Thyroid pathology was assessed 35 days later.
*$p < 0.04$.

6. T Cell Regulation and MHC Restriction

As summarized in Section 1, EAT susceptibility and resistance are categorized by their MHC class II gene differences. Since susceptible mice have naturally existing CD4⁺CD25⁺ T cells (Morris and Kong, 2004), it was of interest to determine if CD4⁺CD25⁺ T cells also influence EAT induction in resistant strains.

In pilot experiments using two EAT-resistant strains, prior depletion of CD4+CD25+ T cells in both BALB/c ($H2^d$) mice (Wei et al., 2004) and B10 ($H2^b$) mice (Morris et al., 2004) enabled the animals to respond to immunization with mTg and LPS. However, mononuclear cell infiltration was much less extensive than in susceptible CBA/J mice immunized with mTg and LPS with no prior depletion of CD4+CD25+ T cells. These data suggest that traditional resistance in EAT may not be solely the result of an insufficiency to present Tg epitopes in the context of class II molecules, but may represent a reduced capacity to overcome peripheral regulation and generate a thyroiditogenic response. Removal of CD4+CD25+ regulatory T cells lowers the threshold for thyroiditis development.

Interestingly, we recently used a novel H2E class II transgenic model, E+B10.Ab^0 (H2A^{b-}E^{b+}), for EAT that would permit us to examine more precisely the degree of stringency for MHC restriction. Unlike conventional susceptible strains, wherein EAT is inducible with both hTg and mTg due to conserved thyroiditogenic epitopes, EAT in H2A$^-$E$^+$ mice is inducible only with hTg, but not with self-mTg (Wan et al., 1999). To determine if E$^+$ transgenic mice possessed CD4+CD25+ regulatory T cells that control this unique phenotype for EAT development, CD25 mAb was administered before immunization with hTg or mTg and LPS (Morris et al., 2004). Thyroiditis severity was markedly enhanced after hTg immunization in CD4+CD25+ T cell–depleted mice, but the CD25 mAb-treated animals remained unresponsive to mTg (Table 15.2). Since EAT-resistant B10 (H2A$^+$E$^-$) mice develop mild thyroiditis after CD25 mAb treatment and mTg/LPS immunization, we tested if such mice also responded to hTg and LPS (Morris et al., 2004). The CD25 mAb-treated B10 mice remained unresponsive to hTg. The data indicate that modulation of EAT susceptibility by CD4+CD25+ regulatory T cells is operative for both hTg and mTg in the respective H2E$^+$ and H2A$^+$ mice. However, this regulation apparently does not supersede MHC restriction for specific Tg epitopes leading to EAT induction. It is unknown at present how this stringent MHC restriction would influence the T cell repertoire.

Table 15.2. *In Vivo* Depletion of CD4+CD25+ T Cells Does Not Alter H2E Class II Restriction for hTg in the E+B10.Ab^0 Transgenic Model

		Thyroiditis					
		Number of mice with % thyroid involvement					
	Antigen	0	> 0–10%	> 10–20%	> 20–40%	> 40–80%	> 80%
Rat IgG	hTg	—	4	1	1	1*	—
CD25 antibody	hTg	—	—	—	—	2	4*
Rat IgG	mTg	5	1	—	—	—	—
CD25 antibody	mTg	3	1	—	—	—	—

Mice were preinjected i.v. with rat CD25 antibody or control antibody at days −14 and −10. EAT was induced i.v. with thyroglobulin (40 μg), followed 3 hr later with LPS (20 μg) at days 0 and 7; thyroid pathology was assessed on day 28.
*$p < 0.01$.

7. Conclusion

The increased use of immunotherapy for autoimmune diseases and for tumors in the past decade has led to a resurgence of interest in regulatory (suppressor) T cells and in recognition of their importance. The existence of a clonal balance between regulatory and autoreactive T cells, which can be tilted toward heightened resistance or autoimmune response by Tg-specific, tolerogenic or immunogenic signals (Kong *et al.*, 1982), has been borne out by subsequent studies. We further suggested that this regulatory T cell control is operative at two levels (Kong *et al.*, 1989) and have now provided some evidence, using the CD25 marker on regulatory T cells, and CD25 depleting-mAb for *in vivo* studies. To show the first level of control of peripheral tolerance by naturally existing regulatory T cells, $CD4^+CD25^+$ T cells were depleted before EAT induction. In susceptible mice, depletion enabled the induction of EAT with a weaker immunization protocol of mTg without adjuvant. In contrast, in resistant mice depleted of $CD4^+CD25^+$ T cells, EAT induction still required the inclusion of an adjuvant, indicating that while $CD4^+CD25^+$ regulatory T cells can modulate EAT development, MHC class II restriction remains the primary influence on susceptibility and resistance. However, in strains where only H2E or H2A molecules are expressed and present only hTg or mTg, respectively, MHC restriction is very stringent even after $CD4^+CD25^+$ regulatory T cells have been depleted.

For the second level of control by $CD4^+CD25^+$ regulatory T cells, peripheral tolerance in susceptible mice can be strengthened by increasing circulatory mTg levels; the mice then withstand challenge even with mTg and adjuvant. This induced resistance lasts for many days and may explain why autoimmune diseases are not more prevalent in individuals with class II susceptibility alleles, despite the continuous presence of self-antigen and autoreactive T cells. However, even in mice with induced resistance, we have shown that the strong tolerance can be circumvented and moderated by signaling through T cell costimulatory molecules, such as those from the TNFR family. It is clear that the clonal balance between regulatory and autoreactive cells can be influenced by T cell manipulation.

Acknowledgment

This research was supported by Grant DK45960 from the National Institute of Diabetes and Digestive and Kidney Diseases.

References

Beisel, K.W., David, C.S., Giraldo, A.A., Kong, Y.M., and Rose, N.R. (1982). Regulation of experimental autoimmune thyroiditis: Mapping of susceptibility to the *I-A* subregion of the mouse *H-2*. *Immunogenetics*, **15**, 427–431.

Cannons, J.L., Lau, P., Ghumman, B., Debenedette, M.A., Yagita, H., Okumura, K., and Watts, T.H. (2001). 4-1bb ligand induces cell division, sustains survival, and enhances effector function of CD4 and CD8 T cells with similar efficacy. *J. Immunol.*, **167**, 1313–1324.

Caspi, R.R. (1998). IL-12 in autoimmunity. *Clin. Immunol. Immunopathol.*, **88**, 4–13.

Dinarello, C.A. (1989). Interleukin-1 and its biologically related cytokines. *Adv. Immunol.*, **44**, 153–205.

Elrehewy, M., Kong, Y.M., Giraldo, A.A., and Rose, N.R. (1981). Syngeneic thyroglobulin is immunogenic in good responder mice. *Eur. J. Immunol.*, **11**, 146–151.

Fuller, B.E., Okayasu, I., Simon, L.L., Giraldo, A.A., and Kong, Y.M. (1993). Characterization of resistance to murine experimental autoimmune thyroiditis: Duration and afferent action of thyroglobulin- and TSH-induced suppression. *Clin. Immunol. Immunopathol.*, **69**, 60–68.

Germann, T., Szeliga, J., Hess, H., Störkel, S., Podlaski, F.J., Gately, M.K., Schmitt, E., and Rüde, E. (1995). Administration of interleukin 12 in combination with type II collagen induces severe arthritis in DBA/1 mice. *Proc. Natl. Acad. Sci. USA*, **92**, 4823–4827.

Kong, Y.M. (1999). In R. Volpe (ed.), *Contemporary Endocrinology: Autoimmune Endocrinopathies*. Humana Press, Totowa, New Jersey. pp. 91–111.

Kong, Y.M. and David, C.S. (2000). New revelations in susceptibility to autoimmune thyroiditis by the use of *H2* and *HLA* class II transgenic models. *Int. Rev. Immunol.*, **19**, 573–585.

Kong, Y.M., David, C.S., Lomo, L.C., Fuller, B.E., Motte, R.W., and Giraldo, A.A. (1997). Role of mouse and human class II transgenes in susceptibility to and protection against mouse autoimmune thyroiditis. *Immunogenetics*, **46**, 312–317.

Kong, Y.M., Flynn, J.C., Wan, Q., and David, C.S. (2003). *HLA* and *H2* class II transgenic mouse models to study susceptibility and protection in autoimmune thyroid disease. *Autoimmunity*, **36**, 397–404.

Kong, Y.M., Giraldo, A.A., Waldmann, H., Cobbold, S.P., and Fuller, B.E. (1989). Resistance to experimental autoimmune thyroiditis: L3T4+ cells as mediators of both thyroglobulin-activated and TSH-induced suppression. *Clin. Immunol. Immunopathol.*, **51**, 38–54.

Kong, Y.M., Lomo, L.C., Motte, R.W., Graldo, A.A., Baisch, J., Strauss, G., Hämmerling, G.J., and David, C.S. (1996). HLA-DRB1 polymorphism determines susceptibility to autoimmune thyroiditis in transgenic mice: Definitive association with HLA-DRB1*0301 (DR3) gene. *J. Exp. Med.*, **184**, 1167–1172.

Kong, Y.M., Okayasu, I., Giraldo, A.A., Beisel, K.W., Sundick, R.S., Rose, N.R., David, C.S., Audibert, F., and Chedid, L. (1982). Tolerance to thyroglobulin by activating suppressor mechanisms. *Ann. NY Acad. Sci.*, **392**, 191–209.

Leonard, J.P., Waldburger, K.E., Schaub, R.G., Smith, T., Hewson, A.K., Cuzner, M.L., and Goldman, S.J. (1997). Regulation of the inflammatory response in animal models of multiple sclerosis by interleukin-12. *Crit. Rev. Immunol.*, **17**, 545–553.

Lewis, M., Fuller, B.E., Giraldo, A.A., and Kong, Y.M. (1992). Resistance to experimental autoimmune thyroiditis is correlated with the duration of raised thyroglobulin levels. *Clin. Immunol. Immunopathol.*, **64**, 197–204.

Lewis, M., Giraldo, A.A., and Kong, Y.M. (1987). Resistance to experimental autoimmune thyroiditis induced by physiologic manipulation of thyroglobulin level. *Clin. Immunol. Immunopathol.*, **45**, 92–104.

McHugh, R.S. and Shevach, E.M. (2002). Cutting edge: Depletion of CD4+CD25+ regulatory T cells is necessary, but not sufficient, for induction of organ-specific autoimmune disease. *J. Immunol.*, **168**, 5979–5983.

Melero, I., Johnston, J.V., Shufford, W.W., Mittler, R.S., and Chen, L. (1998). NK1.1 cells express 4-1BB (CDw137) costimulatory molecule and are required for tumor immunity elicited by anti-4-1BB monoclonal antibodies. *Cell. Immunol.*, **190**, 167–172.

Mignon-Godefroy, K., Brazillet, M.P., Rott, O., and Charreire, J. (1995). Distinctive modulation by IL-4 And IL-10 of the effector function of murine thyroglobulin-primed cells in "transfer-experimental autoimmune thyroiditis." *Cell. Immunol.*, **162**, 171–177.

Morris, G.P., Chen, L., and Kong, Y.M. (2003). CD137 signaling interferes with activation and function of CD4+CD25+ regulatory T cells in induced tolerance to experimental autoimmune thyroiditis. *Cell. Immunol.*, **226**, 20–29.

Morris, G.P. and Kong, Y.M. (2004). Naturally-existing CD4+CD25+ regulatory T cells: A peripheral barrier to autoimmunity to thyroiditis. Turkish J. Endocrinol. Metab. 8(Suppl. 1):28.

Morris, G.P., Yan, Y., David, C.S., and Kong, Y.M. (2004). CD4+CD25+ regulatory T cells modulate autoimmune thyroiditis, but do not supersede MHC class II restriction in the novel H2a$^-$E+ transgenic model. *Clin. Invest. Med.*, **27**, 93b.

Nabozny, G.H. and Kong, Y.M. (1992). Circumvention of the induction of resistance in murine experimental autoimmune thyroiditis by recombinant IL-1β. *J. Immunol.*, **149**, 1086–1092.

Pennline, K.J., Roque-Gaffney, E., and Monahan, M. (1994). Recombinant human IL-10 prevents the onset of diabetes in the nonobese diabetic mouse. *Clin. Immunol. Immunopathol.*, **71**, 169–175.

Racke, M.K., Bonomo, A., Scott, D.E., Cannella, B., Levine, A., Raine, C.S., Shevach, E.M., and Rocken, M. (1994). Cytokine-induced immune deviation as a therapy for inflammatory autoimmune disease. *J. Exp. Med.*, **180**, 1961–1966.

Rapoport, M.J., Jaramillo, A., Zipris, D., Lazarus, A.H., Serreze, D.V., Leiter, E.H., Cyopick, P., Danska, J.S., and Delovitch, T.L. (1993). Interleukin 4 reverses T cell proliferative unresponsiveness and prevents the onset of diabetes in nonobese diabetic mice. *J. Exp. Med.*, **178**, 87–99.

Rott, O., Fleischer, B., and Cash, E. (1994). Interleukin-10 prevents experimental allergic encephalomyelitis in rats. *Eur. J. Immunol.*, **24**, 1434–1440.

Sakaguchi, S., Sakaguchi, N., Asano, M., Itoh, M., and Toda, M. (1995). Immunologic self-tolerance maintained by activated T cells expressing IL-2 receptor A-chains (CD25): Breakdown of a single mechanism of self-tolerance causes various autoimmune diseases. *J. Immunol.*, **155**, 1151–1164.

Shimizu, J., Yamazaki, S., Takahashi, T., Ishida, Y., and Sakaguchi, S. (2002). Stimulation of CD25$^+$CD4$^+$ regulatory T cells through GITR breaks immunological self-tolerance. *Nat. Immunol.*, **3**, 135–142.

Staruch, M.J. and Wood, D.D. (1983). The adjuvanticity of interleukin 1 *in vivo*. *J. Immunol.*, **130**, 2191–2194.

Sun, Y., Lin, X., Chen, H.M., Wu, Q., Subudhi, S.K., Chen, L., and Fu, Y.-X. (2002). Administration of agonistic anti-4-1BB monoclonal antibody leads to the amelioration of experimental autoimmune encephalomyelitis. *J. Immunol.*, **168**, 1457–1465.

Thornton, A.M. and Shevach, E.M. (2000). Suppressor effector function of CD4$^+$CD25$^+$ immunoregulatory T cells is antigen nonspecific. *J. Immunol.*, **164**, 183–190.

Tone, M., Tone, Y., Adams, E., Yates, S.F., Frewin, M.R., Cobbold, S.P., and Waldmann, H. (2003). Mouse glucocorticoid-induced tumor necrosis factor receptor ligand is costimulatory for T cells. *Proc. Natl. Acad. Sci. USA*, **100**, 15059–15064.

Trembleau, S., Penna, G., Bosi, E., Mortara, A., Gately, M.K., and Adorini, L. (1995). Interleukin 12 administration induces T helper type 1 cells and accelerates autoimmune diabetes in NOD mice. *J. Exp. Med.*, **181**, 817–821.

Vladutiu, A.O. and Rose, N.R. (1971). Autoimmune murine thyroiditis: Relation to histocompatibility (H-2) type. *Science*, **174**, 1137–1139.

Wan, Q., Shah, R., Mccormick, D.J., Lomo, L.C., Giraldo, A.A., David, C.S., and Kong, Y.M. (1999). *H2-E* transgenic class II-negative mice can distinguish self from nonself in susceptibility to heterologous thyroglobulins in autoimmune thyroiditis. *Immunogenetics*, **50**, 22–30.

Wan, Q., Shah, R., Panos, J.C., Giraldo, A.A., David, C.S., and Kong, Y.M. (2002). HLA-DR and HLA-DQ polymorphism in human thyroglobulin-induced autoimmune thyroiditis: DR3 and DQ8 transgenic mice are susceptible. *Hum. Immunol.*, **63**, 301–310.

Wei, W.-Z., Morris, G.P., and Kong, Y.M. (2004). Anti-tumor immunity and autoimmunity: A balancing act of regulatory T cells. *Cancer Immunol. Immunother.*, **53**, 73–78.

Zhang, W., Flynn, J.C., and Kong, Y.M. (2001). IL-12 prevents tolerance induction with mouse thyroglobulin by priming pathogenic T cells in experimental autoimmune thyroiditis: Role of IFN-γ and the costimulatory molecules CD40L and CD28. *Cell. Immunol.*, **208**, 52–61.

Zhang, W. and Kong, Y.M. (1998). Noninvolvement of IL-4 and IL-10 in tolerance induction to experimental autoimmune thyroiditis. *Cell. Immunol.*, **187**, 95–102.

16

The Role of Autoimmunity in Multiple Sclerosis

Monika Bradl and Hans Lassmann

1. Introduction

Multiple sclerosis (MS), a chronic inflammatory disease of the central nervous system (CNS), is one of the most common neurological diseases of young adults in developed countries. Hallmarks of this disease are focal plaques of demyelination in the white matter of the CNS. Evidence for autoimmune processes in this disease is mostly circumstantial. First, autoreactive T cells are found in the blood of most patients. This by itself does not yet point to a role of these cells in the disease process, since CNS antigen–specific T cells are also normal components in the immune system of healthy individuals. However, experiments in animal models of autoimmune encephalomyelitis clearly demonstrated that such cells can initiate CNS inflammation once they are activated. Second, autoreactive T cells and antibodies have been found in MS lesions, and immunotherapies targeting CNS antigen–specific T cells may delay or ameliorate the disease progression in MS patients. However, the situation is much more complex. To date, no MS specific autoimmune response has been described. The triggers, which might set off or maintain autoimmunity in a chronic disease course lasting years or decades, remain unknown. Autoimmune responses in the CNS of MS patients are not invariably detrimental, but may even be beneficial. Finally, diffuse alterations indicating general neurodegenerative processes are present throughout the brain of many MS patients, and these alterations are not necessarily due to immune-mediated mechanisms. Hence, there are still many gaps to fill in the "autoimmune hypothesis" of MS.

Monika Bradl and Hans Lassmann • Medical University of Vienna, Center for Brain Research, Division of Neuroimmunology, Spitalgasse 4, A-1090 Wien, Austria.

Molecular Autoimmunity: In commemoration of the 100th anniversary of the first description of human autoimmune disease, edited by Moncef Zouali. Springer Science+Business Media, Inc., New York, 2005.

2. The "Autoimmune Hypothesis" of Multiple Sclerosis

MS is one of the most common neurological diseases within developed countries. It usually starts in young adults and leads in a chronic relapsing or progressive course over many years to major neurological disability (Noseworthy *et al.*, 2000). Genetic factors are important in determining the susceptibility to develop this disease, but within families of MS patients no Mendelian pattern of inheritance is found (Kalman and Lublin, 1999). Furthermore, studies on identical twins show that in addition to genetic factors other pathogenetic triggers have to be involved in disease induction. It is suspected that infections may act as triggers of disease, but so far no MS specific agent has been identified.

MS is believed to be an autoimmune disease, since it can in part be ameliorated or controlled by immunosuppressive or immunomodulatory treatments, and since autoreactive T lymphocytes or antibodies can be found in most patients (Wekerle, 1998; Noseworthy *et al.*, 2000). Furthermore, a chronic inflammatory demyelinating disease with close similarities to MS can be induced in many different animal species by active sensitization with brain tissue or specific brain antigens (Wekerle, 1998). However, the concept of autoimmunity in MS is questioned frequently, since no unique, MS-specific autoimmune response has been identified to date. In addition, recent studies show that the disease process in MS is much more complex than previously thought. It includes an interindividual heterogeneity in the pathogenetic mechanisms of lesion formation, and the presence of a chronic neurodegenerative component with unknown relations to immune-mediated processes (Owens, 2003). In this chapter, we summarize current knowledge about the immunology and immunopathology of MS and discuss arguments for or against the "autoimmune hypothesis" of MS.

3. The Multiple Facets of Multiple Sclerosis

3.1. The Clinical Spectrum of Multiple Sclerosis

The clinical course of MS is highly unpredictable (Noseworthy *et al.*, 2000). It ranges from benign disease, in which patients only suffer from some relapses without the development of permanent clinical deficit, to acute fulminate courses, which lead to death of the patients within a few months after disease onset. In more than 80% of patients the disease starts with a relapsing course, characterized by episodes of exacerbations followed by remission and recovery (relapsing–remitting multiple sclerosis, RRMS). After several years this relapsing course develops into a disease with a steady and uninterrupted increase of neurological disability, the so-called secondary progressive course of the disease (secondary progressive multiple sclerosis, SPMS). In 15–20% of the patients the relapsing phase of the disease is missing, and the patients follow a progressive course from the onset of disease. This form of MS is called primary progressive MS (PPMS). Using magnetic resonance imaging to follow the lesion development during these different phases of MS, it was shown that during the relapsing phase of the disease new lesions appear in the brain, which are associated with blood–brain barrier damage (Werring *et al.*, 2000). This suggests that these lesions are triggered by new waves of inflammatory cells, reaching the CNS tissue from the circulation by their passage through cerebral vessels. This is in marked contrast to the situation in the progressive phase of the disease (both in SPMS and PPMS), where new lesions are only rarely formed in the brain white matter, and where blood–brain barrier damage is not detected. Instead, profound abnormalities are found in the so-called "normal" white matter. This suggests a diffuse injury affecting the brain as a whole. Such diffuse changes can already be seen in early stages of RRMS and seem to gradually increase with time (Filippi et al., 2003).

3.2. The Pathological Spectrum of Multiple Sclerosis

In pathology MS is defined as an inflammatory demyelinating disease (Lassmann et al., 2001), since focal plaques with primary demyelination, relative axonal sparing, and reactive astroglial scaring are formed in the course of a chronic inflammatory reaction. The inflammatory infiltrates are mainly composed of lymphocytes and activated macrophages or microglia cells. Within the lymphocyte population of chronic inflammatory lesions, MHC class I–restricted $CD8^+$ T cells dominate (Gay et al., 1997). In addition, and more prominent in the chronic progressive stage of the disease, B lymphocytes and plasma cells accumulate, mainly in the meninges and the perivascular spaces. These B cells and plasma cells are responsible for a continuous intrathecal production of immunoglobulins.

The formation of new and active demyelinating plaques in the white matter (Figure 16.1) is mainly found in the relapsing stage of the disease. This process is associated with pronounced inflammatory infiltration of the whole tissue, accompanied by a profound blood–brain barrier disturbance. Both the structural and the immunopathological features of active MS plaques are different between different patient subpopulations, suggesting an interindividual heterogeneity in the mechanisms of tissue damage (Lucchinetti et al., 2000). In some patients demyelination and tissue injury is mainly associated with cytotoxic T cells and activated macrophages/microglia cells, while in others immunoglobulins and activated complement are deposited at sites of active myelin destruction. In another subgroup of patients the patterns of tissue damage closely resemble hypoxic tissue injury. Furthermore, a small cohort of patients show unusually severe damage of myelin, oligodendrocytes, and/or axons, suggesting problems of the target tissue to cope with the inflammatory insult (Lucchinetti et al., 2000).

In contrast to the relapsing stage of the disease, the formation of new active white matter lesions is rare in the progressive disease. Here, the typical finding is a slow and gradual enlargement of preexisting plaques (Prineas et al., 2001). In addition, a mild but completely diffuse inflammatory process in the whole brain and the meninges is evident. This is associated with diffuse microglia activation and ongoing axonal injury in the "normal" white matter, and with the formation of large areas of cortical demyelination (Kidd et al., 1999) topically related to the inflammatory process in the meninges. It is not clear yet whether this diffuse inflammatory process is the cause of global cortical and white matter damage in progressive MS or whether it occurs as a secondary reaction to the degenerative process in the brain. Is there clear evidence for immune reactions directed against CNS myelin, neurons, or glia cells, i.e., autoimmune reactions, in MS patients?

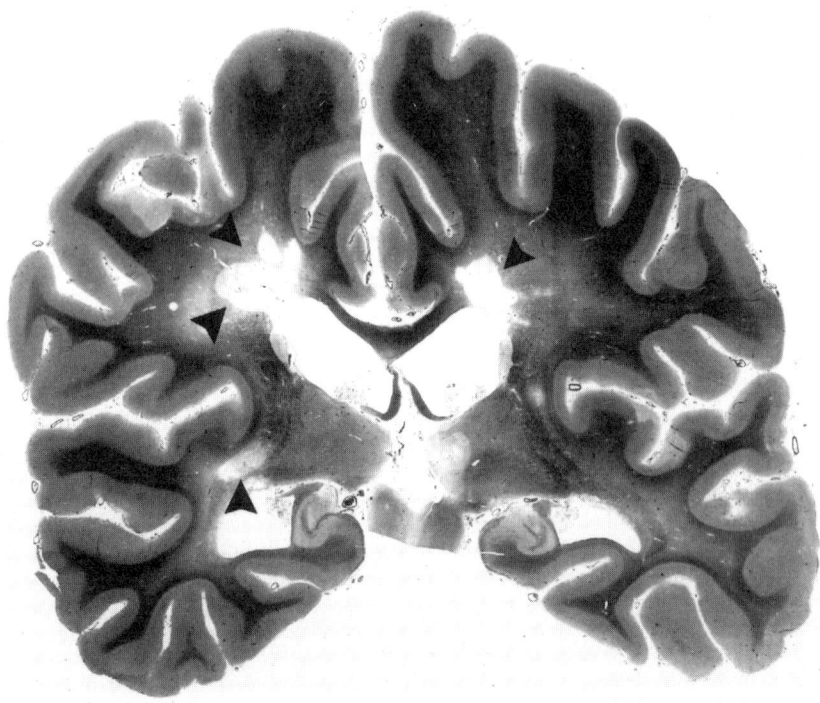

Figure 16.1. Brain lesions of an multiple sclerosis (MS) patient. The arrows point to large demyelinated plaques in central nervous system (CNS) white matter.

3.3. Evidence for T Cell–Mediated Autoimmunity

Since MS is an inflammatory demyelinating disease, which is, at least in northern Europe, often associated with the expression of the MHC class II antigen HLA-DR2, the search for autoimmunity was mainly a search for myelin antigens possibly recognized by brain-infiltrating $CD4^+$, MHC class II–restricted T cells. And indeed, T cells specifically recognizing myelin basic protein (MBP), proteolipid protein (PLP), or myelin oligodendrocyte glycoprotein (MOG) (Pette et al., 1990; Kerlero de Rosbo et al., 1997) could be readily isolated from the blood of MS patients. However, such cells can be isolated from the blood of healthy individuals just as well, indicating that they are normal components of the immune system (Wekerle, 1998). Then, the search was on for quantitative differences (Olsson et al., 1990b) in the antimyelin response of T lymphocytes found in the blood of MS patients and healthy controls. Here, simple assays such as testing for primary proliferation responses turned out to be insufficiently sensitive to detect such differences, with the exception of a single study showing stronger proliferative responses against MOG in MS patients than in controls (Kerlero de Rosbo et al., 1993). The more sensitive approach using ELISPOT techniques proved to be more rewarding, pointing to increased T cell responses also to PLP and MBP

(Olsson *et al.*, 1990b). Then, qualitative differences in the antimyelin response became the focus of intense research. It has been observed that monozygotic twins differ in their anti-MBP response. When both of the twins developed MS, they had a similar epitope recognition pattern, but when only one twin had MS, they differed more in epitope recognition (Utz *et al.*, 1993). Other studies revealed that the autoimmune response against MBP can be focused to a narrow, dominant peptide segment in some selected patients—a finding that contrasts with the broad peptide recognition pattern in controls. This narrow epitope pattern may be remarkably stable for several years over the course of the disease (Goebels *et al.*, 2000). However, in other patients the epitope recognition pattern radically changes over time. This was best demonstrated in human patients progressing from an isolated monosymptomatic demyelinating syndrome (IMDS) to clinically definite MS (Tuohy *et al.*, 1998). Coinciding with, or appearing soon after the diagnosis of IMDS, PLP-specific T cells from these patients recognized a single, specific PLP peptide. In the course of the disease, the response to this peptide regressed, and responses to other PLP epitopes emerged. This change in T cell reactivity concurred with relapses and the diagnosis of clinically definite MS.

Under normal circumstances, CNS antigen–specific T cells are resting components of the immune system. However, once activated, they can trigger autoimmune reactions within the human CNS. Evidence for such a disease scenario comes from a series of neuroparalytic accidents in humans receiving anti-rabies treatments with killed carbolized virus isolated from infected animal brains. Since these vaccines were contaminated with brain material, CNS antigen–specific T cells of the treated patients were activated and initiated CNS inflammation (for review see Bradl and Hohlfeld, 2003). This discovery paved the way for the development of experimental autoimmune encephalomyelitis (EAE), an animal model widely used to decipher certain aspects of CNS inflammation and demyelination seen in MS patients.

Based on this animal model it is undisputed that activated T cells recognizing antigens from myelin, oligodendrocytes, astrocytes, as well as neurons can induce inflammatory brain disease (Berger *et al.*, 1997). However, it became also clear that not every activated CNS antigen–specific T cell will be autoaggressive and hence pathogenic. For example, C57Bl/6 mice harbor PLP-specific T cells in their immune repertoire, but activation of these cells by immunization with PLP fails to provoke CNS inflammation. This is in marked contrast to the situation in SJL mice. With the same treatment regimen, PLP-specific T cells of these animals initiate massive inflammatory lesions in the CNS. The differences between these two strains of mice can be ascribed to variations in the intrathymic expression of tissue-specific self-antigens and tolerance induction (Klein *et al.*, 2000).

Another example is provided by MOG-reactive T cell lines isolated from Lewis rats. While all of these T cell lines vigorously proliferated in response to "their" specific peptides, only cells reactive to MOG_{1-20} or MOG_{35-55} were encephalitogenic and hence able to induce CNS inflammation, while T cells specific for MOG_{60-79} were incapable of mediating an inflammatory response within the CNS (Adelmann *et al.*, 1995). Based on this information, reports about the presence, expansion, or proliferation of myelin-specific T cells in the repertoire of MS patients should be critically evaluated, as long as the encephalitogenic potential of the described T cells remains unclear. There have been attempts to clarify this issue.

Human myelin-specific T cells have been transferred into SCID mice or rhesus monkeys, but the results of these studies were inconclusive (for review see Bradl and Flügel, 2002). At around the same time, a humanized mouse model has been created using transgenic animals expressing three human components involved in T cell recognition: (a) a T cell receptor derived from an MBP-specific T cell clone of an MS patient, which recognizes an MBP epitope that is conserved between mouse and man, (b) an MHC class II antigen associated with MS (HLA-DR2), and (c) the human coreceptor CD4. When the humanized MBP-specific T cells in these animals were activated, CNS inflammation ensued (Madsen *et al.*, 1999).

As seen above, it turned out to be extremely difficult to prove that the myelin-specific T cells found in the immune repertoire of MS patients are directly involved in the initiation and/or progression of an autoimmune reaction within the CNS. Other groups did not use the "*in vivo* approach" to solve this problem. Instead, they concentrated on the inflammatory lesions of MS patients and reasoned that they should be able to find CNS antigen–specific T cells in the plaques, provided that the lesions were initiated by these cells. And indeed, characterization of the T cell receptor usage of T cells found in MS plaques revealed the presence of T cells with similar complementary-determining region-3 sequences to those found in MBP-reactive T cell lines (Oksenberg *et al.*, 1993). In addition, MBP–MHC class II complexes have been found on antigen-presenting cells within MS plaques, indicating that MBP epitopes could be presented by local antigen-presenting cells (Krogsgaard *et al.*, 2000).

Taken together, while several reports describe the presence and the reactivity/receptor usage of $CD4^+$, myelin-specific T cells in MS patients (a recent search in Medline revealed more than 140 entries), there is little evidence for an active role of such CNS antigen–specific T cells in the induction of MS (see above).

Even less is known about the dominant T cell population in the MS plaque, the $CD8^+$ T cells. According to a recent publication, $CD8^+$ T cells in MS lesions represent an oligoclonal population (Babbe *et al.*, 2000). $CD8^+$ T cells are enriched at sites of actively demyelinating lesions, and there is evidence for a direct cytotoxic interaction between $CD8^+$ T lymphocytes and target cells in the CNS (Neumann *et al.*, 2002). However, it remains unclear whether these cells actually recognize myelin proteins and mount an autoimmune response to CNS antigens, or whether they recognize other, probably viral, target structures. The situation is further complicated by the fact that there are only few studies available so far that show that class I–restricted myelin-specific T cells can induce brain inflammation in experimental animals (Huseby *et al.*, 2001; Sun *et al.*, 2001). Unfortunately, these models are not developed to such an extent, that their validity as models for MS can be judged.

Moreover, most, if not all CNS-specific proteins, and even glycolipids (De Libero, 2004) may be targets of encephalitogenic, autoreactive T cells. Hence, T cell–mediated autoimmunity in MS patients may be directed against many different CNS antigens, which may be different between individual patients. It may, thus, not be surprising that a common MS-specific autoimmune response has so far not been identified.

3.4. Evidence for B Cell– or Antibody-Mediated Autoimmunity

There is no doubt about chronic B cell responses and a persistent immunoglobulin production in the CNS of MS patients. This is best exemplified by the diagnostically useful determination of intrathecal immunoglobulin synthesis, and by the presence of oligoclonal immunoglobulin in the cerebrospinal fluid. More recent studies, using new molecular technologies, showed extensive somatic mutations in the immunoglobulin variable genes of B cells isolated from the lesions or the cerebrospinal fluid of affected patients (Owens *et al.*, 1998; Qin *et al.*, 1998). These data also suggest that the B cell and plasma cell populations in the brain of MS patients are derived from the expansion of few individual B cell clones, possibly driven by local antigenic stimulation.

Are the antibodies synthesized intrathecally by these B cell populations directed against autoantigens? This question is still unresolved, since detailed studies aiming to define the antigen specificity of oligoclonal immunoglobulins in MS patients were so far inconclusive. Nevertheless, antibodies in the circulation and the cerebrospinal fluid (CSF) of MS patients may be directed against a large number of different autoantigens (Archelos and Hartung, 2000). Most extensively studied are antibodies against MBP and MOG (Berger *et al.*, 2003). They are found in higher frequencies in MS patients than in controls, and their serum titers at the time of the first clinical presentation may predict how fast the disease will convert to definite MS. However, the presence of antibodies against MOG or MBP is not unique for MS patients, but can be also found in patients with other inflammatory or noninflammatory CNS diseases. Hence, such antibodies may rather represent a secondary response of the immune system to brain damage, rather than being pathogenic by themselves.

Is there evidence for pathogenic autoantibodies responsible for the induction of demyelination or tissue injury? One requirement of an autoantibody to be directly pathogenic is that its target antigen is accessible in the intact tissue. Thus, the respective target antigen has to be located either on the cell surface or on extracellular matrix components. In contrast, most antibodies detected in MS patients are reactive with intracellular antigens such as MBP or neurofilament. The situation is different with MOG, which is a small immunoglobulin-like molecule expressed on the surface of myelin sheaths or oligodendrocytes (Linington *et al.*, 1988). Most anti-MOG antibodies detected in MS patients are pathogenetically irrelevant, since they are directed against linear epitopes hidden within the folded molecular structure (Brehm *et al.*, 1999). However, in a small subgroup of MS patients the antibodies recognize the conformational epitope of MOG, which is accessible on the surface of intact oligodendrocytes (Haase *et al.*, 2001) and myelin sheaths. Since such antibodies drive demyelination in animal models of EAE (Linington *et al.*, 1988), it is likely that they are also truly pathogenic in humans, leading to antibody- and complement-mediated demyelination characteristically found in active lesions in a subset of MS patients (Lucchinetti *et al.*, 2000).

Other antibodies that may be pathogenetically relevant are those directed against AN-2, a surface molecule expressed on glial progenitor cells (Niehaus *et al.*, 2000). Again, these antibodies are only found in a subset of MS patients. Since these antibodies may initiate the destruction of glial progenitor cells, they could be responsible for the failure of myelin repair in some patients. Taken together, autoantibodies directed against cell surface antigens of CNS cells or components of the extracellular matrix of the brain are potentially pathogenic in MS. The target structures of these antibodies may vary between individual patients, but a common, MS-specific autoantibody has not been described yet.

3.5. Evidence for Autoimmunity from Immunotherapies of Multiple Sclerosis

As described above, there is strong, but always indirect, evidence for an immunopathogenesis of MS (Figure 16.2), and different autoimmune mechanisms might operate in different patients or at different stages of the disease. Several immunotherapies provided further evidence for the presence of autoimmune reactions in MS patients, in particular on the level of MBP-specific T lymphocytes. For example, MBP-reactive T cells suspected to be involved in the pathogenesis of MS were depleted by subcutaneous inoculations with irradiated autologous MBP-reactive T cells ("T cell vaccination"). This treatment regimen resulted in the induction of $CD8^+$ T cells that specifically suppressed the MBP-reactive T cells (Zang *et al.*, 2000). It led, at least for an observation interval of 24 months, to a 40% reduction of the relapse rate (as compared to the pretreatment rate), and to a stabilization in lesion activity (Zhang *et al.*, 2002). Other immunotherapies used altered peptide ligands for MBP in MS patients. In these studies, T cells recognizing the putative target epitope for an autoimmune response, MBP_{83-99}, were targeted with peptides differing from the peptide normally recognized by one or two amino acids, in the hope to modulate the T cell response to the native peptide antigen. The outcome of these trials was inconclusive. One group described the induction of regulatory type 2 helper T cells, and a reduction in the volume and the number of lesions (Kappos *et al.*, 2000). Another group found exacerbations of MS directly linked to the activation of MBP-specific T cells in a subgroup of patients (Bielekova *et al.*, 2000). In sum, both of these immunotherapies provide evidence for a role of myelin-specific T cells in the disease. However, there is no evidence that these cells are directly involved in the initiation of MS.

4. The Triggers for Autoimmune Reactions in MS Patients

The studies described above have clearly shown that autoimmune reactions occur and that they appear to be more abundant and pronounced in MS patients compared to controls. Furthermore, some studies indicate that MS may be asso-

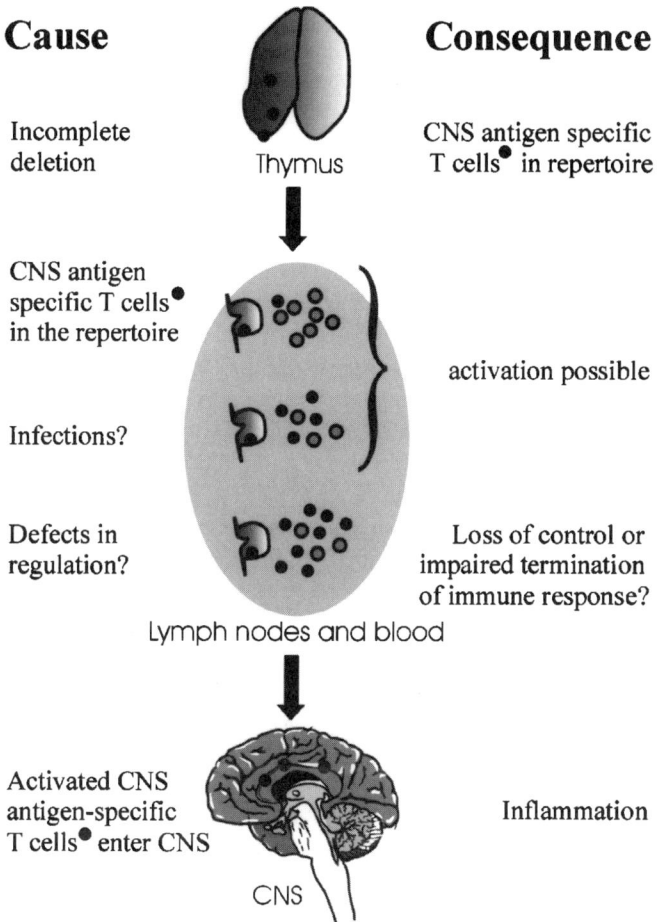

Figure 16.2. Possible causes and consequences of the presence of central nervous system (CNS) antigen–specific T cells in the immune repertoire of multiple sclerosis (MS) patients.

ciated with an increased risk to develop other autoimmune diseases (Henderson et al., 2000). What triggers autoimmune reactions in MS patients?

4.1. Autoimmune Reactions Caused by a Defect in Immune Regulation

A small number of transgenic mice carrying an expanded population of encephalitogenic T cells in the immune repertoire spontaneously develop EAE. If, however, other lymphocytes are genetically ablated, the disease incidence increases to 100% (Lafaille et al., 1994). These data point to the presence of regulatory T cells in the immune system, which n:ay prevent or terminate autoimmune reactions. In MS patients, immune-regulatory processes might act in at least two different situations: in the termination of acute disease episodes or in the global control of autoimmunity.

Early studies showed reduced suppressor activity of peripheral leucocytes in MS patients during active disease (Antel et al., 1979), which was related to reduced numbers of $CD8^+$ T cells in the circulation (Reinherz et al., 1980). This implicated that "suppressor" cells might play a major role in regulating the relapsing remitting course of MS. However, subsequent studies found a much less clearcut correlation between disease activity and loss of suppressor function (Antel et al., 1985) or changes in peripheral $CD8^+$ T cell numbers (Hughes et al., 1988). Reduced numbers of NK1 T cells with putative regulatory function, observed within inflammatory infiltrates in MS lesions, also suggest a defect in the termination of local immune responses (Illés et al., 2000).

A more global defect of immune regulation in MS patients was postulated following the observation that the numbers of regulatory $CD4^+CD25^{high}$ T cells in the circulation of MS patients are lower than in controls (Viglietta et al., 2004), possibly provoked by an impaired release of these cells from the thymus (Hug et al., 2003). This finding is not typical for all MS patients, since another study failed to detect such differences (Putheti et al., 2004).

Finally, the low incidence of asthma in MS patients (Tramlett et al., 2002) suggests a general bias of the immune response toward Th1 cells or cytotoxic T lymphocytes, which are polarized to the production of Th1 cytokines. Indeed, therapeutic strategies, which are believed to shift the immune response from Th1 to Th2 reactions, show some beneficial effect, in particular in patients with acute or relapsing disease (Neuhaus et al., 2000). Taken together, there are only few studies addressing the question of possible regulatory defects in MS patients, and the results of these studies are, at present, far from being conclusive. Hence, the concept of a general defect in immune regulation in MS patients remains an attractive hypothesis, which has to be proven in future large-scale clinical studies.

4.2. Autoimmune Reactions Caused by Infections

Based on extensive studies it became clear that chronic inflammatory demyelinating diseases reminiscent of MS can be induced in experimental animals by several different viruses causing, for example, Theiler's virus-induced encephalomyelitis (Drescher et al., 1997), corona virus–induced demyelinating encephalitis (Nagashima et al., 1978), and canine distemper virus encephalitis (Summers and Appel, 1994). In most of these models the disease starts with a panencephalitis, affecting both the gray and the white matter, which leads to injury of glia cells and neurons. While the infection is gradually cleared from the CNS gray matter, it may persist in the white matter and may cause a slowly progressive inflammatory demyelinating disease culminating in the formation of focal plaques of demyelination (Dethlefs et al., 1997). Tissue damage is mediated by both $CD4^+$ and $CD8^+$ T cells: while class I–restricted T cells appear to be particularly important in the induction of axonal damage and clinical deficit, both T cell populations seem to be involved in the induction and progression of demyelination (Murray et al., 1998). An important lesson from these animal models is that viral infections can induce autoimmune reactions against components of the CNS tissue, which may be involved in the propagation of chronic disease, possibly by antigen or determinant spreading (Croxford et al., 2002). This implicates a contribution of autoimmune T cells to the chronicity of the inflammatory reaction and to the induction of tissue injury. Since the virus is not completely eliminated from the CNS of the experimental animals, it remains unclear whether virus-induced autoimmunity alone, in the absence of viral persistence in the CNS, could maintain a long-lasting chronic inflammatory demyelinating disease.

There is little doubt that infections are unusual precursors to the *onset* of MS, but that they frequently precede the *relapses and progression* of MS. Moreover, oligoclonal IgG bands in the CSF, a diagnostic hallmark of MS, are otherwise almost exclusively found in infectious diseases of the nervous system (e.g., mumps, meningitis, neurosyphilis, cryptococcal meningitis, subacute sclerosing panencephalitis, and chronic rubella panencephalitis) (Gilden, 2002). Do such infections trigger autoimmune reactions, or do stress reactions associated with infections lead to the reactivation of human viruses from sites of latency in the CNS, which may in turn start immune reactions directed against viral structures? Research programs addressed both these scenarios.

The first line of research followed observations in transgenic mice carrying murine, encephalitogenic T cell receptors. The fact that these animals remained healthy when they were kept under specific pathogen-free conditions but developed spontaneous CNS inflammation when they lived in a normal environment (Goverman et al., 1993) led to the suggestion that the activation of T cells before their entry into the CNS may result from the action or recognition of bacterial/viral proteins. A similar, systemic trigger was also postulated for patients with early relapsing/remitting MS, since lesions appear concurrently in the brain and the spinal cord (Thorpe et al., 1996). Clear evidence for an activation of CNS antigen–specific T cells by infectious agents came from the pioneering work of Wucherpfennig and Strominger (1995). They demonstrated that MBP-specific T cell clones derived from MS patients can be activated by peptides derived from herpes simplex virus, Epstein–Barr virus, influenza virus, or Pseudomonas—and the list of such viral or bacterial peptides increased ever since (Hemmer et al., 1997).

The second line of research was stimulated by the isolation of an MS-associated retrovirus from the choroid plexus or from B lymphocytes of MS patients (Perron et al., 1997). This virus was related to a novel family of human endogenous retroviruses, human endogenous retrovirus type W (HERV-W) (Perron et al., 1997). A complete HERV-W provirus is present on chromosome 7, in a region associated with susceptibility to MS (Charmley et al., 1991). We recently detected HERV-W retroviral antigens in neurons, axons, and endothelial cells of active MS lesions (unpublished observation). However, their expression in inactive lesions was low or absent. It will have to be determined in the future whether this virus is activated in response to the stressful environment of an MS lesion or whether immune responses to such retroviruses cause the development of MS and/or the propagation of lesions. Thus, attractive as a viral/bacterial trigger of autoimmune reactions in MS patients may be, there is no known infection that provokes the onset of the disease.

5. Protective Autoimmunity

Autoreactive T cells and autoantibodies directed against components of the CNS are part of the normal immune repertoire. Since MBP-specific T cells can also be activated by viral/bacterial peptides, and because the response to such peptides might be as high, or even higher than the response to MBP (Wucherpfennig and Strominger, 1995), it seems fair to assume that the risk to develop autoimmune diseases is the price to pay for an efficient and fast immune response against pathogens. Moreover, inflammation should not always be considered an undesirable reaction. In fact, research in animal models of traumatic or excitotoxic brain lesions convincingly demonstrated that inflammatory T cells may serve beneficial functions by limiting tissue injury and stimulating tissue repair, an observation that even led to the concept of "protective autoimmunity" (Moalem et al., 1999).

Activated lymphocytes and macrophages are potent sources of growth factors and neurotrophins (Moalem et al., 2000), e.g., brain-derived neurotrophic factor (BDNF), which is able to rescue neurons *in vitro* (Kerschensteiner et al., 1999). Interestingly, neurons and glia cells at the active edge of demyelinating inflammatory MS lesions express receptors for BDNF (Stadelmann et al., 2002). Moreover, leukocyte-derived neurotrophins are also essential in the recruitment of oligodendrocyte progenitor cells and the induction of remyelination (Franklin, 2002).

Protective autoimmunity is not only a feature of self-antigen-specific T cells; it is also observed with autoantibodies. For example, murine and human antibodies of the IgM isotype, which recognize antigens on the surface of oligodendrocytes, may promote remyelination in the CNS (Asakura et al., 1996). These observations should be a point of concern when extensive immunosuppression in MS patients is planned: Such a treatment could not only reduce or limit possible deleterious consequences of CNS inflammation, but could also impair remyelination and repair.

6. What Remains of the "Autoimmune Hypothesis" of Multiple Sclerosis?

From all the data discussed above it seems that MS patients have just one feature in common: they share the presence of autoreactive T cells in the immune repertoire. However, they share this with healthy subjects just as well. All other features of the disease seem to be rather patient-specific: the trigger of the disease, the disease course, the pathological characteristics of the MS plaque, the contribution of the immune system to the pathological changes in the CNS, the (auto?)antigens recognized by T cells and/or antibodies, and even the response to treatments. There is evidence that autoimmune processes could be involved in every single aspect mentioned above. Bringing all these aspects together to a uniform picture of MS remains a challenge for the future.

References

Adelmann, M., Wood, J., Benzel, I., Fiori, P., Lassmann, H., Matthieu, J.-M., Gardinier, M.V., Dornmair, K., and Linington, C. (1995). The N-terminal domain of the myelin oligodendrocyte glycoprotein (MOG) induces acute demyelinating experimental autoimmune encephalomyelitis in Lewis rats. *J. Neuroimmunol.*, **63**, 17–27.

Antel, J.P., Arnason, B.G.W., and Medof, M.E. (1979). Suppressor cell function in multiple sclerosis: Correlation with clinical disease severity. *Ann. Neurol.*, **5**, 338–342.

Antel , J.P., Reder, A.T., and Noronha, A.B. (1985). Cellular immunity and immune regulation in multiple sclerosis. *Semin. Neurol.*, **5**, 117–126.

Archelos, J.J. and Hartung, H.P. (2000). Pathogenetic role of autoantibodies in neurological disease. *Trends Neurosci.*, **23**, 317–327.

Asakura, K., Miller, D.J., Murray, K., Bansal, R., Pfeiffer, S.E., and Rodriguez, M. (1996). Monoclonal autoantibody SCH94.03, which promotes central nervous system remyelination, recognizes an antigen on the surface of oligodendrocytes. *J. Neurosci. Res.*, **43**, 273–281.

Ascherio, A. and Munch, M. (2001). Epstein-Barr virus and multiple sclerosis. *Epidemiology*, **11**, 220–224.

Babbe, H., Roers, A., Waisman, A., Lassmann, H., Goebels, N., Hohlfeld, R., Friese, M., Schröder, R., Deckert, M., Schmidt, S., Ravid, R., and Rajewsky, K. (2000). Clonal expansion of CD8$^+$ T cells dominate the T cell infiltrate in active multiple sclerosis lesions as shown by micromanipulation and single cell polymerase chain reaction. *J. Exp. Med.*, **192**, 393–404.

Berger, T., Rubner, P., Schautzer, F., Egg, R., Ulmer, H., Mayringer, I., Dilitz, E., Deisenhammer, F., and Reindl, M. (2003). Antimyelin antibodies as a predictor of clinically definite multiple sclerosis after the first demyelinating event. *N. Engl. J. Med.*, **349**, 139–145.

Berger, T., Weerth, S., Kojima, K., Linington, C., Wekerle, H., and Lassmann, H. (1997). Experimental autoimmune encephalomyelitis: The antigen specificity of T-lymphocytes determines the topography of lesions in the central and peripheral nervous system. *Lab. Invest.*, **7**, 355–364.

Bielekova, B., Goodwin, B., Richert, N., Cortese, I., Kondo, T., Afshar, G., Gran, B., Eaton, J., Antel, J., Frank, J.A., McFarland, H.F., and Martin, R. (2000). Encephalitogenic potential of the myelin basic protein peptide (amino acids 83–99) in multiple sclerosis: Results of a phase II clinical trial with an altered peptide ligand. *Nat. Med.*, **6**, 1167–1175.

Bjartmar, C., Wujek, J.R., and Trapp, B.D. (2003). Axonal loss in the pathology of MS: Consequences for understanding the progressive phase of the disease. *J. Neurol. Sci.*, **206**, 165–171.

Bradl, M. and Flügel, A. (2002). The role of T cells in brain pathology. In B. Dietzschold and J.A. Richt (eds) Protective and pathological immune responses in the CNS. *Curr. Top. Microbiol. Immunol.*, **265**, 141–162.

Bradl, M. and Hohlfeld, R. (2003). Molecular pathogenesis of neuroinflammation. *J. Neurol. Neurosurg. Psychiatry*, **74**, 1364–1370.

Brehm, U., Piddlesden, S.J, Gardinier, M.V., and Linington, C. (1999). Epitope specificity of demyelinating monoclonal autoantibodies directed against the human myelin oligodendrocyte glycoprotein. *J. Neuroimmunol.*, **97**, 9–15.

Charmley, P., Beal, S.S., Concannon, P., Hood, L., and Gatti, R.A. (1991). Further localization of multiple sclerosis susceptibility gene on chromosome 7q using a new T cell receptor specific beta chain DANN polymorphism. *J. Neuroimmunol.*, **32**, 231–240.

Croxford, J.L., Olson, J.K., and Miller, S.D. (2002). Epitope spreading and molecular mimicry as triggers of autoimmunity in the Theiler's virus-induced demyelinating disease model of multiple sclerosis. *Autoimmun. Rev.*, **1**, 251–260.

De Libero, G. (2004). Immunology. The Robin Hood of antigen presentation. *Science*, **303**, 485–487.

Dethlefs, S., Brahic, M., and Larsson-Sciard, E.L. (1997). An early, abundant cytotoxic T-lymphocyte response against Theiler's virus is critical for preventing viral persistence. *J. Virol.*, **71**, 8875–8878.

Drescher, K.M., Pease, L.R., and Rodriguez, M. (1997). Antiviral immune responses modulate the nature of central nervous system (CNS) disease in a murine model of multiple sclerosis. *Immunol. Rev.*, **159**, 177–193.

Filippi, M., Bozzali, M., Rovaris, M., Gonen, O., Kesavadas, C., Ghezzi, A., Martinelli, V., Grossman, R., Scotti, G., Comi, G., and Falini, A. (2003). Evidence for widespread axonal damage at the earliest clinical stage of multiple sclerosis. *Brain*, **126**, 433–437.

Franklin, R.J. (2002). Why does remyelination fail in multiple sclerosis? *Nat. Rev. Neurosci.*, **3**, 705–714.

Gay, F.W., Drye, G.W., Dick, G.W.A., and Esiri, M.M. (1997). The application of multifactorial cluster analysis in the staging of plaques in early multiple sclerosis: Identification and characterization of the primary demyelinating lesion. *Brain*, **120**, 1461–1483.

Gilden, D.H. (2002). Multiple sclerosis exacerbations and infection. *Lancet Neurol.*, **1**, 145.

Goebels, N., Hofstetter, H., Schmidt, S., Brunner, C., Wekerle, H., and Hohlfeld, R. (2000). Repertoire dynamics of autoreactive T cells in multiple sclerosis patients and healthy subjects. Epitope spreading versus clonal persistence. *Brain*, **123**, 508–518.

Goverman, J., Woods, A., Larson, L., Weiner, L.P., Hood, L., and Zaller, D.M. (1993). Transgenic mice that express a myelin basic protein-specific T cell receptor develop spontaneous autoimmunity. *Cell*, **72**, 551–560.

Haase, C.G., Guggenmos, J., Brehm, U., Andersson, M., Olsson, T., Reindl, M., Schneidewind, J.M., Zettl, U.K., Heidenreich, F., Berger, T., Wekerle, H., Hohlfeld, R., and Linington, C. (2001). The fine specificity of the myelin oligodendrocyte glycoprotein autoantibody response in patients with multiple sclerosis and normal healthy controls. *J. Neuroimmunol.*, **114**, 220–225.

Hemmer, B., Fleckenstein, B.T., Vergelli, M., Jung, G., McFarland, H., Martin, R., and Wiesmuller, K.H. (1997). Identification of high potency microbial and self ligands for a human autoreactive class II-restricted T cell clone. *J. Exp. Med.*, **185**, 1651–1659.

Henderson, R.D., Bain, C.J., and Pender, M.P. (2000). The occurrence of autoimmune diseases in patients with multiple sclerosis and their families. *J. Clin. Neurosci.*, **7**, 434–437.

Hohlfeld, R. (1997). Biotechnological agents for the immunotherapy of multiple sclerosis. Principles, problems and perspectives. *Brain*, **120**, 865–916.

Hug, A., Korporal, M., Schroder, I., Haas, J., Gratz, K., Storch-Hagenlocher, B., and Wildemann, B. (2003). Thymic export function and T cell homeostasis in patients with relapsing remitting multiple sclerosis. *J. Immunol.*, **171**, 432–437.

Hughes, P.J., Kirk, P.F., Dyas, J., Munro, J.A., Welsh, K.I., and Compston, D.A.S. (1988). Factors influencing circulating OKT8 cell phenotypes in patients with multiple sclerosis. *J. Neurol. Neurosurg. Psychiatry*, **50**, 1156–1159.

Huseby, E.S., Liggitt, D., Brabb, T., Schnabel, B., Ohlen, C., and Goverman, J. (2001). A pathogenic role for myelin-specific CD8(+) T cells in a model for multiple sclerosis. *J. Exp. Med.*, **194**, 669–676.

Illés, Z., Kondo, T., Newcombe, J., Oka, N., Tabira, T., and Yamamura, T. (2000). Differential expression of NK T cells Vγ24JδQ invariant TCR chain in the lesions of multiple sclerosis and chronic inflammatory demyelinating polyneuropathy. *J. Immunol.*, **164**, 4375–4381.

Johnston, J.B., Silva, C., Holden, J., Warren, K.G., Clark, A.W., and Power, C. (2001). Monocyte activation and differentiation augment human endogenous retrovirus expression: Implications for inflammatory brain diseases. *Ann. Neurol.*, **50**, 434–442.

Jolivet Reynaud, C., Perron, H., Ferrante, P., Becquart, L., Dalbon, P., and Mandrand, B. (1999). Specificities of multiple sclerosis cerebrospinal fluid and serum antibodies against mimotopes. *Clin. Immununol.*, **93**, 283–293.

Kalman, B. and Lublin, F.D. (1999). The genetics of multiple sclerosis. A review. *Biomed. Pharmacother.*, **53**, 358–370.

Kappos, L., Comi, G., Panitch, H., Oger, J., Antel, J., Conlon, P., and Steinman, L. (2000). Induction of a non-encephalitogenic type 2 T helper-cell autoimmune response in multiple sclerosis after administration of an altered peptide ligand in a placebo-controlled, randomized phase II trial. The altered peptide ligand in relapsing MS study group. *Nat. Med.*, **6**, 1176–1182.

Kerlero de Rosbo, N., Hoffman, M., Mendel, I., Yust, I., Kaye, J., Bakimer, R., Flechter, S., Abramsky, O., Milo, R., Karni, A., and Ben-Nun, A. (1997). Predominance of the autoimmune response to myelin oligodendrocyte glycoprotein (MOG) in multiple sclerosis: Reactivity to the extracellular domain of MOG is directed against three main regions. *Eur. Immunol.*, **27**, 3059–3069.

Kerschensteiner, M., Gallmeier, E., Behrens, L., Leal, V.V., Misgeld, T., Klinkert, W.E., Kolbeck, R., Hoppe, E., Oropeza-Wekerle, R.L., Bartke, I., Stadelmann, C., Lassmann, H., Wekerle, H., and Hohlfeld, R. (1999). Activated human T cells, B cells, and monocytes produce brain-derived neurotrophic factor in vitro and in inflammatory brain lesions: A neuroprotective role of inflammation? *J. Exp. Med.*, **189**, 865–870.

Kidd, T., Barkhof, F., McConnell, R., Algra, P.R., Allen, I.V., and Revesz, T. (1999). Cortical lesions in multiple sclerosis. *Brain*, **122**, 17–26.

Klein, L., Klugmann, M., Nave, K.A., Tuohy, V.T., and Kyewski, B. (2000). Shaping of the autoreactive T-cell repertoire by a splice variant of self protein expressed in thymic epithelial cells. *Nat. Med.*, **6**, 56–61.

Krogsgaard, M., Wucherpfennig, K.W., Cannella, B., Hansen, B.E., Svejgaard, A., Pyrdol, J., Ditzel, H., Raine, C., Engberg, J., Fugger, L., and Canella, B. (2000). Visualization of myelin basic protein (MBP) T cell epitopes in multiple sclerosis lesions using a monoclonal antibody specific for the human histocompatibility leukocyte antigen (HLA)-DR2-MBP 85-99 complex. *J. Exp. Med.*, **191**, 1395–1412.

Lafaille, J.J., Nagashima, K., Katsuki, M., and Tonegawa, S. (1994). High incidence of spontaneous autoimmune encephalomyelitis in immunodeficient anti-myelin basic protein T cell receptor transgenic mice. *Cell*, **78**, 399–408.

Lassmann, H., Brück, W., and Lucchinetti, C. (2001). Heterogeneity of multiple sclerosis pathogenesis: Implications for diagnosis and therapy. *Trends Mol. Med.*, **7**, 115–121.

Linington, C., Bradl, M., Lassmann, H., Brunner, C., and Vass, K. (1988). Augmentation of demyelination in rat acute allergic encephalomyelitis by circulating mouse monoclonal antibodies directed against a myelin/oligodendrocyte glycoprotein. *Am. J. Pathol.*, **130**, 443–454.

Lucchinetti, C., Brück, W., Parisi, J., Scheithauer, B., Rodriguez, M., and Lassmann, H. (2000). Heterogeneity of multiple sclerosis lesions: Implications for the pathogenesis of demyelination. *Ann. Neurol.*, **47**, 707–717.

Madsen, L.S., Andersson, E.C., Jansson, L., Krogsgaard, M., Andersen, C.B., Engberg, J., Strominger, J.L., Svejgaard, A., Hjorth, J.P., Holmdahl, R., Wucherpfennig, K.W., and Fugger, L. (1999). A humanized model for multiple sclerosis using HLA-DR2 and a human T cell receptor. *Nat. Genet.*, **23**, 343–347.

Moalem, G., Gdalyahu, A., Shani, Y., Otten, U., Lazarovici, P., Cohen, I.R., and Schwartz, M. (2000). Production of neurotrophins by activated T cells: Implications for neuroprotective autoimmunity. *J. Autoimmun.*, **15**, 331–345.

Moalem, G., Leibowitz-Amit, R., Yoles, E., Mor, F., Cohen, I.R., and Schwartz, M. (1999). Autoimmune T cells protect neurons from secondary degeneration after central nervous system axotomy. *Nat. Med.*, **5**, 49–55.

Murray, P.D., Pavelko, K.D., Leibowitz, J., Lin, X., and Rodriguez, M. (1998). CD4(+) and CD8(+) T cells make discrete contributions to demyelination and neurologic disease in a viral model of multiple sclerosis. *J. Virol.*, **72**, 7320–7329.

Musette, P., Bequet, D., Delarbre, C., Gachelin, G., Kourilsky, P., and Dormont, D. (1996). Expansion of a recurrent Vβ 5.3$^+$ T cell population in newly diagnosed and untreated HLA-DR2 multiple sclerosis patients. *Proc. Natl. Acad. Sci. USA*, **93**, 12461–12466.

Nagashima, K., Wege, H., and ter Meulen, V. (1978). Early and late CNS-effects of corona virus infection in rats. *Adv. Exp. Med. Biol.*, **100**, 395–409.

Neuhaus, O., Farina, C., Yassouridis, A., Wiendl, H., Then Bergh, F., Dose, T., Wekerle, H., and Hohlfeld, R. (2000). Multiple sclerosis: Comparison of copolymer-1-reactive T cell lines from treated and untreated subjects reveals cytokine shift from T helper 1 to T helper 2 cells. *Proc. Natl. Acad. Sci. USA*, **97**, 7452–7457.

Neumann, H., Medana, I., Bauer, J., and Lassmann, H. (2002). Cytotoxic T lymphocytes in autoimmune and degenerative CNS diseases. *Trend Neurosci.*, **25**, 313–319.

Niehaus, A., Shi, J., Grzenkowski, M., Diers-Fenger, M., Hartung, H.P., Toyka, K., Bruck, W., and Trotter, J. (2000). Patients with active relapsing-remitting multiple sclerosis synthesize antibodies recognizing oligodendrocyte progenitor cell surface protein: Implications for remyelination. *Ann. Neurol.*, **48**, 362–371.

Noseworthy, J.H., Lucchinetti, C., Rodriguez, M., and Weinshenker, B.G. (2000). Multiple sclerosis. *N. Engl. J. Med.*, **343**, 938–952.

Oksenberg, J.R., Panzara, M.A., Begovich, A.B., Mitchell, D., Erlich, H.A., Murray, R.S., Shimonkevitz, R., Sherritt, M., Rothbard, J., and Bernard, C.C. (1993). Selection for T-cell receptor V beta-D beta-J beta gene rearrangements with specificity for a myelin basic protein peptide in brain lesions of multiple sclerosis. *Nature*, **362**, 68–70.

Olsson, T., Baig, S., Höjeberg, B., and Link, H. (1990a). Anti-myelin basic protein and anti-myelin antibody-producing cells in multiple sclerosis. *Ann. Neurol.*, **27**, 132–136.

Olsson, T., Zhi, W.W., Höjeberg, B., Kostulas, V., Yu-Ping, J., Anderson, G., Ekre, H.-P., and Link, H. (1990b). Autoreactive T lymphocytes in multiple sclerosis determined by secretion of interferon-γ. *J. Clin. Invest.*, **86**, 981–985.

Owens, T., (2003). The enigma of multiple sclerosis: Inflammation and neurodegeneration causes heterogenous dysfunction and damage. *Curr. Opin. Neurol.*, **16**, 259–265.

Owens, G.P., Kraus, H., Burgoon, M.P., Smith-Jensen, T., Devlin, M.E., and Gilden, D.H. (1998). Restricted use of V_H4 germline segments in an acute multiple sclerosis brain. *Ann. Neurol.*, **43**, 236–243.

Pette, M., Fujita, K., Wilkinson, D., Altmann, D.M., Trowsdale, J., Giegerich, G., Hinkkanen, A., Epplen, J.T., Kappos, L., and Wekerle, H. (1990). Myelin autoreactivity in multiple sclerosis: Recognition of myelin basic protein in the context of HLA-DR2 products by T lymphocytes of multiple sclerosis patients and healthy donors. *Proc. Natl. Acad. Sci. USA*, **87**, 7968–7972.

Perron, H., Garson, J.A., Bedin, F., Beseme, F., Paranhos-Baccala, G., Komurian-Pradel, F., Mallet, F., Tuke, P.W., Voisset, C., Blond, J.L., Lalande, B., Seigneurin, J.M., and Mandrand, B. (1997). Molecular identification of a novel retrovirus repeatedly isolated from patients with multiple sclerosis. The collaborative research group on multiple sclerosis. *Proc. Natl. Acad. Sci. USA*, **94**, 7583–7588.

Prineas, J.W., Kwon, E.E., Cho, E.S., Sharer, L.R., Barnett, M.H., Oleszak, E.L., Hoffman, B., and Morgan, B.P.(2001). Immunopathology of secondary-progressive multiple sclerosis. *Ann. Neurol.*, **50**, 646–657.

Putheti, P., Pettersson, A., Soderstrom, M., Link, H., and Yuang, Y.M. (2004). Circulating $CD4^+CD25^+$ T regulatory cells are not altered in multiple sclerosis and unaffected by disease-modulating drugs. *J. Clin. Immunol.*, **24**, 155–161.

Qin, Y., Duquette, P., Zhang, Y., Poole, R., and Antel, J.P. (1998). Clonal expansion and somatic hypermutation of V_H genes of B cells from cerebrospinal fluid in multiple sclerosis. *J. Clin. Invest.*, **102**, 1045–1050.

Reinherz, E.L., Weiner, H.L., Hauser, S.L., Cohen, J.A., DiStaso, J.A., and Schlossman, S.F. (1980). Loss of suppressor cells in active multiple sclerosis. *N. Engl. J. Med.*, **303**, 125–129.

Stadelmann, C., Kerschensteiner, M., Misgeld, T, Brück, W., Hohlfeld, R., and Lassmann, H. (2002). BDNF and gp145trkB in multiple sclerosis brain lesions: Neuroprotective interactions between immune cells and neuronal cells? *Brain*, **125**, 75–85.

Summers, B.A. and Appel, M.J. (1994). Aspects of canine distemper virus and measles virus encephalomyelitis. *Neuropath. Appl. Neurobiol.*, **20**, 525–534.

Sun, D., Whitaker, J.N., Huang, Z., Liu, D., Coleclough, C., Wekerle, H., and Raine, C.S. (2001). Myelin antigen-specific CD8+ T cells are encephalitogenic and produce severe disease in C57BL/6 mice. *J. Immunol.*, **166**, 7570–7587.

Thorpe, J.W., Kidd, D., Moseley, I.F., Kendall, B.E., Thompson, A.J., MacManus, D.G., McDonald, W.I., and Miller, D.H. (1996). Serial gadolinium-enhanced MRI of the brain and spinal cord in early relapsing-remitting multiple sclerosis. *Neurology*, **46**, 373–378.

Tremlett, H.L., Evans, J., Wiles, C.M., and Luscombe, D.K. (2002). Asthma and multiple sclerosis: An inverse association in a case-control general practice population. *Q. J. Med.*, **95**, 753–756.

Tuohy, V.K., Yu, M., Yin, L., Kawczak, J.A., Johnson, J.A., Mathisen, P.M., Weinstock-Guttman, B., and Kinkel, R.P. (1998). The epitope spreading cascade during experimental autoimmune encephalomyelitis and multiple sclerosis. *Immunol. Rev.*, **164**, 93–100.

Utz, U., Biddison, W.E., McFarland, H.F., McFarlin, D.E., Flerlage, M., and Martin, R. (1993). Skewed T-cell receptor repertoire in genetically identical twins correlates with multiple sclerosis. *Nature*, **364**, 243–247.

Viglietta, V., Baecher-Allan, C., Weiner, H.L., and Hafler, D.A. (2004). Loss of functional suppression by CD4$^+$CD25$^+$ regulatory T cells in patients with multiple sclerosis. *J. Exp. Med.*, **199**, 971–979.

Wekerle, H. (1998). Immunology. In D.A.S. Compston (ed.) Mc Alpine's Multiple Sclerosis, 3rd edn. Churchill Livingstone, London. pp. 379–407.

Werring, D.J., Brassat, D., Droogan, A.G., Clark, C.A., Symms, M.R., Barker, G.J., MacManus, D.G., Thompson, A.J., and Miller, D.H. (2000). The pathogenesis of lesions and normal-appearing white matter changes in multiple sclerosis: A serial diffusion MRI study. *Brain*, **123**, 1667–1676.

Wucherpfennig, K.W. and Strominger, J.L. (1995). Molecular mimicry in T cell-mediated autoimmunity: Viral peptides activate human T cell clones specific for myelin basic protein. *Cell*, **80**, 695–705.

Zang, Y.C., Hong, J., Rivera, V.M., Killian, J., and Zhang, J.Z. (2000). Preferential recognition of TCR hypervariable regions by human anti-idiotypic T cells induced by T cell vaccination. *J. Immunol.*, **164**, 4011–4017.

Zhang, J.Z., Rivera, V.M., Tejada-Simon, M.V., Yang, D., Hong, J., Li, S., Hang, H., Killian, J., and Zang, Y.C. (2002). T cell vaccination in multiple sclerosis: Results of a preliminary study. *J. Neurol.*, **249**, 212–218.

17

Crippled B Lymphocyte Signaling Checkpoints in Systemic Autoimmunity

Moncef Zouali

1. Introduction

A remarkable feature of the immune system is that the potential to mount a response to self-antigens (Ags) is rooted in the genome of all normal individuals. Yet, pathogenic autoimmunity remains a relatively rare event that affects approximately 5% of the population. Despite intensive scrutiny, the origin of autoimmune reactivity remains an enigma, and a number of theories have been formulated to account for the initiation and persistence of an "aggressive" autoimmune reaction in diseased subjects, including a role for autoantigen, pathogen-related antigens, molecular mimicry, increased expression of major histocompatibility complex (MHC) class II antigens, polyclonal activation, altered antigen processing, and/or presentation. None of these mechanisms fully accounts for all the findings in autoimmune diseases, and it is conceivable that multiple processes may act simultaneously or in temporal succession in a single autoimmune disorder and that distinct processes may operate in a single autoimmune subject. It also is generally accepted that autoimmune diseases require contributions from inherited and environmental factors, and several cell types can be recruited to mount an aggressive autoimmune response. In systemic autoimmune disease, however, B lymphocytes play a paramount role. Focusing on systemic lupus erythematosus (SLE), this chapter reviews data suggesting that crippled B cell receptor (BcR)–mediated signaling checkpoints provide a biochemical and molecular background for the observed diverse abnormalities involved in disease progression.

Moncef Zouali • Institut National de Santé et de Recherche Médicale (INSERM U 430), Immunopathologie Humaine, Paris, France.

Molecular Autoimmunity: In commemoration of the 100th anniversary of the first description of human autoimmune disease, edited by Moncef Zouali. Springer Science+Business Media, Inc., New York, 2005.

2. B Lymphocytes Participate in Both innate and Adaptive Immunity

B lymphocytes are generated throughout life by differentiation from hematopoietic stem progenitors. Cells that engage in the B cell lineage in the bone marrow execute a programmed development, first sequentially rearranging immunoglobulin (Ig) heavy (H)-chain genes at the pro–B cell stage, then undergoing multiple rounds of clonal expansion at the pre–B cell stage and, finally, assembling Ig light (L)-chain genes to give rise to newly formed surface IgM$^+$ immature B cells. However, the random Ig variable (V) gene rearrangements can produce autoreactive lymphocytes that must be tolerized by negative selection before emigrating to peripheral lymphoid organs. In the bone marrow, immature self-reactive B lymphocytes are eliminated by apoptosis (clonal deletion), are functionally silenced (anergy), or undergo secondary V gene rearrangements to extinguish their autoreactivity (receptor editing). After undergoing these selection processes, newly formed B cells can migrate to the periphery, express IgD, and give rise to mature naive B cells, which can be categorized in three functionally and phenotypically distinct populations: conventional FO (or B-2) B cells, MZ B cells of the spleen, and B-1a B cells in the peritoneal cavity. Upon new encounters with self-Ags not present in the bone marrow, B cells must again be tolerized in the periphery before they enter the long-lived pool of B cells (reviewed in Radic and Zouali, 1996; Nemazee *et al.*, 2000).

At the mature stage, B cells possess a system that can sense the presence of microorganisms and contribute to their destruction. In addition to secreting Igs, B cells are able to present Ags (Roosnek and Lanzavecchia, 1991), upregulate costimulatory molecules, express antimicrobial activity by producing reactive oxygen intermediates and other inflammatory cytokines, and secrete factors that can directly mediate microbial destruction (Yi *et al.*, 1996; Lee and Koretzky, 1998). More recently, it was recognized that B cells can express toll-like receptors (TLRs) (Applequist *et al.*, 2002; Bourke *et al.*, 2003). This expression is augmented following engagement of the BcR or the costimulatory molecule CD40 or by stimulation with *S. aureus* Cowan I bacteria or unmethylated CpG DNA. Since TLR9 recognizes unmethylated CpG motifs characteristic of bacterial DNA and is involved in the immediate response to a wide range of microbes, B cells may, in addition to their role as Ab-producing cells during the adaptive immune response, respond to pathogens in a manner associated with the innate branch of immune defense. Inducible expression of TLRs in B cells may provide a link between the innate and adaptive branches of the immune system. Additionally, two B cell subpopulations with innate-like functions are present in the peripheral lymphocyte compartment: B-1 and MZ B cells. Because of their anatomical location and their functional properties, these two B cell subsets are involved in T cell–independent, innate-like immunity and represent an immune mechanism of first-line defense against pathogens.

3. The Critical Role of B Cells in Autoimmunity

The main immunological event in the pathogenesis of SLE is B cell hyperactivity, and several lines of evidence demonstrate that B cells are essential for development of autoimmunity and disease expression (Lu and Cyster, 2002; Zouali, 2005). Transfer of cultured pre-B cells derived from (NZB × NZW) F_1 fetal liver into SCID mice is sufficient to generate a lupus-like syndrome (Reininger *et al.*, 1992). Since T cells do not develop from these donor cells, it appears that B-lineage-intrinsic defects play a primary role in the pathogenesis of the autoimmune disorder. It is also clear that, before disease development, (NZB × NZW) F_1 mice have large numbers of B cells spontaneously producing low-affinity, IgM anti-DNA antibodies (Steward and Hay, 1976). In human lupus, the number of B cells that secrete Igs spontaneously is dramatically increased (Zouali *et al.*, 1991). Characterization of autoantibody genes shows that pathogenic, high-affinity IgG anti-DNA antibody (Ab) result from an Ag-driven process (Demaison *et al.*, 1994; Radic and Weigert, 1994), implying that the adaptive immune branch underlies their production. However, other evidence suggests that B cells with innate-like functions may play a role in autoimmunity (Viau and Zouali, 2005). In humans, increased numbers of B-1 cells are found in patients with certain autoimmune diseases. In experimental models, early studies recognized that B-1 cells exhibit binding to self-Ags and that they can potentially produce high-affinity autoantibodies typical of autoimmune disease. The autoimmune NZB mouse has an expanded B-1 population that could be linked to the *Sle2* lupus susceptibility locus. In mice homozygous for the *lpr* mutation, B-1 cells produce anti-erythrocyte autoantibodies responsible for autoimmune hemolytic anemia and their elimination reverses autoimmunity. Studies of B-1 cell trafficking also support their role in autoimmunity. While B-1 cells are known to preferentially migrate toward the chemoattractant CXCL13 (BLC), they fail to home to the peritoneal cavity in aged (NZB × NZW) F_1 mice, developing lupus nephritis (Ishikawa *et al.*, 2001). When injected intravenously, they are preferentially recruited to the target organs that show ectopic expression of CXCL13, namely the kidney, the lung, and the thymus. This aberrant homing of B-1 cells in aged (NZB × NZW) F_1 mice may favor autoantibody production.

Studies of MZ B cells also suggest that they can potentially play a role in the spontaneous development of autoantibodies (Viau and Zouali, 2005). In humans, infiltrating cells exhibiting a MZ B cell phenotype have been described in Grave's disease and in Sjögren's syndrome (Segundo *et al.*, 2001; Groom *et al.*, 2002). In the autoimmune NZB mouse model, several findings suggest that MZ B cell function is abnormal. There is an enhanced proliferation following stimulation with anti-Ig and -MHC class II Abs. Additionally, NZB mice have an increased proportion of MZ B cells exhibiting an "activated" phenotype, with increased levels of costimulatory molecules, as compared with nonautoimmune mouse strains. Like its parental NZB strain, the lupus-prone (NZB × NZW) F_1 mouse model demonstrates increased numbers of $CD1^{high}$

B cells, a phenotype of MZ B cells. Importantly, this MZ B cell expansion is detectable as early as 4 weeks of age and is responsible for production of large amounts of anti-DNA Abs, as compared with FO B cells (Zeng et al., 2000; Schuster et al., 2002).

Studies of longevity factors (APRIL and BAFF) that influence B cell maturation and survival at several levels also point to a link of MZ B cells with autoimmunity. Mice overexpressing BAFF spontaneously develop an SLE-like syndrome associated with a dramatic increase in MZ B cells (Mackay et al., 1999; Batten et al., 2000). More recently, it was found that the salivary glands of BAFF transgenic mice contain a subpopulation of B cells with an MZ-like phenotype (B22O$^+$HAS$^+$ CD21^{high}CD1^{high}) that could derive from the expanded MZ population present in the spleen of BAFF transgenic mice (Groom et al., 2002). Importantly, MZ-like B cells also have been detected in the thyroid gland of patients with Graves's disease. It is possible that these cells have aberrantly acquired trafficking receptors, enabling them to circulate and home to other lymphoid locations, as occurs in aged (NZB × NZW) F_1 mice whose B-1 cells abnormally migrate to ectopic target organs (Ito et al., 2004).

4. B Cell Receptor–Mediated Signaling Checkpoints

Throughout B cell development, lymphocyte survival and selection are determined by the specificity of the hetero-oligomeric BcR, a module that interacts with coreceptors, protein tyrosine kinases (PTKs) and phosphatases (PTPs), adapters, and other proteins to propagate signaling cascades (Reth and Wienands, 1997; Benschop and Cambier, 1999). Since the BcR is responsible for Ag recognition, the transduction events that occur after its ligation mediate interpretation of the magnitude and duration of BcR signaling. In addition to orchestrating the efficient development and subsequent activation of B lymphocytes at several discrete stages during B lineage differentiation, the BcR plays a central role in self-tolerance. During B lymphocyte development, potentially aggressive autoreactive B cells must be tolerized at various checkpoints. While studies over the past few years have provided a wealth of new information regarding the selection processes, systemic autoimmune disease is often considered to result from a deficiency in tolerance induction, resulting in failure to eliminate self-reactive lymphocytes. In this scheme, B cells that slip through the mesh of the tolerance safety nets would attack and destroy peripheral tissues. Lupus, for example, would result from rogue B cells reactive to various tissues. Yet, it is clear, from recent work with transgenic systems in which most B cells express receptors directed against self-antigens, that additional mechanisms allow the immune system to control potentially devastating lymphocytes and avoid pathology. Notably, transgenic expression or inactivation of genes encoding certain signaling molecules was found to lead to autoimmune phenomena and to disease, suggesting that a tight balance in BcR signaling pathways is required to prevent pathogenic autoimmune reactions (Hasler and Zouali, 2001; Kammer et al., 2002; Yu et al., 2003; Zouali and Sarmay, 2004).

5. Critical Regulators of B Cell Receptor Signaling

Cumulative evidence indicates that the CD19 coreceptor can induce positive signals that could enhance B cell responses. By regulating Src kinases and PI3K activity, it lowers the threshold of BcR-mediated signaling. Strikingly, study of $CD19^{-/-}$ mice revealed an apparent tight regulation of CD19 cell surface density during B cell development (Saito *et al.*, 2002). In contrast to mice that overexpress CD19, $CD19^{-/-}$ mice have a markedly elevated BcR signaling threshold compared with wild-type mice. Reversibly, transgenic expression of low levels of human CD19 with normal levels of mouse CD19 resulted in hyperactive B cells and loss of tolerance to nuclear Ags (Sato *et al.*, 2000). Since the product of PI3K, phosphatidyl inositol 3,4,5 trisphosphate, activates protein kinase B (PKB), which in turn promotes B cell survival, CD19 participates positively in B cell activation. Its positive effect on PI3K activity also may result in survival of immature B cells with low-affinity-binding BcRs, a potential source of the autoreactivity seen in mice that have only a 15–30% increase in CD19 expression. The CD19 transgenic mice exhibit a profoundly different phenotype from normal controls and develop SLE-like manifestations. Like mice that overexpress CD19, the tight skin (TSK) mouse, a genetic model for human systemic sclerosis, also contains spontaneously activated B cells and autoAbs against systemic sclerosis–specific target autoAgs. In TSK mice, CD19 deficiency results in quiescent B cells, with significantly reduced autoAb production and skin fibrosis (Saito *et al.*, 2002). Thus, even subtle increases in CD19 density were sufficient to predispose mice to autoimmune manifestations, suggesting that modest alterations in CD19 expression could contribute to the development of autoAbs in humans.

The transmembrane CD45 PTP is expressed on all nucleated hemopoietic cells and constitutes up to 10% of all membrane proteins in T and B cells. It possesses an intracytoplasmic region bearing two domains, D1 and D2, and an extracellular domain with variable composition and structure, due to alternative splicing of several exons and differential glycosylation. The function of CD45 in immunoreceptor signaling is to regulate the activity of Src family kinases by dephosphorylating their regulatory C terminal tyrosine. It is thought that dimerization of CD45 inhibits PTP activity through symmetrical interactions between its inhibitory structural wedge and its catalytic site. Recently, the phenotype of a gain-of-function mutation in CD45 that may enhance activity of Src family PTKs has been reported (Majeti *et al.*, 2000). Homozygous and heterozygous mice in which a single point mutation, glutamate 613 to arginine, that inactivates the inhibitory wedge of CD45 exhibited polyclonal lymphocyte activation, lymphoproliferation, autoAb production, and severe glomerulonephritis (GN), resulting in death. These findings demonstrate the *in vivo* importance of negative regulation of CD45 and suggest that the level of Src family PTK activity is an important determinant of immune tolerance.

Members of the novel serine/threonine protein kinase C (PKC) family, which includes PKC-δ, -ε, -θ and -η, can phosphorylate a multitude of cellular substrates. It is emerging that the different family members have diverse roles in the immune system and are implicated in various cellular processes, such as growth, differentiation, and death. Among the 12 known isoforms, several PKC members are expressed in B lineage cells and activated by BcR stimulation, suggesting a contribution of PKCs in the transduction of B cell–mediated immune responses. One of them, PKC-δ, is unique in that its overexpression can have a potent negative influence on cell behavior, inhibiting proliferation and growth and enhancing death. This isoform is known to be highly expressed in B cells and to be involved in BcR signaling. It has the unusual property of being tyrosine-phosphorylated and this phosphorylation event takes place within a minute of BcR engagement. To understand its role in B cell immunity, mice deficient for PKC-δ were analyzed (Mecklenbrauker *et al.*, 2002; Miyamoto *et al.*, 2002). The most overt abnormalities were splenomegaly and lymphadenopathy, both attributed to an increase in the number of conventional B 2 cells. This augmentation was not observed in the bone marrow, indicating that it occurred in secondary lymphoid organs. In the absence of Ag stimulation, PKC-δ-deficient mice exhibit B cell expansion, formation of unusually high numbers of numerous germinal centers (GCs), and defective B cell tolerance to self-Ag. With age, the B cell abnormalities had pathological consequences, including increased concentrations of serum IgG_1 and IgA, autoAbs, immune-complex (IC)–type GN, and lymphocyte perivascular infiltration in many organs. Additionally, while acute exposure of PKC-δ-deficient B cells to self-Ag elicits immunity, chronic stimulation with the same Ag failed to induce tolerance via anergy, as it does for wild-type B cells (Mecklenbrauker *et al.*, 2002), further underscoring the importance of PKC-δ in negative regulation of B cell proliferation and in establishing B cell tolerance.

While Lyn is not required to initiate BcR signaling, it is an essential inhibitor of transduction pathways. B cells from mice homozygous for a disruption at the *Lyn* locus had a delayed, but increased, Ca^{2+} flux and an exaggerated negative selection response to Ag, and a spontaneous hyperactivity (Hibbs *et al.*, 1995; Nishizumi *et al.*, 1995). The deficient mice also had circulating autoreactive Abs and severe GN caused by the deposition of ICs in the kidney, a pathology reminiscent of SLE. The role of Lyn in BcR-mediated signaling and in establishing B cell tolerance to self-Ags also stems from the phenotype of Lyn deficiency on a C57BL/6 background. The mutant mice are more susceptible to myelin oligodendrocyte glycoprotein–induced experimental allergic encephalomyelitis (Du and Sriram, 2002). This strain dependence suggests that *Lyn* may represent a genetic susceptibility factor for autoimmune disease.

Further analysis of signaling pathways in $Lyn^{-/-}$ mice demonstrated that regulation of BcR signaling is a complex quantitative trait in which the Lyn activating role is mediated by the phosphorylation of tyrosine residues within immunoreceptor-activating motifs (ITAMs) of Igα, Igβ and CD19, and the subsequent recruitment of signaling enzymes, such as Syk, PLCγ2, and PI-3Kinase (Cornall *et al.*, 1998). These positive effects are balanced by the negative regulatory role of Lyn in B cells through at least two pathways. First, Lyn is responsible for phosphorylating tyrosine residues of the negative BcR coreceptors FcγRIIb, platelet endothelial cell adhesion molecule-1 (PECAM-1), and PD-1 (see below). This activity may account, at least in part, for the enhanced BcR signaling seen in $Lyn^{-/-}$ B cells and for the autoimmune phenotype observed in mice lacking these BcR coreceptors. Second, Lyn was shown to negatively regulate

B cells by opposing the effect of Syk on BcR-mediated activation of Akt/PKB. Deregulation of Akt/PKB correlates with the BcR-mediated hyperresponsiveness of Lyn$^{-/-}$ B cells and might contribute to the autoimmune syndrome that develops in Lyn-deficient animals.

The key role of Lyn in establishing and maintaining peripheral tolerance also comes from studies of the consequences of sustained activation of Lyn *in vivo* using a targeted gain-of-function mutation (Hibbs *et al.*, 2002). The mice, designated Lyn$^{up/up}$, carry a single point mutation (Y508) in a sequence that negatively regulates Lyn activity and express a constitutively activated form of Lyn, allowing study of the consequences of constitutive engagement of both stimulatory and inhibitory signaling pathways. The mutant mice have reduced numbers of conventional B-2 lymphocytes, downregulated surface IgM and costimulatory molecules, and elevated numbers of B-1a cells. *In vitro*, there is a heightened Ca^{2+} flux in response to BcR stimulation and exaggerated positive signaling. While there is a constitutive phosphorylation of negative regulators of BcR signaling (SHP-1 and SHIP-1), Syk and phospholipase Cγ2 are constitutively phosphorylated. Surprisingly, Lyn$^{up/up}$ mice developed circulating autoreactive Abs and lethal autoimmune GN, suggesting that enhanced positive signaling eventually overrides constitutive negative signaling. The breakdown of self-tolerance seen in these mice may result from an imbalance between chronic negative signaling and enhanced positive signaling. It indicates that Lyn plays a central role in maintaining the equilibrium between positive and negative B cell signaling pathways, and in regulating B cell tolerance and development of autoimmunity. Moreover, the fact that both Lyn$^{-/-}$ and Lyn$^{up/up}$ mice show a breakdown in self-tolerance and develop circulating autoreactive autoAbs and severe lupus-like nephritis (Hibbs *et al.*, 1995, 2002; Nishizumi *et al.*, 1995) clearly demonstrates that Lyn is a key regulator of B cell signaling and suggests that any imbalance in signaling, either by deletion on activation, may result in severe autoimmunity.

Also important in BcR signaling and in self-tolerance is SHP-1, a PTP with several targets in immunoreceptor signaling. SHP-1 is reportedly recruited to CD22 or CD72, even in the absence of BcR coaggregation. It also can dephosphorylate the ITAMs of Src kinases, Syk kinases, adapters such as SLP-76, effectors such as Vav and PI-3' kinase, as well as receptors such as CD19. There also is strong indication that SHP-1 can dephosphorylate immunoreceptor inhibitory motif (ITIM)–containing receptors (see below), thereby providing a potential mechanism of autoregulation (Reth and Wienands, 1997; Benschop and Cambier, 1999). A significant clue regarding the role of SHP-1 in immune cell homeostasis was provided by the finding that its gene is mutated in *motheaten* (m^e) and *viable motheaten* (m^{ev}) mice (Shultz *et al.*, 1993; Tsui *et al.*, 1993). The mutant mice exhibit a spontaneous point mutation in SHP-1 that results in a protein with decreased PTP activity (10–20% of normal activity) and immune defects, including expansion of B-1 cells, a low threshold of membrane Ig signaling, hypergammaglobulinemia, autoantibody production, and GN. Their B cells exhibit augmented BcR-induced proliferation, protein tyrosine phosphorylation, and mitogen-activated protein kinase (MAPK) activation. Similar defects were observed in a B cell line expressing a dominant-negative form of SHP-1. Thus, SHP-1 is a critical negative regulator of immunoreceptor signaling in B cells.

6. Negative Regulators of B Cell Receptor–Mediated Signal Transduction

The prototypes of negative coreceptors are those that target the Fc region of IgGs (Fcγ). By virtue of their broad tissue distribution, they have the ability to sense humoral concentrations of Abs. Linking ligands with effector cells of the immune system, they initiate cellular responses useful in host defense (Daeron, 1997; Ravetch and Bolland, 2001). One of them, FcγRIIb1, is exclusively expressed on B cells and has the potential to terminate B cell signal transduction.

The prediction that FcγR may control autoantibody production as well as IC-induced inflammation was amply confirmed in studies of FcγR-deficient models. Ablation of FcγRIIb renders mice susceptible to the induction, or spontaneous development, of various autoimmune disorders, such as collagen-induced arthritis (CIA), Goodpasture syndrome, and GN, indicating that FcγRIIb is a critical component of peripheral tolerance (Takai, 2002). Significantly, while FcγRIIb deficiency results in anti-DNA and antichromatin Abs, and fatal autoimmune GN on the C57BL/6 background, FcγRIIb$^{-/-}$ BALB/c mice maintain tolerance to self and are resistant to the development of autoimmunity. This strain dependence suggests that FcγRIIb is a genetic susceptibility factor for autoimmune disease, a conclusion further supported by studies showing that FcγRIIb deficiency can render a nonpermissive MHC haplotype susceptible to CIA, a murine model of rheumatoid arthritis. While mice expressing a nonpermissive haplotype are resistant to CIA induction, FcγRIIb$^{-/-}$ mice immunized with type II collagen (CII) develop cellular and humoral immunity to collagen, resulting in inflammatory destruction of the joints and arthritis that can be transferred by immune serum. These data indicate that FcγRIIb negatively regulates both humoral immune responses and IC-mediated inflammation, and plays a critical role in suppressing the induction of CIA.

The precise molecular mechanism for the synergistic effect of combinations of susceptibility alleles is poorly defined. MRL-*lpr/lpr* mice, a model for SLE and RA, have a Fas mutation that results in spontaneous development of systemic autoimmune disease and a short life span. Half of them die by 5–6 months of age due to massive progression of systemic autoimmune disease, including lupus GN. However, the C57BL/6 (B6).Fas(*lpr/lpr*) strain does not develop such disorders within the normal life span, indicating that suppressor gene(s) in B6 mice may control the onset and exacerbation of disease. In FcγRIIb$^{-/-}$ mice deficient in Fas (B6.IIB$^{-/-}$–Fas$^{lpr/lpr}$), the combined absence of FcγRIIb and Fas led to the development of systemic autoimmune disease, including anti-DNA and anti-CII autoAbs, and cryoglobulin production (Yajima *et al.*, 2003). The double-mutant mice were short-lived, due to enhanced autoAb production culminating in fatal IC-mediated lupus GN. Thus, FcγRIIb deletion with Fas mutation is sufficient to mediate systemic autoimmunity in B6 mice, implying that the *FcγRIIb* gene is a critical SLE suppressor.

Polymorphisms of FcγR have been shown to impact on the induction and regulation of autoimmunity, and on the pathogenesis of autoimmune disease. In NZB and (NZB × NZW) F$_1$ mice, FcγRIIb1 expression was abnormally downregulated in follicular GC B cells from aged mice (Jiang *et al.*, 2000). This reduced surface expression of FcγRIIb1 may be attributable to DNA polymorphism in the promoter region of the FcγRIIb1 gene (Jiang et al., 2000; Pritchard et al., 2000). Since during T cell–dependent humoral immune responses, Ag-activated B cells accumulate in GCs where Ig class switching, somatic mutation, receptor revision, and generation of high-affinity IgG Abs occur, the FcγRIIb1 promoter allele may well be functioning to maintain and/or upregulate FcγRIIb1 expression levels on GC B cells during the process of T cell–mediated activation in healthy strains of mice. As lupus IgG anti-DNA Abs in SLE also are affinity-selected (Demaison et al., 1994; Radic and Weigert, 1994) and must be, at least in part, generated in GCs, FcγRIIb1 downregulation may represent a mechanism that allows high-affinity IgG autoAbs to be overproduced by GC B cells that escape negative signals for IgG production.

PECAM-1/CD31 represents a new B cell coreceptor that serves to negatively regulate BcR signaling and B cell tolerance *in vivo*. It is a 130-kDa glycoprotein and a member of the Ig superfamily. In humans, it is expressed at the lateral junctions of endothelial cells and on the surface of hematopoietic cells, including monocytes, neutrophils, natural killer (NK) cells, platelets, and naive B and T cells. In the B cell lineage, PECAM-1 is expressed on human naive follicular mantle zone B cells and plasma cells, but not on GC memory B cells. Its cytoplasmic domain contains two ITIMs that, upon phosphorylation of the putative tyrosine residues, recruit and activate the PTPs SHP-1 and SHP-2 (Henshall et al., 2001). By doing so, PECAM-1 might negatively regulate BcR signaling.

The inhibitory effects observed *in vitro* in response to cross-linking suggested that PECAM-1 is an important regulator of Ag-induced lymphocyte activation. Studies *in vivo* demonstrated that PECAM-1 has a functional immunomodulatory role, thereby adding a new B cell coreceptor to the list of known negative regulators of BcR-mediated signaling (Wilkinson et al., 2002). PECAM-1$^{-/-}$ mice exhibit a developmental defect in transition from immature to mature B cells, increased numbers of B-1a cells in the peritoneum, reduced mature recirculating B-2 cells in the periphery, hyperresponsive B cells in response to BcR cross-linking and polyclonal stimulation with lipopolysaccharide (LPS), and elevated antibody responses to T-cell-independent Ags. With age, autoAbs, IC-mediated GN and lupus-like autoimmune disease are seen (Wilkinson et al., 2002). These findings are consistent with the PECAM-1 serving to negatively regulate the signaling threshold of early B cell activation initiated through the BcR complex, which directs downstream effector functions, including B cell proliferation (Wilkinson et al., 2002). It may contribute to the maintenance of peripheral tolerance and protection from autoimmunity.

The recently identified CD28 homolog and costimulatory molecule programmed death-1 (PD-1) and its ligands, PD-L1 and PD-L2, are homologs of B7 (Nishimura and Honjo, 2001). PD-1 is a type 1 transmembrane protein that belongs to the Ig superfamily and is transcriptionally induced in activated T and B cells, and myeloid cells. Its cytoplasmic region contains an ITIM with two tyrosine residues. In human tonsils, PD-1 is expressed on most T cells and on a small subset of centrocytes in the light zone of GCs, where clonal selection of centrocytes takes place, suggesting that PD-1 may play an important role in the GC reaction.

Because the sequence surrounding its N-terminal tyrosine residue fulfills the requirement of ITIM, it has been thought that PD-1 negatively regulates immune responses. Further studies showed that coligation of PD-1 with the BcR inhibited Ca^{2+} mobilization and tyrosine phosphorylation of effector molecules, including Igβ, Syk, phospholipase C-γ2, and ERK1/2, whereas phosphorylation of Lyn and Dok was not affected (Okazaki et al., 2001). PD-1 can inhibit BcR signaling by recruiting SHP-2 to its phosphotyrosine and dephosphorylating key BcR-signaling transducers. Because PD-1 is phosphorylated exclusively by coligation with the BcR and is expressed only on activated cells, it probably functions in the downmodulation of excessive and prolonged activation, and inhibition and/or suppression of inappropriate activation, such as occurs in autoimmunity, by elevating the threshold for restimulation.

These *in vitro* studies suggested that PD-1 acts to downregulate immune responses and plays a critical role in the establishment and/or maintenance of peripheral tolerance. Not surprisingly, ablation of PD-1 led to breakdown of peripheral tolerance and to autoimmune disease (Okazaki et al., 2001). Specifically, lack of PD-1 on a BALB/c background gave rise to a lethal autoAb-mediated dilated cardiomyopathy, a life-threatening disease in humans for which only heart transplantation is effective (Nishimura and Honjo, 2001). This disease is mediated by an autoAb reactive to a heart-specific 30-kDa protein. Because ligands for PD-1 are strongly expressed in the heart and the Ag recognized by the autoAb is strictly restricted to heart, the direct interaction between $PD-1L^+$ heart tissue and $PD-1^+$ preactivated B cells normally prevents this deadly pathology by inhibiting autoreactive B cells at the effector site. By contrast, on a C57BL-6 background, PD-1 deficiency results in spontaneous typical lupus-like GN and destructive arthritis (Nishimura et al., 1999). These studies reinforce the notion that in genetically predisposed individuals, dysfunction of PD-1 may underlie distinct types of autoimmune disease.

Finally, CD22 is a B cell–specific transmembrane protein of the Ig superfamily with seven Ig-like domains and three cytoplasmic ITIMs that function as BcR coreceptors. It first appears intracellularly during the late pro–B cell stage of ontogeny, shifting to the plasma membrane with B cell maturation until plasma cell differentiation. CD22 is expressed at low levels on pre-B cells and at higher levels on mature IgM^+IgD^+ B cells and is absent on terminally differentiated plasma cells. It acts as a homing receptor for recirculating B cells, probably through the affinity of its extracellular domain for 2,6-linked sialic acid–bearing glycans (Nitschke et al., 1999).

CD22 is associated with the BcR both structurally and functionally. Upon BcR cross-linking, its cytoplasmic domain is rapidly tyrosine phosphorylated, resulting in recruitment of a number of signaling molecules, including Lyn, Syk, phospholipase C-γ1, and PI3K. In addition, tyrosine phosphorylated CD22 recruits and activates SH2-domain-containing PTP (SHP-1 and SHIP), leading to activation of a CD22/SHP-1/SHIP regulatory pathway that reduces both Ca^{2+} mobilization and MAPK activation and downregulates CD19 phosphorylation and BcR-mediated signal transduction, culminating in negative regulation of BcR signaling. It is thought that by downmodulating BcR signaling, CD22 sets a threshold for BcR ligation.

The *in vitro* studies suggesting that CD22 may help prevent unwanted activation of low-affinity or autoreactive B cells that could emerge as a result of somatic mutation in response to exogenous Ags are supported by several observations *in vivo*. Targeted disruption of the *cd22* gene results in a hyperactive

B cell phenotype, an augmented thymus-independent immune response, an expanded B-1 cell population, and increased titers of serum autoAbs (Sato et al., 1996; Nitschke et al., 1997; O'Keefe et al., 1999). Interestingly, as the CD22-deficient mice age, a large proportion develop high titers of serum IgG anti-DNA Abs (O'Keefe et al., 1999). Thus, a single gene defect, exclusive to B cells, suffices to trigger autoAb production in mice.

Since deregulated expression of CD22 could lead to excessive activation of B cells and autoAb production, as seen in CD22$^{-/-}$ mice, studies were performed to probe this coreceptor in systemic autoimmunity. Genomewide mapping analysis of lupus susceptibility loci in autoimmune-prone NZW mice revealed that an interval containing the *Cd22* gene on chromosome 7 is linked with autoAb production and lupus-like GN (Wakeland et al., 2001). Remarkably, one of the loci contributing to autoimmunity in New Zealand Mixed (NZM) mice (*Sle3*) also has been mapped to a region of chromosome 7 in the vicinity of *cd22*. These latter autoimmune mice produce aberrant CD22 mRNA species that result from alternative splicing due to an insertion in one intron of the *cd22* gene (Mary et al., 2000). Heterozygous expression of this allele, called *cd22^a*, promoted autoAb production in mice bearing the Y chromosome–linked autoimmune acceleration gene *Yaa* (Mary et al., 2000), implying that even a partial CD22 deficiency may contribute to lupus susceptibility. This gene linkage appears to have functional consequences. Activation of B cells with LPS in the presence of IL-4 upregulated CD22 expression in NZW mice bearing the *Cd22^a* allele, as compared with B6 mice bearing the *Cd22^b* allele (Wakeland et al., 2001), suggesting that defective upregulation of CD22 on potentially autoreactive B cells may favor the production of autoAbs in lupus-prone mice.

7. Disrupted B Cell Signaling Pathways in Human Autoimmunity

As discussed above, experimentally induced modification of receptor expression or alteration of signaling pathways may have a significant impact on B cell tolerance to self. In humans too, there are indications that abnormal B cell signaling may contribute to autoimmune disease. In lupus, stimulation of circulating B cells through their sIgM produced significantly higher Ca^{2+} fluxes compared with similarly induced responses of B cells from patients with other systemic rheumatic diseases (Liossis et al., 1996). The overall level of sIgM-initiated protein tyrosyl phosphorylation also was significantly enhanced, and correlated with the augmented BcR-mediated free Ca^{2+} responses. This aberrant BcR-mediated signaling process was not associated with disease activity, medications used, or specific clinical manifestations. It was disease-specific, suggesting a possible intrinsic SLE B cell defect that may have pathogenic implications. In further studies, the content of lupus B cells in Lyn, CD45, and SHP-1 was significantly altered in patients (Huck et al., 2001; Liossis et al., 2001). Investigation of gene polymorphisms in the Japanese population suggested that *cd22* could be considered a candidate susceptibility gene for autoimmune disease (Hatta et al., 1999). Although the function of the CD22/Lyn signaling-inhibitor complex was not addressed, it is likely that decreased Lyn in SLE B cells may contribute to the B cell overactivity. In line with studies in rodents showing that CD19 overexpression by 20% induces autoAb production in normal mice, expression of CD19 and CD21 levels are 20% higher on B cells from patients with systemic sclerosis compared with healthy individuals (Saito *et al.*, 2002). In Japanese patients with SLE, an CD19 single nucleotide polymorphism (SNP), which was rare in Caucasians, was increased (Kuroki *et al.*, 2002).

A particular feature of human autoimmune disease is the wide variety of affected organs, the diversity of immune responses involved, the distinctive phenotypes in time course, severity, and response to medication. One possibility is that genetic predisposition contributes to determination of the pathogenic process, a view supported by the strain dependency of disease manifestations observed in mice deficient in inhibitory BcR coreceptors. Initial support to a pivotal role of FcγRIIb polymorphisms as susceptibility alleles in the pathogenesis of SLE came from mapping of a region in the telomere of human chromosome 1 (Tsao *et al.*, 1997). This region is syntetic to a region on mouse chromosome 1 in the vicinity of the *FcγRIIb1* locus (Wakeland *et al.*, 2001) and encodes a variety of immunologically relevant molecules, including FcγRIIb and FcγRIII, and poly(ADP-ribose) polymerase (PARP). More recently, the frequency of an SNP (695T/C) coding for a nonsynonymous $232^{Ile/Thr}$ substitution within the transmembrane domain of FcγRIIb was significantly increased in Japanese SLE patients compared with healthy individuals (Kyogoku *et al.*, 2002). It is of further interest that another allele called *FCGR2B-187T*, which mediates a higher level of CD19 dephosphorylation and a greater degree of Ca^{2+} response when coengaged with the BcR than does *FCGR2B-187I*, is not associated with SLE in both African Americans and Caucasians (Li *et al.*, 2003). This lack of association raises the interesting possibility that *FCGR2B-187T* may be interacting epistatically with background genes that differ between Japanese patients and both African American and Caucasian patients.

In mouse, PD-1 appears to inhibit immune responses *in vivo*, and abrogation of this inhibition could result in development of autoimmune disease. This conclusion is supported by gene mapping studies of patients wherein an intronic SNP in *PD-1* was reportedly associated with SLE in 12% of Europeans and 7% of Mexicans (Prokunina *et al.*, 2002). The identified allele alters a binding site for a transcription factor located in an intronic enhancer and could lead to aberrant regulation of PD-1 and, hence, deregulated self-tolerance. These data suggest a mechanism through which PD-1 can contribute to the development of SLE in humans (Prokunina *et al.*, 2002).

Overall, studies of mouse models revealed that subtle alterations in overlapping signaling pathways that influence B cell responses to transmembrane and intracellular signals are sufficient to predispose mice to autoAb production. It is important to probe additional signal transduction pathways in autoimmune patients. For example, given that the lack of PKC-δ leads to autoimmunity in mice, it is possible that the activity of PKC-δ is reduced in the cells of humans with certain autoimmune diseases. Moreover, strikingly similar abnormalities of Ag-receptor signaling have previously been reported in T cells from patients with SLE (Brundula *et al.*, 1999; Kammer *et al.*, 2002) pointing toward potentially unifying Ag-receptor-mediated signaling defects in lupus lymphocytes (Zouali, 1998; Kammer et al., 2002). Accordingly, the signaling abnormalities encountered in SLE lymphocytes may provide a biochemical and molecular background for such diverse abnormalities as lymphocyte activation, anergy, and cell death (Hasler and Zouali, 2001; Kammer et al., 2002; Zouali and Sarmay, 2004).

8. Conclusions

Studies of B cells in systemic autoimmune diseases have provided important clues. Their role in autoimmunity is more important than previously thought. Only recently was it realized that B cells can be subject to positive selection generated and maintained on the basis of their autoreactivity and that B cells are essential in promoting systemic autoimmunity. Their depletion has been used in treating a number of autoimmune conditions, including autoimmune thrombocytopenic, rheumatoid arthritis, lupus, autoimmune hemolytic anemia, cold agglutinin disease, mixed cryoglobulinemia, autoimmune neuropathies, myasthenia gravis, Wegener's granulomatosis, and dermatomyositis. In many of these conditions major improvement is seen in a good proportion of cases, particularly in rheumatoid arthritis, lupus, dermatomyositis, autoimmune neuropathies, immune thrombocytopenic purpura, and hemolytic anemia (Patel, 2002).

While it seems clear that immune receptor signaling checkpoints are involved in the progression of systemic autoimmunity, their origin remains unclear. Even though genetic factors are important for disease development, the environmental contribution to clinical expression cannot be ignored, and it is likely that different mechanisms could lead to loss of self-tolerance characteristic of SLE. The observation that different SLE patients produce different spectra of autoantibodies suggests that more than one factor could play a role in a single patient and it is conceivable that the combination of factors varies throughout the disease. Environmental factors include infectious agents, endogenous retroviruses, pollutants, and hormones (Hasler and Zouali, 2003; Zouali, 2005). While immune receptor signaling checkpoints may not prove to be the single key to elucidating all aspects of autoimmunity, it is likely that further studies of these and related pathways may lead to novel approaches of more specific therapeutic intervention in human systemic autoimmunity.

References

Applequist, S.E., Wallin, R.P., and Ljunggren, H.G. (2002). Variable expression of toll-like receptor in murine innate and adaptive immune cell lines. *Int. Immunol.*, **14**, 1065–1074.

Batten, M., Groom, J., Cachero, T.G., Qian, F., Schneider, P., Tschopp, J., Browning, J.L., and Mackay, F. (2000). BAFF mediates survival of peripheral immature B lymphocytes. *J. Exp. Med.*, **192**, 1453–1466.

Benschop, R.J. and Cambier, J.C. (1999). B cell development: Signal transduction by antigen receptors and their surrogates. *Curr. Opin. Immunol.*, **11**, 143–151.

Bourke, E., Bosisio, D., Golay, J., Polentarutti, N., and Mantovani, A. (2003). The toll-like receptor repertoire of human B lymphocytes: Inducible and selective expression of TLR9 and TLR10 in normal and transformed cells. *Blood*, **102**, 956–963. Epub 2003 Apr 2010.

Brundula, V., Rivas, L.J., Blasini, A.M., Paris, M., Salazar, S., Stekman, I.L., and Rodriguez, M.A. (1999). Diminished levels of T cell receptor zeta chains in peripheral blood T lymphocytes from patients with systemic lupus erythematosus. *Arthritis Rheum.*, **42**, 1908–1916.

Cornall, R.J., Cyster, J.G., Hibbs, M.L., Dunn, A.R., Otipoby, K.L., Clark, E.A., and Goodnow, C.C. (1998). Polygenic autoimmune traits: Lyn, CD22, and SHP-1 are limiting elements of a biochemical pathway regulating BCR signaling and selection. *Immunity*, **8**, 497–508.

Daeron, M. (1997). Fc receptor biology. *Annu. Rev. Immunol.*, **15**, 203–234.

Demaison, C., Chastagner, P., Theze, J., and Zouali, M. (1994). Somatic diversification in the heavy chain variable region genes expressed by human autoantibodies bearing a lupus-associated nephritogenic anti-DNA idiotype. *Proc. Natl. Acad. Sci. U S A*, **91**, 514–518.

Du, C. and Sriram, S. (2002). Increased severity of experimental allergic encephalomyelitis in lyn$^{-/-}$ mice in the absence of elevated proinflammatory cytokine response in the central nervous system. *J. Immunol.*, **168**, 3105–3112.

Groom, J., Kalled, S.L., Cutler, A.H., Olson, C., Woodcock, S.A., Schneider, P., Tschopp, J., Cachero, T.G., Batten, M., Wheway, J., Mauri, D., Cavill, D., Gordon, T.P., Mackay, C.R., Mackay, F. (2002). Association of BAFF/BLyS overexpression and altered B cell differentiation with Sjögren's syndrome. *J. Clin. Invest.*, **109**, 59–68.

Hasler, P. and Zouali, M. (2001). B cell receptor signaling and autoimmunity. *FASEB J.*, **15**, 2085–2098.

Hasler, P. and Zouali, M. (2003). Subversion of immune receptor signaling by infectious agents. *Genes Immun.*, **4**, 95–103.

Hatta, Y., Tsuchiya, N., Matsushita, M., Shiota, M., Hagiwara, K., and Tokunaga, K. (1999). Identification of the gene variations in human CD22. *Immunogenetics*, **49**, 280–286.

Henshall, T.L., Jones, K.L., Wilkinson, R., and Jackson, D.E. (2001). Src homology 2 domain-containing protein-tyrosine phosphatases, SHP-1 and SHP-2, are required for platelet endothelial cell adhesion molecule-1/CD31-mediated inhibitory signaling. *J. Immunol.*, **166**, 3098–3106.

Hibbs, M.L., Harder, K.W., Armes, J., Kountouri, N., Quilici, C., Casagranda, F., Dunn, A.R., and Tarlinton, D.M. (2002). Sustained activation of Lyn tyrosine kinase *in vivo* leads to autoimmunity. *J. Exp. Med.*, **196**, 1593–1604.

Hibbs, M.L., Tarlinton, D.M., Armes, J., Grail, D., Hodgson, G., Maglitto, R., Stacker, S.A., and Dunn, A.R. (1995). Multiple defects in the immune system of Lyn-deficient mice, culminating in autoimmune disease. *Cell*, **83**, 301–311.

Huck, S., Le Corre, R., Youinou, P., and Zouali, M. (2001). Expression of B cell receptor-associated signaling molecules in human lupus. *Autoimmunity*, **33**, 213–224.

Ishikawa, S., Sato, T., Abe, M., Nagai, S., Onai, N., Yoneyama, H., Zhang, Y., Suzuki, T., Hashimoto, S., Shirai, T., Lipp, M., Matsushima, K. (2001). Aberrant high expression of B lymphocyte chemokine (BLC/CXCL13) by C11b+CD11c+ dendritic cells in murine lupus and preferential chemotaxis of B1 cells towards BLC. *J. Exp. Med.*, **193**, 1393–1402.

Ito, T., Ishikawa, S., Sato, T., Akadegawa, K., Yurino, H., Kitabatake, M., Hontsu, S., Ezaki, T., Kimura, H., and Matsushima, K. (2004). Defective B1 cell homing to the peritoneal cavity and preferential recruitment of B1 cells in the target organs in a murine model for systemic lupus erythematosus. *J. Immunol.*, **172**, 3628–3634.

Jiang, Y., Hirose, S., Abe, M., Sanokawa-Akakura, R., Ohtsuji, M., Mi, X., Li, N., Xiu, Y., Zhang, D., Shirai, J., Hamano, Y., Fuji, H., Shirai, T. (2000). Polymorphisms in IgG Fc receptor IIB regulatory regions associated with autoimmune susceptibility. *Immunogenetics*, **51**, 429–435.

Kammer, G.M., Perl, A., Richardson, B.C., and Tsokos, G.C. (2002). Abnormal T cell signal transduction in systemic lupus erythematosus. *Arthritis Rheum.*, **46**, 1139–1154.

Kuroki, K., Tsuchiya, N., Tsao, B.P., Grossman, J.M., Fukazawa, T., Hagiwara, K., Kano, H., Takazoe, M., Iwata, T., Hashimoto, H., and Tokunaga, K. (2002). Polymorphisms of human CD19 gene: Possible association with susceptibility to systemic lupus erythematosus in Japanese. *Genes Immun.*, **3** (Suppl. 1), S21–S30.

Kyogoku, C., Dijstelbloem, H.M., Tsuchiya, N., Hatta, Y., Kato, H., Yamaguchi, A., Fukazawa, T., Jansen, M.D., Hashimoto, H., van de Winkel, J.G., Kallenberg, C.G., Tokunaga, K. (2002). Fcgamma receptor gene polymorphisms in Japanese patients with systemic lupus erythematosus: Contribution of FCGR2B to genetic susceptibility. *Arthritis Rheum.*, **46**, 1242–1254.

Lee, J.R. and Koretzky, G.A. (1998). Production of reactive oxygen intermediates following CD40 ligation correlates with c-Jun N-terminal kinase activation and IL-6 secretion in murine B lymphocytes. *Eur. J. Immunol.*, **28**, 4188–4197.

Li, X., Wu, J., Carter, R.H., Edberg, J.C., Su, K., Cooper, G.S., and Kimberly, R.P. (2003). A novel polymorphism in the Fcgamma receptor IIB (CD32B) transmembrane region alters receptor signaling. *Arthritis Rheum.*, **48**, 3242–3252.

Liossis, S.N., Kovacs, B., Dennis, G., Kammer, G.M., and Tsokos, G.C. (1996). B cells from patients with systemic lupus erythematosus display abnormal antigen receptor-mediated early signal transduction events. *J. Clin. Invest.*, **98**, 2549–2557.

Liossis, S.N., Solomou, E.E., Dimopoulos, M.A., Panayiotidis, P., Mavrikakis, M.M., and Sfikakis, P.P. (2001). B-cell kinase lyn deficiency in patients with systemic lupus erythematosus. *J. Invest. Med.*, **49**, 157–165.

Lu, T.T. and Cyster, J.G. (2002). Integrin-mediated long-term B cell retention in the splenic marginal zone. *Science*, **297**, 409–412.

Mackay, F., Woodcock, S.A., Lawton, P., Ambrose, C., Baetscher, M., Schneider, P., Tschopp, J., and Browning, J.L. (1999). Mice transgenic for BAFF develop lymphocytic disorders along with autoimmune manifestations. *J. Exp. Med.*, **190**, 1697–1710.

Majeti, R., Xu, Z., Parslow, T.G., Olson, J.L., Daikh, D.I., Killeen, N., and Weiss, A. (2000). An inactivating point mutation in the inhibitory wedge of CD45 causes lymphoproliferation and autoimmunity. *Cell*, **103**, 1059–1070.

Mary, C., Laporte, C., Parzy, D., Santiago, M.L., Stefani, F., Lajaunias, F., Parkhouse, R.M., O'Keefe, T.L., Neuberger, M.S., Izui, S., and Reininger, L. (2000). Dysregulated expression of the Cd22 gene as a result of a short interspersed nucleotide element insertion in Cd22a lupus-prone mice. *J. Immunol.*, **165**, 2987–2996.

Mecklenbrauker, I., Saijo, K., Zheng, N.Y., Leitges, M., and Tarakhovsky, A. (2002). Protein kinase Cdelta controls self-antigen-induced B-cell tolerance. *Nature*, **416**, 860–865.

Miyamoto, A., Nakayama, K., Imaki, H., Hirose, S., Jiang, Y., Abe, M., Tsukiyama, T., Nagahama, H., Ohno, S., Hatakeyama, S., and Nakayama, K.I. (2002). Increased proliferation of B cells and auto-immunity in mice lacking protein kinase Cdelta. *Nature*, **416**, 865–869.

Nemazee, D., Kouskoff, V., Hertz, M., Lang, J., Melamed, D., Pape, K., and Retter, M. (2000). B-cell-receptor-dependent positive and negative selection in immature B cells. *Curr. Top. Microbiol. Immunol.*, **245**, 57–71.

Nishimura, H. and Honjo, T. (2001). PD-1: An inhibitory immunoreceptor involved in peripheral tolerance. *Trends Immunol.*, **22**, 265–268.

Nishimura, H., Nose, M., Hiai, H., Minato, N., and Honjo, T. (1999). Development of lupus-like autoimmune diseases by disruption of the PD-1 gene encoding an ITIM motif-carrying immunoreceptor. *Immunity*, **11**, 141–151.

Nishizumi, H., Taniuchi, I., Yamanashi, Y., Kitamura, D., Ilic, D., Mori, S., Watanabe, T., and Yamamoto, T. (1995). Impaired proliferation of peripheral B cells and indication of autoimmune disease in lyn-deficient mice. *Immunity*, **3**, 549–560.

Nitschke, L., Carsetti, R., Ocker, B., Kohler, G., and Lamers, M.C. (1997). CD22 is a negative regulator of B-cell receptor signalling. *Curr. Biol.*, **7**, 133–143.

Nitschke, L., Floyd, H., Ferguson, D.J., and Crocker, P.R. (1999). Identification of CD22 ligands on bone marrow sinusoidal endothelium implicated in CD22-dependent homing of recirculating B cells. *J. Exp. Med.*, **189**, 1513–1518.

Okazaki, T., Maeda, A., Nishimura, H., Kurosaki, T., and Honjo, T. (2001). PD-1 immunoreceptor inhibits B cell receptor-mediated signaling by recruiting Src homology 2-domain-containing tyrosine phosphatase 2 to phosphotyrosine. *Proc. Natl. Acad. Sci. U S A*, **98**, 13866–13871.

O'Keefe, T.L., Williams, G.T., Batista, F.D., and Neuberger, M.S. (1999). Deficiency in CD22, a B cell–specific inhibitory receptor, is sufficient to predispose to development of high affinity autoantibodies. *J. Exp. Med.*, **189**, 1307–1313.

Patel, D.D. (2002). B cell–ablative therapy for the treatment of autoimmune diseases. *Arthritis Rheum.*, **46**, 1984–1985.

Pritchard, N.R., Cutler, A.J., Uribe, S., Chadban, S.J., Morley, B.J., and Smith, K.G. (2000). Autoimmune-prone mice share a promoter haplotype associated with reduced expression and function of the Fc receptor FcgammaRII. *Curr. Biol.*, **10**, 227–230.

Prokunina, L., Castillejo-Lopez, C., Oberg, F., Gunnarsson, I., Berg, L., Magnusson, V., Brookes, A.J., Tentler, D., Kristjansdottir, H., Grondal, G., Bolstad, A.I., Svenungsson, E., Lundberg, I., Sturfelt, G., Jonssen, A., Truedsson, L., Lima, G., Alcocer-Varela, J., Jonsson, R., Gyllensten, U.B., Harley, J.B., Alarcon-Segovia, D., Steinsson, K., Alarcon-Riquelme, M.E. (2002). A regulatory polymorphism in PDCD1 is associated with susceptibility to systemic lupus erythematosus in humans. *Nat. Genet.*, **32**, 666–669.

Radic, M.Z. and Weigert, M. (1994). Genetic and structural evidence for antigen selection of anti-DNA antibodies. *Annu. Rev. Immunol.*, **12**, 487–520.

Radic, M.Z. and Zouali, M. (1996). Receptor editing, immune diversification and self-tolerance. *Immunity*, **5**, 505–511.

Ravetch, J.V. and Bolland, S. (2001). IgG Fc receptors. *Annu. Rev. Immunol.*, **19**, 275–290.

Reininger, L., Radaszkiewicz, T., Kosco, M., Melchers, F., and Rolink, A.G. (1992). Development of autoimmune disease in SCID mice populated with long-term "in vitro" proliferating (NZB × NZW)F1 pre-B cells. *J. Exp. Med.*, **176**, 1343–1353.

Reth, M. and Wienands, J. (1997). Initiation and processing of signals from the B cell antigen receptor. *Annu. Rev. Immunol.*, **15**, 453–479.

Roosnek, E. and Lanzavecchia, A. (1991). Efficient and selective presentation of antigen–antibody complexes by rheumatoid factor B cells. *J. Exp. Med.*, **173**, 487–489.

Saito, E., Fujimoto, M., Hasegawa, M., Komura, K., Hamaguchi, Y., Kaburagi, Y., Nagaoka, T., Takehara, K., Tedder, T.F., and Sato, S. (2002). CD19-dependent B lymphocyte signaling thresholds influence skin fibrosis and autoimmunity in the tight-skin mouse. *J. Clin. Invest.*, **109**, 1453–1462.

Sato, S., Hasegawa, M., Fujimoto, M., Tedder, T.F., and Takehara, K. (2000). Quantitative genetic variation in CD19 expression correlates with autoimmunity. *J. Immunol.*, **165**, 6635–6643.

Sato, S., Miller, A.S., Inaoki, M., Bock, C.B., Jansen, P.J., Tang, M.L., and Tedder, T.F. (1996). CD22 is both a positive and negative regulator of B lymphocyte antigen receptor signal transduction: Altered signaling in CD22-deficient mice. *Immunity*, **5**, 551–562.

Schuster, H., Martin, T., Marcellin, L., Garaud, J.C., Pasquali, J.L., and Korganow, A.S. (2002). Expansion of marginal zone B cells is not sufficient for the development of renal disease in NZB × NZW F1 mice. *Lupus*, **11**, 277–286.

Segundo, C., Rodriguez, C., Garcia-Poley, A., Aguilar, M., Gavilan, I., Bellas, C., and Brieva, J.A. (2001). Thyroid-infiltrating B lymphocytes in Graves' disease are related to marginal zone and memory B cell compartments. *Thyroid*, **11**, 525–530.

Shultz, L.D., Schweitzer, P.A., Rajan, T.V., Yi, T., Ihle, J.N., Matthews, R.J., Thomas, M.L., and Beier, D.R. (1993). Mutations at the murine motheaten locus are within the hematopoietic cell protein-tyrosine phosphatase (Hcph) gene. *Cell*, **73**, 1445–1454.

Steward, M.W. and Hay, F.C. (1976). Changes in immunoglobulin class and subclass of anti-DNA antibodies with increasing age in N/ZBW F1 hybrid mice. *Clin. Exp. Immunol.*, **26**, 363–370.

Takai, T. (2002). Roles of Fc receptors in autoimmunity. *Nat. Rev. Immunol.*, **2**, 580–592.

Tsao, B.P., Cantor, R.M., Kalunian, K.C., Chen, C.J., Badsha, H., Singh, R., Wallace, D.J., Kitridou, R.C., Chen, S.L., Shen, N., Song, Y.W., Isenberg, D.A., Yu, C.L., Hahn, B.H., Rotter, J.I. (1997). Evidence for linkage of a candidate chromosome 1 region to human systemic lupus erythematosus. *J. Clin. Invest.*, **99**, 725–731.

Tsui, H.W., Siminovitch, K.A., de Souza, L., and Tsui, F.W. (1993). Motheaten and viable motheaten mice have mutations in the haematopoietic cell phosphatase gene. *Nat. Genet.*, **4**, 124–129.

Viau, M. and Zouali, M. (2005). Innate-like B cells and autoimmunity. *Clin. Immunol.*, **114**, 17–26.

Wakeland, E.K., Liu, K., Graham, R.R., and Behrens, T.W. (2001). Delineating the genetic basis of systemic lupus erythematosus. *Immunity*, **15**, 397–408.

Wilkinson, R., Lyons, A.B., Roberts, D., Wong, M.X., Bartley, P.A., and Jackson, D.E. (2002). Platelet endothelial cell adhesion molecule-1 (PECAM-1/CD31) acts as a regulator of B-cell development, B-cell antigen receptor (BCR)-mediated activation, and autoimmune disease. *Blood*, **100**, 184–193.

Yajima, K., Nakamura, A., Sugahara, A., and Takai, T. (2003). FcgammaRIIB deficiency with Fas mutation is sufficient for the development of systemic autoimmune disease. *Eur. J. Immunol.*, **33**, 1020–1029.

Yi, A.K., Klinman, D.M., Martin, T.L., Matson, S., and Krieg, A.M. (1996). Rapid immune activation by CpG motifs in bacterial DNA. Systemic induction of IL-6 transcription through an antioxidant-sensitive pathway. *J. Immunol.*, 157, 5394–5402.

Yu, C.C., Mamchak, A.A., and DeFranco, A.L. (2003). Signaling mutations and autoimmunity. *Curr. Dir. Autoimmun.*, **6**, 61–88.

Zeng, D., Lee, M.K., Tung, J., Brendolan, A., and Strober, S. (2000). Cutting edge: A role for CD1 in the pathogenesis of lupus in NZB/NZW mice. *J. Immunol.*, **164**, 5000–5004.

Zouali, M. (1998). Signaling in human lupus T lymphocytes. *Lupus*, **7**, 499–502.

Zouali, M. (2005). Taming Lupus. *Scientific American*, **292**, 70–77.

Zouali, M., Fournie, G.J., and Theze, J. (1991). Quantitative clonal analysis of the B cell repertoire in human lupus. *Cell. Immunol.*, **133**, 161–177.

Zouali, M. and Sarmay, G. (2004). B lymphocyte signaling pathways in systemic autoimmunity: Implications for pathogenesis and treatment. *Arthritis Rheum.*, **50**, 2730–2741.

18

Disrupted T Cell Receptor Signaling Pathways in Systemic Autoimmunity

Ana M. Blasini and Martín A. Rodríguez

1. Introduction

The homeostasis of adaptive immune responses is determined by the integrity of postreceptor signaling in its two main cellular components: T and B cells. The last 2 decades have brought up mounting information on the different intracellular routes triggered by receptor binding in T and B cells, with the identification of key signaling molecules and their genes. Today we understand better the role of different signaling molecules that drive intrathymic differentiation and post-thymic function of immune cells and we are identifying the defects underlying the triggering and perpetuation of autoimmune responses.

Tolerance to autoantigens depends predominantly on negative selection of autoreactive cells in the thymus and the neutralization of a small, but pathogenetically significant, surviving population of self-reactive T and B cells. There is no evidence of abnormal thymic negative selection as the basis for human systemic autoimmune diseases. Therefore, autoimmunity develops by the failure of down-regulatory mechanisms needed to check activation of autoreactive cells. The mechanisms checking postthymic autoreactive cells are tolerization to self-antigens, deletion by apoptosis, and peripheral restriction by regulatory $CD4^+CD25^+$ cells. In most autoimmune diseases, abnormalities of these mechanisms combine to a different degree. For example, in systemic lupus erythematosus (SLE), diminished numbers of $CD4^+CD25^+$ cells (Liu *et al.*, 2004), increased apoptosis (Emlen *et al.*, 1994), and loss of tolerance to nucleosomes (Datta, 2003) probably all contribute to a variable extent.

In recent years the importance of regulated postmembrane cell signaling in the maintenance of immune homeostasis has become evident. Some biological processes are closely related to autoimmunity, among them apoptosis, a complex

---

Ana M. Blasini and Martín A. Rodríguez • Centro Nacional de Enfermedades Reumáticas, Servicio de Reumatología, Hospital Universitario de Caracas, Caracas, Venezuela.

Molecular Autoimmunity: In commemoration of the 100th anniversary of the first description of human autoimmune disease, edited by Moncef Zouali. Springer Science+Business Media, Inc., New York, 2005.

set of biochemical events executed by diverse signaling routes and needed to delete autoreactive cells. Faulty regulation of apoptosis and defective cytokine expression have been proposed as pathogenetic mechanisms in SLE and both mechanisms may be intertwined. Thus, antiapoptotic cytokine signaling may influence the deregulation of cell death in lupus lymphocytes since IL-2, IL-4, IL-7, and IL-15 are known to induce a pronounced increment of Bcl-2 and a concomitant reduction of T cell death (Graninger et al., 2000). Also, abnormal assembly of the T cell receptor (TCR)/CD3 complex in lupus T cells, due to diminished expression of ζ chains (Liossis et al., 1998; Brundula et al., 1999), may alter the routes needed for regulated apoptosis (Combadière et al., 1996) and normal interleukin 2 (IL-2) production (Solomou et al., 2001). Combined genetic defects and abnormal metabolic processing may alter the configuration of the signaling cascade and favor the development of autoimmune responses. In Figure 18.1 we show a composite of mechanisms involving defective signaling in T cells that may potentially contribute to generation of sustained autoimmune responses.

2. Signaling Pathways in T Cells

The molecular basis for signal transduction in T cells has been the focus of intense and productive research in the last several years. Signal transduction after antigen binding occurs through the TCR-associated CD3 complex and ζ-ζ homodimers. Upon antigen-receptor engagement Src family kinases Lck and Fyn phosphorylate immunoreceptor tyrosine-based activation motifs (ITAMs) on the intracellular portions of the invariant chains of the antigen receptor (Figure 18.2). This leads to the recruitment, phosphorylation, and activation of ζ-chain-associated protein 70 (ZAP-70). Active ZAP-70 phosphorylates SH2-domain-containing leukocyte protein of 76 kDa (SLP-76), the small dual-specific phosphatase VHR (*Vaccinia* virus H1 protein-related phosphatase), and an adaptor protein essential for T cell signaling, the membrane-associated adaptor protein linker of activated T cells (LAT). Phosphorylation of LAT at multiple tyrosine residues creates binding sites for Gads (Gads/GrpL) adaptor proteins and growth factor receptor-bound protein 2 (Grb2), leading to nucleation of multiprotein signaling complexes. Lck and Fyn also phosphorylate and activate Tec family kinases and phosphoinositide 3-kinase (PI3K). Together, these families of kinases phosphorylate several downstream adaptors, leading to the recruitment of signal transducers essential for mobilizing Ca^{2+}, activating mitogen-activated protein kinases (MAPKs) and phospho-inositide 3 kinase (PI3K), regulating the actin cytoskeleton, and inducing transcription factor activity (Cannons and Schwartzberg, 2004). Lck activity is regulated by the opposing actions of the Csk kinase and protein tyrosine phosphatase (PTP) CD45, which maintains Src kinases in a dephosphorylated state, ready to participate in signal transduction (Mustelin et al., 2004).

These early signaling events lead to the formation of multiprotein complexes that activate the Ras signaling cascade (Figure 18.2). Ras signaling requires the formation of upstream intermolecular complexes including adaptors Shc and Grb2; Grb2 couples to hSos and positions it in the proximity of Ras,

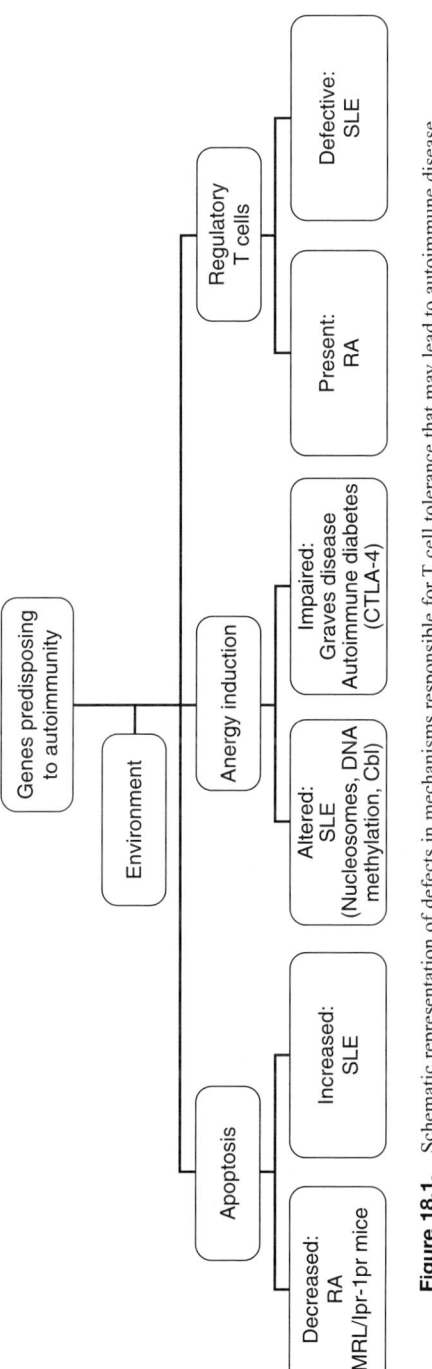

Figure 18.1. Schematic representation of defects in mechanisms responsible for T cell tolerance that may lead to autoimmune disease.

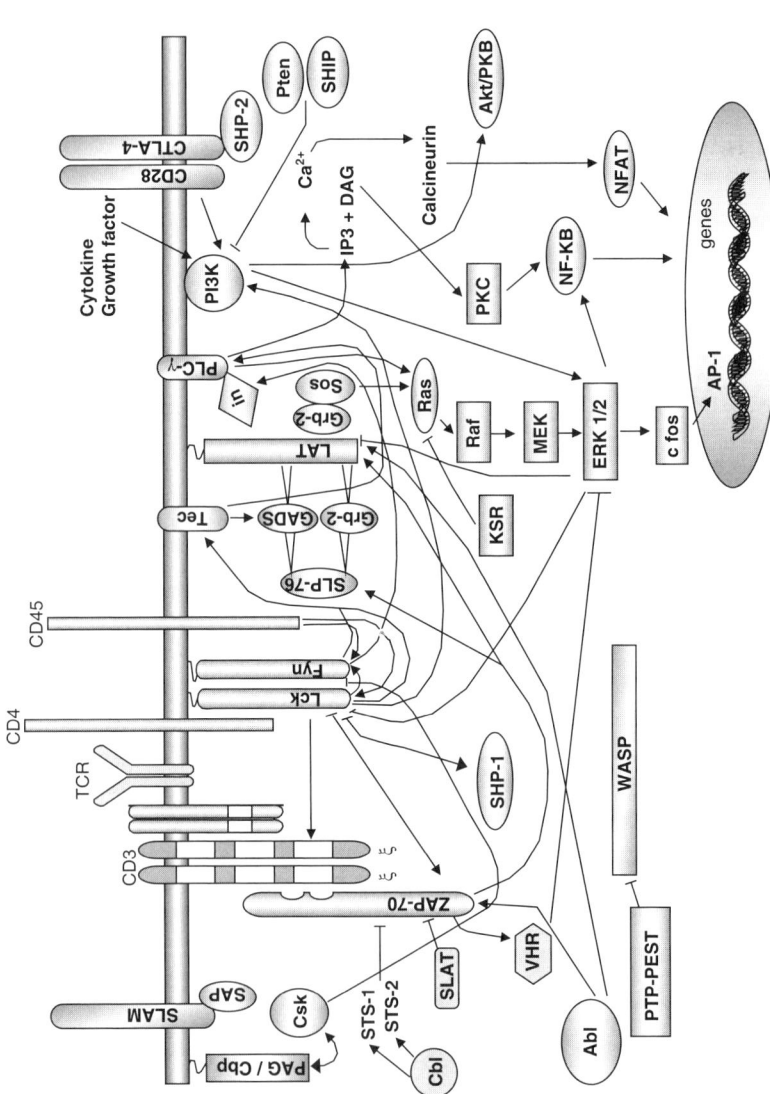

Figure 18.2. Signaling molecules triggered by TCR/CD3 ligation. Only those molecules relevant to pathogenesis of autoimmunity are shown.

allowing exchange of GDP for GTP, which switches Ras from an inactive to an active form (Altman and Deckert, 1999). Ras engagement induces activation of the serine kinase Raf, which activates the dual specific kinases, MEK-1 and MEK-2, that in turn trigger activation of MAPKs ERK-1 and ERK-2. The duration of ERK activation affects c-fos expression, a component of the AP-1 transcription complex. Transient or sustained ERK activation acts as a signal interpreter of the strength of TCR engagement and directs cellular responses (Schade and Levine, 2004). ERK activation leads to nuclear translocation of various transcription factors involved in enhancing IL-2 gene promoter activity and cell proliferation (Altman and Deckert, 1999). The MAPK cascade is essential for signaling of various extracellular stimuli from the membrane to the nucleus and is involved in a variety of cellular responses including proliferation, differentiation, and apoptosis (Treisman, 1996). Also, blocked signaling through the Ras/Raf/MAPK pathway ending up in inactivation of transcription factor AP-1 is critical for induction of anergy in T cells (Fields et al., 1996; Li et al., 1996). Anergic T cells are associated with distinctive kinetics, amplitude, and localization of MAPK signaling (Adams et al., 2004), and thus this signaling route is key for T cell tolerance.

Antigen binding by TCR signaling leads to the formation of a multimolecular complex at the TCR antigen-presenting cell (APC) contact face, an area known as the immunological synapse, in which signaling is initiated before a mature synapse is established with stable centralized and peripheral supramolecular activating complexes (SMACs) (Jacobelli et al., 2004). Lipid rafts are areas in cell membranes enriched in sphyngolipids and cholesterol, forming gel-like microdomains partitioned into liquid-ordered phase, and representing ~40% of immune cell membranes (Thomas et al., 2004). GM1 and GM3, the main ganglioside constituents of the plasma cell membrane, concentrate at lipid rafts. These domains, which are enriched in the synapse, congregate or segregate signaling molecules immediately after coupling of antigen/MHC complexes by TCR, and are the physical platform for the assembly of the signaling cascade at the very early steps of the T cell activation process.

Signal transduction is a complex choreography of phosphatases counterbalancing kinases, synergistic and antagonistic cross-talk between pathways, and coordinated engagement of multiple adaptor molecules, leading to a coherent functional outcome. Alterations of this delicate balance may contribute to loss of tolerance in autoreactive T cells and to triggering of autoimmune responses. Cytotoxic T lymphocyte antigen-4 (CTLA-4) accumulates in a two-step process at the immunological synapse and inhibits T cell activation, either by increasing the threshold for activation or by dampening existing immune responses (Chikuma and Bluestone, 2002). Another inhibitor, SH2-containing PTPase (SHP-1), helps T cells to discriminate between agonist and antagonist stimulation. Weakly binding ligands induce a reduced number of intracellular signals, including diminished binding of ZAP-70 to partially phosphorylated ζ chains, phosphorylation of SHP-1 by activated Lck, and inactivation of Lck by SHP-1. In contrast, stimulatory ligands activate ERK, which in turn phosphorylates Lck on Ser^{59}, interfering with SHP-1 recruitment and precluding Lck inactivation

(Stefanova *et al.*, 2003). Further downstream, VHR phosphatase translocates from the cytosol to the immunological synapse where it becomes a target of ZAP-70-induced phosphorylation and exerts a negative regulation of ERK activation (Alonso *et al.*, 2003). TCR signaling is also downregulated by the induction of the PTPase CD148, which attenuates activation by dephosphorylating LAT and PLCγ. Downregulation of TCR membrane expression is another mechanism involved in attenuation of signaling in T lymphocytes.

3. T Cell Signaling Abnormalities in Systemic Autoimmune Disease

The critical contribution of T cells to the generation and perpetuation of systemic autoimmune disease is beyond question. In experimental models they can transfer disease, and genetically engineered antigen-specific autoreactive T cells can promote sustained autoimmunity. In addition, treatment modalities that interfere with T or B cell function ameliorate disease. The discovery by two independent groups (Fields *et al.*, 1996; Li *et al.*, 1996) of the role of the route leading to transcription factor AP-1, as a pivotal mechanism responsible for the maintenance of T cell tolerance, illustrates the importance of postreceptor cell signaling in immune homeostasis. Recent work, showing bidirectional signaling, triggered by the interaction of cells responsible for innate and adaptive immunity, adds a new perspective to the understanding of the interplay between the environment and genes, a subject fundamental to the pathogenesis of autoimmune diseases. Thus, a full understanding of signaling abnormalities in T cells relevant to autoimmunity pathogenesis is not possible without looking at innate immunity cells.

3.1. Signaling Abnormalities in Antigen-Presenting Cells and Autoimmune Disease

Whereas stimulation by immature dendritic cells (DCs) induces tolerization, mature DCs promote full activation and effector functions of T cells (Ohashi and DeFranco, 2002). This is due to the expression of costimulatory molecules in mature DCs, which promote a stable immunological synapse between DCs and adaptive immune cells, allowing a continuous stimulation of T cells (Jacobelli *et al.*, 2004). Some of the critical postreceptor signals needed for efficient antigen-presenting function by APCs have been identified. For example, the coupling of the CD40/CD40L pair triggers bidirectional signaling affecting APCs and T cells (Fujii *et al.*, 2004). The full immunostimulant capability of APC involves augmented expression of MHC and CD80 molecules. Postreceptor signaling in APC influences immune outcomes and may have an impact on immune homeostasis. For example, disruption of STAT3, an adaptor molecule coupling downstream signaling triggered by activation of Janus kinase (JAK) in APCs can break tolerance to antigen in anergic antigen-specific CD4[+] cells (Cheng *et al.*, 2003). In other cases, signaling abnormalities may shift the direction of immune responses back to homeostasis. Thus, defective C-Rel expression interfering with normal IL-12 and IL-23 gene expression limits the APC capacity for induction of the Th1 phenotype in T cells, shifting immune responses away from organ-specific autoimmune disease in the genetically susceptible individual (Hilliard et al., 2002). Subtle stimulatory changes, such as those induced by TCR binding of altered peptide ligand presented by APC, induce anergy in autoreactive T cell clones (Quaratino et al., 2000) by aborting the activation process at intermediate steps. The biochemical basis is formation of the p21 (with only two tyrosine-phosphorylated ITAMs) instead of p23 (with all three tyrosine-phosphorylated ITAMs) ζ chains (Sloan-Lancaster and Allen, 1996). DC presenting a self-epitope with an altered peptide can induce anergy in a human autoreactive T cell clone (Quaratino et al., 2000), and short-lived peptides can also have a tolerogenic effect, by diminishing the density of peptide/MHC complexes (Mirshahidi et al., 2004). The basis for these effects is the completion of postreceptor signaling in T cells, from membrane receptors to the nucleus. A better knowledge of these very first steps after TCR ligation may be the basis for future therapeutic approaches in autoimmune diseases (Pugliese, 2003).

Recently, the role of pattern recognition receptors (PRRs) serving as links between APCs and adaptive immune cells has shed new light on our understanding of the interplay between innate and adaptive responses, a subject fundamental to the pathogenesis of autoimmune disease. Defective expression or altered molecular configuration of toll-like receptors (TLRs), a highly conserved family of 10 PRRs expressed in innate and adaptive immune cells, may foster sustained responses to self-antigens. For example, the expression of TLR2 by antigen-activated T cells (Komai-Koma et al., 2004) may render them susceptible to activation by bacterial lipopeptide, a ligand for TLR2, and partly explain the triggering of autoimmune disease or its reactivation in the presence of bacterial infection. Another example is the breaking of tolerance to proteolipid protein in mice naturally resistant to experimental encephalomyelitis, when T cells are stimulated in the presence of APCs that have been activated via TLR4 or TLR9, with unmethylated CpG oligodeoxy-nucleotides (CpG ODNs) (Waldner et al., 2004). Also, TLRs participate in the interaction between DCs and B cells needed for T cell responses to host-derived DNA (Seibl et al., 2004). Signaling through TLRs is necessary for full maturation of DCs, a role in part played by endogenous ligands HSP60 and HSP70. Ligation of TLRs by these endogenous molecules may precipitate autoimmunity, as occurs with HSP70 (Millar et al., 2003) or with HSPg96, when this molecule is genetically engineered for surface expression (Liu et al., 2003). Recently, the hypothesis of a third signal participating in the regulation of T cell responses has been proposed (Thomas, 2004). Lack of a third signal conveyed by IL-12 and IL1-β may induce anergy in $CD8^+$ T cells (Curtsinger et al., 2003). On the contrary, signals delivered by IL-1β stimulate IL-12 production by DCs, and contribute to the induction of autoimmune myocarditis (Thomas, 2004). Recently, Scheinecker et al. (2000) showed a decreased stimulatory capacity of DCs on T cells from SLE patients, which was attributed to a reduction of the CD11-myeloid-related subset, the mature CD phenotype. Since DCs expressing CD11 correspond to the mature subset (Turley, 2002), a predominance of the tolerogenic CD11-negative DC in SLE, along with intrinsic signaling abnormalities, may further contribute to the paradoxal "anergic" phenotypic profile shown by lupus T cells (Blasini and Rodríguez, 2004).

3.2. Signaling Abnormalities in T Cells and Autoimmune Disease

To understand abnormalities in T cell signaling that may condition sustained autoimmune responses we need to identify the critical checkpoints in postmembrane signaling responsible for the maintenance of tolerance (Figure 18.2).

3.2.1. Early Signaling Abnormalities of T Cells in Systemic Autoimmunity

The first postmembrane signaling abnormality identified in lupus T cells is altered metabolism of cyclic AMP and defective phosphorylation of protein kinase A (PKA-1), due to a deficiency of the PKA-I and PKA-I isoenzyme activity present in approximately 70% and 37% of SLE patients, respectively (reviewed in Kammer *et al.*, 2004). Another relevant signaling aberration of T cells related to systemic autoimmunity involves disregulated PKB (Akt) activation (Ohashi and DeFranco, 2002). The increased activity of this kinase in transgenic mice expressing T cells with a constitutively active form of Akt was sufficient for the development of splenomegaly, lymphadenopathy, and autoimmune glomerulonephritis (Rathemell *et al.*, 2003), by a mechanism involving prevention of Fas-mediated death in T cells. Defects in the expression or activity of tyrosine phosphatases may also cause loss of tolerance. Mice heterozygous for Pten and SHIP develop a lymphoproliferative autoimmune syndrome akin to human Sjögren' syndrome, by a mechanism involving overproduction of IL-4 by $CD4^+$ cells (Moody and Jirik, 2004). Another downregulatory mechanism involving tyrosine phosphatases is the coupling of tyrosine phosphatase SHP-2 by the intracytoplasmatic tail of CTLA-4 (Lee *et al.*, 1998). In addition, a molecule expressed after T cell activation and also displaying coinhibitor capabilities is the program death-1 (PD-1) molecule, the receptor for two B7 family members, B7-H1 and B7-DC (Moretta and Bottino, 2004). Like CTLA-4, PD-1 is also able to recruit SHP-2. The importance of CTLA-4 in the maintenance of tolerance is underscored by the predisposition to Graves' disease, autoimmune hypothyroidism, and type I diabetes in patients expressing lower messenger RNA levels due to an allelic variation of *CTLA-4* gene (Ueda *et al.*, 2003). The nonobese diabetic (NOD) mouse also exhibits defective CTLA-4 function (Salojin *et al.*, 1998). Besides its role as a negative regulator of downstream signaling in activated T cells, CTLA-4 is also needed to activate $CD4^+CD25^+$ regulatory T cells (Takahashi *et al.*, 2000).

Current studies are showing peculiar accommodations of signaling molecules in lipid rafts of T cells from patients with systemic autoimmune diseases. In normal T cells Lck is enriched within, whereas tyrosine phosphatase CD45 is excluded out of these domains upon antigen binding by the TCR complex (Cannon and Schwartzberg, 2004). This pattern is reversed in T cells from patients with SLE (Jury *et al.*, 2004). In fact, upon activation lupus T cells show diminished levels of raft-associated Lck, persistence of CD45 within these microdomains (Jury *et al.*, 2004), and augmented amounts of GM1 ganglioside in the membrane of lupus T cells (Jury *et al.*, 2004; Krishnan *et al.*, 2004), indicating a higher density of lipid rafts in T cells from SLE patients. One potential consequence of the latter abnormality is accelerated actin polymerization following T cell activation, probably explaining the increased capping induced by anti-CD3 stimulation in lupus T cells (Krishnan *et al.*, 2004). The diminished expression of Lck in rafts of lupus T cells is apparently due to increased ubiquitination, an effect probably related to the effect of sustained oxidative stress *in vivo* (Jury *et al.*, 2004). We previously reported enhanced activation of p^{59} Fyn, a src kinase constitutively associated to the CD3 complex in lupus T cells (Blasini *et al.*, 1998a). Enhancement of Fyn activity has also been observed in T cells from NOD mice (Salojin *et al.*, 1997). Fyn is an integral part of signaling from the coreceptor signaling lymphocyte activation molecule (SLAM), a glycoprotein expressed on activated T lymphocytes and APCs that acts as a coregulator of antigen-driven T cell responses that, after coupling to adaptor protein serum amyloid P component (SAP), fulfills a costimulatory function for T cell activation (Veillete, 2004). This pathway is involved in the pathogenesis of autoimmunity, as evidenced by the protection from hypergammaglobulinemia, production of anti-DNA antibodies, and glomerulonephritis in mice bearing a targeted mutation in the SAP gene (Hron *et al.*, 2004). These defects in critical src kinases, acting at very early stages of the T cell activation process, may explain the abnormal pattern of tyrosine phosphorylation observed in lupus T cells (Matache *et al.*, 1996; Blasini *et al.*, 1998b; Liossis *et al.*, 1998). The abnormal assembly of molecules in the early signalosome formed after TCR ligation may be in part contributed for by the defective expression of TCR ζ chains (Liossis *et al.*, 1998; Brundula *et al.*, 1999). These molecules act as a scaffold for downstream signaling by bearing three ITAMs that are tyrosine-phosphorylated upon T cell activation. Tyrosine-phosphorylated ITAMs allow the coupling of ZAP-70 (and its subsequent phosphorylation by Lck), along with SLP-76 and LAT, a complex that activates PLCγ1 (Leo *et al.*, 2002). This complex triggers calcium mobilization by generating IP3 from PIP_2 in membranes (Figure 18.2). Interestingly, lipid rafts of lupus T cells show absence of ZAP-70 (Krishnan *et al.*, 2004), a fact that may explain the abnormal pattern of calcium mobilization in lupus T cells, possibly due to signaling through an alternate route in the "rewired" cascade in which TCR ζ chains are substituted by FCεγRI (Tsokos *et al.*, 2003). The importance of ZAP-70 in the maintenance of tolerance is illustrated by the increased susceptibility to autoimmune arthritis in mice bearing a spontaneous single point mutation of ZAP-70 SH2 domain, which interferes with its coupling to ζ chains (Sakaguchi *et al.*, 2003). T cells from ZAP-70 mutants show an altered threshold for T cell activation that hampers thymic deletion of potentially autoreactive clones (Sakaguchi *et al.*, 2003). We have recently observed defective expression of LAT in lipid rafts of lupus T cells (unpublished data), another molecule critical for assembly of the calcium activation complex in T lymphocytes. Displacement of LAT from the signalosome and abnormal calcium mobilization has been observed in synovial T cells from patients with rheumatoid arthritis (RA) (Gringhuis *et al.*, 2000; Cope, 2002).

3.2.2. Intermediate and Late Signaling Abnormalities of T Cells in Systemic Autoimmunity

The role of the Ras/Raf/MAPK signaling cascade in the maintenance of T cell anergy is well established (Williams, 1996). Anergic T cells show diminished activity of Ras that impairs activation of downstream ERK-1 and ERK-2, leading to defective transactivation of AP-1 and inhibition of IL-2 production (Fields *et al.*, 1996). The signals responsible for Ras inactivation in anergic cells are not clear as yet, but possibly involve the triggering of alternate signaling routes when T cells are stimulated in the absence of a second or may be a third signal. Interestingly, altered signaling in T cells from patients with autoimmune diseases resembles the pattern observed in anergic normal T cells (Blasini and Rodriguez, 2004), suggesting that both induction of anergy and rupture of tolerance involve deviation from usual signaling routes.

We have recently shown diminished activation of ERK-1 and ERK-2 in T cells from SLE patients, possibly due to abnormal coupling of Ras nucleotide exchange factor hSos to adaptor protein Grb2 in cells activated through the TCR/CD3 complex *in vitro* (Cedeño *et al.*, 2003). Also, diminished ERK activation was demonstrated by Deng *et al.* (2003) in lupus T cells, and was proposed as the mechanism for diminished DNA methylation and enhanced gene activation in lupus T cells (Oelke and Richardson, 2004). In a model of *in vitro* induced anergy, Yi *et al.* (2000) showed upregulated expression of CD40-L that was caused by lack of phosphorylation of Cbl/Cbl-b, and sustained phosphorylation of ERK. Cbl-b, an ubiquitin ligase, functions as a negative regulator of receptor clustering and raft aggregation in T cells. Loss of the molecular adaptor Cbl-b frees antigen receptor–triggered receptor clustering, lipid raft aggregation, and sustained tyrosine phosphorylation from the requirement for CD28 costimulation, diminishing the threshold for antigen responses in T cells. Introduction of the Cbl-b mutation into a $Vav1^{-/-}$ background relieved the functional defects of $Vav1^{-/-}$ T cells and caused spontaneous autoimmunity (Krawczyk *et al.*, 2000). Cbl-b-deficient mice develop anti-DNA antibodies (Bachmaier *et al.*, 2000), further illustrating the importance of this downregulatory molecule in the maintenance of tolerance. Finally, Cbl-interacting proteins Sts-1 and Sts-2 negatively regulate TCR signaling. T cells from mice lacking Sts-1/2 are hyperresponsive to TCR stimulation, show increased ZAP-70 activity, augmentation of cytokine production, and increased susceptibility to autoimmunity in a mouse model of multiple sclerosis (Carpino *et al.*, 2004; Kowanetz *et al.*, 2004). These data suggest abnormal signaling involving the Ras/Raf/MAPK pathway in T cells from SLE and other autoimmune conditions. Clearly, further studies are needed to understand the status of this critical pathway in the breakdown of immune homeostasis.

Late effects of upstream signaling abnormalities in autoimmune T cells are defective activation of transcription factors. T cells from SLE patients show defective nuclear translocation of NFkB, due to abnormal expression of the p^{65} subunit (Wong *et al.*, 1999). Two other transcription factors, AIRE and FOXP3, are involved in induction of tolerance in humans and their mutation could induce severe autoimmune disease (Ramsdell and Ziegler, 2003).

One of the most relevant consequences of impaired signaling in T cells is abnormal regulation of calcium mobilization, a critical signal for IL-2 gene activation. Defective IL-2 production is a mechanism for disruption of tolerance. Responses mediated through the IL-2 receptor β are needed to maintain homeostasis and prevent autoimmunity (Suzuki et al., 1995; Malek et al., 2000), and gene-targeted mice lacking the IL-2 gene develop various forms of autoimmune manifestations, including autoimmune colitis (Ludviksson et al., 1997). T cells from SLE patients have diminished production of IL-2 due to limited transcriptional activity of the IL-2 gene promoter (Wong et al., 1999; Solomou et al., 2001). Also, IL-2 deficiency favors autoimmunity by compromising the generation of regulatory $CD4^+CD25^+$ T cells (Shevach, 2000; De Lafaille and Lafaille, 2002). In addition, SLE patients show diminished numbers of peripheral blood $CD4^+CD25^+$ regulatory T cells (Liu et al., 2004) and mice prone to lupus show early diminishment of $CD4^+CD25^+$ cells (Wu and Staines, 2004). The generation and function of $CD4^+CD25^+$ regulatory T cells depend on normal signaling involving STAT molecules, including STAT5 (Antov et al., 2003) and STAT1. STAT1-deficient mice become susceptible to experimental autoimmune encephalomyelitis (Nishibori et al., 2004) and STAT5A/5B-deficient mice show autoimmunity affecting multiple organs (Snow et al., 2003).

Some of the above-discussed signaling abnormalities can alter cell cycle control and disrupt biological responses in the T cell compartment that could promote autoimmunity. For instance, resistance to Fas-mediated apoptosis related to high expression of the cyclin kinase inhibitor p21 (WAF-1/CIP-1) can be observed in T and B cells from lupus-prone mice (Lawson et al., 2004). This defect may explain the accumulation of the activated/memory type $CD44^{high}CD4^+$ cells arrested in the G0/G1 phase of the cell cycle. Table 18.1 summarizes several identified abnormalities of key signaling molecules in T lymphocytes that may be involved in the pathogenesis of autoimmune disease.

4. Conclusions

Many of the above-discussed abnormalities in T cell responses seen in patients with systemic autoimmunity can be related to identifiable signaling abnormalities. Some have a genetic basis; others are induced by stressing environment and thus, could be potentially reversible. The whole picture, which includes genes encoding regulatory molecules responsible for immune homeostasis and the environmental endogenous and exogenous agents that trigger and perpetuate autoimmune responses, is far from complete. The possibility of therapeutic intervention to correct critical T and B cell signaling defects by pharmacological, or even genetic means, has accelerated the pace of the scientific endeavor to unveil the immune cell signaling machinery and open new frontiers in the management of patients with systemic autoimmunity.

Table 18.1. Signaling Abnormalities of T cells in autoimmune conditions

Molecule	Abnormality	Autoimmune condition	References
TCR ζ	Decreased amounts: transcriptional defects, increased ubiquitination	SLE, RA	Liossis et al., 1998; Brundula et al., 1999
PKA	Decreased activity: transcriptional defects PKAI PKAII isoenzymes	SLE	Kammer et al., 2004
PKB, Akt	Increased activity	Autoimmune glomerulonephritis and lymphoproliferation	Rathemell et al., 2003
Lck	Decreased location in rafts, decreased activity	SLE, RA	Jury et al., 2004, Krishnan et al., 2004
Fyn	Increased activity	SLE, NOD mice	Blasini et al., 1998, Salojin et al., 1997
ZAP-70	Decreased location in rafts, decreased activity, mutation	SLE, CIA	Krishnan et al., 2004; Sakaguchi et al., 2003
mSos	Decreased coupling to adaptor Grb2	SLE	Cedeño et al., 2003
ERK-1-2	Decreased phosphorylation and activity	SLE	Deng et al., 2003, Cedeño et al., 2003
Cbl-b	Decreased activity or deficiency	Systemic autoimmunity, SLE	Krawczyk et al., 2000; Bachmaier et al., 2000; Yi et al., 2000
Pten	Deregulated activity	Autoimmune lymphoproliferative disease, NOD mice	Moody and Jirik, (2004); Salojin et al., 1998
CTLA-4	Diminished mRNA, polymorphism of *CTLA-4* gene	Graves, hypothyroidism, type I diabetes	Ueda et al., 2003
STAT-1	Deficiency	EAE	Nishibori et al., 2004
STAT-3	Disruption	Systemic autoimmunity	Cheng et al., 2003
NF*k*B p65	Abnormal expresión	SLE	Wong et al., 1999

SLE, systemic lupus erythematosus; RA, rheumatoid arthritis; NOD, non-bese diabetic; CIA, collagen-induced arthritis; EAE, experimental autoimmune encephalomyelitis.

Acknowledgments

We thank Mrs. Mayra Mayora for efficient secretarial work and design of figures. Supported by Fondo Nacional de Innovación y Tecnología (Fonacit) No. S1-200000440.

References

Adams, C.L., Grierson, A.M., Mowat, A.M., Harnett, M.M., and Garside, P. (2004). Differences in the kinetics, amplitude, and localization of ERK activation in anergy and priming revealed at the level of individual primary T cells by laser scanning cytometry. *J. Immunol.*, **173**, 1579–1586.

Alonso, A., Rahmouni, S., Williams, S., van Stipdonk, M., Jaroszeqski, L., Godzik, A., Abraham, R.T., Schoenberg, S.P., and Mustelin, T. (2003). Tyrosine phosphorylation of VHR phosphatase by ZAP-70. *Nat. Immunol.*, **4**, 44–48.

Altman, A. and Deckert, M. (1999). The function of small GTPases in signaling by immune recognition and other leukocyte receptors. *Adv. Immunol.*, **72**, 1–99.

Antov, A., Yang, L., Vig, M., Baltimore, D., and Van Parijs, L. (2003). Essential role for STAT5 signaling in CD4+CD25+ regulatory cell homeostasis and the maintenance of self-tolerance. *J. Immunol.*, **171**, 3435–3441.

Bachmaier, K., Krawczyk, C., Kozieradzki, I., Kong, Y., Sasaki, T., Oliveira-Santos, A., Mariathasan, S., Bouchard, D., Wakeham, A., Itie, A., Le, J., Ohash-Sarosi, I., Nishina, H., Lipkowitz, S., and Penninger, J. (2000). Negative regulation of lymphocyte activation and autoimmunity molecular adaptor Cbl-b. *Nature*, **403**, 211–216.

Blasini, A.M., Brundula, V., Paris, M., Rivas, L., Salazar, S., Stekman, I.L., and Rodríguez, M.A. (1998a). Protein tyrosine kinase in T lymphocytes from patients with systemic lupus erythematosus. *J. Autoimmun.*, **11**, 387–393.

Blasini, A.M., Chacon, R., Riera, R., Stekman, I.L., and Rodriguez, M.A. (1998b). Abnormal pattern of tyrosine phosphorylation in unstimulated peripheral blood T lymphocytes from patients with systemic lupus erythematosus. *Lupus*, **7**, 515–523.

Blasini, A.M. and Rodríguez, M.A. (2004). Altered signaling triggered by ligation of CR/CD3 receptor in T lymphocytes from patients with systemic lupus erythematosus: The road from anergy to autoimmunity. *Int. Rev. Immunol.*, **23**, 265–272.

Brundula, V., Rivas, L., Blasini, A.M., Paris, M., Salazar, S., Stekman, I.L., and Rodríguez, M.A. (1999). Diminished levels of TCR ζ chain in peripheral blood (PB) T lymphocytes from patients with systemic lupus erythematosus (SLE). *Arthritis Rheum.*, **42**, 1908–1916.

Cannons, J.L. and Schwartzberg, P.L. (2004). Fine-tuning lymphocyte regulation: What's new with tyrosine kinases and phosphatases? *Curr. Opin. Immunol.*, **16**, 296–303.

Carpino, N., Turner, S., Mekala, D., Takahashi, Y., Zang, H., Geiger, T.L., Doherty, P., and Ihle, J.N. (2004). Regulation of ZAP-70 activation and TCR signaling by two related proteins, Sts-1 and Sts-2. *Immunity*, **20**, 37–46.

Cedeño, S., Cifarelli, D.F., Blasini, A.M., Paris, M., Placeres, F., Alonso, G., and Rodriguez, M.A. (2003). Defective activity of ERK-1 and ERK-2 mitogen-activated protein kinases in peripheral blood T lymphocytes from patients with systemic lupus erythematosus. Potential role of altered coupling of Ras nucleotide exchange factor hSos to adaptor protein Grb2. *Clin. Immunol.*, **106**, 41–49.

Cheng, F., Wang, H.W., Cuenca, A., Huang, M., Ghansah, T., Brayer, J., Kerr, W., Takeda, K., Akira, S., Schoenberger, S., Yu, H., Jove, R., and Sotomayor, E. (2003). A critical role for Stat3 signaling in immune tolerance. *Immunity*, **19**, 425–436.

Chikuma, S. and Bluestone, J.A. (2002). CTLA-4: Acting at the synapse. *Mol. Intervention*, **2**, 205–208.

Combadière, B., Freedman, M., Chen, L., Shores, E.W., Love, P., and Lenardo, M.J. (1996). Qualitative and quantitative contributions of the T cell receptor ζ chain to mature T cell apoptosis. *J. Exp. Med.*, **183**, 2109–2117.

Cope, A.P. (2002). Studies of T-cell activation in chronic inflammation. *Arthritis Res.*, **4**, S197–S211.
Curtsinger, J.M., Lins, D.C., and Mescher, M. (2003). Signal 3 determines tolerance versus full activation of naïve CD8 T cells: Dissociating proliferation and development of effector function. *J. Exp. Med.*, **197**, 1141–1151.
Datta, S.K. (2003). Major peptide autoepitopes for nucleosome-centered T and B cell interaction in human and murine lupus. *Ann. NY Acad. Sci.*, **987**, 79–90.
De Lafaille, M.A.C. and Lafaille, J.J. (2002). CD4+ regulatory T cells in autoimmunity and allergy. *Curr. Opin. Immunol.*, **14**, 771–778.
Deng, C., Lu, Q., Zhang, Z., Rao, T., Attwood, J., Yung, R., and Richardson, B. (2003). Hydralazine may induce autoimmunity by inhibiting extracellular-regulated signal-regulated kinase pathway signaling. *Arthritis Rheum.*, **48**, 746–756.
Emlen, W., Niebur, J., and Richard, K. (1994). Accelerated in vitro apoptosis of lymphocytes from patients with systemic lupus erythematosus. *J. Immunol.*, **152**, 3685–3692.
Fields, P.E., Gajewski, T.F., and Fitch, F.W (1996). Blocked Ras activation in anergic $CD4^+$ T cells. *Science*, **271**, 1276–1278.
Fujii, S., Liu, K., Smith, C., Bonito, A., and Steinman, R. (2004). The linkage of innate to adaptive immunity via maturing dendritic cells in vivo requires CD40 ligation in addition to antigen presentation and CD80/86 costimulation. *J. Exp. Med.*, **199**, 1607–1618.
Graninger, W.B., Steiner, C.W., Graninger, M.T., Aringer, M., and Smolen, J.S. (2000). Cytokine regulation of apoptosis and Bcl-2 expression in lymphocytes of patients with systemic lupus erythematosus. *Cell Death Differ.*, **7**, 966–972.
Gringhuis, S., Leow, A., Papendrecht-van der Voort, E., Remans, P., Breedveld, F., and Verweij, C. (2000). Displacement of linker for activation of T cells from the plasma membrane due to redox balance alterations results in hyporesponsiveness of synovial fluid T lymphocytes in rheumatoid arthritis. *J. Immunol.*, **164**, 2170–2179.
Hilliard, B.A., Mason, N., Xu, L., Sun, J., Lamhamedi-Cherradi, S.E., Liou, H.C., Hunter, C., and Chen, Y.H. (2002). Critical roles of c-Rel in autoimmune inflammation and helper T cell differentiation. *J. Clin. Invest.*, **110**, 843–850.
Hron, J.D., Caplan, L., Gerth, A.J., Schwartzberg, P.L., and Peng, S.L. (2004). SH2D1A regulates T-dependent humoral autoimmunity. *J. Exp. Med.*, **200**, 261–266.
Jacobelli, J., Andres, P.G., Boisvert, J., and Krummel, M.F. (2004). New views of the immunological synapse: Variations in assembly and function. *Curr. Opin. Immunol.*, **16**, 345–352.
Jury, E.C., Kabouridis, P.S., Flores-Borja, F., Mageed, R.A., and Isenberg, D.A. (2004). Altered lipid raft-associated signaling and ganglioside expression in T lymphocytes from patients with systemic lupus erythematosus. *J. Clin. Invest.*, **113**, 176–187.
Kammer, G.M., Laxminarayana, D., and Khan, I.U. (2004). Mechanisms of deficient type I protein kinase A activity in lupus T lymphocytes. *Int. Rev. Immunol.*, **23**, 225–244.
Komai-Koma, M., Jones, L., Ogg, G., Xu, D., and Liew, F. (2004). TLR2 is expressed on activated T cells as a costimulatory receptor. *Proc. Natl. Acad. Sci. USA*, **101**, 3029–3034.
Kowanetz, K., Cresette, N., Haglund, K., Schmidt, M.H., Heldin, C.H., and Dikie, I. (2004). Suppressors of T-cell receptor signaling sts-1 and sts-2 bind to cbl and inhibit endocytosis of receptor tyrosine kinases. *J. Biol. Chem.*, **279**, 32786–32795.
Krawczyk, C., Bachmaier, K., Sasaki, T., Jones, G.R., Snapper, B.S., Bouchard, D., Kozieradzki, I., Ohashi, S.P., Alt, W.F., and Penninger, M.J. (2000). Cbl-b is a negative regulator of receptor clustering and raft aggregation in T cells. *Immunity*, **13**, 463–473.
Krishnan, S., Nambiar, M.P., Warke, V.G., Fisher, C.U., Mitcell, J., Delaney, N., and Tsokos, G.C. (2004). Alterations in lipid raft composition and dynamics contribute to abnormal T cell responses in systemic lupus erythematosus. *J. Immunol.*, **172**, 7821–7831.
Lawson, B.R., Baccala, R., Song, J., Croft, M., Kono, D.H., and Theofilopoulos, A.N. (2004). Deficiency of cyclin kinase inhibitor p21 (WAF-1/CIP-1) promotes apoptosis of activated/memory T cells and inhibits spontaneous systemic autoimmunity. *J. Exp. Med.*, **199**, 547–557.
Lee, K.M., Chuang, E., Griffin, M., Khattri, R., Hong, D.K., Zhang, W., Straus, D., Samelson, L., Thompson, C., and Bluestone, J. (1998). Molecular basis of T cell inactivation by CTLA-4. *Science*, **282**, 2263–2266.
Leo, A., Wienands, J., Baier, G., Horejsi, V., and Schraven, B. (2002). Adapters in lymphocyte signaling. *J. Clin. Invest.*, **109**, 301–309.

Li, W., Whaley, C.D., Mondino, A., and Mueller, D.L. (1996). Blocked signal transduction to the ERK and JNK protein kinases in anergic CD4+ T cells. *Science*, **271**, 1272–1275.

Liossis, S.N.C., Ding, X.Z., Dennis, G.J., and Tsokos, G.C. (1998). Altered pattern of TCR/CD3-mediated protein-tyrosyl phosphorylation in T cells from patients with systemic lupus erythematosus. Deficient expression of the T cell receptor γ chain. *J. Clin. Invest.*, **101**, 1448–1457.

Liu, B., Dai, J., Zheng, H., Stoilova, D., Sun, S., and Li, Z. (2003). Cell surface expression of an endoplasmic reticulum resident heat shock protein gp96 triggers Myd88-dependent systemic autoimmune diseases. *Proc. Natl. Acad. Sci. USA*, **100**, 15842–15829.

Liu, M.F., Wang, C.R., Fung, L.L., and Wu, C.R. (2004). Decreased CD4+CD25+ T cells in peripheral blood of patients with systemic lupus erythematosus. *Scand. J. Immunol.*, **59**, 198–202.

Ludviksson, B.R., Gray, B., Strober, W., and Ehrhardt, R.O. (1997). Dysregulated intra-thymic development in the IL-2-deficient mouse leads to colitis-inducing thymocytes. *J. Immunol.*, **158**, 104–111.

Malek, T.R., Porter, B.O., Codias, E.K., Sciberlli, P., and Yu, A. (2000). Normal lymphoid homeostasis and lack of lethal autoimmunity in mice containing mature T cells with severely impaired IL-2 receptors. *J. Immunol.*, **15**, 2905–2914.

Matache, M., Stefanescu, M., Onu, A., Szegli, G, Barel, M., Tanseanu, S., Matei, I., Boullie, S., and Frade, R. (1996). Tyrosine phosphorylation in peripheral lymphocytes from patients with systemic lupus erythematosus. *Autoimmunity*, **24**, 217–228.

Millar, D.G., Garza, K.M., Odermat, B., Elford, A.R., Ono, N., Li, Z., and Ohashi, P.S. (2003). Hsp 70 promotes antigen-presenting cell function and converts cell tolerance to autoimmunity in vivo. *Nat. Med.*, **9**, 1469–1476.

Mirshahidi, S., Ferris, L.C., and Sadegh-Nasseri, S. (2004). The magnitude of TCR engagement is a critical predictor of T cell anergy or activation. *J. Immunol.*, **172**, 5346–5355.

Moody, J.L. and Jirik, F.R. (2004). Compound heterozygosity for Pten and SHIP augments T-cell dependent humoral immune responses and cytokine production by CD4+ T cells. *Immunology*, **112**, 404–412.

Moretta, A. and Bottino, C. (2004). Regulated equilibrium between opposite signals: A general paradigm for T cell function? *Eur. J. Immunol.*, **34**, 2084–2088.

Mustelin, T., Alonso, A., Bottini, N., Huynh, H., Rahmouni, S., Nika, K., Louis-dit-Sully, C., Tautz, L., Togo, S., Bruckner, S, Mena-Duran, A., and al-Khouri, A.M. (2004). Protein tyrosine phosphatases in T cell physiology. *Mol. Immunol.*, **41**, 687–700.

Nishibori, T., Tanabe, Y., Su, L., and David, M. (2004). Impaired development of CD4+CD25+ regulatory T cells in the absence of STA1: Increased susceptibility to autoimmune disease. *J. Exp. Med.*, **199**, 25–34.

Oelke, K. and Richardson, B. (2004). Decreased T cell ERK pathway signaling may contribute to the development of lupus through effects on DNA methylation and gene expression. *Int. Rev. Immunol.*, **23**, 315–331.

Ohashi, P.S. and DeFranco, A.L. (2002). Making and breaking tolerance. *Curr. Opin. Immunol.*, **14**, 744–759.

Pugliese, A. (2003). Peptide-based treatment for autoimmune diseases: Learning how to handle a double-edge sword. *J. Clin. Invest.*, **111**, 1280–1282.

Quaratino, S., Duddy, L., and Londei, M. (2000). Fully competent dendritic cells as inducers of T cell anergy in autoimmunity. *Proc. Natl. Acad. Sci. USA*, **97**, 10911–10916.

Ramsdell, F. and Ziegler, S.F. (2003). Transcription factors in autoimmunity. *Curr. Opin. Immunol.*, **15**, 718–724.

Rathemell, J.C., Elstrom, R.L., Cinalli, R.M., and Thompson, C.B. (2003). Activated Akt promotes increased resting T cell size, CD28-independent T cell growth, and development of autoimmunity and lymphoma. *Eur. J. Immunol.*, **33**, 2223–2232.

Sakaguchi, N., Takahashi, T., Hata, H., Nomura, T., Tagami, T., Yamazaki, S., Sahikama, T., Matsutani, T., Negishi, I., Nakatsuru, S., and Sakaguchi, S. (2003). Altered thymic T-cell selection due to a mutation of the ZAP-70 gene causes autoimmune arthritis in mice. *Nature*, **426**, 454–460.

Salojin, K.V., Zhang, J., Cameron, M., Gill, B., Arreaza, G., Ochi, A., and Delovitch, T.L. (1997). Impaired plasma membrane targeting of Grb2-murine son of sevenless (mSos) complex and differential activation of the Fyn-T cell receptor (TCR)-ζ-Cbl pathway mediate T cell hyporesponsiveness in autoimmune nonobese diabetic mice. *J. Clin. Invest.*, **186**, 887–897.

Salojin, K.V., Zhang, J., Madrenas, J., and Delovitch, T.L. (1998). T-cell anergy and altered T-cell receptor signaling: Effects on autoimmune disease. *Immunol. Today*, **19**, 468–473.

Schade, A.E. and Levine, A.D. (2004). Cutting edge: Extracellular signal-regulated kinases 1/2 function as integrators of TCR signal strength. *J. Immunol.*, **172**, 5828–5832.

Scheinecker, C., Zwölfer, B., Köller, M., Männer, G., and Smolen, J.S. (2000). Alterations of dendritic cells in systemic lupus erythematosus. *Arthritis Rheum.*, **44**, 856–865.

Seibl, R., Kyburz, D., Lauener, R.P., and Gay, S. (2004). Pattern recognition receptors and their involvement in the pathogenesis of arthritis. *Curr. Opin. Rheumatol.*, **16**, 411–418.

Shevach, E.M. (2000). Regulatory T cells in autoimmunity. *Annu. Rev. Immunol.*, **18**, 423–449.

Sloan-Lancaster, J. and Allen, P. (1996). Altered peptide ligand-induced partial T cell activation: Molecular mechanisms and role in T cell biology. *Annu. Rev. Immunol.*, **14**, 1–27.

Snow, J.W., Abraham, N., Na, M.C., Herndier, B.G., Pastuszak, A.W., and Goldsmith, M.A. (2003). Loss of tolerance and autoimmunity affecting multiple organ STAT5A/5B-deficient mice. *J. Immunol.*, **171**, 5042–5050.

Solomou, E.E., Juang, Y.T., Gourley, M.F., Kammer, G.M., and Tsokos, G.C. (2001). Molecular basis of deficient IL-2 production in T cells from patients with systemic lupus erythematosus. *J. Immunol.*, **166**, 4216–4222.

Stefanova, I., Hemmer, B., Vergami, M., Martin, R., Biddison, W.E., and Germain, R.N. (2003). TCR ligand discrimination is enforced by competing ERK positive and SHP-1 negative feedback pathways. *Nat. Immunol.*, **4**, 248–254.

Suzuki, H., Kundig, T.M., Furlonger, C., Wakeman, A., Timms, E., Matsuyama, T., Schmits, R., Simard, J.J., Ohashi, P.S., Griesser, H., Taniguchi, T., Paige, C.J., and Mak, T.W. (1995). Deregulated T cell activation and autoimmunity in mice lacking interleukin-2 receptor β. *Science*, **9**, 1472–1476.

Takahashi, T., Tanagmi, T., Yamazaki, S., Uede, T., Shimizu, J., Sakaguchi, N., Mak, T, and Sakaguchi, S. (2000). Immunological self-tolerance maintained by CD4+CD25+ regulatory T cells constitutively expressing cytotoxic T lymphocyte-associated antigen 4. *J. Exp. Med.*, **192**, 303–310.

Thomas, R. (2004). Signal 3 and its role in autoimmunity. *Arthritis Res. Ther.*, **6**, 26–27.

Thomas, S., Preda-Pais, A., Casares, S., and Brumeanu, T.D. (2004). Analysis of lipid rafts in T cells. *Mol. Immunol.*, **41**, 399–409.

Treisman, R. (1996). Regulation of transcription by MAP kinase cascades. *Curr. Opin. Cell Biol.*, **8**, 205–215.

Tsokos, G.C., Nambiar, M.P., Tenbrock, K., and Juang, Y.T. (2003). Rewiring the T-cell: Signaling defects and novel prospects for the treatment of SLE. *Trends Immunol.*, **24**, 259–263.

Turley, S.J. (2002). Dendritic cells: Inciting and inhibiting autoimmunity. *Curr. Opin. Immunol.*, **14**, 765–770.

Ueda, H., Howson, J.M.M., Esposito, L., Heward, J., Snook, H., Chamberlain, G., Rainbow, D., Hunter, K., Smith, A., Di Genova, G., Herr, M., Dahlman, I., Payne, F., Smyth, D., Lowe, C., Twells, R., Howlett, S., Healy, B., Nutland, S., Rance, H., Everett, V., Smink, L., Lam, A., Cordell, H., Walker, N., Bordin, C., Hulme, J., Motzo, C., Cucca, F., Hess, J., Metzker, M., Rogers, J., Gregory, S., Allahabadia, A., Nithiyananthan, R., Tuomilehto-Wolf, E., Tuomilheto, J., Bingley, P., Gillespie, K., Undlien, D., Ronningen, K., Guja, C., Ionescu-Tirgoviste, C., Savage, D., Maxwell, A., Carson, D., Patterson, C., Franklyn, J., Clayton, D., Peterson, L., Wicker, L., Todd, J., and Gough, S. (2003). Association of the T-cell regulatory gene *CTLA4* with susceptibility to autoimmune disease. *Nature*, **423**, 506–511.

Veillete, A. (2004). SLAM family receptors regulate immunity with and without SAP-related adaptors. *J. Exp. Med.*, **199**, 1175–1178.

Waldner, H., Collins, M., and Kuchroo, V.K. (2004). Activation of antigen-presenting cells by microbial products breaks self tolerance and induces autoimmune disease. *J. Clin. Invest.*, **113**, 990–997.

Williams, N. (1996). T cell inactivation linked to Ras block. *Science*, **271**, 1234.

Wong, H.K., Kammer, G., Dennis, G., and Tsokos, G.C. (1999). Abnormal NF-*k*B activity in T lymphocytes from patients with systemic lupus erythematosus is associated with decreased p65-RelA protein expression. *J. Immunol.*, **163**, 1682–1689.

Wu, H.Y. and Staines, N.A. (2004). A deficiency of CD4$^+$CD25$^+$ T cells permits the development of spontaneous lupus-like disease in mice, and can be reversed by induction of mucosal tolerance to histone peptide autoantigens. *Lupus*, **13**, 192–200.

Yi, Y., McNerney, M., and Datta, S.K. (2000). Regulatory defects in Cbl and mitogen-activated protein kinase (extracellular signal-related kinase) pathways cause persistent hyperexpression of CD40 ligand in human lupus T cells. *J. Immunol.*, **165**, 6627–6634.

19

Immune Cell Signaling and Gene Transcription in Human Systemic Lupus Erythematosus

Christina G. Katsiari and George C. Tsokos

1. Introduction

Interruption of self-tolerance is believed to represent the central event underlying autoimmune disease pathogenesis. Loss of tolerance to multiple self-antigens is a prerequisite for systemic autoimmunity. Systemic lupus erythematosus (SLE) represents by far the best example of generalized loss of tolerance. Recent information discussed herein stipulates that primary immune cell biochemical abnormalities underwrite the autoreactive nature of T and B cells in SLE. Obviously, the use of the term "primary" in this context does not imply that these abnormalities are not the result of multiple unidentified genetic effects, the expression of which has been conditioned by environmental and/or hormonal factors (Tsokos, 1999).

Lymphocytes recognize and respond to antigens through specialized surface receptors, namely the T cell and B cell receptors (TCRs and BCRs, respectively). This unique property as well as the fact that putative autoantigens interact with autoreactive lymphocytes through their antigen receptors (AgRs) has led investigators to characterize molecular events that take place in SLE lymphocytes in response to the engagement of their AgRs (Kammer et al., 2002). In normal lymphocytes, AgR interaction results in activation of protein tyrosine kinases (PTKs), subsequent protein phosphorylation of additional cytosolic proteins, calcium mobilization, and activation of transcription factors. Activated transcription factors translocate to the nucleus to modify the expression of genes that regulate cellular operations such as lymphocyte activation, cellular proliferation, secretion of soluble factors, phenotypic changes, anergy or programmed cell death, and effector

---

Christina G. Katsiari and George C. Tsokos • Department of Cellular Injury, Walter Reed Army Institute of Research, Silver Spring, Maryland 20190 and Department of Medicine, Uniformed Services University, Bethesda, Maryland 20814.

Molecular Autoimmunity: In commemoration of the 100th anniversary of the first description of human autoimmune disease, edited by Moncef Zouali. Springer Science+Business Media, Inc., New York, 2005.

functions. Because ligation of other surface receptors to either costimulatory molecules or cytokines is important for the regulation of the immune response, the respective biochemical pathways have also attracted research interest. During the last 10 years a significant amount of information has been generated that has helped investigators gain insights into the molecular and biochemical events that underlie the aberrant lymphocyte function in SLE patients. Lymphocytes from patients with SLE present enormous complexity, which has led investigators to coin the term "enigma" to the SLE T cell abnormal function (Dayal and Kammer, 1996). Namely, although T cells from patients with SLE are known to provide excessive cognate help to B cells to produce autoantibodies, they cannot mount proper cytotoxic responses to fend off infected cells and altered self cells (Tsokos, 1999). Also, at the cytokine production levels, whereas the production of IL-6 and IL-10 has been reported to be increased, SLE T cells do not produce sufficient amounts of IFN-γ and IL-2 (Froncek and Horwitz, 1999). In this chapter we summarize recent data that unravel the biochemical abnormalities that underwrite the diverse SLE immune cell abnormalities.

2. Altered Pattern of Tyrosine Phosphorylation and Calcium Responses

Under physiological conditions, the signal initiated by the TCR/CD3 complex triggers phosphorylation of phospholipase Cγ on Tyr and Ser residues, hydrolysis of phosphatidylinositol (4, 5)-biphosphate to inositol (1, 4, 5)-triphosphate, and a rapid rise in the free intracellular calcium concentration $[Ca^{2+}]_i$. In SLE T cells, the CD3-initiated rise in $[Ca^{2+}]_i$ is significantly higher and prolonged compared to the response of T cells from normal individuals, and that of patients with other systemic rheumatic diseases (Vassilopoulos *et al.*, 1995). The magnitude of the calcium response does not correlate with overall disease activity, specific disease manifestations, or therapeutic interventions. The enhanced $[Ca^{2+}]_i$ responses are not restricted to a particular T cell subpopulation and they extend to short-term T cell lines established from patients with SLE and to autoantigen-specific T cell lines (Liossis *et al.*, 1998), suggesting that the defective response and/or the molecular events that lead to it are inherent to SLE T cells. It is possible that the increased response is accentuated by additional contributing factors such as nitric oxide, which is produced at increased amounts by SLE monocytes and T cells. Nitric oxide contributes to increase calcium release from mitochondria (Nagy *et al.*, 2004).

Stimulation of T cells from patients with SLE results also in significantly enhanced production of tyrosine-phosphorylated cellular proteins, which represent the most proximal TCR-mediated biochemical event. Of interest, anti-CD3-mediated T cell stimulation displays abnormal kinetics of phosphotyrosine production characterized by a rapid and abrupt increase, followed by a steep decrease to baseline levels, whereas protein tyrosine phosphorylation in normal T cells gradually increases during the same period of time. Analysis of the TCR/CD3-initiated tyrosine-phosphorylated proteins revealed the absence or deficiency of a protein band, which was always present in T cell lysates obtained from normal subjects and from patients with rheumatic disorders other than SLE. Immunoblotting, immunoprecipitation, and flow-cytometric analysis revealed that the deficient band was the CD3ζ chain (Liossis et al., 1998).

3. TCR ζ Chain Deficiency

The TCR consists of α and β chains that are responsible for antigen recognition. For the initiation of signal transduction TCR joins the CD3 complex consisting of the ε, γ, δ, and ζ chains. While expression of α, β, and ε chains is preserved, ζ chain expression is impaired or absent in approximately three quarters of SLE patients (Nambiar et al., 2003). Deficient ζ chain expression in SLE T cells was also reported by other laboratories, suggesting that this defect is not geographically restricted (Takeuchi et al., 1998; Brundula et al., 1999). TCR ζ chain deficiency does not correlate with disease activity, therapy, age, race, or gender, and in one study it was found to persist over a 3-year period of follow-up (Nambiar et al., 2003). The TCR ζ chain gene is located on chromosome 1q23. Genomewide scans of multiplex SLE families indicate that genes on chromosome 1q may contribute to genetic predisposition and susceptibility to the disease. Genetic linkage of the ζ chain gene to the FcγRII and RIII gene cluster, a candidate locus implicated in genetic susceptibility and deficiency of TCR ζ chain in SLE, suggests that ζ chain loss might play an important role in genetic predisposition to this autoimmune disease (Jensen et al., 1992).

Although impaired TCR ζ chain has also been reported in other disorders such as rheumatoid arthritis, melanoma, tumor-infiltrating lymphocytes, viral infections, and leprosy, many aspects of the molecular mechanisms of TCR ζ chain downregulation in human SLE are discrete and characterize the disease. It appears that multiple mechanisms lead to decreased expression of TCR ζ chain in SLE T cells (Table 19.1).

Table 19.1. Mechanisms of TCR ζ Chain Deficiency in SLE T cells

1. Impaired TCR ζ chain gene transcription
    Reduced formation of 98-kDa Elf-1 (type I defect)
    Decreased binding of Elf-1 to IL-2 gene promoter (type II defect)
2. Impaired translation and posttranscription events
    Polymorphisms, mutations, and alternative splicing in the nucleotide sequence of ζ chain transcripts → abnormal RNA editing?
    Abnormal ζ chain transcripts (absence of exon VII, shorter 3′ untranslated region) → abnormal translation?
3. Impaired posttranslational functions
    Protein degradation: increased amounts of ubiquitinated forms of ζ chain and enhanced activity of caspases
    Protein phosphorylation: possible defect due to LCK deficiency
    Oxidative stress: elevated mitochondrial transmembrane potential and reactive oxygen species

3.1. Impaired TCR ζ Chain Gene Transcription

The molecular basis of ζ chain deficiency in SLE T cells lies in part at the level of transcription since a significant number of SLE patients display decreased amounts of ζ chain messenger RNA (mRNA) (Liossis *et al.*, 1998; Takeuchi *et al.*, 1998). Indeed, the activity of ζ chain gene promoter in SLE T cells is low compared to that of normal T cells. Since no mutations or polymorphisms of the TCR ζ chain promoter were detected to explain this finding, it was assumed that either the production and/or activation of the transcription factors that regulate the activity of the TCR ζ chain promoter could be defective. The TCR ζ chain promoter is notably simple in that only two Elf-1 binding sites have been identified in the proximal region and shown to regulate its activity (Rellahan *et al.*, 1998). Elf-1 is an Ets family transcription factor that encodes for a 619–amino acid polypeptide. Elf-1 is subject to posttranslational modifications, mainly glycozylation and phosphorylation, resulting in the 80- and 98-kDa forms of the protein, respectively (Juang *et al.*, 2002). Each form displays a different distribution inside the cell, the 80-kDa form being located in the cytoplasm and the 90-kDa form in the nucleus. Fully phosphorylated and glycozylated Elf-1 (98 kDa) binds to the TCR ζ chain gene promoter and enhances its activity (Juang *et al.*, 2002).

In SLE T cells, the levels of Elf-1 mRNA and the expression of the cytoplasmic 80-kDa form are comparable to those of normal T cells, suggesting that transcription and translation of Elf-1 gene are intact. In contrast, the form of Elf-1 that binds to the ζ chain promoter was decreased in SLE T cells. Specifically, in some SLE patients the 98-kDa Elf-1 is not formed (type I defect), whereas in others although it is produced it does not bind to DNA (type II defect) (Juang *et al.*, 2002). Thus, defects in expression or function of the 98-kDa Elf-1 result in decreased binding to the TCR ζ chain gene promoter and, consequently, decreased transcription of TCR ζ chain.

3.2. Impaired Translation and Posttranscription Events

TCR ζ chain transcripts expressed at low levels in SLE T cells display frequent and heterogeneous polymorphisms, mutations, and alternative splicing in their nucleotide sequence. Because respective findings in the genomic DNA were not found, the mechanism of RNA editing may have an additional role in ζ chain deficiency in SLE T cells. Defective RNA editing has been shown for protein kinase A I (PKA-I) in SLE T cells (Laxminarayana *et al.*, 2002) and it may represent an additional mechanism that contributes to the diverse defects in SLE T cells. In addition, a novel ζ chain transcript was found to be increased in SLE T cells, pointing toward a defect at the level of transcription. Specifically, ζ chain transcripts missing exon VII have been reported to be increased in SLE T cells (Nambiar *et al.*, 2002; Takeuchi *et al.*, 1998) and to contribute to the decreased assembly of TCRs in SLE T cells and possibly to the decreased production of IL-2. In addition, another ζ chain transcript has been described in SLE

T cells, which has a shorter 3′ untranslated region (Nambiar et al., 2001). This alternatively spliced transcript is unstable and produces little protein. Summarily, it appears that the production of alternatively spliced ζ chain transcripts with either defective ζ chain products or unstable ζ chain mRNA contributes to decreased expression of ζ chain protein in SLE T cells.

3.3. Impaired Posttranslational Functions

Proteolytic cleavage constitutes a major mechanism through which the cell regulates its protein content. The TCR ζ chain carries multiple putative caspase cleavage sites (Gastman et al., 1999). Incubation of SLE T cells with caspase inhibitors upregulates ζ chain expression, suggesting that proteolysis by caspases plays a role in ζ chain downregulation.

The final disposal apparatus in eucaryocytes is the proteasome, an abundant cylinder-shaped, ATP-dependent protease complex dispersed throughout the cytosol and the nucleus. With few exceptions, the proteasome acts on proteins that have been specifically marked for degradation by the attachment of multiple copies of the small protein ubiquitin. Ubiquitination and subsequent proteolysis aim to confer short half-lives on specific normal proteins whose concentrations must change promptly, serving as a homeostatic mechanism during cell activation. Indeed, following phosphorylation, the TCR ζ chain undergoes ubiquitination and proteasome-mediated degradation (Valitutti et al., 1997). Depending on the intensity of this process, the ζ chain exists in multiple-sized ubiquitinated forms in T cells at a steady state (Cenciarelli et al., 1992). In SLE T cells, increased amounts of ubiquitinated forms of ζ chain were found (Nambiar et al., 2002). Increased ubiquitination and/or decreased rate of proteasome-mediated degradation could explain this finding. Because in SLE T cells caspase-mediated proteolytic activity appears to be potent, it is more likely that increased ubiquitination of the ζ chain is the consequence of an aberrant ubiquitination process or increased ζ chain turnover rather than a defective proteolysis.

Increased ubiquitination has also been implicated for the decreased levels of lymphocyte-specific protein kinase (LCK) observed in SLE T cells (Jury et al., 2003). Moreover, because LCK is important for the phosphorylation of ζ chain during cell activation, it has been proposed that LCK deficiency may contribute even further to the reduction of ζ chain activity in SLE T cells.

3.4. Oxidative Stress

In normal T cells, oxidative stress induces the downregulation of the ζ chain through the action of heat shock proteins (Nambiar et al., 2002), a mechanism that accounts for the decrease of ζ chain in the majority of pathologies other than SLE. Because in SLE T cells the mitochondrial transmembrane potential and reactive oxygen species are elevated (Gergely et al., 2002), oxidative stress may represent an additional mechanism that contributes to the downregulation of the ζ chain.

3.5. Role of IFNγ

Baniyash et al. have developed an in vivo experimental system that mimics conditions of chronic immune cell activation and have shown that sustained exposure of mice to bacterial antigens is sufficient to induce TCR ζ chain downregulation and impair T cell function. In this model, the decrease of TCR ζ chain seems to be dependent on the presence of a robust IFNγ response because chronic infection in IFNγ-deficient mice failed to decrease the expression of TCR ζ chain (Baniyash, 2004).

4. Mechanisms of Increased TCR/CD3–Mediated $[Ca^{2+}]_i$ Response in SLE T Cells

The reduced TCR ζ chain expression described above could account for both the decreased activation-induced cell death and the diminished natural killer cell activity observed in T cells from patients with SLE (Horwitz et al., 1999). Conversely, TCR ζ chain deficiency and augmented phosphorylation of cytoplasmic proteins and increased intracellular calcium responses in SLE T cells appear to be contradicting findings. Current investigation, however, has revealed several mechanisms that can explain, at least in part, the TCR/CD3-mediated calcium response in SLE T cells despite the deficiency of ζ chain. They include increased expression of FcRγ chain that takes the place of ζ chain in the TCR/CD3 complex, and rewires the downstream signaling events and alterations in the composition and function of lipid rafts, favoring the hyperexcitability of SLE T cells.

4.1. FcRγ Chain Substitutes for Defective ζ Chain

To explain amplified calcium responses in SLE T cells following TCR ligation despite the presence of small amounts or even the absence of the TCR ζ chain, it was hypothesized that other members of the ζ chain family may take its place in the CD3 complex. Immunoprecipitation and immunoblotting experiments demonstrated that in contrast to controls, a large proportion of SLE T cells express very high levels of FcRγ chain that are functionally associated with components of the CD3 and participate in antigen receptor–mediated signal transduction (Enyedy et al., 2001). Unlike the ζ chain that associates with ZAP-70 to initiate its downstream signaling, FcRγ associates with Syk kinase. The FcRγ–Syk complex is able to mediate 100 times more potent signals (Krishnan et al., 2003). Interestingly, forced expression of FcRγ chain in normal T cells resulted in signaling events that were similar to those observed in SLE T cells. Specifically, normal T cells overexpressing FcRγ chain displayed increased CD3-mediated protein tyrosine phosphorylation and calcium responses (Nambiar et al., 2003). Conversely, reconstitution of the missing TCR ζ chain in SLE T cells resulted in normalized CD3-mediated early signaling events (Nambiar et al., 2003). The generated information suggests that the presence of FcRγ chain in the CD3 complex renders T cells overexcitable, whereas the presence of the TCR ζ chain in the CD3 complex is required for a "normal" response. The presence of the FcRγ chain on the surface of SLE T cells indicates that they may become easily stimulated in the presence of autoantigens that have low affinity for TCR. In contrast, cells that express normal levels of the TCR ζ chain do not respond to autoantigens and thus an autoimmune response is averted.

4.2. Altered Composition and Dynamics of Lipid Rafts

It has become clear that cell membrane structure plays a dynamic role in the intracellular signaling process. Because inside its lipid by-layer, the van der Waals attractive forces between adjacent fatty acid tails are in most cases unable to hold these molecules together, each cell membrane lipid monolayer is a random combination of different types of lipid molecules. On the other hand, because some lipid molecules, such as sphingolipids, have longer and saturated fatty acid tails, the attractive forces are strong enough to hold neighboring molecules together, at least transiently, forming discrete microdomains called lipid rafts. Like scaffold proteins, lipid rafts, which are rich in both sphingolipids and cholesterol, can be thought of as lipid scaffolds where signaling molecules can assemble and interact (Simons and Ikonen, 1997). In normal T cells, TCR ligation induces rapid clustering of lipid rafts that results in the accumulation of signaling proteins at the area of contact between T and antigen-presenting cells, known as the immunological synapse. The integrity of the immune synapse is essential for TCR signaling and T cell activation (Alonso and Millan, 2001).

Several factors can influence the strength of signals controlled by lipid rafts, such as lipid raft pool size, its membrane distribution, and protein content as well as the kinetics of cytoskeletal rearrangements following T cell stimulation. It was recently shown that GM1 ganglioside, a lipid raft marker, is present in higher amounts in SLE T cells compared with that in normal T cells, while stimulation induced further overexpression of GM1 in SLE T cells (Jury *et al.*, 2004; Krishnan *et al.*, 2004), which was due to *de novo* production of the molecule (Krishnan *et al.*, 2004). These results indicate that SLE T cells posses more extensive basal levels of lipid rafts as well as greater capacity to generate lipid rafts than normal T cells in response to stimulation.

In the same studies the expression and function of different components of lipid rafts were assessed. Using confocal microscopy, it was shown that while the TCR ζ chain is uniformly distributed on the normal T cell membrane, in SLE T cells residual ζ chain is preferentially colocalized with lipid rafts (Krishnan *et al.*, 2004), confirming previous results where the bulk of residual ζ chain in SLE T cells was found to be limited to detergent-insoluble fractions (Nambiar *et al.*, 2002). More importantly, lipid rafts from SLE T cells contained FcRγ-chain and Syk kinase, which were absent in normal T cells. In addition, ZAP-70 present in normal lipid rafts was not detected in SLE lipid rafts (Krishnan *et al.*, 2004). These findings were furthermore confirmed by the analysis of lipid raft fractions following sucrose density gradient ultracentrifugation. Other qualitative alterations of lipid raft composition have been recently reported where the majority of SLE patients studied presented an increased raft association of the protein tyrosine phosphatase (PTP) CD45 (Jury *et al.*, 2004) and a reduced raft expression of the LCK (Jury *et al.*, 2004; Krishnan *et al.*, 2004). Furthermore, in response to stimulation, SLE T cells displayed an altered pattern of CD45 phosphatase raft expression kinetics characterized by a rapid and more pronounced recruitment of CD45 within the lipid raft clusters followed by their faster exclusion from the site of activation (Jury *et al.*, 2004). It should be noted that although the overall amount of LCKs localized in lipid rafts was reduced, the active form of the enzyme was increased, in particular in T cells from patients with SLE with an increased expression of CD45 in raft domains. Because the PTP CD45 regulates LCK function through the dephosphorylation of an inhibitory tyrosine residue, thus allowing LCK to assume an open/active conformation, and because immunoprecipitation experiments performed in T cell lysates revealed increased association of PTP CD45 with LCK in patients with SLE, the authors postulate that in SLE T cells LCK may be differentially regulated through the altered association of PTP CD45 with lipid rafts (Jury *et al.*, 2004).

Another finding that characterized SLE T cells was that in the absence of any activation stimulus, TCR–CD3 complexes were clustered in contrast to normal T cells in which they were evenly allocated throughout the cell membrane. Upon stimulation, the kinetics of receptor capping was accelerated and more intense in SLE T cells. Because CD3 clusters and caps colocalized with GM1, it is suggested that SLE T cells are also characterized by faster kinetics of lipid raft clustering and polarization. In response to stimulation and preceding TCR–CD3 capping, it was shown that the kinetics of actin polymerization was enhanced in SLE T cells. Since actin polymerization represents a central event in cytoskeleton remodeling, which is essential for the receptor cap formation, altered actin dynamics in SLE T cells seem to have a role in the kinetic performance of TCR–CD3 complexes and lipid rafts (Krishnan *et al.*, 2004).

An important aspect of the study by Krishnan *et al.* (2004) was that it provided a link between the altered lipid rafts function and the abnormal T cell responses in SLE. TCR ligation with anti-CD3 antibodies induced augmented calcium responses in SLE T cells, as described above. Concomitant treatment with methyl-β-cyclodextrin (MbCD), a compound that disrupts lipid rafts by depleting cholesterol, led to a dose-dependent reduction of intracellular calcium concentrations in both normal and SLE T cells, although with the latter needing a higher dose to achieve the same decrease observed in normal T cells. In the same set of experiments it was documented that PLCγ1, a crucial factor in the biochemical pathway that leads to intracellular calcium influx, exhibited a significant shift from the cytoplasm to the lipid rafts in response to T cell activation, pointing toward a direct association between PLCγ1 and lipid raft performance in regard to calcium homeostasis. Overall, these results show that augmented calcium responses in SLE T cells are mediated by, at least in part, a lipid raft–dependent mechanism, which remains to be defined. It should be noted that the aberrations described in these studies do not correlate with disease activity or medication intake. Future studies aiming to thoroughly characterize the full spectrum of such alterations and to describe in detail the direct effects on T cell signaling will provide new insight into the T cell "rewired" signaling pathways in SLE.

5. Protein Kinase A Function

PKA is a cAMP-dependent serine/threonine kinase. Several years ago Kammer et al., who were the first to search for biochemical defects in SLE T cells, identified abnormalities in cAMP metabolism and PKA-catalyzed protein phosphorylation (Mandler et al., 1982; Hasler et al., 1990). In the inactive state, PKA forms a tetramer consisting of two regulatory and two catalytic subunits (R_2C_2). Depending on the isoform of the R-subunit, PKA exists as either PKA-I or PKA-II isoenzyme, localized to the plasma membrane and the cytoplasm, respectively. Binding of cAMP to the regulatory subunits alters their confirmation, causing them to dissociate from the PKA complex (R_2C_2 + 4 cAMP ↔ R_2 $cAMP_4$ + 2C). The released catalytic subunits are thereby activated to phosphorylate a diverse panel of target proteins both in the cytoplasmic and the nuclear compartments (Torgersen et al., 2002).

Initial studies revealed that cAMP-primed SLE T cells display impaired PKA-catalyzed protein phosphorylation mainly due to decreased activity of the isoenzyme PKA-I (Kammer et al., 1994). First-order enzyme kinetics studies subsequently showed that in SLE T cells the binding capacity of cAMP for the R-I subunit was significantly impaired (Kammer et al., 1996). A mutated form of R-I subunit with lower cAMP-binding activity could provide a logical explanation for this finding. Investigators found that SLE T cells have decreased protein expression of the R-I subunits (α and β isoforms) that paralleled decreased amounts of R-I transcripts, which may reflect defects at the transcriptional level (Laxminarayana et al., 1999). Lower levels of functional R-I possibly denote lower levels of PKA-I (RI_2C_2) complexes and thus a smaller pool of PKA-I for cAMP to act on, leading to impaired performance of this signaling pathway. On the other hand, since the expression of the catalytic subunits appears to be intact, it is logical to expect increased amounts of unrelated (and thus possibly active) C subunits that could spontaneously promote PKA-mediated actions. Nevertheless, when R-I transcripts from SLE T cells were analyzed several mutations were found in the absence of any genetic mutations, implying defective mRNA editing. Whether altered R-I subunits from SLE T cells display reduced ability to bind cAMP remains to be addressed. Along with a better characterization of the nature of PKA-I deficiency, a more thorough understanding of its contribution to cell and cytokine defects in SLE T cells is anticipated with great interest.

Moreover, a series of experiments indicate that the β isoform of the R-subunit (RIIß) of PKA-II isoenzyme could mediate impaired cAMP-initiated responses in the nucleus of lupus T cells (Kammer, 2002). In the nucleus, transcriptional regulation by cAMP is mediated by a family of cAMP-responsive nuclear factors, which bind to and regulate the expression of genes containing the cAMP-responsive element (CRE) consensus in their promoters. Phosphorylation of these CRE-binding proteins (CREBs) by the C subunit of PKA that translocates into the nucleus following activation by cAMP modulates their activity (Skalhegg and Tasken, 2000). In SLE T cells, deficient PKA-II activity is characterized by spontaneous dissociation of the cytosolic $RIIβ_2C_2$ holoenzyme, aberrant RIIβ translocation to the nucleus from the cytosol, and retention of RIIβ in the nucleus. Experiments addressing a possible interplay with extracellular stimuli in the lupus microenvironment showed that RIIβ disorder was primary to the SLE T cell (Mishra et al., 2000). Interestingly, forced expression of RIIβ in Jurkat cells revealed a novel, specific, and direct interaction between RIIβ and CREB that led to inhibition of CRE-dependent expression of the proto-oncogene c-fos (Elliott et al., 2003). In addition, Jurkat cells overexpressing a phosphorylated form of RIIβ were shown to suppress the production of IL-2 in parallel with upregulation of the costimulatory molecule CD40L (Elliott et al., 2004). Taken together these results indicate that accumulation of RIIβ in the nucleus of SLE T cells could contribute to some phenotypic abnormalities observed in these cells, such as reduced production of IL-2 and overexpression of CD40L.

Importantly, PKA-I and -II abnormalities encountered in SLE T cells do not correlate with disease activity, therapy, or demographic characteristics, and are sustained over time (Kammer, 2002). In addition, the fact that the genes encoding RIß and RIIß map to 7p22 and 7q22 regions, which have been identified as susceptibility loci for SLE (Gaffney et al., 2000), underlines the importance of the study of PKA function in SLE.

6. Regulation of Transcription Determines Interleukin 2 Deficiency in SLE T Cells

Interleukin-2 (IL-2) represents a crucial cytokine in the activation of lymphocytes. IL-2 exerts multiple functions, which serve both the promotion and the suppression of the immune response (Nelson, 2004). Fine-tuning of the production of this cytokine is the result of numerous interrelated events, which make its study a complex but intriguing task. Reduction in IL-2 production in SLE T cells upon stimulation was first documented in the 1980s, when PHA-primed cells from lupus patients produced less IL-2 compared to normal T cells (Linker-Israeli et al., 1983), and it was later on associated with the decreased immune responses to infectious agents encountered in patients with SLE.

IL-2 deficiency in SLE T cells is the result of transcriptional repression of the IL-2 gene. Study of transcriptional regulation of IL-2 in normal T cells has shown that the decisive elements for the transcription of the molecule lie within 300 bp upstream of the start codon of the IL-2 gene promoter (Jain et al., 1995). Within these 300 bp several AP-1 sites are found, which are closely related to adjacent NFAT sites, as well as an NFκB, a C-responsive element (RE) and a CD28-RE site. Currently, defects leading to decreased IL-2 production in SLE T cells include changes in the expression and function of the transcription factors NFκB, AP-1, CREM, and CREB (Figure 19.1).

In SLE T cells NFκB activity is decreased due to deficiency of its p65 subunit (Wong et al., 1999). Under physiologic conditions, the p65/p50 heterodimer accounts for the NFκB-mediated IL-2 upregulation (Jain et al., 1995). Conversely, SLE T cells express adequate amounts of p50, which in the absence of p65 homodimerize, conveying a repressor effect (Wong et al., 1999). The importance of p65 deficiency is furthermore supported by experiments showing

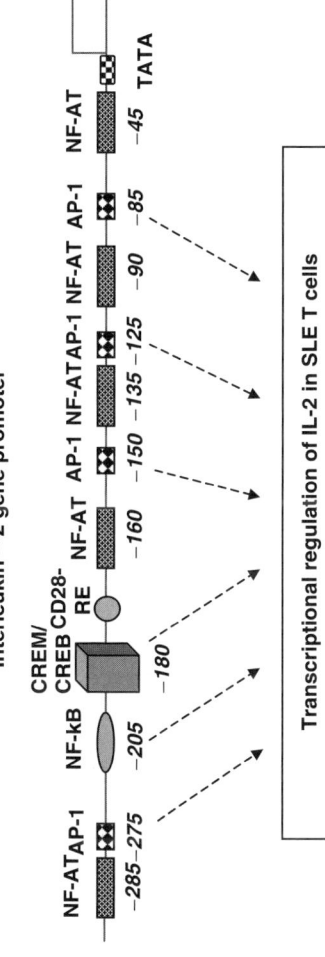

Figure 19.1. Structure/function of the IL-2 promoter. Several *cis* sites have been identified and the *trans* factors that bind to them have been studied extensively. In SLE T cells, decreased occupancy of the IL-2 promoter by transcriptional enhancers such as NFκB, CREB, and AP1, along with increased occupancy by the repressor CREM, are responsible for decreased IL-2 promoter activity and, ultimately, decreased production of IL-2.

that forced expression of p65 in SLE T cells increased IL-2 gene promoter activity (Herndon et al., 2002). The cause underlying p65 deficiency is currently unknown although it is postulated that caspase-8, which display increased activity in SLE T cells, might bind and digest the p65 chain (Horiuchi et al., 2000).

The CRE site (−180) of the IL-2 promoter represents the binding site of CREB and CREM (Powell et al., 1998). In normal T cells, upon activation-initiated phosphorylation by protein kinase C and other downstream kinases, phosphorylated CREB (pCREB) enters the nucleus and binds to the −180 site in complex with CRE-binding protein (CBP) and p300 where it enhances IL-2 promoter activity (Mayr and Montminy, 2001). Subsequent return to basal state is achieved by the removal of pCREB along with the recruitment of CREM to the same binding site to suppress IL-2 promoter transcriptional activity. In SLE T cells, transcriptionally upregulated CREM was found to be associated with decreased IL-2 production (Solomou et al., 2001). Restoration of IL-2 production in SLE T cells following transfection of an antisense CREM plasmid that suppressed the expression of CREM confirmed the importance of this transcriptional repressor (Tenbrock et al., 2003). In addition, CREM binding to the c-fos promoter in unstimulated live T cells from SLE patients is significantly increased. Increased binding results in decreased transcription of c-fos gene, decreased production of c-fos protein and AP-1 binding in the nuclei of SLE T cells (Kyttaris et al., 2004). Therefore, CREM contributes to the decreased expression of IL-2 by both binding directly to the IL-2 promoter and limiting the AP-1 binding through its action on the c-fos promoter.

Although several defects underlying lupus T cell IL-2 deficiency have been defined, many other issues remain to be addressed. Apart from the ongoing study of other transcription factors as putative contributors to IL-2 deficiency, the pursuit for the "missing link" between a particularly robust calcium response and a phenotype resembling that of anergic cells with deficient responses to recall antigens and deficient production of IL-2 upon activation lies in the center of our research.

7. Conclusions

SLE T cells have been known to display diverse cellular and cytokine abnormalities. The studies reviewed in this chapter have systematically shed light on the biochemical and molecular aberrations responsible for the aberrant SLE T cell biology. We now understand that rewiring of the SLE T cell receptor by the replacement of the ζ chain by the FcRγ chain and the subsequent engagement of downstream molecules leads to an overexcitable T cell phenotype that may account for the presence of increased numbers of T cells in the peripheral blood of SLE patients, providing increased help to B cells to produce autoantibodies. We recently described increased spontaneous aggregation of lipid rafts on the surface membrane of SLE T cells. We found that these aggregates are dynamic both in terms of lateral mobility on the surface of the T cells and in terms of biosynthesis. We believe that these preaggregated lipid rafts contribute to the well-established overexcitable T cell phenotype. We now understand an array of mechanisms that are responsible for the decreased expression of the ζ chain in SLE T cells. It should be noted that the decreased ζ chain, which we and others have observed in effector T cells, is due mostly to increased degradation rather than to decreased transcription and to expression of various forms of alternatively spliced transcripts that are either unstable or display a poor translational rate.

But if the initial response of the SLE T cells is strong, then why are a number of effector functions, such as the production of IL-2, decreased? Our studies have shown that a number of downstream events "hijack" an otherwise robust response. Specifically, we have found that SLE T cells express increased amounts of CREM, a transcriptional repressor that binds to the IL-2 promoter and limits its expression. Obviously, our studies and those of others have only started describing the many elements of the puzzle. Yet, we are encouraged by the fact that we are already able to put pieces together and have started understanding the origin of certain "aberrant" T cell functions.

References

Alonso, M.A. and Millan, J. (2001). The role of lipid rafts in signaling and membrane trafficking in T lymphocytes. *J. Cell Sci.*, **114**, 3957–3965.

Baniyash, M. (2004). TCR ζ-chain down regulation: Curtailing an excessive inflammatory immune response. *Nat. Rev. Immunol.*, **4**, 675–687.

Brundula, V., Rivas, L.J., Blasini, A.M., Paris, M., Salazar, S., Stekman, I.L., and Rodriguez, M.A. (1999). Diminished levels of T cell receptor zeta chains in peripheral blood T lymphocytes from patients with systemic lupus erythematosus. *Arthritis Rheum.*, **42**, 1908–1916.

Cenciarelli, C., Hou, D., Hsu, K.-C., Rellahan, B.L., Wiest, D.L., Smith, H.T., Fried, V.A., and Weissman, A.M. (1992). Activation-induced ubiquitination of the T cell antigen receptor. *Science*, **257**, 795–797.

Dayal, A.K. and Kammer, G.M. (1996). The T cell enigma in lupus. *Arthritis Rheum.*, **39**, 23–33.

Jury, E.C., Kabouridis, P.S., Abba, A., Mageed, R.A., and Isenberg, D.A. (2003). Increased ubiquitination and reduced expression of LCK in T lymphocytes from patients with systemic lupus erythematosus. *Arthritis Rheum.*, **48**, 1343–1354.

Elliott, M.R., Shanks, R.A., Khan, I.U., Brooks, J.W., Burkett, P.J., Nelson, B.J., Kyttaris, V., Juang, Y.T., Tsokos, G.C., and Kammer, G.M. (2004). Down-regulation of IL-2 production in T lymphocytes by phosphorylated protein kinase A-RII{beta}. *J. Immunol.*, **172**, 7804–7812.

Elliott, M.R., Tolnay, M., Tsokos, G.C., and Kammer, G.M. (2003). Protein kinase A regulatory subunit type II{beta} directly interacts with and suppresses CREB transcriptional activity in activated T cells. *J. Immunol.*, **171**, 3636–3644.

Enyedy, E.J., Nambiar, M.P., Liossis, S.N., Dennis, G., Kammer, G.M., and Tsokos, G.C. (2001). Fc epsilon receptor type I gamma chain replaces the deficient T cell receptor zeta chain in T cells of patients with systemic lupus erythematosus. *Arthritis Rheum.*, **44**, 1114–1121.

Froncek, M.C. and Horwitz, D. (1999). Cytokines in the pathogenesis of systemic lupus erythematosus. In R.G. Lahita (ed.), *Systemic Lupus Erythematosus*. Academic Press, New York. pp. 187–203.

Gaffney, P.M., Ortmann, W.A., Selby, S.A., Shark, K.B., Ockenden, T.C., Rohlf, K.E., Walgrave, N.L., Boyum, W.P., Malmgren, M.L., Miller, M.E., Kearns, G.M., Messner, R.P., King, R.A., Rich, S.S., and Behrens, T.W. (2000). Genome screening in human systemic lupus erythematosus: Results from a second Minnesota cohort and combined analyses of 187 sib-pair families. *Am. J. Hum. Genet.*, **66**, 547–556.

Gastman, B.R., Johnson, D.E., Whiteside, T.L., and Rabinowich, H. (1999). Caspase-mediated degradation of T-cell receptor zeta-chain. *Cancer Res.*, **59**, 1422–1427.

Gergely, P., Jr., Grossman, C., Niland, B., Puskas, F., Neupane, H., Allam, F., Banki, K., and Perl, A. (2002). Mitochondrial hyperpolarization and ATP depletion in patients with systemic lupus erythematosus. *Arhtritis Rheum.*, **46**, 175–190.

Hasler, P., Schultz, L.A., and Kammer, G.M. (1990). Defective cAMP-dependent phosphorylation of intact T lymphocytes in active systemic lupus erythematosus. *Proc. Natl. Acad. Sci. USA*, **87**, 1978–1982.

Herndon, T.M., Juang, Y.T., Solomou, E.E., Rothwell, S.W., Gourley, M.F., and Tsokos, G.C. (2002). Direct transfer of p65 into T lymphocytes from systemic lupus erythematosus patients leads to increased levels of interleukin-2 promoter activity. *Clin. Immunol.*, **103**, 145–153.

Horiuchi, T., Himeji, D., Tsukamoto, H., Harashima, S., Hashimoto, C., and Hayashi, K. (2000). Dominant expression of a novel splice variant of caspase-8 in human peripheral blood lymphocytes. *Biochem. Biophys. Res. Commun.*, **272**, 877–881.

Horwitz, D., Stohl, W., and Dixon Gray, J (1999). T lymphocytes, natural killer cells, and immune regulation. In R.G. Lahita (ed), *Systemic Lupus Erythematosus*. Academic Press, New York. pp. 157–185.

Jain, J., Loh, C., and Rao, A. (1995). Transcriptional regulation of the IL-2 gene. *Curr. Opin. Immunol.*, **7**, 333–342.

Jensen, J.P., Hou, D., Ramsburg, M., Taylor, A., Dean, M., and Weissman, A.M. (1992). Organization of the human T cell receptor zeta/eta gene and its genetic linkage to the Fc gamma RII-Fc gamma RIII gene cluster. *J. Immunol.*, **148**, 2563–2571.

Juang, Y.-T., Solomou, E., Rellahan, B., and Tsokos, G.C. (2002a). Phosphorylation and O-linked glycosylation of Elf-1 leads to its translocation to the nucleus and binding to the promoter of the T cell receptor ζ chain. *J. Immunol.*, **168**, 2865–2871.

Juang, Y.-T., Tenbrock, K., Nambiar, M.P., Gourley, M.F., and Tsokos, G.C. (2002b). Defective production of the 98 kDa form of Elf-1 is responsible for the decreased expression of TCR zeta chain in patients with systemic lupus erythematosus. *J. Immunol.*, **169**, 6048–6055.

Jury, E.C., Kabouridis, P.S., Flores-Borja, F., Mageed, R.A., and Isenberg, D.A. (2004). Altered lipid raft-associated signaling and ganglioside expression in T lymphocytes from patients with systemic lupus erythematosus. *J. Clin. Invest.*, **113**, 1176–1187.

Kammer, G.M. (2002). Deficient protein kinase A in systemic lupus erythematosus: A disorder of T lymphocyte signal transduction. *Ann. NY Acad. Sci.*, **968**, 96–105.

Kammer, G.M., Khan, I.U., Kammer, J.A., Olorenshaw, I., and Mathis, D. (1996). Deficient type I protein kinase A isozyme activity in systemic lupus erythematosus T lymphocytes: II. Abnormal isozyme kinetics. *J. Immunol.*, **157**, 2690–2698.

Kammer, G.M., Khan, I.U., and Malemud, C.J. (1994). Deficient type I protein kinase A isozyme activity in systemic lupus erythematosus T lymphocytes. *J. Clin. Invest.*, **94**, 422–430.

Kammer, G.M., Perl, A., Richardson, B.C., and Tsokos, G.C. (2002). Abnormal T cell signal transduction in systemic lupus erythematosus. *Arthritis Rheum.*, **46**, 1139–1154.

Krishnan, S., Nambiar, M.P., Warke, V.G., Fisher, C.U., Mitchell, J., Delaney, N., and Tsokos, G.C. (2004). Alterations in lipid raft composition and dynamics contribute to abnormal T cell responses in systemic lupus erythematosus. *J. Immunol.*, **172**, 7821–7831.

Krishnan, S., Warke, V.G., Nambiar, M.P., Tsokos, G.C., and Farber, D.L. (2003). The FcRγ subunit and Syk kinase replace CD3ζ and ZAP-70 in the TCR/CD3 signaling complex of human effector CD4 T cells. *J. Immunol.*, **170**, 4189–4195.

Kyttaris, V.C., Juang, Y.T., Tenbrock, K., Weinstein, A., and Tsokos, G.C. (2004). Cyclic adenosine 5′-monophosphate response element modulator is responsible for the decreased expression of c-fos and activator protein-1 binding in T cells from patients with systemic lupus erythematosus. *J. Immunol.*, **173**, 3557–3563.

Laxminarayana, D., Khan, I.U., and Kammer, G. (2002). Transcript mutations of the alpha regulatory subunit of protein kinase A and up-regulation of the RNA-editing gene transcript in lupus T lymphocytes. *Lancet*, **360**, 842–849.

Laxminarayana, D., Khan, I.U., Mishra, N., Olorenshaw, I., Tasken, K., and Kammer, G.M. (1999). Diminished levels of protein kinase A RI alpha and RI beta transcripts and proteins in systemic lupus erythematosus T lymphocytes. *J. Immunol.*, **162**, 5639–5648.

Linker-Israeli, M., Bakke, A.C., Kitridou, R.C., Gendler, S., Gillis, S., and Horwitz, D.A. (1983). Defective production of interleukin 1 and interleukin 2 in patients with systemic lupus erythematosus (SLE). *J. Immunol.*, **130**, 2651–2655.

Liossis, S.N., Ding, D.Z., Dennis, G.J., and Tsokos, G.C. (1998a). Altered pattern of TCR/CD3-mediated protein-tyrosyl phosphorylation in T cells from patients with systemic lupus erythematosus. Deficient expression of the T-cell receptor zeta chain. *J. Clin. Invest.*, **101**, 1448–1457.

Liossis, S.N., Hoffman, R.W., and Tsokos, G.C. (1998b). Abnormal early TCR/CD3-mediated signaling events of a snRNP-autoreactive lupus T cell clone. *Clin. Immunol. Immunopathol.*, **88**, 305–310.

Mandler, R., Birch, R.E., Polmar, S.H., Kammer, G.M., and Rudolph, S.A. (1982). Abnormal adenosine-induced immunosuppression and cAMP metabolism in T lymphocytes of patients with systemic lupus erythematosus. *Proc. Natl. Acad. Sci USA*, **79**, 7542–7546.

Mayr, B. and Montminy, M. (2001). Transcriptional regulation by the phosphorylation-dependent factor CREB. *Nat. Rev. Mol. Cell Biol.*, **2**, 599–609.

Mishra, N., Khan, I.U., Tsokos, G.C., and Kammer, G.M. (2000). Association of deficient type II protein kinase A activity with aberrant nuclear translocation of the RII{beta} subunit in systemic lupus erythematosus T lymphocytes. *J. Immunol.*, **165**, 2830–2840.

Nagy, G., Barcza, M., Gonchoroff, N., Phillips, P.E., and Perl, A. (2004). Nitric oxide-dependent mitochondrial biogenesis generates Ca2+ signaling profile of lupus T cells. *J. Immunol.*, **173**, 3676–3683.

Nambiar, M.P., Enyedy, E.J., Fisher, C.U., Krishnan, S., Warke, V.G., Gilliland, W.R., Oglesby, R.J., and Tsokos, G.C. (2002a). Abnormal expression of various molecular forms and distribution of T cell receptor zeta chain in patients with systemic lupus erythematosus. *Arthritis Rheum.*, **46**, 163–174.

Nambiar, M.P., Enyedy, E.J., Warke, V.G., Krishnan, S., Dennis, G., Kammer, G.M., and Tsokos, G.C. (2001). Polymorphisms/mutations of TCR-zeta-chain promoter and 3' untranslated region and selective expression of TCR zeta-chain with an alternatively spliced 3' untranslated region in patients with systemic lupus erythematosus. *J. Autoimmun.*, **16**, 133–142.

Nambiar, M.P., Fisher, C., Enyedy, E., Warke, V., Kumar, A., and Tsokos, G.C. (2002b). Oxidative stress is involved in the heat stress-induced downregulation of TCR zeta chain expression and TCR/CD3-mediated [Ca(2+)](i) response in human T-lymphocytes. *Cell. Immunol.*, **215**, 151–161.

Nambiar, M.P., Fisher, C.U., Kumar, A., Tsokos, C.G., Warke, V.G., and Tsokos, G.C. (2003a). Forced expression of the Fc receptor γ chain renders human T cells hyperresponsive to antigen stimulation. *J. Immunol.*, **170**, 2871–2876.

Nambiar, M.P., Fisher, C.U., Warke, V.G., Krishnan, S., Mitchell, J.P., Delaney, N., and Tsokos, G.C. (2003b). Reconstitution of deficient T cell receptor zeta chain restores T cell signaling and augments T cell receptor/CD3-induced interleukin-2 production in patients with systemic lupus erythematosus. *Arthritis Rheum.*, **48**, 1948–1955.

Nambiar, M.P., Mitchell, J.P., Ceruti, M.A., Malloy, M.A., and Tsokos, G.C. (2003c). Prevalence of T cell receptor zeta chain deficiency in systemic lupus erythematosus. *Lupus*, **12**, 46–51.

Nelson, B.H. (2004). IL-2, regulatory T cells, and tolerance. *J. Immunol.*, **172**, 3983–3988.

Powell, J.D., Ragheb, J.A., Kitagawa-Sakakida, S., and Schwartz, R.H. (1998). Molecular regulation of interleukin-2 expression by CD28 co-stimulation and anergy. *Immunol. Rev.*, **165**, 287–300.

Rellahan, B.L., Jensen, J.P., Howcroft, T.K., Singer, D.S., Bonvini, E., and Weissman, A.M. (1998). Elf-1 regulates basal expression from the T cell antigen receptor {zeta}-chain gene promoter. *J. Immunol.*, **160**, 2794–2801.

Simons, K. and Ikonen, E. (1997). Functional rafts in cell membranes. *Nature*, **387**, 569–572.

Skalhegg, B.S. and Tasken, K. (2000). Specificity in the cAMP/PKA signaling pathway. Differential expression, regulation, and subcellular localization of subunits of PKA. *Front. Biosci.*, **5**, D678–D693.

Solomou, E.E., Juang, Y.T., Gourley, M.F., Kammer, G.M., and Tsokos, G.C. (2001). Molecular basis of deficient IL-2 production in T cells from patients with systemic lupus erythematosus. *J. Immunol.*, **166**, 4216–4222.

Takeuchi, T., Tsuzaka, K., Pang, M., Amano, K., Koide, J., and Abe, T. (1998). TCR zeta chain lacking exon 7 in two patients with systemic lupus erythematosus. *Int. Immunol.*, **10**, 911–921.

Tenbrock, K., Juang, Y.-T., Tolnay, M., and Tsokos, G.C. (2003). The cAMP response element modulator suppresses IL-2 production in T cells by a chromatin-dependent mechanism. *J. Immunol.*, **170**, 2971–2976.

Torgersen, K.M., Vang, T., Abrahamsen, H., Yaqub, S., and Tasken, K. (2002). Molecular mechanisms for protein kinase A-mediated modulation of immune function. *Cell. Signal.*, **14**, 1–9.

Tsokos, G.C. (1999). Overview of cellular immune function in systemic lupus erythematosus. In R.G. Lahita (ed), *Systemic Lupus Erythematosus*. Academic Press, New York. pp. 17–54.

Valitutti, S., Muller, S., Salio, M., and Lanzavecchia, A. (1997). Degradation of T cell receptor (TCR)-CD3-zeta complexes after antigenic stimulation. *J. Exp. Med.*, **185**, 1859–1864.

Vassilopoulos, D., Kovacs, B., and Tsokos, G.C. (1995). TCR/CD3 complex-mediated signal transduction pathway in T cells and cell lines from patients with systemic lupus erythematosus. *J. Immunol.*, **155**, 2269–2281.

Wong, H.K., Kammer, G.M., Dennis, G., and Tsokos, G.C. (1999). Abnormal NF-kappaB activity in T lymphocytes from patients with systemic lupus erythematosus is associated with decreased p65-relA protein expression. *J. Immunol.*, **163**, 1682–1689.

20

Accumulation of Self-Antigens in Systemic Lupus Erythematosus

Koji Yasutomo

1. Introduction

The immune system has evolved to combat infectious organisms. During the evolution of immune systems, T and B cells with antigen-specific receptors emerged and an acquired immune system was established (Germain, 2001). During the early stage of development, both types of lymphocytes carry out repertoire selection, a mechanism that allows cells to distinguish self from non-self antigens (Sebzda et al., 1999; Germain, 2001). For instance, immature T cells expressing high-affinity receptors for self-antigens presented by major histocompatibility complex (MHC) molecules are deleted in the thymus (negative selection) (Robey and Fowlkes, 1994). In contrast, immature T cells with low-affinity receptors for self-peptides presented by MHC molecules can develop further (positive selection) (Robey and Fowlkes, 1994). These two selection steps are critical to maintaining a strong T cell response directed against non-self antigens while avoiding self-reactivity.

Defective regulation of self and non-self discrimination (thymic selection) may cause autoimmune diseases, generally classified as either systemic or organ-specific. The systemic autoimmune disease, systemic lupus erythematosus (SLE), is caused by aberrant autoreactive T and B cells (Davidson and Diamond, 2001; Yasutomo, 2003). This is clearly supported by clinical laboratory findings such as high titer of a variety of serum autoantibodies, including anti-dsDNA and anti-Sm antibodies in SLE patients (Davidson and Diamond, 2001). Although the contribution of such antinuclear autoantibodies to the progression of SLE is unclear, the level of serum anti-dsDNA antibody is well correlated with disease activity (Davidson and Diamond, 2001). In addition, although SLE is controlled by treatment with immunosuppressive drugs including steroids or cyclosporine A,

---

Koji Yasutomo • Department of Immunology and Parasitology, Institute of Health Biosciences, The University of Tokushima Graduate School, 3-18-15 Kuramoto, Tokushima 770-8503, Japan.

Molecular Autoimmunity: In commemoration of the 100th anniversary of the first description of human autoimmune disease, edited by Moncef Zouali. Springer Science+Business Media, Inc., New York, 2005.

controlling the disease progression of SLE without the side effects of such drugs is difficult. For these reasons, many studies have tried to identify candidate genes that are responsible for the progression of SLE. For example, mouse models have been established by targeting genes that are associated with cell death or lymphocyte activation threshold. As a result, several genes have been discovered that are related to the development of an SLE-like syndrome in mice (Shull *et al.*, 1992; Tivol *et al.*, 1995; Lachmann, 1996; O'Keefe *et al.*, 1996; Taylor *et al.*, 1996; Botto *et al.*, 1998; Bickerstaff *et al.*, 1999; Nishimura *et al.*, 1999; Wang *et al.*, 1999; Bachmaier *et al.*, 2000; Balomenos *et al.*, 2000; Chiang *et al.*, 2000; Majeti *et al.*, 2000; Napirei *et al.*, 2000; Brunkow *et al.*, 2001; Chui *et al.*, 2001; Demetriou *et al.*, 2001; Le *et al.*, 2001; Scott *et al.*, 2001; Yasutomo *et al.*, 2001). However, genes responsible for causing typical human SLE have not been identified.

Therefore, the search for genes that cause SLE continues. For example, a candidate locus on chromosome 1q was identified by linkage study (Tsao, 2000). Several hundreds to 1000 sibpairs or transmission disequilibrium testing may be required to document linkage at this position. And future analysis of a separate group of lupus families will be necessary to confirm the existence of a locus at chromosome 1q, with unusually strong contributions to disease. The chromosomal intervals initially mapped in a linkage analysis of murine crosses or affected sibpairs are usually 10–20 centiMorgans, and therefore may contain up to 500 genes, most of which are unknown. Since in that locus, tumor necrosis factor receptor 2 (TNFR2), complement component C1q, Fc gamma receptors, T cell receptor (TCR) zeta chain, interleukin-10, poly(ADP-ribose) polymerase, and HRES-1 are located, it would be necessary to check the role of each gene in susceptibility to or progression of SLE. The recent development of genomewide scanning techniques should facilitate the discovery of candidate genes affecting SLE.

Recent studies, including ours, have provided evidence that most mouse $CD4^+$ or $CD8^+$ T cells have a potential to respond to self-antigens (Dorfman *et al.*, 2000). This new finding suggests that human T cells also have the same ability. Accordingly, we, and others, found that the *in vivo* persistence of self-antigens can induce persistent activation of mature T cells, resulting in the development of human SLE or SLE-like syndrome in mice (Kirschfink *et al.*, 1993; Botto *et al.*, 1998; Bickerstaff *et al.*, 1999; Napirei *et al.*, 2000; Yasutomo *et al.*, 2001). We originally found two DNase1-deficient SLE patients who were defective in clearance of nucleosomal antigens because of low DNase1 activity. These two patients had very high frequencies of T and B cells against nucleosomal antigens and high serum titers of anti-dsDNA antibodies (Yasutomo *et al.*, 2001). These recent findings in basic and clinical immunology would lead us to reconsider the importance of antigen clearance and persistence as a cause of SLE in addition to the regulation of T cells by molecules expressed on and in T cells.

2. T Cell in Human Lupus

The immune system responds to antigens that engage specialized receptors present on the T cell surface (TCR) (Yablonski and Weiss, 2001). This engagement induces various intracellular biochemical interactions that culminate in the transmission of an extracellular signal that can trigger cell activation, proliferation, secretion of soluble mediators, phenotypic changes, acquisition of effector functions, anergy, and programmed cell death (Lanzavecchia et al., 1999; Germain, 2001; Yablonski and Weiss, 2001). The TCR stimulation outcome can vary considerably depending on the degree and extent of integration of other membrane receptor–initiated specific accessory signals. In particular, CD28–CD80/86 and CD40–CD40 ligand interactions are important for complete T cell and B cell activation, respectively. Because TCR-mediated signaling events direct these diverse but equally important outcomes, it is likely that the diverse cellular aberrations described in lupus patients reflect signaling defect(s) responsible for the disease pathogenesis. Such defects might be due to expression of a signaling molecule(s) stemming from either a defective gene(s) or defective gene expression regulation.

Tsokos et al. studied TCR/CD3-mediated signaling events to identify lupus T cell primary abnormalities (Vassilopoulos et al., 1995; Liossis et al., 1998). Using anti-human CD3 monoclonal antibody (mAb), they showed that lupus T cells exhibited significantly increased Ca^{2+} responses compared with the T cell responses from patients with non-SLE autoimmune disease (Vassilopoulos et al., 1995). In particular, anti-CD3 mAb enhanced Ca^{2+} fluxes in lupus but not in nonlupus T cell clones. Furthermore, they also demonstrated that lupus TCR/CD3 signaling is defective in tyrosine phosphorylation, which is necessary for proper T cell activation (Liossis et al., 1998). Specifically, T cell stimulation in SLE patients results in significantly enhanced production of tyrosine-phosphorylated cellular proteins and consists of an early elevated response as well as a steep decrease to baseline phosphorylation levels. In contrast, protein tyrosine phosphorylation in normal T cells gradually increases during the same time period (Liossis et al., 1998). Taken together, these results suggest that proximal signaling by the TCR of lupus T cells is more sensitive in terms of peak intensity or TCR signaling duration, which would elicit abnormally high T cell activation that could cause tissue damage.

T cell activation is regulated by a number of cell surface and intracellular molecules as well as other cell types. Filaci et al. (2001) showed that the inhibitory function in CD8-positive cells in lupus patients is low, which would contribute to the development of lupus (Filaci et al., 2001). Recent suppressor T cell studies revealed that $CD4^+CD25^+$ T cells play a pivotal role in the development of autoimmune diseases in mice (Shevach et al., 2001). Several groups reported that human blood also possesses $CD4^+CD25^+$ T cells that have suppressive function (Sakaguchi, 2004). So far, the contribution of $CD4^+CD25^+$ T cells to lupus has not been reported, but much attention will be paid to the possible role of this unique population in the defective T cell response of lupus patients.

3. Antigen Clearance and Autoimmunity

3.1. DNASE1-Deficient Patients: Gene Mutation and Clinical Features

Naiperi et al. (2000) reported that DNASE1-deficient mice develop an SLE-like syndrome even if the DNASE1 mutation is heterozygous (Napirei et al., 2000). Like humans with SLE, these mice possess very high titers of anti-dsDNA antibodies and develop severe glomerulonephritis. When the DNASE1-coding sequence and exon–intron boundaries were scanned for mutations, an A to G transversion in exon 2 at position 172 of the cDNA sequence (replacing a lysine (AAG) with a stop signal (TAG) at residue 5 (K5X)) was found in two SLE patients. These patients (both girls), one 13 (patient 1) and the other 17 (patient 2) years old, were diagnosed for SLE on the basis of both clinical features and high serum titers of anti-dsDNA and SS-A antibodies.

3.2. DNASE1-Deficient Patients: Laboratory Findings

In these two patients the gene mutation is heterozygous, and their transformed B cells had a significantly lower level of DNASE1 activity, showing that this mutation reduces the total activity of this enzyme. Previous studies demonstrated low serum DNASE1 activity in SLE patients (Chitrabamrung *et al.*, 1981; Tew *et al.*, 2001), but the DNASE1 gene was not examined. Our unpublished analysis indicates that many SLE patients without any DNASE1 mutation also have low levels of serum DNASE1 activity.

3.3. DNASE1-Deficient Patients: Effect on Autoreactivity

Because of the importance of knowing whether low DNASE1 activity has a specific effect on any of the defining pathological characteristics of SLE, we examined antibody titers against nucleosomal antigens and dsDNA in the patients with DNASE1 mutation. The titer of IgG against nucleosomal antigens was 7–8 times greater in patients with the DNASE1 mutation than in other SLE patients, and 70–80 times greater than in normal controls. The IgG titer against dsDNA in two patients was also higher than that in other SLE patients. Furthermore, the precursor frequencies of T and B cells responsive to nucleosomal antigens were about 10 times greater in the patients with DNASE1 mutation than in other SLE patients, and 100 times greater than in normal controls. These results strongly suggest that the low activity of DNASE1 in these patients contributes to the expansion of autoreactive T and B cells against nucleosomal antigens.

4. Defective Clearance of Self-Antigens in SLE

4.1. Evidence from Knockout Mice

Aberrant rates of apoptosis and increased levels of free-circulating chromatin have been reported in human lupus (Amoura *et al.*, 1997), and humans with DNASE1 and C1q gene mutations develop SLE (Kirschfink *et al.*, 1993; Yasutomo *et al.*, 2001), suggesting that efficient removal of chromatin or chromatin–protein complex is crucial to prevent SLE. For example, serum amyloid P component (SAP)-deficient mice develop an SLE-like syndrome (Bickerstaff *et al.*, 1999). Two models are likely to explain the connection between SAP deficiency and SLE. First, SAP participates in dissolution of chromatin, activates complement components, C1–C4, and potentiates hepatic clearance through complement receptors. Thus,

SAP deficiency leads to apoptotic debris accumulation, which activates preexisting autoreactive T and B cells. Second, SAP helps binding of C4b to chromatin and these complexes bind to bone marrow stromal cells by the C4b receptor. In the presence of SAP, immature autoreactive B cells acquire tolerance against chromatin through clonal deletion or anergy or both. In SAP deficiency, chromatin cannot be recruited to the bone marrow, which results in autoreactive B cell escape into the periphery and antigen-driven B cell expansion.

Boes *et al.* (2000) established mutant mice that cannot secrete IgM but that are able to express surface IgM and IgD, and to secrete other classes of immunoglobulins (Boes *et al.*, 2000). They crossed these mutant mice with lupus-prone lymphoproliferative (lpr) mice and observed elevated levels of IgG autoantibodies to double-stranded DNA and histones, and increased deposition of immune complexes in the glomeruli. Furthermore, the absence of secreted IgM also resulted in accelerated development of IgG autoantibodies, even in normal mice. The accelerated autoantibody responses in mice deficient in secreted IgM resemble the effect of complement deficiency on the development of autoimmune disease. In humans and mice, deficiencies in the early components of the complement cascade, including C1q, C2, and C4, are associated with a high incidence of SLE (Kirschfink *et al.*, 1993; Botto *et al.*, 1998). The complement promotes the removal of autoantigens and, hence, could reduce the chance that autoreactive B cells are activated. Intravenous injection of syngeneic apoptotic cells into normal mice induces a rapid response against nuclear antigens (Mevorach *et al.*, 1998b), suggesting that the source of autoantigens for the autoantibody response may be the apoptotic cells (Casciola-Rosen *et al.*, 1994). C1q can bind apoptotic blebs directly, and the activation of complement is required for the clearance of apoptotic cells by macrophages (Mevorach *et al.*, 1998a), suggesting a critical role of complement in the clearance of apoptotic cells.

Mer (a member of the Axl/Mer/Tyro3 receptor tyrosine kinase family)-deficient mice also possess high serum titers of anti-DNA antibodies, although this chapter does not mention whether they develop a pathological autoimmune response (Scott *et al.*, 2001). Since Mer is required for the engulfment and efficient clearance of apoptotic cells, the defective clearance of chromatin in this deficient mouse may induce overt activation of autoreactive lymphocytes.

Hanayama *et al.* (2004) found that milk fat globule epidermal growth factor 8 (MFG-E8)–deficient mice develop glomerulonephritis as a result of autoantibody production (Hanayama *et al.*, 2004). MFG-E8 binds to apoptotic cells by recognizing phosphatidylserine, and this binding enhances the engulfment of apoptotic cells by macrophages. This finding demonstrates that MFG-E8 has a critical role in removing apoptotic cells and thereby in the accumulation of self-antigens.

Taken together, the defective clearance of nucleosomal antigens can be a causative factor for the generation of anti-dsDNA antibodies, resulting in tissue damage. Since a significant number of autoreactive T cells against self-ligands including nucleosomal antigens are present in the normal body, I would like to postulate that the degree of T cell response against self-ligands is dependent on the duration or amount of TCR signaling as well as on the affinity of interaction between the TCRs and their ligands (Figure 20.1). Normally, because the TCR

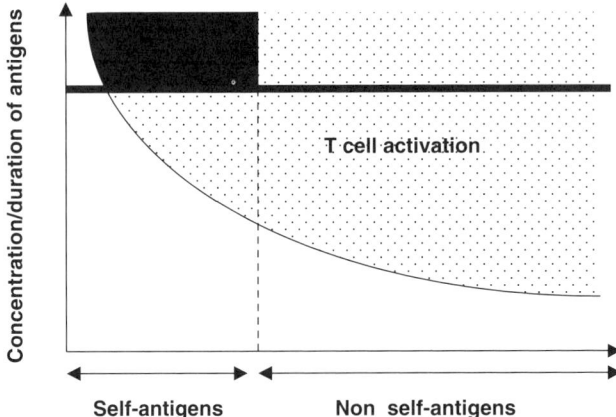

Figure 20.1. Accumulation of self-antigens triggers T cell autoreactivity. Mature T cells generally do not respond to self-antigens while proliferating against non-self antigens. However, defective clearance accumulation of self-antigens would allow mature T cells to reach the threshold for activation (black area).

affinity is generally low and self-ligands are rapidly cleared from the body, the duration or amount of TCR signaling does not reach the threshold for activation of sufficient T cells to cause self-damage. However, when accumulation of self-ligands is abnormal, T cells are exposed to them for enough time or in amounts sufficient to induce the differentiation of pathological T cells.

4.2. Mechanisms of Accumulation of Self-Antigens in SLE

The issue of why lymphocytes reactive against nucleosomal antigens generally inside the cells are generated and induce autoimmunity deserves consideration. The role of anti-dsDNA antibodies in the pathophysiology of SLE remains unclear. Over the past decades, the antinuclear specificities of disease-associated autoantibodies have lured investigators to one of the likely sources of autoreactive stimulation in SLE. There are two possible explanations. The first is that cells (either inadvertently damaged [ultraviolet burns] or intentionally targeted [programmed cell death]) die, the condensed chromatin is broken down into nucleosomes, and the nuclear envelope fragments within vesicles filled with nucleosomal components. This regimented process of apoptosis is often distinguished histologically by pyknotic nuclei and vacuolated cytoplasm. Casciola-Rosen *et al.* (1994) demonstrated that in the course of death by apoptosis, cells translocate the SLE autoantigens from the nuclear compartment to the cell surface, displaying them to the extracellular environment in membranous blebs, called "apoptotic bodies" (Casciola-Rosen *et al.*, 1994). The physiologic display of nuclear antigens in apoptotic bodies offers a plausible mechanism whereby the autoantigens prominent in SLE are exposed to the immune system, possibly stimulating an autoimmune response.

Another explanation provided by DeGiorgio *et al.* (2001) is that a subset of anti-DNA antibodies cross-reacts with the *N*-methyl-D-aspartate (NMDA) subtype of glutamate receptors and induces neuronal cell injury (more generally, that anti-DNA antibody cross-reacts with self-antigens thereby eliciting tissue damage). However, the authors do not directly show that this autoantibody is really responsible for the development of the central nervous damage seen in SLE patients. It would also be very important to identify target antigens expressed in other organs that are recognized by anti-dsDNA antibodies.

A recent study showed that self-IgG2a-reactive B cells are activated by the recognition of immune complexes composed of IgG2a and nucleosomes by cell surface IgM and toll-like receptor (TLR) 9 present on B cells (Leadbetter *et al.*, 2002, 2003). Autoreactive B cells are present in the lymphoid tissues of healthy individuals, but normally remain inactive. Over the decades, it has been demonstrated that to become fully activated, naive B cells must receive signals through Ig receptors and accessory receptors from helper T cells that recognize the same antigen (Fagarasan and Honjo, 2000). However, recent reports have indicated that various constituents of microbes can induce B cells to produce antibody in the absence of such helper T cells (Poeck *et al.*, 2004). For example, lipopolysaccharide (LPS), a major constituent of the cell wall of Gram-negative bacteria, is recognized by a surface receptor present on all B cells, TLR4 (Poltorak *et al.*, 1998). The simultaneous recognition of the LPS-containing bacteria by antigen receptors and TLR4 induces a synergistic signal, and triggers proliferation and production of antibodies by bacteria-specific B cells without T cell help (Poeck *et al.*, 2004). Another toll-like receptor, TLR9, usually serves as a pathogen sensor by detecting unmethylated CpG base pairs that are more common in bacterial than in mammalian DNA, although unmethylated CpG base pairs are found in certain parts of mammalian genes. Leadbetter *et al.* (2002) have recently reported that immune IgG–nucleosome complexes could activate self-IgG-specific B cells through both the antigen receptor and TLR9, which indicates that TLR9 signaling facilitates the production of rheumatoid factors (Leadbetter *et al.*, 2002). Immune complexes composed of nucleosomes or other self-antigens might have the same effect. Therefore, autoreactive B cells, which recognize nucleosomes directly or indirectly with the help of TLR9 signaling, may not need T cell help to proliferate and produce a variety of antibodies.

Regarding the role of TLR in the induction of autoimmune diseases, Eriksson *et al.* (2003) reported that dendritic cell activation through both CD40 and TLR4, but not CD40 alone, was needed to induce autoimmune disease in mouse recipients of dendritic cells pulsed with self-antigens. This result indicates that the activation of antigen-presenting cells through TLR plays a central role in the induction of autoimmune diseases. On the basis of these reports, we propose the following model to explain the role of antinucleosome-specific lymphocytes in the induction of SLE. Nucleosomal antigen accumulation due to the defect of DNASE1 increases the formation of immune complexes with methylated DNA and a variety of self-antigens. These immune complexes activate antigen-presenting cells (e.g., dendritic cells and B cells), which, in turn, greatly increase the sensitivity of autoreactive T cells to self-antigens as well as directly stimulate B cell production of autoantibodies.

4.3. Clearance of Self-Antigens as a Therapeutic Strategy

Macanovic *et al.* (1966) have reported that recombinant DNASE1 reduces the autoimmune response of lupus-prone mice. However, in clinical trials of recombinant DNASE1, the SLE symptoms and laboratory data of patients with SLE did not significantly improve (Davis *et al.*, 1999). According to our studies, DNASE1 mutation could only be detected in 2 out of 100 SLE patients. Thus, it is important to determine whether treatment of recombinant DNASE1 is effective for SLE patients with DNASE1 mutation. Furthermore, in advanced SLE, T and B cells have many nucleosome specificities, whereas, initially, lymphocytes with only a single nucleosomal antigen specificity trigger the disease. Therefore, the treatment using recombinant DNASE1 may have a benefit only at a very early stage of SLE.

Regarding the therapeutic potential of nucleosomal antigens, Lu *et al.* (1999) reported that CD4$^+$ T cells from lupus patients responded strongly to certain histone peptides (H2B10-33, H416-39, H471-94, H391-105, H2A34-48, and H449-63). These same peptides overlap with major epitopes for helper CD4$^+$ T cells that induce anti-DNA autoantibodies and nephritis in lupus mice. The altered peptide ligands can induce anergy in T cells (Sloan-Lancaster *et al.*, 1993, 1994; Ryan and Evavold, 1998). On the basis of this finding, Kaliyaperumal *et al.* (1999) tried to inhibit murine lupus using altered histone peptides and succeeded in partially suppressing SLE-like syndrome (Kaliyaperumal *et al.*, 1999). The ability of each peptide associated with human SLE to affect T cell proliferation or cytokine secretion profiles is different, suggesting that each epitope plays a specific role in autoimmune T cell response induction or suppression. Thus, caution should be exercised when applying this technique in the clinic because it would be dangerous to inhibit T cell responses that are inhibiting SLE. Furthermore, because human HLA differs from person to person, it is very difficult to identify the altered ligands in each SLE patient. To overcome this problem, it is essential to establish therapeutic strategies that apply to all SLE patients.

References

Amoura, Z., Piette, J.C., Chabre, H., Cacoub, P., Papo, T., Wechsler, B., Bach, J.F., and Koutouzov, S. (1997). Circulating plasma levels of nucleosomes in patients with systemic lupus erythematosus: Correlation with serum antinucleosome antibody titers and absence of clear association with disease activity. *Arthritis Rheum.*, **40**, 2217–2225.

Bachmaier, K., Krawczyk, C., Kozieradzki, I., Kong, Y.Y., Sasaki, T., Oliveira-dos-Santos, A., Mariathasan, S., Bouchard, D., Wakeham, A., Itie, A., Le, J., Ohashi, P.S., Sarosi, I., Nishina, H., Lipkowitz, S., and Penninger, J.M. (2000). Negative regulation of lymphocyte activation and autoimmunity by the molecular adaptor Cbl-b. *Nature*, **403**, 211–216.

Balomenos, D., Martin-Caballero, J., Garcia, M.I., Prieto, I., Flores, J.M., Serrano, M., and Martinez, A.C. (2000). The cell cycle inhibitor p21 controls T-cell proliferation and sex-linked lupus development. *Nat. Med.*, **6**, 171–176.

Bickerstaff, M.C., Botto, M., Hutchinson, W.L., Herbert, J., Tennent, G.A., Bybee, A., Mitchell, D.A., Cook, H.T., Butler, P.J., Walport, M.J., and Pepys, M.B. (1999). Serum amyloid P component controls chromatin degradation and prevents antinuclear autoimmunity. *Nat. Med.*, **5**, 694–697.

Boes, M., Schmidt, T., Linkemann, K., Beaudette, B.C., Marshak-Rothstein, A., and Chen, J. (2000). Accelerated development of IgG autoantibodies and autoimmune disease in the absence of secreted IgM. *Proc. Natl. Acad. Sci. USA*, **97**, 1184–1189.

Botto, M., Dell'Agnola, C., Bygrave, A.E., Thompson, E.M., Cook, H.T., Petry, F., Loos, M., Pandolfi, P.P., and Walport, M.J. (1998). Homozygous C1q deficiency causes glomerulonephritis associated with multiple apoptotic bodies. *Nat. Genet.*, **19**, 56–59.

Brunkow, M.E., Jeffery, E.W., Hjerrild, K.A., Paeper, B., Clark, L.B., Yasayko, S.A., Wilkinson, J.E., Galas, D., Ziegler, S.F., and Ramsdell, F. (2001). Disruption of a new forkhead/winged-helix protein, scurfin, results in the fatal lymphoproliferative disorder of the scurfy mouse. *Nat. Genet.*, **27**, 68–73.

Casciola-Rosen, L.A., Anhalt, G., and Rosen, A. (1994). Autoantigens targeted in systemic lupus erythematosus are clustered in two populations of surface structures on apoptotic keratinocytes. *J. Exp. Med.*, **179**, 1317–1330.

Chiang, Y.J., Kole, H.K., Brown, K., Naramura, M., Fukuhara, S., Hu, R.J., Jang, I.K., Gutkind, J.S., Shevach, E., and Gu, H. (2000). Cbl-b regulates the CD28 dependence of T-cell activation. *Nature*, **403**, 216–220.

Chitrabamrung, S., Rubin, R.L. and Tan, E.M. (1981). Serum deoxyribonuclease I and clinical activity in systemic lupus erythematosus. *Rheumatol. Int.*, **1**, 55–60.

Chui, D., Sellakumar, G., Green, R., Sutton-Smith, M., McQuistan, T., Marek, K., Morris, H., Dell, A., and Marth, J. (2001). Genetic remodeling of protein glycosylation in vivo induces autoimmune disease. *Proc. Natl. Acad. Sci. USA*, **98**, 1142–1147.

Davidson, A. and Diamond, B. (2001). Autoimmune diseases. *N. Engl. J. Med.*, **345**, 340–350.

Davis, J.C., Jr., Manzi, S., Yarboro, C., Rairie, J., McInnes, I., Averthelyi, D., Sinicropi, D., Hale, V.G., Balow, J., Austin, H., Boumpas, D.T., and Klippel, J.H. (1999). Recombinant human Dnase I (rhDNase) in patients with lupus nephritis. *Lupus*, **8**, 68–76.

DeGiorgio, L.A., Konstantinov, K.N., Lee, S.C., Hardin, J.A., Volpe, B.T., and Diamond, B. (2001). A subset of lupus anti-DNA antibodies cross-reacts with the NR2 glutamate receptor in systemic lupus erythematosus. *Nat. Med.*, **7**, 1189–1193.

Demetriou, M., Granovsky, M., Quaggin, S., and Dennis, J.W. (2001). Negative regulation of T-cell activation and autoimmunity by Mgat5 N-glycosylation. *Nature*, **409**, 733–739.

Dorfman, J.R., Stefanova, I., Yasutomo, K., and Germain, R.N. (2000). CD4+ T cell survival is not directly linked to self-MHC-induced TCR signaling. *Nat. Immunol.*, **1**, 329–335.

Fagarasan, S. and Honjo, T. (2000). T-Independent immune response: New aspects of B cell biology. *Science*, **290**, 89–92.

Filaci, G., Bacilieri, S., Fravega, M., Monetti, M., Contini, P., Ghio, M., Setti, M., Puppo, F., and Indiveri, F. (2001). Impairment of CD8+ T suppressor cell function in patients with active systemic lupus erythematosus. *J. Immunol.*, **166**, 6452–6457.

Germain, R.N. (2001). The T cell receptor for antigen: Signaling and ligand discrimination. *J. Biol. Chem.*, **276**, 35223–35226.

Hanayama, R., Tanaka, M., Miyasaka, K., Aozasa, K., Koike, M., Uchiyma, Y., and Nagata, S. (2004). Autoimmune disease and impaired uptake of apoptotic cells in MFG-E8-deficient mice. *Science*, **304**, 1147–1150.

Kaliyaperumal, A., Michaels, M.A., and Datta, S.K. (1999). Antigen-specific therapy of murine lupus nephritis using nucleosomal peptides: Tolerance spreading impairs pathogenic function of autoimmune T and B cells. *J. Immunol.*, **162**, 5775–5783.

Kirschfink, M., Petry, F., Khirwadkar, K., Wigand, R., Kaltwasser, J.P., and Loos, M. (1993). Complete functional C1q deficiency associated with systemic lupus erythematosus (SLE). *Clin. Exp. Immunol.*, 94, 267–272.

Lachmann, P.J. (1996). The in vivo destruction of antigen – A tool for probing and modulating an autoimmune response. *Clin. Exp. Immunol.*, **106**, 187–189.

Lanzavecchia, A., Lezzi, G., and Viola, A. (1999). From TCR engagement to T cell activation: A kinetic view of T cell behavior. *Cell*, **96**, 1–4.

Le, L.Q., Kabarowski, J.H., Weng, Z., Satterthwaite, A.B., Harvill, E.T., Jensen, E.R., Miller, J.F., and Witte, O.N. (2001). Mice lacking the orphan G protein-coupled receptor G2A develop a late-onset autoimmune syndrome. *Immunity*, **14**, 561–571.

Leadbetter, E.A., Rifkin, I.R., Hohlbaum, A.M., Beaudette, B.C., Shlomchik, M.J., and Marshak-Rothstein, A. (2002). Chromatin-IgG complexes activate B cells by dual engagement of IgM and toll-like receptors. *Nature*, **416**, 603–607.

Leadbetter, E.A., Rifkin, I.R., and Marshak-Rothstein, A. (2003). Toll-like receptors and activation of autoreactive B cells. *Curr. Dir. Autoimmun.*, **6**, 105–122.

Liossis, S.N., Ding, X.Z., Dennis, G.J., and Tsokos, G.C. (1998). Altered pattern of TCR/CD3-mediated protein-tyrosyl phosphorylation in T cells from patients with systemic lupus erythematosus. Deficient expression of the T cell receptor zeta chain. *J. Clin. Invest.*, **101**, 1448–1457.

Majeti, R., Xu, Z., Parslow, T.G., Olson, J.L., Daikh, D.I., Killeen, N., and Weiss, A. (2000). An inactivating point mutation in the inhibitory wedge of CD45 causes lymphoproliferation and autoimmunity. *Cell*, **103**, 1059–1070.

Mevorach, D., Mascarenhas, J.O., Gershov, D., and Elkon, K.B. (1998a). Complement-dependent clearance of apoptotic cells by human macrophages. *J. Exp. Med.*, **188**, 2313–2320.

Mevorach, D., Zhou, J.L., Song, X., and Elkon, K.B. (1998b). Systemic exposure to irradiated apoptotic cells induces autoantibody production. *J. Exp. Med.*, **188**, 387–392.

Napirei, M., Karsunky, H., Zevnik, B., Stephan, H., Mannherz, H.G., and Moroy, T. (2000). Features of systemic lupus erythematosus in Dnase1-deficient mice. *Nat. Genet.*, **25**, 177–181.

Nishimura, H., Nose, M., Hiai, H., Minato, N., and Honjo, T. (1999). Development of lupus-like autoimmune diseases by disruption of the PD-1 gene encoding an ITIM motif-carrying immunoreceptor. *Immunity*, **11**, 141–151.

O'Keefe, T.L., Williams, G.T., Davies, S.L., and Neuberger, M.S. (1996). Hyperresponsive B cells in CD22-deficient mice. *Science*, **274**, 798–801.

Poeck, H., Wagner, M., Battiany, J., Rothenfusser, S., Wellisch, D., Hornung, V., Jahrsdorfer, B., Giese, T., Endres, S., and Hartmann, G. (2004). Plasmacytoid dendritic cells, antigen, and CpG-C license human B cells for plasma cell differentiation and immunoglobulin production in the absence of T-cell help. *Blood*, **103**, 3058–3064.

Poltorak, A., He, X., Smirnova, I., Liu, M.Y., Van Huffel, C., Du, X., Birdwell, D., Alejos, E., Silva, M., Galanos, C., Freudenberg, M., Ricciardi-Castagnoli, P., Layton, B., and Beutler, B. (1998). Defective LPS signaling in C3H/HeJ and C57BL/10ScCr mice: Mutations in Tlr4 gene. *Science*, **282**, 2085–2088.

Robey, E. and Fowlkes, B.J. (1994). Selective events in T cell development. *Annu. Rev. Immunol.*, **12**, 675–705.

Ryan, K.R. and Evavold, B.D. (1998). Persistence of peptide-induced CD4+ T cell anergy in vitro. *J. Exp. Med.*, **187**, 89–96.

Sakaguchi, S. (2004). Naturally arising CD4+ regulatory T cells for immunologic self-tolerance and negative control of immune responses. *Annu. Rev. Immunol.*, **22**, 531–562.

Scott, R.S., McMahon, E.J., Pop, S.M., Reap, E.A., Caricchio, R., Cohen, P.L., Erp, H.S., and Matsushima, G.K. (2001). Phagocytosis and clearance of apoptotic cells is mediated by MER. *Nature*, **411**, 207–211.

Sebzda, E., Mariathasan, S., Ohteki, T., Jones, R., Bachmann, M.F., and Ohashi, P.S. (1999). Selection of the T cell repertoire. *Annu. Rev. Immunol.*, **17**, 829–874.

Shevach, E.M., McHugh, R.S., Piccirillo, C.A., and Thornton, A.M. (2001). Control of T-cell activation by CD4+CD25+ suppressor T cells. *Immunol. Rev.*, **182**, 58–67.

Shull, M.M., Ormsby, I., Kier, A.B., Pawlowski, S., Diebold, R.J., Yin, M., Alle, R., Sidman, C., Proetzel, G., Calvin, D., and *et al.* (1992). Targeted disruption of the mouse transforming growth factor-beta 1 gene results in multifocal inflammatory disease. *Nature*, **359**, 693–699.

Sloan-Lancaster, J., Evavold, B.D., and Allen, P.M. (1993). Induction of T-cell anergy by altered T-cell-receptor ligand on live antigen-presenting cells. *Nature*, **363**, 156–159.

Sloan-Lancaster, J., Evavold, B.D., and Allen, P.M. (1994). Th2 cell clonal anergy as a consequence of partial activation. *J. Exp. Med.*, **180**, 1195–1205.

Taylor, G.A., Carballo, E., Lee, D.M., Lai, W.S., Thompson, M.J., Patel, D.D., Schenkman, D.I., Gilkeson, G.S., Broxmeyer, H.E., Haynes, B.F., and Blackshear, P.J. (1996). A pathogenetic role for TNF alpha in the syndrome of cachexia, arthritis, and autoimmunity resulting from tristetraprolin (TTP) deficiency. *Immunity*, **4**, 445–454.

Tew, M.B., Johnson, R.W., Reveille, J.D., and Tan, F.K. (2001). A molecular analysis of the low serum deoxyribonuclease activity in lupus patients. *Arthritis Rheum.*, **44**, 2446–2447.

Tivol, E.A., Borriello, F., Schweitzer, A.N., Lynch, W.P., Bluestone, J.A., and Sharpe, A.H. (1995). Loss of CTLA-4 leads to massive lymphoproliferation and fatal multiorgan tissue destruction, revealing a critical negative regulatory role of CTLA-4. *Immunity*, **3**, 541–547.

Tsao, B.P. (2000). Lupus susceptibility genes on human chromosome 1. *Int. Rev. Immunol.*, **19**, 319–334.

Vassilopoulos, D., Kovacs, B., and Tsokos, G.C. (1995). TCR/CD3 complex-mediated signal transduction pathway in T cells and T cell lines from patients with systemic lupus erythematosus. *J. Immunol.*, **155**, 2269–2281.

Wang, J., Zheng, L., Lobito, A., Chan, F.K., Dale, J., Sneller, M., Yao, X., Puck, J.M., Straus, S.E., and Lenardo, M.J. (1999). Inherited human caspase 10 mutations underlie defective lymphocyte and dendritic cell apoptosis in autoimmune lymphoproliferative syndrome type II. *Cell*, **98**, 47–58.

Yablonski, D. and Weiss, A. (2001). Mechanisms of signaling by the hematopoietic-specific adaptor proteins, SLP-76 and LAT and their B cell counterpart, BLNK/SLP-65. *Adv. Immunol.*, **79**, 93–128.

Yasutomo, K. (2003). Pathological lymphocyte activation by defective clearance of self-ligands in systemic lupus erythematosus. *Rheumatology*, **42**, 214–222.

Yasutomo, K., Horiuchi, T., Kagami, S., Tsukamoto, H., Hashimura, C., Urushihara, M., and Kuroda, Y. (2001). Mutation of DNase1 in people with systemic lupus erythematosus. *Nat. Genet.*, **28**, 313–314.

21

B Lymphocyte Depletion Therapy in Autoimmune Disorders: Chasing Trojan Horses

Jonathan C. W. Edwards, Geraldine Cambridge, and Maria J. Leandro

1. Introduction

Since the anti-CD20 therapeutic monoclonal antibody rituximab became available for clinical use in 1997 the effects of B lymphocyte depletion (BLyD) have been explored in a range of autoimmune disorders (Edwards *et al.*, 2002). The results of a formal controlled trial in rheumatoid arthritis (RA) confirm that clinical benefit can be substantial (Edwards *et al.*, 2004). Perhaps more significantly in the long term, the opportunity afforded by BLyD to study mechanisms underlying pathogenesis of many human autoimmune diseases is unique and exciting.

This chapter reviews the rationale for BLyD in the context of changing ideas about the role of B cells in autoimmunity, and explores the emerging concept of autoimmune B cells as immunologic Trojan horses. The potential for clinical application, the logistics employed, and the clinical results obtained with rituximab so far are reviewed. Finally, evolving immunodynamic studies are described, which highlight correlations between clinical state and autoantibody profiles, and which may elucidate mechanisms of action and point to means of optimization of therapeutic benefit and development of further targeted approaches to therapy.

2. Human Autoimmunity: An Abnormality of B Cell Function

The original concept of autoimmunity was based on the presence of antibodies to self: B cell autoreactivity (Landsteiner, 1904). One hundred years later, there is a broad body of evidence for abnormalities of B cell function in a wide

---

Jonathan C. W. Edwards, Geraldine Cambridge, and Maria J. Leandro • University College London Centre for Rheumatology, Arthur Stanley House, London W1T 4NJ, England.

Molecular Autoimmunity: In commemoration of the 100th anniversary of the first description of human autoimmune disease, edited by Moncef Zouali. Springer Science+Business Media, Inc., New York, 2005.

range of human autoimmune diseases. Although T cell responses to autoantigens are documented in a number of murine models of autoimmunity (Busser *et al.*, 2003), evidence for T cell autoreactivities that might be responsible for production of autoantibodies such as rheumatoid factors and antibodies to a range of nuclear antigens in the human is limited. It has often proved difficult to differentiate consistently between T cell responses in autoimmune subjects and normal individuals. Recent studies on removal of regulatory T cell populations further confirm that the presence of T cells with autoreactive potential is a normal phenomenon, not itself sufficient to explain disease (Danke *et al.*, 2004).

The concept that human autoimmune disease relates primarily to an abnormality of B cell function formed the rationale behind the use of BLyD at University College London (UCL) (Edwards *et al.*, 1999). The aim of this chapter is to review the basis for that rationale and to assess whether clinical and immunological responses to BLyD support the hypothesis.

2.1. A Brief Review of Investigation of B and T Cell Autoreactivity in Human Autoantibody-Associated Diseases

Although it is clear that both T and B cells are involved in human autoimmune disease, the particular abnormality of cell function that initiates the process in each disease is not. Identification of such abnormalities should inform the development of therapies capable of modifying disease at a more fundamental level. Unfortunately, critical analysis of evidence with respect to the underlying role of T and B cells in many human diseases is often colored by concepts of disease dynamics based on animal models. These models, although vitally important for understanding basic cellular and molecular mechanisms, have, by nature, different dynamics, determined by experimental manipulation.

In this context, the following is a brief summary of the available evidence "one way or another" for T and B cells in human autoantibody-associated disease. First, the association of specific autoantibodies with disorders such as RA, systemic lupus erythematosus (SLE), myasthenia gravis, and pernicious anemia is well documented. The correlation between autoantibody specificities and clinical conditions or organ distribution within conditions (e.g., anti-Jo-1 antibodies and lung involvement in patients with myositis; anti-DNA antibodies and renal involvement in lupus) suggest a pathogenic role.

Second, corresponding T cell responses to self-antigens are often not consistently found. Anti-IgG Fc T cell responses have been reported in only 0–50% of patients with RA (Lang *et al.*, 1999). T cell responses to ANA-associated antigens such as La (Davies *et al.*, 2002) and topoisomerase-1 (Kuwana *et al.*, 2001) were no greater in patients with autoantibodies to these antigens than in controls. T cell reactivity to acetylcholine receptor does not seem to exist outside the thymus in myasthenia—there is no T cell–mediated myositis. T cell responses to nonprotein antigens such as DNA and cardiolipin in SLE and peanut agglutinin receptor in sarcoidosis (Pilatte *et al.*, 1990) would not generally be predicted.

The production of anti-DNA antibodies in SLE is probably not T cell–independent. T cell responses to histone peptide are increased in patients (Lu *et al.*, 1999). However, such responses may be driven by unusual costimulatory signals from antigen-presenting cells from antigens bound to DNA, perhaps through toll-like receptors (Boule et al., 2004), which may in turn be dependent on modulating effects of anti-DNA antibodies on DNA clearance pathways.

Third, both cellular and antibody responses to a wide range of autoantigens can be induced in animals by the use of adjuvants or T cell receptor (TCR) transgenes (Matsumoto et al., 2002), but nothing equivalent is recognized in humans. Clinically significant human autoimmunity is restricted to very few autoantigens and occurs stochastically, without evidence of an environmental trigger, at a relatively low lifetime frequency in genetically susceptible individuals.

Fourth, T cell–targeted therapy (antibodies to CD3, CD4, and CD52) has produced little benefit in classical autoantibody-associated disease. Juvenile onset diabetes mellitus, in which autoantibodies have only relatively recently been recognized, may be an important exception to the general argument (Glandt et al., 2003). In those conditions not traditionally associated with specific autoantibody responses, such as psoriasis or ankylosing spondylitis, in which T cell–mediated effector mechanisms seem probable, it is not clear whether an autoreactive adaptive immune response is involved. Inflammation may arise from responses to foreign antigens or nonspecific innate immune signals that are poorly controlled. It may not be appropriate to call it autoimmunity.

Our conclusion is that although abnormalities of B cell function in systemic autoimmune diseases are easy to find, they may not be associated with loss of T cell tolerance. Given the sophistication of current *in vitro* techniques, this can no longer be attributed to practical difficulties of measuring human T cell responses to autoantigens (Davies et al., 2002). An alternative conclusion is that we are not finding such antigen-specific T cells as they are not there. We would therefore suggest that T cells have been "duped" into providing inappropriate help for renegade autoreactive B cells (Edwards et al., 1999).

2.2. Generation of Autoreactive T Cells

There may be good reasons why specific B cell, but not T cell, autoreactivities should occur. No clear reason has emerged to explain why the interaction between TCR and antigen should go wrong for just one self-antigen. The generation of diversity of the TCR relies on gene rearrangement and recombination, which if negative or positive selection processes in the thymus break down could potentially produce autoreactive T cells. However, effective peripheral tolerance mechanisms, including regulatory T cells, appear to be available to control autoreactive T cells that have escaped deletion. The TCR–antigen interaction is tightly controlled by the necessity of coengagement of accessory molecules. The output is a stereotyped signal through the TCR. Problems might be expected to occur on a broad front if signaling thresholds are changed but an isolated mistake makes no sense. An autoimmune response resulting from cross-reactivity, or "molecular mimicry," if possible at all, should happen every day, given the similarity between many exogenous and endogenous proteins. The complexity of the coreceptor, cytokine, and chemokine requirements for a productive autoimmune response resulting from "molecular mimicry" type TCR–antigen interactions probably prohibits such mistakes.

2.3. Autoantibodies as Effector Molecules

In contrast, it is relatively easy to see how an interaction between antibody and antigen could create problems. As the interaction can occur in any functional context and antibody can potentially mimic any protein, a great number of unusual effects can and do result. Aberrant effects from antibody–antigen interactions are commonplace. In thyrotoxicosis, for example, the antibody becomes a pseudohormone, and in myasthenia a neurotransmitter antagonist; crossing of the placenta by anti-Ro antibodies results in fetal cardiac dysfunction.

2.4. Autoantibodies as Trojan Horse Immunomodulators

The generation of autoreactive B cells results from a stochastic process based on immunoglobulin gene rearrangement and mutation. Although receptor editing, deletion, and anergy mechanisms are present to prevent B cell autoreactivity, autoantibodies (produced by daughter plasma cells from autoreactive B cells) are readily detectable in low levels in normal individuals. The question in autoimmunity is not why autoreactive B cells are generated but why they survive to expand and undergo affinity maturation.

They could survive with appropriate autoreactive T cell help. As already discussed, evidence for this in human systemic autoimmune disease is poor. There is an alternative, however. B cells may be able to engineer their own survival signals through their unique antibody products. To escape programmed cell death all they require is to produce antibodies that, bound to antigens, act as aberrant immunomodulators, and specifically as aberrant B cell growth factors. Once autoantibody production is initiated it enters a vicious cycle (Edwards *et al.*, 1999). The first soldiers in the Trojan horse open the gates to let the army in (see Figure 21.1). The T cells act only as gullible gatekeepers.

All antibody production is designed to be a vicious cycle of amplification, which wanes only as antigen is removed. Antibody–antigen interaction normally drives this cycle for foreign but not for self-antigens. However, for certain autoantigens it is predictable that things may go wrong. The most obvious is IgG Fc—the dominant antigen in RA. An IgG antibody to IgG Fc (IgG rheumatoid factor) is its own antigen. IgG Fc is a potent signaling ligand in regulation of B cell survival either as a binding site for complement or as a coligand for Fc receptors. If antigen takes on the function of antibody, problems can be expected.

The potential self-perpetuation mechanism for IgG rheumatoid factor (RhF)–secreting B cells (Edwards *et al.*, 1999) cannot be covered here in detail, but the essential steps can be summarized as follows:

First, IgG or possibly IgA RhF-secreting B cells arise through several possible mechanisms including a replacement mutation in a B cell antigen receptor specific for another antigen or through class switching of IgM RhF-B cells. B cell surface RhF will bind IgG bound to foreign antigens, mediating endocytosis and presentation of antigenic peptides to appropriate T cells. The RhF B cell can in this way gain T cell help as if it recognized a foreign antigen (Roosnek and Lanzavecchia, 1991) and autoreactive T cells are not required.

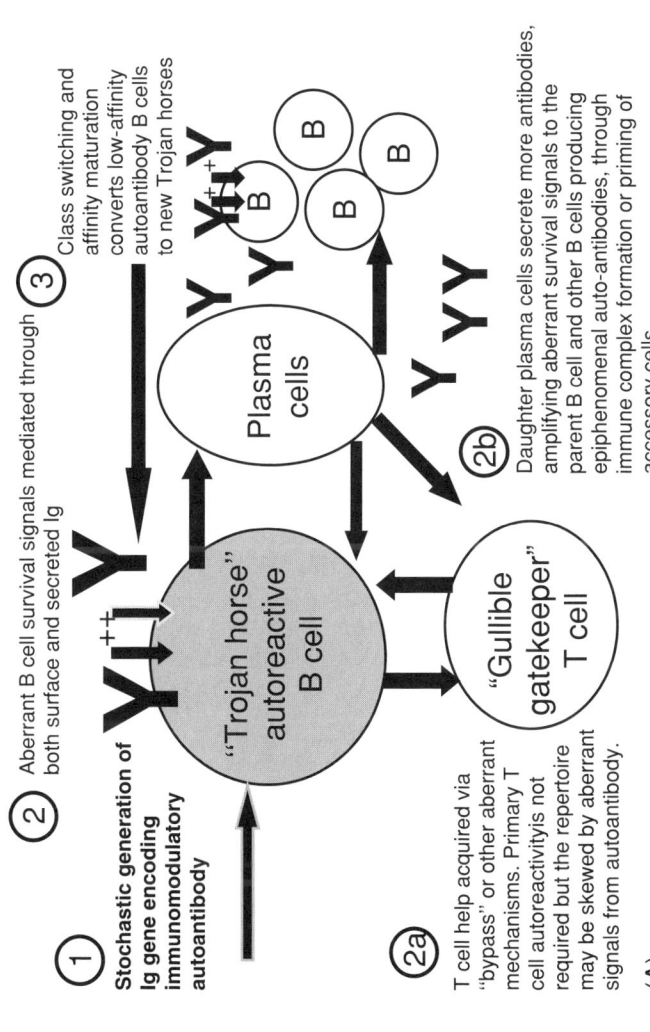

Figure 21.1. (A) The Trojan horse concept of self-perpetuating autoimmune B cells. T cells are necessary to the process but only as "gullible gatekeepers." *Continued*

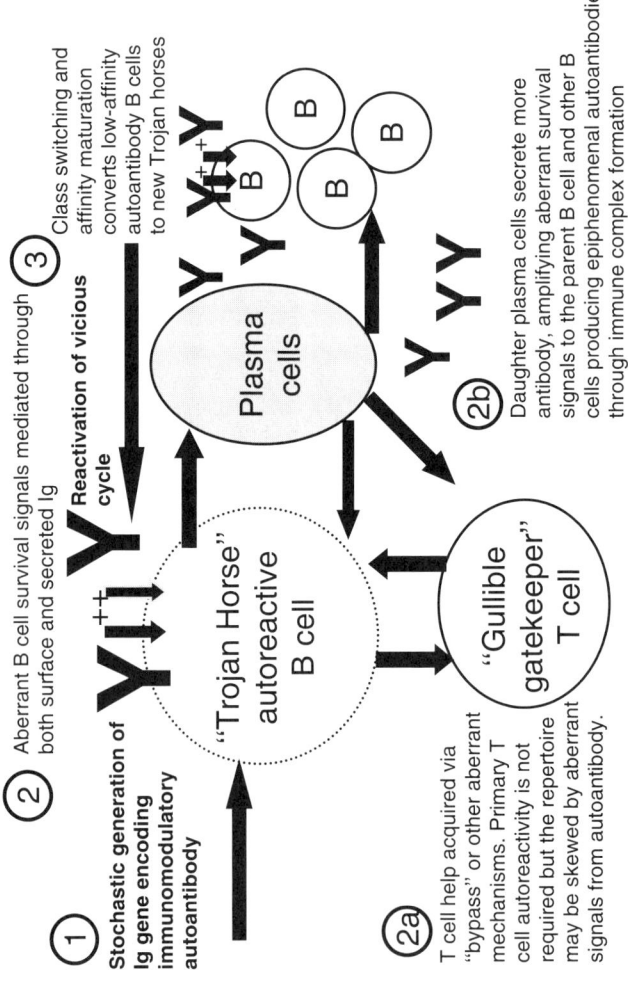

Figure 21.1. *continued* (B) Removal of pathogenic Trojan horse B cells does not necessarily remove the disease because autoantibody from plasma cells can recruit new Trojan horses.

Second, soluble IgG RhF from daughter plasma cells locally self-associate into polymers that bind and activate complement. The resulting lattice of IgG and C3d can provide the second survival signal required by the parent B cell. Since the lattice will always have free antigenic Fc epitopes the normal situation in which a soluble antibody competes with cell surface antibody for binding, thereby limiting the ability of a B cell to receive afferent signals from its own antibody products, will fail to operate. Binding of complexes to B cell receptor (i.e., surface RhF) may also outcompete the negative effect of binding to FcγRIIb.

IgG rheumatoid factor is the paradigmatic Trojan horse autoantibody, but others, such as antibodies to C1q (Davies *et al.*, 1990) or autoantibodies derived from the IgG VH 4–34 germ line gene, may plausibly fit into this scenario (Pugh-Bernard *et al.*, 2001). This mechanism also implies that multiple diverse T and B cell interactions, and not clonally restricted T cells of limited specificity, can be responsible for the propagation of the autoantibody response. Moreover, the Trojan horse concept is in many ways in keeping with the prevailing view of anti-DNA antibodies, in that their interaction with DNA may, by altering clearance, lead to aberrant signals to both B cells and other cells able to prime T cells reactive with nucleosomal proteins (Lu *et al.*, 1999).

Perhaps the most attractive feature of the principle is that it explains why autoantibodies are formed to a restricted but heterogeneous set of antigens. Each may represent a Trojan horse capable of generating an aberrant B cell survival signal through a different loophole in the regulatory pathways. There are likely to be a limited number of ways to make such a Trojan horse. The range of B cells to receive the survival signal may vary widely, from cells of the same specificity (e.g., RhF), through a broad subset of autoreactive, class-switched, VH4-34 positive cells, to all B cells, which might shed light on the multiplicity of autoantibodies produced in SLE and the general B cell expansion seen in Sjögren's syndrome.

A potential disadvantage of the Trojan horse concept is that it may be difficult to model in animals, although transferable self-perpetuating autoimmune B cells could provide an interesting proof of concept. This disadvantage may be less serious if new therapeutic agents allow human disease to be explored directly.

The Trojan horse model of autoimmunity outlined above may or may not prove the ideal viewpoint. Moreover, it may represent only one of many layers in a complex regulatory disturbance. There will be exceptions. For instance, type I diabetes mellitus does not seem to fit the paradigm, and it is unclear where antibodies figure in the pathogenesis of human multiple sclerosis and Crohn's disease. Nevertheless, it may illustrate an abnormality of regulatory dynamics common to many of the classical autoantibody-associated syndromes.

3. Clinical Significance of the Trojan Horse Concept

The clinical importance of the Trojan horse concept is that it predicts that BLyD may have the potential to produce permanent remission in a range of autoimmune disorders (Edwards *et al.*, 1999).

If autoimmunity is based on T cell autoreactivity then removing B cells is unlikely to produce long-term benefits. Even assuming that clinical disease is antibody-mediated, as in myasthenia, an episode of BLyD would not be expected to produce anything more than a temporary reduction in disease severity before more T cell–driven, autoantibody-producing B cell clones would expand. If the role of B cells is primarily to present antigen to autoreactive T cells, which then mediate disease directly, then it is doubtful that anything would happen at all. B cells are potent antigen presenters, but there is little evidence that they are uniquely necessary. For example, agammaglobulinaemic, B cell–deficient patients supported with soluble immunoglobulins appear to have protective T cell responses. There seems no particular reason why removing B cells should restore T cell tolerance to self (assuming it is breached).

If, on the other hand, autoimmunity is due to the stochastic emergence of Trojan horse autoreactive B cells then the treatment strategy is as for a B cell lymphoproliferative disorder, except that it may be necessary to clear the whole set of pathogenic clones rather than a single neoplastic clone. Permanent cure in B lymphoproliferative disorders is not easy to achieve but is regularly seen in childhood lymphoblastic leukemia and Hodgkin's disease. Rituximab was designed precisely for the purpose of removing unwanted clones of B cells (Mclaughlin, 2001) so its use in autoimmunity to attempt long-term abrogation is entirely logical. That is the reason why it was tried in RA.

3.1. Effector Mechanisms in RA

The above discussion may puzzle the reader familiar with the concept of T cell effector mechanism for tissue pathology in RA. For autoantibody production to be largely independent of T cell autoreactivity is one thing, but for tissue pathology to be independent of T cell effector mechanisms is another. The plausibility of a T cell effector mechanism in RA has been a matter of debate for over 20 years (Janossy *et al.*, 1981; Firestein and Zvaifler, 1990).

Recent evidence from our laboratory has suggested that an antibody-based effector mechanism can explain tissue pathology. Briefly, small, non–complement fixing complexes of self-associated IgG-RhF (Mannik and Nardella, 1985) are ideally suited to generate macrophage activation not only in synovium but also in the other eight target tissues in RA (Nardella *et al.*, 1983; Bhatia *et al.*, 1998; Edwards *et al.*, 1999). The triggering of FcγRIIIa, the favored ligand of such small complexes, on macrophages has been shown to result in the production of TNFα and IL-1 (Abrahams *et al.*, 2002), the major mediators of inflammation in RA (Maini *et al.*, 1997). Others have shown that macrophage activation in the joint precedes lymphocyte infiltration in RA (Kraan *et al.*, 1998). The ability of circulating antibodies to trigger arthritis has also been modeled in the K/BxN transgenic mouse (Matsumoto *et al.*, 2002).

Support for an antibody-mediated effector mechanism has also now come from the use of B lymphocyte–targeted therapy in RA and other autoantibody-associated diseases (Edwards *et al.*, 2002). The clinical responses of remission and relapse have been shown to follow autoantibody rather than B cell levels

(Cambridge *et al.*, 2003). No responses have been seen in clinically similar but antibody-negative conditions (Edwards *et al.*, 2004).

3.2. Logistics of B Cell Depletion

Significant B cell depletion is achieved with traditional immunoablative therapy with high-dose cyclophosphamide, as used before bone marrow transplantation. This type of regimen has been reported to produce clinical benefit in autoimmunity. In SLE long-lasting remissions may be achieved (Traynor *et al.*, 2002). Nevertheless, toxicity is not trivial. Moreover, it is relatively difficult to use evidence from broad-spectrum ablative therapy to probe pathogenesis. Selective B cell depletion became a practical option with the emergence of the anti-CD20 monoclonal antibody rituximab (McLaughlin, 2001).

3.3. Anti-CD20 Therapeutic Agents

CD20 is expressed on the surface of B cells. It is also present at a low level on a small proportion of $CD3^+$ T cells (0–10%, mean 3%, mostly $CD4^-$, $CD8^-$) and a very small number of $CD56^+CD3^-$ (natural killer phenotype) cells, which are also depleted by rituximab (M. J. Leandro, unpublished data; Figure 21.2). CD20 is a calcium channel associated with lipid rafts (Li *et al.*, 2004), but whether its function is involved in the action of therapeutic antibodies is not established.

CD20 has a low rate of cycling in the cell membrane and is thus an attractive target for directly cytolytic antibodies. Several anti-CD20 antibodies have been produced, recent examples being either humanized or fully human (generated in mice transgenic for human Ig), several of which are in preclinical and early clinical development (Teeling *et al.*, 2004). However, only rituximab is in

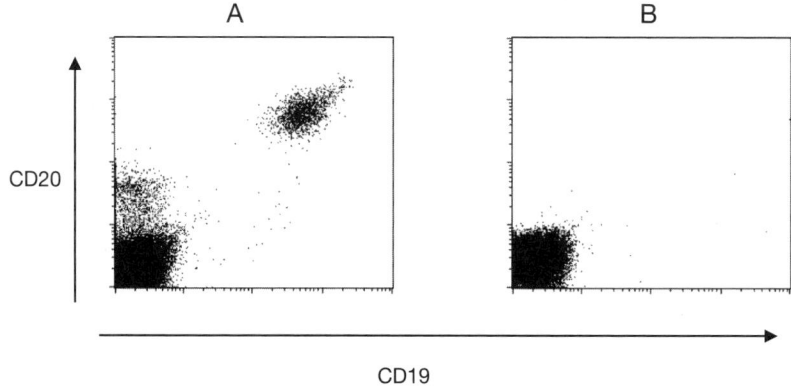

Figure 21.2. FACS plot showing CD19 versus CD20 staining of peripheral blood cells in the lymphocyte gate (A) before and (B) after treatment with rituximab. $CD19^+/CD20^+$ cells are B lymphocytes. $CD19^+CD20^-$ (low) cells are plasmablasts. $CD19^-/CD20^+$ (low) cells are T and NK cells. All CD20 expressing cells are depleted following rituximab. Plasmablasts also disappear from the circulation.

regular clinical use and has provided virtually all information on specific B cell depletion in autoimmune disease (Edwards *et al.*, 2002).

3.4. Rituximab

Rituximab is derived from the C2B8 murine clone chimerized with human IgG1 constant regions (McLaughlin, 2001; Grillo-Lopez *et al.*, 2002). It has *in vitro* cytolytic activity in complement-dependent, antibody-dependent cellular cytotoxicity (ADCC) and apoptotic assays. The relative importance of these mechanisms *in vivo* has not been resolved. Recent *in vivo* studies using mice transgenic for human CD20 suggest that cytolysis may be predominantly complement-mediated (Di Gaetano *et al.*, 2003). There is a suggestion that efficacy of depletion in humans is dependent on FcγRIIIa allotype, implying a contribution from ADCC (Anolik *et al.*, 2003). However, no correlation between FcγRIIIa allotype and either depletion or clinical benefit was observed in RA or SLE patients at UCL (unpublished data). Other antibodies, such as B1, show different spectra of activity in the three types of *in vitro* assays (Teeling *et al.*, 2004). Clarification of *in vivo* cytolytic mechanisms is likely to be crucial for further development of anti-CD20 reagents.

Doses of rituximab used clinically achieve virtually complete depletion of circulating B cells (Edwards *et al.*, 2004) (Figure 21.2). The level of depletion in solid tissues is less clear. Anecdotal reports suggest that splenic B cells may be cleared completely but primate studies suggest that depletion in lymph nodes is incomplete (Reff *et al.*, 1994). Recent studies of bone marrow examined 3 months after BLyD with rituximab suggest that B cell depletion is complete in some but not in others (M. J. Leandro, manuscript in preparation). An interesting finding is the absence of pro-B cells in one case 3 months after treatment. Pro-B cells do not express CD20 and their absence following rituximab might not be expected, unless transit to the pre-B cell compartment is accelerated and residual rituximab prevents further maturation. This might shed some light on the prolonged B lymphopenia the drug induces. Nonselective bone marrow ablation regimens tend to produce 3 months of B cytopenia, whereas with rituximab the period is often 7–9 months and can exceed 2 years.

Rituximab is given by intravenous (IV) infusion. The oncological dosing regimen of 375 mg per square meter of body surface area on four occasions at weekly intervals has been compressed and standardized into two infusions of 1 g for a number of studies in autoimmunity. Little is known about optimum dose, but in RA this dose routinely achieves >98% depletion of circulating B cells (M. J. Leandro, manuscript in preparation). In RA the two doses are given 1 week apart. In SLE, UCL patients receive a 750 mg dose of IV cyclophosphamide the day after each rituximab infusion and for this reason a 2-week interval is used (Leandro *et al.*, 2002a). The use of cyclophosphamide in SLE is partly because its use is standard practice and partly because B cell depletion is much less reliably achieved in SLE for reasons that are not understood but might be related to the complement dysfunction in this condition. It is not known whether differences in the timing of administration affect the efficiency of B cell depletion.

3.5. Efficacy

Much of the evidence on the efficacy of BLyD in autoimmune disease comes from open studies. However, in many conditions comparable results have been reported from more than one study and changes in objective variables suggest benefits are real. In RA the first open study initiated at UCL in 1998 suggested that a single cycle of BLyD with combination therapy could induce almost complete clinical remission for periods up to 3 years (Edwards and Cambridge, 2001). A further dose-ranging study provided plausible evidence for dose response (Leandro et al., 2002b). A further open study from De Vita et al. (2002) produced comparable results using rituximab alone. A formal randomized double-blind controlled trial has now confirmed substantial efficacy (Edwards et al., 2004).

In SLE open studies set up by Looney and by Eisenberg in the USA, by Edwards at UCL, and by van Vollenhoven in Sweden have reported evidence for significant efficacy and dose response (Anolik et al., 2001; Edwards et al., 2002; Leandro et al., 2002a). Problems such as infusion reactions, failure of depletion, and human antichimeric antibody (HACA) responses have been more evident in SLE, but responses of life-threatening problems such as thrombocytopenia and nephritis have been at least as substantial and persistent as in RA.

In immune thrombocytopenia two open studies observed improvement in more than half of the cases and complete remission in about a third (Cooper et al., 2004). Several small open studies of hemolytic anemia have reported good results in a proportion of cases, but with more heterogeneity (Edwards et al., 2002). An early study in IgM-associated peripheral neuropathy reported major benefits (Levine and Pestronk, 1999). A study in myositis reported benefits, and other studies have been set up but no formal publication is available. An open study in Wegener's granulomatosis reported benefits (Specks et al., 2001) and a formal trial in ANCA-associated vasculitis is now under way. Case reports have mostly described good results in Goodpasture's syndrome, other forms of vasculitis, antiphospholipid syndrome, myasthenia gravis, and bullous skin disorders (Edwards et al., 2002). Although there have been no open studies reported, at least one formal trial has been set up in multiple sclerosis.

The conclusion from reports published to date is that in a proportion of cases BLyD therapy is associated with partial or complete remission of disease. However, responses have not been universal, with a quarter to a third of cases showing little or no evidence of response. This immediately raises the question of intersubject variables in pathogenic mechanisms that determine responsiveness.

The second main observation from these studies is that improvement following a brief cycle of BLyD (8–22 days) is prolonged for months or years. In most cases it continues at least for the period of peripheral depletion, in the order of 8 months. However, in perhaps a third to a half of cases, it continues for up to another 2–3 years. In RA the longest response so far is 42 months. Responses of over 2 years in immune thrombocytopenia and SLE are not uncommon (Leandro et al., 2002a; Cooper et al., 2004).

3.6. Failure of Seronegative Disease to Respond

Perhaps the simplest qualitative observation to tease out of the immunodynamic data from BLyD so far is that disorders not associated with autoantibodies show no benefit (Edwards *et al.*, 2004). Most interestingly, this includes the seven reported seronegative cases of what would otherwise be considered RA. In these cases symptoms did not improve and C-reactive protein levels did not fall. In at least some of these cases, despite the absence of the nail abnormalities normally considered the most reliable discriminant, psoriatic arthropathy may have been the correct diagnosis. What remains to be seen is whether all RhF-negative "RA" is really allied to the psoriatic/spondarthropathy group, or whether there is a "true seronegative RA", and if so whether it responds to BLyD.

Coexistent psoriasis in patients treated for RA with BLyD at UCL has shown no evidence of response and may have got worse. A patient treated for immune thrombocytopenia with coexistent Crohn's disease experienced worsening of Crohn's disease (Papadakis *et al.*, 2003).

These observations appear to suggest that B cells play a very different role in RA from that in the conditions associated with spondarthritis, including psoriasis and Crohn's disease. It supports the suggestion that RhF may have a more direct relationship to pathogenesis than has been assumed. If B cells contributed to RA simply as antigen-presenting cells one might expect broadly similar responses in the two groups of conditions. The complete absence of response in seronegative conditions so far suggests that the dependence of RA on B cells is more specific. Antigen presentation by B cells is integral to antibody production, and by implication autoantibody production, whereas other accessory cells can support T cell–mediated inflammation.

3.7. Adverse Events Associated with BLyD

Adverse events attributable to BLyD with rituximab have been uncommon (Edwards *et al.*, 2002, 2004). Infusion reaction seen in lymphoma patients, thought to relate to cytokine release and other events associated with cytolysis, is not a major problem in autoimmune patients. Sensitivity reactions and hemodynamic disturbances occur only occasionally. A common feature of first rituximab infusions is a transient pricking sensation in the throat, which we have interpreted as indicating penetration of rituximab into Waldeyer's ring.

BLyD might be expected to induce susceptibility to infection, but this has not proved to be a common problem, probably owing to the relative preservation of antimicrobial antibody levels. One of 121 patients to receive rituximab in the RA phase II study subsequently died of a cardiac event following an episode of pneumonia (Edwards *et al.*, 2004). He had preexisting rheumatoid lung disease and was known to have a poor prognosis. Two other respiratory infections occurred in the trial. The occurrence of febrile episodes with lower respiratory symptoms, including cough with purulent sputum and pleuritic pain, and constitutional features suggestive of infection is also the one issue of concern in the UCL cohort. Following 120 cycles of BLyD, ten significant lower respiratory episodes have occurred, probably more than expected by chance. All these may be infective, and some may represent hospital-acquired infection. However, four have occurred within a few days of rituximab treatment, suggesting some sort of noninfective late immunological reaction.

HACA responses to rituximab are infrequent in RA, estimated at 3% in the phase II study (Edwards *et al.*, 2004), but much higher rates have been seen in SLE. In initial studies with low doses (Anolik *et al.*, 2001) rates as high as 50% were reported but evidence of a blocking response has only been seen on two occasions in 21 patients at UCL.

3.8. Repeated Cycles of B Cell Depletion

All 40 RA patients treated with B cell depletion over a 5-year period at UCL have been considered for re-treatment whenever clinically necessary. A total of 75 treatment cycles have been given. Twenty-seven patients have continued on the program. Those that have not continued have had either poor or brief responses, have been maintained on standard drugs following one or more cycles of BLyD or have been withdrawn because of falling IgM levels.

Patients in the repeat BLyD program at UCL have been followed for a total of 125 patient years. Average length of clinical benefit for all cycles was 15 months. Patients have gained a similar degree and length of response from subsequent treatments as from the first cycle. Secondary failure of clinical response or depletion was not observed in up to four cycles. Two patients have had significantly longer responses to subsequent treatments than to the first, and none have had shorter responses, but there is no evidence as yet that repeated treatment can induce long-term remission where a single treatment cannot.

4. Do Data from BLyD Support the Trojan Horse Concept?

The use of BLyD has, therefore, provided us with a powerful new tool with which to treat a number of autoimmune conditions and, perhaps more importantly for the future, to probe underlying pathogenic mechanisms. It was initially thought by some that BLyD would not work in such autoantibody-associated conditions, since rituximab-based ablation of circulating B cells in lymphoma for a period of many months was not associated with major falls in immunoglobulin levels (McLaughlin, 2001; Grillo-Lopez *et al.*, 2002). Clinical responses seen in a number of conditions showed this prediction to be wrong.

4.1. Autoantibody Levels Fall Selectively Compared with Antimicrobial and Total Immunoglobulin Levels

Early experience showed that significant, apparently selective falls in autoantibody levels could occur (Edwards and Cambridge, 2001; Specks *et al.*, 2001). This is confirmed by immunodynamic studies in RA and SLE (Leandro *et al.*, 2002; Cambridge *et al.*, 2003; Edwards *et al.*, 2004). It appears that B cell depletion may be more useful than predicted because of a differential sensitivity of autoantibody-secreting cells. Correlations between decline in C-reactive protein and RhF levels are almost linear in many RA patients. In SLE, the drop in anti-DNA antibodies mirrors renal disease most closely (Leandro *et al.*, 2002a). In other conditions antibodies are more difficult to quantify. In contrast, levels of antimicrobial antibodies, to pneumococcal capsular polysaccharide (PCP) and tetanus toxoid, in most patients with RA and lupus did not decrease significantly, even after several cycles of treatment (Cambridge *et al.*, 2003).

Since circulating antibody comes chiefly from plasma cells rather than B cells, it seems likely that differential attrition of *auto*antibody levels, compared with antimicrobial responses, following BLyD reflects relatively short half-lives for at least some autoantibody-secreting plasma cells. Plasma cells do not themselves express significant levels of CD20 and would be unlikely to be ablated by the protocols used, although cyclophosphamide may be modestly plasmacytolytic.

There is also a suggestion that marginal zone B cells in the spleen, which secrete antibodies to T cell–independent, often carbohydrate, antigens, may be relatively resistant to rituximab. This could explain preservation of levels of antibodies to PCP seen in RA and lupus (Cambridge *et al.*, 2003; G. Cambridge, M. J. Leandro, J. C. W. Edwards, manuscript in preparation). In fact there is sometimes a transient rise in such levels after B cell depletion, suggesting that marginal zone–derived plasma cells may take over space made available by plasma cells derived from rituximab-sensitive B cells (G. Cambridge, unpublished data).

4.2. Total Immunoglobulin Levels May Fall after Repeat Cycles

An observation of particular interest is that after three courses of BLyD based on rituximab serum IgM in several RA patients has fallen to undetectable levels (M. J. Leandro, manuscript in preparation). During the first cycle of depletion, falls of up to 40% are seen. It seems that certain subpopulations of B cells, and their daughter plasma cells, may not be regenerated and that repeated cycles may deplete any remaining relatively long-lived IgM-secreting plasma cell populations. IgM is thought to provide a frontline defense against infection through "natural antibody," capable of recognizing, with low affinity, microbes to which the host does not have an adaptive immune response, and facilitating scavenging and presentation of microbial antigen by macrophages and dendritic cells to T cells. Thus IgM is important for T cell–mediated immunity, quite apart from its role in humoral immunity. A defect in both systems following B cell depletion is of potential concern. Nevertheless, in the experience at UCL, and also apparently elsewhere, there is little or no indication that B cell depletion makes patients susceptible to the classical opportunistic infections such as pneumocystis carinii, tuberculosis, commensal fungi, or viruses. To date, none of the patients with absent IgM levels in our cohort have suffered significant infective episodes.

Serum IgG levels tend to fall by a mean of 20–30%, but remain within the normal range, even following repeated cycles of rituximab (Cambridge *et al.*, 2003). Interestingly, following rituximab patients with both RA and SLE may normalize their IgG levels from either very high or abnormally low levels. This can be dramatic and encourages the hope that some form of restoration of immune regulation or removal of inhibitory or cytotoxic antibodies is occurring.

The broad conclusion, therefore, is that the simplest explanation for the benefit of BLyD is the reduction in levels of pathogenic autoantibodies. Our recent immunodynamic monitoring of patients with autoimmune disease undergoing BLyD has, however, produced a number of findings that suggest that the relationship between B cells and disease is even more complex (Cambridge et al., 2003).

4.3. Clinical Response Follows Serological Response, Not B Cell Numbers

If the therapeutic action of BLyD was through the removal of B cells responsible for presenting autoantigens to T cells, benefit would be expected only for as long as B cells were depleted. B lymphocytes are often sparse in synovium and even when present form a minority of antigen-presenting cells, but this would not preclude the possibility that they were functioning in extrasynovial sites. Observations following BLyD make this explanation unlikely. B lymphocyte numbers fall within days of rituximab therapy but clinical improvement occurs over a period of several months (Cambridge et al., 2003). If B cells are functioning as the key providers of antigen presentation to T cells capable of inducing cytokine production from synovial or other macrophage populations, one would expect a rapid and dramatic response to BLyD, as observed with TNFα antagonists. This does not occur.

4.4. The Kinetics of Relapse Follow Autoantibody Rises Rather than B Cell Return

Although evidence to date suggests that B cell depletion as currently performed is unlikely to produce long-term remission in RA it is not uncommon for patients to remain in remission for 1–2 years following the return of circulating B cells (Cambridge et al., 2003). In almost exactly half of cases relapse occurs within a few weeks of B cell return, but in the other half there is a delay of several months (see Figure 21.3). B cell return is therefore necessary but not sufficient for relapse. However, return of circulating autoantibodies does appear to be both necessary and sufficient in that a return to pretreatment levels is consistently associated with relapse. This supports the view that autoantibodies are directly involved both in the propagation of the autoimmune response and in clinical disease expression, as in the Trojan horse model (see Figure 21.1).

The long delay between B cell return and clinical relapse indicates the presence of a rate-limiting step (or steps) in the returning pathological immune response. The often-protracted period of increasing autoantibody production, which is also known to precede the initial onset of clinical RA in many cases, suggests that this is a rerun of an early stage of disease propagation. These observations are consistent with a role for infrequent stochastic generation of antibody species with aberrant signaling capacity, as initially envisaged (Edwards et al.,

1999). An important question is whether the continued production of autoantibody from plasma cells may determine relapse by providing afferent immunoregulatory signals favoring reemergence of a self-amplification loop of autoreactive B cell survival (Figure 21.1B). Agents targeting different windows of activity in B cell life history may help answer this question.

4.5. Why Are There Two Patterns of Relapse?

In RA there appear to be two distinct temporal patterns of relapse (Figure 21.3) immediately following B cell return or at random over a period of many months. The first pattern might reflect the resistance of some pathogenic memory B cell populations to rituximab due perhaps to their geographic situation, e.g., in the marginal zone of the spleen or in synovial tissue. The significance of survival of B cells in the synovium may be not so much that it is the target tissue but that it is a solid tissue that may be difficult to penetrate. Relapse at B cell return might therefore reflect involvement of "original" pathogenic clones in reestablishment of a vicious cycle and reattaining of clinically relevant levels of proinflammatory autoantibodies. Initial studies of bone marrow from RA patients and of peripheral blood in RA and SLE patients supports the idea that inadequate depletion of B cells may be associated both with early B cell return and an immediate autoantibody rise associated with relapse, indicating that doses used for B cell depletion may at times be suboptimal.

In the delayed pattern of relapse, it is possible that qualitative as well as quantitative factors are involved. IgG RhF from (CD20$^-$) plasma cells can, in theory, recruit newly emerging IgM-RhF B cells to enter into a vicious cycle (Figure 21.1B). Such B cells would then become the focus of increased and inappropriate T cell help, resulting in germinal center formation, affinity maturation, and class switching. The pattern of RhF return in patients would support this in that IgM-RhF is usually the first class to rise before relapse.

5. Conclusions

BLyD therapy has made a significant impact in autoimmune disease. However, the original hope of achieving long-term remission by ablation of pathogenic B cell clones remains to be realized. More potent anti-B cell agents may achieve the goal. Longer-term remission may only be achievable if BLyD is combined with a plasma cell depletion strategy. At present, no safe and effective antiplasma cell agents are available, but elucidation of the survival signals required by plasma cells may open new therapeutic avenues. Immunodynamic studies tend to support the original hypothesis that Trojan horse B cells may control disease through both afferent and inflammatory effector signals mediated by autoantibody. Further immunodynamic studies may be particularly useful in identifying mechanisms of relapse and how these may be blocked. In the meantime, progress is highly encouraging, and, used with care, BLyD promises to be an important option for severe or refractory autoimmune disease.

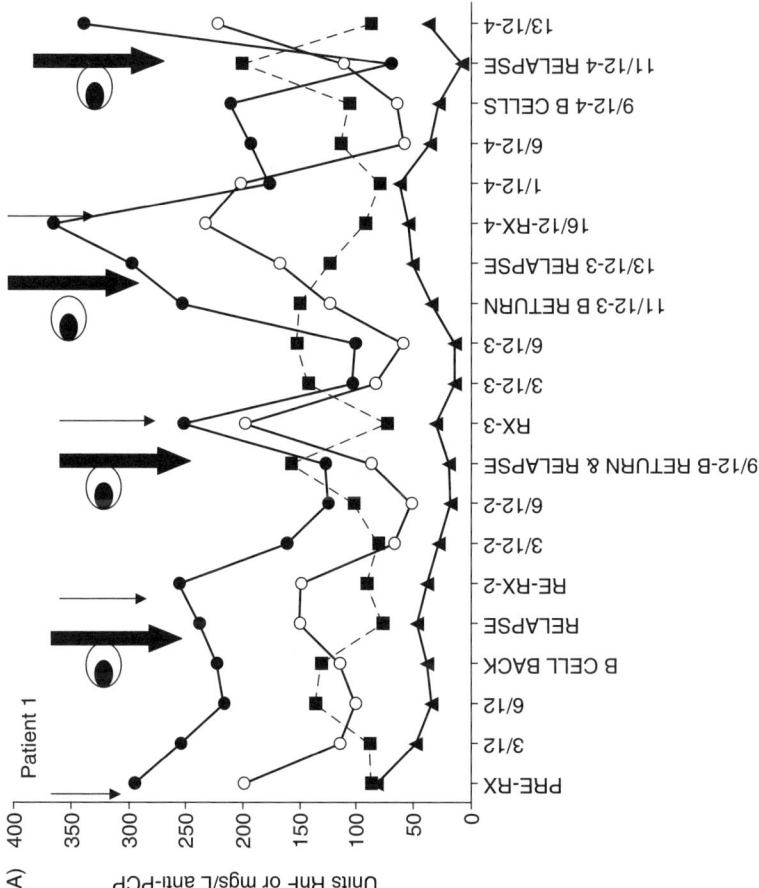

Figure 21.3. Serial rheumatoid factor (RhF) and antipneumococcal polysaccharide (anti-CCP) antibody levels in two patients with RA undergoing repeated cycles of BLyD. (A) Patient 1 relapsed on B cell return.

Continued

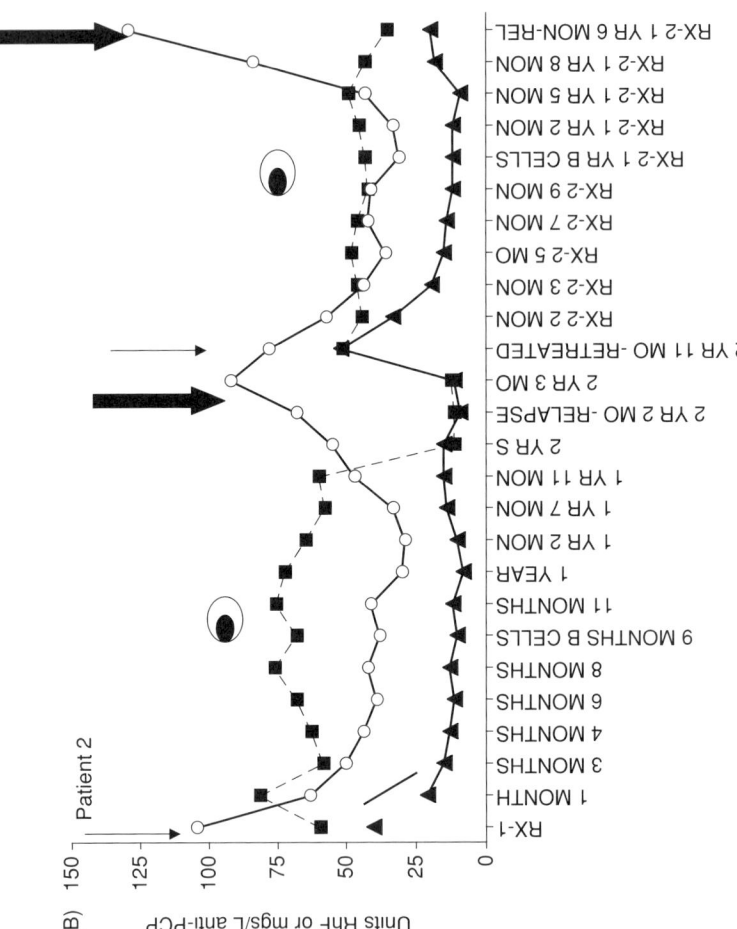

Figure 21.3. *continued* (B) Patient 2 relapsed months after B cell return.

References

Abrahams, V.M., Cambridge, G., and Edwards, J.C.W. (2002). Induction of tumour necrosis factor alpha production by human monocytes: A key role for FcγRIIIa in rheumatoid arthritis. *Arthritis Rheum.*, **43**, 608–616.

Anolik, J.H., Campbell, D., Felgar, R.E., Young, F., Sanz, I., Rosenblatt, J., and Looney, R.J. (2003). The relationship of FcγRIIIa genotype to degree of B cell depletion by rituximab in the treatment of systemic lupus erythematosus. *Arthritis Rheum.*, **48**, 455–459.

Anolik, J.H., Campbell, D., Ritchlin, C., Looney, J.R., and Sanz, I. (2001). B lymphocyte depletion as a novel treatment for systemic lupus erythematosus: Phase I/II trial of rituximab in SLE. *Arthritis Rheum.*, **46** (Abstr. Suppl.), S2009.

Bhatia, A., Blades, S., Cambridge, G., and Edwards, J.C.W. (1998). Differential distribution of FcγRIIIa in normal human tissues and co-localization with DAF and fibrillin-1: Implications for immunological microenvironments. *Immunology*, **94**, 65–638.

Boule, M.W., Broughton, C., Mackay, F., Akira, S., Marshak-Rothstein, A., and Rifkin, I.R. (2004). Toll-like receptor 9-dependent and -independent dendritic cell activation by chromatin–immunoglobulin G complexes. *J. Exp. Med.*, **199**, 1631–1640.

Busser, B.W., Adair, B.S., Erikson, J., and Laufer, T.M. (2003). Activation of diverse repertoires of autoreactive T cells enhances the loss of anti-dsDNA B cell tolerance. *J. Clin. Invest.*, **112**, 1361–1371.

Cambridge, G., Leandro, M.J., Edwards, J.C.W., Ehrenstein, M.R., Salden, M., Bodman-Smith, M., and Webster, A.D.B. (2003). Serological changes following B lymphocyte depletion therapy for rheumatoid arthritis. *Arthritis Rheum.*, **48**, 2146–2154.

Cooper, N., Stasi, R., Cunningham-Rundles, S., Feuerstein, M.A., Leonard, J.P., Amadori, S., and Bussel, J.B. (2004). The efficacy and safety of B-cell depletion with anti-CD20 monoclonal antibody in adults with chronic immune thrombocytopenic purpura. *Br. J. Haematol.*, **125**, 232–239.

Danke, N.A., Koelle, D.M., Yee, C., Beheray, S., and Kwok, W.W. (2004). Autoreactive T cells in healthy individuals. *J. Immunol.*, **172**, 5967–5972.

Davies, K.A., Hird, V., Stewart, S., Sivolpenko, G.B., Jose, P., Epentos, A.A., and Walport, M.J. (1990) A study of in vivo immune complex formation and clearance in man. *J. Immunol.*, **144**, 4613–4620.

Davies, M.L., Taylor, E.J., Gordon, C., Young, S.P., Welsh, K., Bunce, M., Wordsworth, B.P., Davidson, B., and Bowman, S.J. (2002). Candidate T cell epitopes of the human La/SSB autoantigen. *Arthritis Rheum.*, **46**, 209–214.

De Vita, S., Zaja, F., Sacco, S., De Candia, A., Fanin, R., and Ferraccioli, G. (2002). Efficacy of selective B cell blockade in the treatment of rheumatoid arthritis: Evidence for a pathogenetic role of B cells. *Arthritis Rheum.*, **46**, 2029–2033.

Di Gaetano, N., Cittera, E., Nota, R., Vecchi, A., Grieco, V., Scanziani, E., Botto, M., Introna, M., and Golay, J. (2003). Complement activation determines the therapeutic activity of rituximab in vivo. *J. Immunol.*, **171**, 1581–1587.

Donath, J. and Landsteiner K. (1904). Ucber paroxysmale Hämoglobinurie. *Muench. med. Wochenschr.*, **51**, 1590–1593.

Edwards, J.C.W., Cambridge, G., and Abrahams, V.M. (1999). Do self-perpetuating B lymphocytes drive human autoimmune disease? *Immunology*, **97**, 1868–1876.

Edwards, J.C.W. and Cambridge, G. (2001). Sustained improvement in rheumatoid arthritis following a protocol designed to deplete B lymphocytes. *Rheumatology*, **40**, 205–211.

Edwards, J.C.W., Leandro, M.J., and Cambridge, G. (2002). B lymphocyte depletion therapy in rheumatoid arthritis and other autoimmune disorders. *Biochem. Soc. Trans.*, **30**, 824–828.

Edwards, J.C.W., Leandro, M.J., and Cambridge, G. (2004). B lymphocyte depletion therapy with rituximab in rheumatoid arthritis. *Rheum. Dis. Clin. North Am.*, **30**, 393–404.

Edwards, J.C., Szczepanski, L., Szechinski, J., Filipowicz-Sosnowska, A., Close, D.R., Stevens, R.M., and Shaw, T. (2004). Efficacy of B cell targeted therapy with rituximab, in rheumatoid arthritis. *N. Engl. J. Med.*, **350**, 2572–2581.

Firestein, G.S. and Zvaifler, N.J. (1990). How important are T cells in chronic rheumatoid synovitis? *Arthritis Rheum.*, **33**, 768–773.

Glandt, M., Hagopian, W., and Herold, K.C. (2003). Treatment of type 1 diabetes with anti-CD3 monoclonal antibody. *Rev. Endocr. Metabol. Disord.*, **4**, 361–368.
Grillo-Lopez, A.J., Hedrick, E., Rashford, M., and Benyunes, M. (2002). Rituximab: Ongoing and future clinical development. *Semin. Oncol.*, **29** (Suppl. 2), 105–112.
Janossy, G., Duke, O., Poulter, L.W., Panayi, G., Bofill, M., and Goldstein, G. (1981). Rheumatoid arthritis: A disease of T lymphocyte-macrophage immunoregulation. *Lancet*, **ii**, 839–842.
Kraan, M.C., Versendaal, H., Jonker, M., Bresnihan, B., Post, W.J., Hart, B.A., Breedveld, F.C., and Tak, P.P. (1998). Asymptomatic synovitis precedes clinically manifest arthritis. *Arthritis Rheum.*, **41**, 1481–1488.
Kuwana, M., Feghali, C.A., Medsger, T.A., Jr., and Wright, T.M. (2001). Autoreactive T cells to topoisomerase I in monozygotic twins discordant for systemic sclerosis. *Arthritis Rheum.*, **44**, 1654–1659.
Lang, A.K., Macht, L.M., Kirwan, J.R., Wraith, D.C., and Elson, C.J. (1999). Ability of T cells from patients with rheumatoid arthritis to respond to immunoglobulin G. *Immunology*, **98**, 116–122.
Leandro, M.J., Edwards, J.C.W., Cambridge, G., Ehrenstein, M.R., and Isenberg, D.A. (2002a). An open study of B lymphocyte depletion in systemic lupus erythematosus. *Arthritis Rheum.*, **46**, 2673–2677.
Leandro, M.J., Edwards, J.C.W., and Cambridge, G. (2002b). Clinical outcome in 22 patients with RA treated with B lymphocyte depletion. *Ann. Rheum. Dis.*, **61**, 1–5.
Levine, T.D. and Pestronk, A. (1999). IgM antibody-related polyneuropathies: B-cell depletion chemotherapy using rituximab. *Neurology*, **52**, 1701–1704.
Li, H., Ayer, L.M., Polyak, M.J., Mutch, C.M., Petrie, R.J., Gauthier, L., Shariat, N., Hendzel, M.J., Shaw, A.R., Patel, K.D., and Deans, J.P. (2004). The CD20 calcium channel is localized to microvilli and constitutively associated with membrane rafts; antibody binding increases the affinity of the association through an epitope-dependent crosslinking-independent mechanism. *J. Biol. Chem.*, **279**, 19893–19901.
Lu, L., Kaliyaperumal, A., Boumpas, D.T., and Datta, S.K. (1999). Major peptide autoepitopes for nucleosome-specific T cells of human lupus. *J. Clin. Invest.*, **104**, 345–355.
Maini, R.N., Elliott, M., Brennan, F.M., Williams, R.O., and Feldmann, M. (1997). TNF blockade in rheumatoid arthritis: Implications for therapy and pathogenesis. *Acta Pathol. Microbiol. Immunol. Scand.*, **105**, 257–263.
Mannik, M. and Nardella, F.A. (1985). IgG rheumatoid factors and self-association of these antibodies. *Clin. Rheum. Dis.*, **11**, 551–572.
Matsumoto, I., Maccioni, M., Lee, D.M., Maurice, M., Simmons, B., Brenner, M., Mathis, D., and Benoist, C. (2002). How antibodies to a ubiquitous cytoplasmic enzyme may provoke joint-specific autoimmune disease. *Nat. Immunol.*, **3**, 360–365.
McLaughlin, R. (2001). Rituximab: Perspective on single agent experience, and future directions in combination trials. *Crit. Rev. Oncol. Haematol.*, **40**, 3–16.
Nardella, F.A., Dayer, J.M., Roelke, M., Krane, S.M., and Mannik, M. (1983). Self-associating IgG rheumatoid factors stimulate monocytes to release prostaglandins and mononuclear cell factor that stimulates collagenase and prostaglandin production by synovial cells. *Rheumatol. Int.*, **3**, 183–186.
Papadakis, K.A., Rosenbloom, B., and Targan, S.R. (2003). Anti-CD20 chimeric monoclonal antibody (rituximab) treatment of immune-mediated thrombocytopenia associated with Crohn's disease. *Gastroenterology*, **124**, 583.
Pilatte, Y., Tisserand, E.M., Greffard, A., Bignon, J., and Lambre, C.R. (1990). Anticarbohydrate autoantibodies to sialidase-treated erythrocytes and thymocytes in serum from patients with pulmonary sarcoidosis. *Am. J. Med.*, **88**, 486–492.
Pugh-Bernard, A.E., Silverman, G.J., Cappione, A.J., Villano, M.E., Ryan, D.H., Insel, R.A., and Sanz, I. (2001). Regulation of inherently autoreactive VH4-34 B cells in the maintenance of human B cell tolerance. *J. Clin. Invest.*, **108**, 1061–1070.
Reff, M.E., Carner, K., Chambers, K.S., Chinn, P.C., Leonard, J.E., Raab, R., Newman, R.A., Hanna, N., and Anderson, D.R. (1994). Depletion of B cells in vivo by a chimeric mouse human monoclonal antibody to CD20. *Blood*, **83**, 435–445.
Roosnek, E. and Lanzavecchia, A. (1991). Efficient and selective presentation of antigen-antibody complexes by rheumatoid factor B cells. *J. Exp. Med.*, **173**, 487–489.

Specks, U., Fervenza, F.C., McDonald, T.J., and Hogan, M.C.E. (2001). Response of Wegener's granulomatosis to anti-CD20 chimeric monoclonal antibody therapy. *Arthritis Rheum.*, **44**, 2836–2840.

Teeling, J.L., French, R.R., Cragg, M.S., van den Brakel, J., Pluyter, M., Huang, H., Chan, C., Parren, P.W., Hack, C.E., Dechant, M., Valerius, T., van de Winkel, J.G.J., and Glennie, M.J. (2004). Characterisation of new human CD20 monoclonal antibodies with potent cytolytic activity against non-Hodgkin's lymphomas. *Blood*, **104**, 1793–1800.

Traynor, A.E., Barr, W.G., Rosa, R.M., Rodriguez, J., Oyama, Y., Baker, S., Brush, M., and Burt, R.K. (2002). Hematopoietic stem cell transplantation for severe and refractory lupus. Analysis after five years and fifteen patients. *Arthritis Rheum.*, **46**, 2917–2923.

22

B Lymphocyte Stimulator (BLyS) and Autoimmune Rheumatic Diseases

William Stohl

1. Introduction

Patients with autoimmune rheumatic diseases and their physicians recognize all too well that present-day therapy for these conditions remains inadequate and complicated by unacceptable serious toxicities. The recent discovery of B lymphocyte stimulator (BLyS) and its receptors provides cautious optimism that a key contributor and facilitator to autoimmune rheumatic diseases has now been identified for therapeutic targeting. A variety of BLyS antagonists have been developed, and some of these have already entered phase I and phase II clinical trials. One may be cautiously optimistic that BLyS antagonists will collectively hold a prominent place in the armamentarium of the clinician who treats patients with autoimmune rheumatic diseases.

2. BLyS and Its Receptors

2.1. General Biology

BLyS is a 285-amino acid member of the tumor necrosis factor (TNF) ligand superfamily. Other names for this factor are B cell–activating factor belonging to the TNF family (BAFF); TNF- and ApoL-related leukocyte-expressed ligand 1 (TALL-1); a TNF homolog that activates apoptosis, NF-κB, and JNK (THANK); TNF superfamily member 20 (TNFSF20), subsequently renamed TNFSF13B; and zTNF4 (Moore *et al.*, 1999; Mukhopadhyay *et al.*, 1999; Schneider *et al.*, 1999; Shu *et al.*, 1999; Tribouley *et al.*, 1999; Gross *et al.*, 2000). Polymorphisms within the human *Blys* gene have been identified, but no association with any distinct polymorphism has yet been appreciated among patients with autoimmune rheumatic diseases (i.e., systemic lupus erythematosus [SLE] or rheumatoid arthritis [RA]) (Kawasaki *et al.*, 2002).

William Stohl • Division of Rheumatology, University of Southern California Keck School of Medicine, Los Angeles, California 90033.

Molecular Autoimmunity: In commemoration of the 100th anniversary of the first description of human autoimmune disease, edited by Moncef Zouali. Springer Science+Business Media, Inc., New York, 2005.

Expression of BLyS is largely (but not exclusively) restricted to myeloid lineage cells (i.e., monocytes, macrophages, dendritic cells, neutrophils) (Moore et al., 1999; Schneider et al., 1999; Shu et al., 1999; Tribouley et al., 1999; Nardelli et al., 2001; Scapini et al., 2003). BLyS is also expressed to some degree by T cells (Schneider et al., 1999), and although not expressed by peripheral blood B cells, BLyS expression can be detected in several tonsillar B cell subsets (He et al., 2004).

BLyS is a type II transmembrane protein that is cleaved at the cell surface by a furin protease, resulting in release of a soluble, biologically active 17-kDa molecule (Schneider et al., 1999; Nardelli et al., 2001). In vivo, BLyS circulates in trimeric form (Schneider et al., 1999; Kanakaraj et al., 2001). Some laboratories have induced BLyS to assemble into virus-like clusters of 60 monomers in vitro (Liu, et al., 2002, 2003; Kim et al., 2003), but other laboratories have not detected multimeric self-assembly (Karpusas et al., 2002; Oren et al., 2002). The multimeric self-assembly of BLyS may simply be an artifact of the manner in which BLyS is "tagged" for in vitro experimentation (Zhukovsky et al., 2004), so whether virus-like clusters of BLyS can actually form in vivo either in the circulation or locally in tissues remains uncertain.

In addition to its full-length isoform, a naturally produced shorter isoform of BLyS (called ΔBAFF) has been identified (Gavin et al., 2003). Although it can be expressed on the cell surface of ΔBAFF-transfectants, ΔBAFF is not released into the surrounding culture medium. ΔBAFF does not bind to cells expressing BLyS receptors and, not surprisingly, is biologically inactive. Moreover, since ΔBAFF can form heterotrimers with full-length BLyS, ΔBAFF can actually block BLyS activity. How ΔBAFF production and degradation are regulated and what determines production of one isoform rather than the other are questions that remain to be addressed.

Not only is the expression of BLyS restricted, but expression of the three BLyS receptors (B cell maturation antigen [BCMA]; transmembrane activator and calcium-modulator and cyclophilin ligand-interactor [TACI]; and BAFF receptor [BAFFR], also known as BLyS receptor 3 [BR3]) is also highly restricted. Receptor expression is largely limited to B cells, although activated T cells also express TACI and BAFFR to some degree (Laabi et al., 1994; von Bülow and Bran, 1997; Thompson et al., 2001; Yan et al., 2001a; Ng et al., 2004). Accordingly, BLyS binds strongly to B cells, weakly to T cells, and not at all to NK cells or monocytes (Moore et al., 1999; Xia et al., 2000). Most, if not all, of the BLyS that binds to human peripheral blood B cells does so via surface BAFFR and/or TACI, with little, if any, BLyS binding via BCMA (Ng et al., 2004; Novak et al., 2004). However, in vitro generated human plasmablasts upregulate surface BCMA expression and downregulate surface expression of BAFFR and TACI (Avery et al., 2003), and the numbers of antigen-specific long-lived immunoglobulin (Ig)-secreting cells (plasma cells) in the bone marrow of mice genetically deficient in BCMA are much lower than those in the bone marrow of BCMA-intact mice (O'Connor et al., 2004). Thus, it is highly likely that BLyS binds discrete B cell subpopulations in vivo via BCMA as well as via BAFFR and/or TACI.

Binding of BLyS to its receptors on B cells triggers a complex intracellular signaling scheme. Several TNF receptor-associated factors (TRAFs), including TRAF1, TRAF2, TRAF3, TRAF5, and TRAF6, interact with one or more of the three BLyS receptors (Hatzoglou et al., 2000; Shu and Johnson, 2000; Xia et al., 2000; Xu and Shu, 2002). NF-κB1 and NF-κB2 are each activated (Claudio et al., 2002; Kayagaki et al., 2002; Hatada et al., 2003), with activation of the latter predominating (Zarnegar et al., 2004). The intricate network of intracellular and intranuclear events remains to be fully elucidated, but, in any case, BLyS-triggered signals lead to increased B cell survival (Batten et al., 2000; Do et al., 2000; Thompson et al., 2000; Harless et al., 2001; Hsu et al., 2002; Avery et al., 2003; Hatada et al., 2003) and the differentiation of immature B cells to marginal zone (MZ) B cells (Tardivel et al., 2004). Since MZ B cells may be crucial to the early T cell–independent (TI) protective responses against blood-borne pathogens, BLyS may vitally contribute to the host's initial antimicrobial defense (Balázs et al., 2002). Consistent with BLyS playing such role, a genome scan has identified the locus containing the *Blys* gene to highly influence susceptibility to *Ascaris* infection (Williams-Blangero et al., 2003).

BLyS also has a costimulatory role on T cells *in vitro* (Huard et al., 2001, 2004; Ng et al., 2004). BAFFR is the principal BLyS receptor involved in this costimulation (Ng et al., 2004), but the intracellular signaling pathways remain largely unexplored as is the physiologic relevance of the *in vitro* observations. Nonetheless, when assessing the consequences of excessive BLyS or BLyS antagonism on B cell function *in vivo*, one must recognize that indirect T cell–mediated effects may be contributory as well.

2.2. *In Vivo* Deficiency of BLyS or Its Receptors

BLyS is incontrovertibly vital to normal B cell development. BLyS-deficient mice display considerable, albeit incomplete, global reductions in mature "conventional" (B2) B cells (with intact peritoneal B1 B cells and intact immature bone marrow B cells) and in baseline serum Ig levels and Ig responses to T cell–dependent (TD) and TI antigens (Gross et al., 2001; Schiemann et al., 2001; Gorelik et al., 2004). This incomplete B cell depletion points to some BLyS-independent means of B cell survival (and function). *In vivo* studies using the hen egg lysozyme (HEL)/anti-HEL double-transgenic mouse model have suggested that survival of autoreactive B cells may be more BLyS-dependent than that of non-autoreactive B cells (Lesley et al., 2004; Thien et al., 2004). Whether this greater BLyS dependency of autoreactive B cells extends to disease-promoting autoreactive B cells remains to be determined.

The phenotypes of mice genetically deficient in individual BLyS receptors are highly disparate. Although immunized BCMA-deficient mice do not harbor as many antigen-specific long-lived Ig-secreting cells in their bone marrow as do BCMA-intact mice (O'Connor et al., 2004), BCMA-deficient mice otherwise exhibit no discernible phenotypic or functional abnormalities (Schiemann et al., 2001; Xu and Lam, 2001). This suggests that in a normal (non-autoimmune) environment, surface expression of BLyS receptors other than BCMA is sufficient to transmit the requisite BLyS-triggered signals for "normal" B cell survival and ultimate function. Whether BLyS–BCMA interactions play an important contributory role in an autoimmune environment remains to be formally tested. Of note, BCMA is polymorphic in humans. However, no association between SLE or RA and any specific BCMA polymorphism has been detected (Kawasaki *et al.*, 2001), consistent with a limited role (at most) for BCMA in development of clinical autoimmunity.

TACI, a second BAFF receptor, does not appear to be critical for the agonist effects of BLyS on B cells. TACI-deficient mice harbor increased, rather than decreased, numbers of B cells (although they manifest impaired Ig responses to TI, but not TD, antigens) (von Bülow *et al.*, 2001; Yan *et al.*, 2001b). As they age, these mice develop elevated circulating titers of autoantibodies, Ig deposition in their kidneys with concomitant glomerulonephritis (GN), and premature death (Seshasayee *et al.*, 2003). *In vitro* treatment of B cells with anti-TACI monoclonal antibody (mAb) blocks B cell responses to agonists (Seshasayee *et al.*, 2003), strongly suggesting that TACI transmits a negative signal to B cells.

In contrast to the phenotypes of BCMA- or TACI-deficient mice, A/WySnJ mice (which bear a mutated *Baffr* gene and express a mutant BAFFR protein) display deficiencies in mature B cell number and antibody responses qualitatively similar to those of BLyS-deficient mice (Thompson *et al.*, 2001; Yan *et al.*, 2001a). When injected with exogenous BLyS, A/WySnJ mice do not undergo splenic B lymphocytosis (whereas similarly treated A/J control mice do), and BLyS does not enhance survival of B cells from A/WySnJ mice *in vitro*. Moreover, in bone marrow chimeric mice harboring B cells that bear the mutated *Baffr* gene and B cells that bear the wild-type *Baffr* gene, the B cells bearing the mutated *Baffr* gene have decreased *in vivo* survival (Harless *et al.*, 2001). Taken together, these observations strongly point to BLyS–BAFFR interactions as essential for the agonist effects of BLyS on B cells in non-autoimmune-prone mice. Whether BLyS–BAFFR interactions are truly indispensable for development of disease in autoimmune-prone hosts requires further investigation.

2.3. Supranormal Levels of BLyS *In Vivo*

Administration of exogenous BLyS to mice at the time of immunization with antigen enhances *in vivo* antigen-specific antibody production (Do *et al.*, 2000). Repeated administration of BLyS to mice, even without intentional antigenic immunization, results in B cell expansion and polyclonal hypergammaglobulinemia (Moore *et al.*, 1999). Some of the increased Ig production likely is directed to common environmental (foreign) antigens, but some of the increased Ig production could also be directed to self-antigens. Hosts in whom endogenous BLyS production is persistently elevated may be at increased risk for development of clinical autoimmunity. Indeed, BLyS promotes *in vitro* T cell–independent class switching of IgD$^+$ B cells, which, when coupled with cross-linking of B cell surface Ig, leads to secretion of class-switched antibodies (Litinskiy *et al.*, 2002). Although *in vivo* generation of "pathogenic" autoantibodies (e.g., anti-double-stranded [ds]DNA) in SLE is felt to be a helper T cell–dependent process, it remains theoretically possible that production of such autoantibodies (and the ensuing consequences for disease) could be driven by high levels of BLyS even in the absence (marked reduction) of "pathologic" helper T cell function.

In fact, constitutive overproduction of BLyS does lead to clinical autoimmunity (although the degree of its T cell dependence remains to be established). Mice that express a *Blys* transgene (BLyS-Tg mice) frequently develop not just polyclonal hypergammaglobulinemia but also elevated titers of multiple autoantibodies (including anti-dsDNA), circulating immune complexes, and renal Ig deposits (Mackay *et al.*, 1999; Gross *et al.*, 2000; Khare *et al.*, 2000). Of note, SLE-prone (NZB × NZW)F1 (BWF1) and MRL-*lpr/lpr* mice harbor elevated circulating levels of BLyS at the onset of disease (Gross *et al.*, 2000), suggesting that "natural" BLyS overexpression may play a contributory role in the "natural" development of SLE.

Due to ethical constraints, causality between BLyS overexpression and development of clinical autoimmunity cannot be directly tested in humans. Nonetheless, there is considerable inferential evidence pointing to a role for BLyS overexpression in a number of human autoimmune conditions. Cross-sectional studies have demonstrated elevated circulating levels of BLyS in 20–30% of human SLE patients tested at a single point in time (Cheema *et al.*, 2001; Zhang *et al.*, 2001). Weak correlation was observed between circulating BLyS and total IgG levels, and stronger correlation was observed between circulating BLyS levels and anti-dsDNA titers.

A subsequent 12-month longitudinal study of 68 SLE patients (and 20 healthy control subjects) highlighted the differences in BLyS expression between SLE patients and healthy controls. Whereas the control subjects uniformly maintained stable "normal" serum BLyS levels over time, elevated serum levels of BLyS were persistently observed in ~25% of the SLE patients, and intermittent elevations in serum BLyS levels were observed in additional ~25% of patients (Stohl *et al.*, 2003). Although the mechanism underlying BLyS overexpression remains to be determined, it is clear that BLyS overexpression is common among human SLE patients.

Of note, circulating BLyS levels do not overtly correlate with disease activity (measured by the SLE Disease Activity Index [SLEDAI]) for any individual SLE patient followed over time, (Stohl *et al.*, 2003). However, at the population level, circulating BLyS levels do correlate with disease activity. Among 245 SLE patients from 4 different medical centers followed for an average of 15 months, circulating BLyS levels correlated weakly but significantly with disease activity when analyzed across the entire population (Petri *et al.*, 2003). Thus, although BLyS has no known direct proinflammatory properties, its positive effects on B cell survival and/or autoantibody production appear to increase the likelihood of aggravating and/or exacerbating disease.

Circulating BLyS levels are elevated not just in SLE patients but also in a substantial fraction of RA patients (Cheema *et al.*, 2001; Zhang *et al.*, 2001). Furthermore, in patients with RA or other inflammatory arthritides, the BLyS levels in the synovial fluids (SFs) from clinically affected joints are almost always higher than those in corresponding sera (Tan *et al.*, 2003). The latter observation points to local joint BLyS production in inflammatory arthritis. Although speculative, it may be that local production of BLyS in the joint plays a vital role in pathogenesis of RA and related disorders.

Human patients with Sjögren's syndrome (SS) frequently harbor elevated circulating levels of BLyS (Groom *et al.*, 2002; Mariette *et al.*, 2003), reminiscent of the SS-like illness developed late in life by BLyS-Tg mice (Groom *et al.*, 2002). One study documented a correlation between circulating levels of BLyS and anti-Ro/SSA or anti-La/SSB autoantibodies in human SS patients (Mariette *et al.*, 2003), although this finding was not replicated in another study (Groom *et al.*, 2002). T cells infiltrating the salivary glands in SS patients overexpress BLyS (Lavie *et al.*, 2004), but it is not yet certain whether these infiltrating T cells actually produce the BLyS or whether they passively adsorb BLyS produced by other cell types. In any case, the increased local accumulation of BLyS may permit survival of B cells undergoing malignant lymphomatous transformation, perhaps the most ominous and worrisome complication of SS.

Circulating BLyS levels are also often increased in patients infected with HIV (Stohl *et al.*, 2002; Rodriguez *et al.*, 2003). Whether these patients are at greater risk for development of B cell lymphomas relative to the HIV-infected population at large is unknown. In addition, patients in general with non-Hodgkin's B cell lymphomas harbor elevated circulating BLyS levels (Briones *et al.*, 2002). Such patients and those with other hematologic malignancies, including multiple myeloma and chronic lymphocytic leukemia, harbor circulating B cells that elaborate BLyS (Novak *et al.*, 2002, 2004; He *et al.*, 2004; Kern *et al.*, 2004; Moreaux *et al.*, 2004). Since the neoplastic B cells express receptors for BLyS, their elaboration of BLyS may give rise to a clinically important autocrine pathway of survival.

2.4. APRIL and Its Relevance to BLyS

Any discussion of BLyS must include its "cousin," a proliferation-inducing ligand (APRIL; also known as TALL-2, TNF-related death ligand-1 [TRDL-1], and TNFSF13A), a 250-amino acid member of the TNF ligand superfamily that shares substantial homology with BLyS and binds to two of the three BLyS receptors (BCMA and TACI) (Hahne *et al.*, 1998; Shu *et al.*, 1999; Kelly *et al.*, 2000; Marsters *et al.*, 2000; Rennert *et al.*, 2000; Wu *et al.*, 2000; Yu *et al.*, 2000) but not to BAFFR (Thompson *et al.*, 2001). APRIL costimulates B cells *in vitro* and *in vivo* (Marsters *et al.*, 2000; Yu *et al.*, 2000; Litinskiy *et al.*, 2002), although it does so with considerably less potency than that of BLyS (Craxton *et al.*, 2003).

Its ability to bind to BCMA and TACI notwithstanding, neither complete deficiency of APRIL nor its constitutive overexpression has dramatic effects on *in vivo* biology. Mice genetically deficient in APRIL have been reported to be phenotypically normal (Varfolomeev *et al.*, 2004) or to have a modest deficiency in generating IgA responses despite normal B cell numbers and development (Castigli *et al.*, 2004). APRIL-Tg mice, which constitutively overexpress APRIL, manifest only subtle immunologic abnormalities (Stein *et al.*, 2002). Most noteworthy, no serologic or clinical autoimmune features are appreciated in these mice.

The circulating concentrations of APRIL in normal human hosts are considerably greater than those of BLyS (Stohl *et al.*, 2004), raising the likelihood that APRIL exerts meaningful *in vivo* effects. Supporting this notion is the association of a specific *April* polymorphism with SLE in a Japanese cohort (Koyama *et al.*, 2003). Of note, APRIL can complex with BLyS *in vivo* to form BLyS/APRIL heterotrimers (BAHT), which are biologically active *in vitro* (Roschke *et al.*, 2002). Whether BAHT have greater, equal, or lesser bioactivity than does BLyS *in vivo* is not yet known. Regardless, APRIL may counterbalance the autoimmunogenic effects of BLyS overexpression. In our cohort of 68 SLE patients longitudinally studied over a 12-month period, serum APRIL levels inversely correlated with serum anti-dsDNA titers (in anti-dsDNA-positive patients) and inversely correlated with clinical disease activity (as measured by SLEDAI) (Stohl *et al.*, 2004).

3. BLyS Antagonism as a Therapeutic Modality

3.1. Mouse Models

BLyS antagonism has been shown to be an effective therapeutic modality in murine SLE. BWF1 and MRL-*lpr/lpr* mice respond clinically (decreased disease progression and improved survival) to repeated injections of a soluble fusion protein between one of the BLyS receptors (TACI or BAFFR) and IgG Fc (TACI-Ig and BAFFR-Ig, respectively) (Gross *et al.*, 2000; Kayagaki *et al.*, 2002). The ability of TACI-Ig and BAFFR-Ig to similarly inhibit disease in these models indicates that neutralization of BLyS, rather than APRIL, lies at the core of the salutary clinical response, since BAFFR-Ig binds and neutralizes *only* BLyS.

Although the salutary clinical response in BWF1 mice to one BLyS antagonist (BAFFR-Ig) in one study was associated with reduced circulating levels of anti-dsDNA antibodies (Kayagaki *et al.*, 2002), the dramatic *in vivo* clinical response in BWF1 mice to a different BLyS antagonist (TACI-Ig) in another study was not associated with any reduction in circulating anti-dsDNA titers (Gross *et al.*, 2000). It is not known whether these disparate results are due to inherent differences in the BLyS antagonists used. Regardless, they do strongly suggest that effective blockade of clinical autoimmunity by BLyS antagonists may ensue via a pathway independent of anti-dsDNA autoantibodies. Whether this pathway critically depends on some other autoantibody or whether this pathway is actually autoantibody-independent is unknown at present. It may be that physical B cell depletion is required to achieve favorable clinical responses rather than just a reduction in circulating levels of autoantibodies.

An important caveat to the mouse studies is that elicitation of a favorable clinical response by a given BLyS antagonist in one model does not invariably guarantee a favorable response to this BLyS antagonist in a second model. Administration of an adenoviral vector containing a TACI-Ig to MRL-*lpr/lpr* mice promoted persistently elevated circulating levels of TACI-Ig. These hosts demonstrated phenotypic evidence of BLyS neutralization, including substantial amelioration of the renal disease that they would otherwise have developed. However, administration of the identical adenoviral vector to BWF1 mice afforded no clinical protection, likely due to the development of anti-TACI antibodies by the BWF1 hosts (Liu et al., 2004). (Although repeated injections of TACI-Ig to BWF1 mice were clinically efficacious (Gross et al., 2000) and, presumably, did not elicit a clinically significant anti-TACI response, the greater continual exposure to TACI-Ig experienced by the adenovirus-infected hosts may have facilitated development of a biologically more potent anti-TACI response.) Accordingly, the response to any individual BLyS antagonist may be highly variable within a genetically heterogeneous population. Extrapolating to human SLE patients, one single BLyS antagonist may not be effective in all subjects, so multiple different antagonists may need to be developed for human clinical use.

BLyS antagonism has also been shown to be an effective therapeutic modality in murine collagen-induced arthritis (CIA), a model of RA. CIA can be inhibited after its induction by treatment with TACI-Ig. In addition to the marked reduction in joint inflammation and destruction, systemic T cell and B cell responses to the immunogen (collagen) are also markedly decreased (Gross et al., 2001; Wang et al., 2001). Since T cell phenotype, numbers, and function in BLyS-deficient mice are grossly normal despite readily apparent abnormalities in B cells (Schiemann et al., 2001), the therapeutic effects of a BLyS antagonist on a T cell–mediated disease suggest that a factor (such as BLyS) that enhances B cell survival, proliferation, and/or differentiation may play a vital indirect role in propagating a pathologic T cell response. Indeed, accumulation of plasma cells in the affected joints of RA patients is well established and may arise, at least in part, from the increased local levels of BLyS (Tan et al., 2003). The plasma cells might then elaborate antibodies crucial to formation of phlogistic immune complexes and/or cytokines vital to propagating the local destructive process.

3.2. The Human Experience

Given the associations between levels of circulating BLyS and levels of circulating autoantibodies or clinical disease activity in human SLE along with the success of BLyS antagonists in treating murine SLE, clinical trials with BLyS antagonists have been initiated. A phase I clinical trial in SLE patients with a human anti-BLyS mAb (belimumab) has already been completed (Baker et al., 2003; Furie et al., 2003). A total of 70 patients were enrolled in this multicenter double-blind trial, and each patient received either a single infusion of drug at one of four doses (or placebo) or received two infusions of drug at one of the same four doses (or placebo) separated by 3 weeks. Biologic activity of the anti-BLyS mAb was documented by a reduction in circulating B cells among drug-treated (but not placebo-treated) patients, and safety of the anti-BLyS mAb was documented by there being no difference in frequency of adverse events between drug-treated and placebo-treated patients. Phase II clinical trials with this anti-BLyS mAb are currently under way in SLE and RA patients.

Agents other than anti-BLyS mAb are also being evaluated for use in humans. TACI-Ig and BAFFR-Ig are currently undergoing phase I evaluation. These two fusion proteins differ in their abilities to neutralize APRIL, with only TACI-Ig being an effective neutralizer of APRIL. Given the inverse correlation between circulating APRIL levels and disease activity among SLE patients (Stohl et al., 2004), it may be therapeutically judicious to choose a BLyS antagonist that does not antagonize APRIL. Clinical experience will ultimately determine the legitimacy of this concern.

In addition to direct antagonism of BLyS, agents that target specific BLyS receptors could, in principle, be therapeutically successful. Such agents include radiolabeled BLyS or BLyS conjugated to a toxin, nonagonist anti-BAFFR mAb, and agonist anti-TACI mAb that could deliver a negative signal to the B cell and overwhelm any positive signals delivered through BAFFR.

An exciting novel approach has recently been reported for inactivation of TNF signaling. A computational structure-based design strategy was utilized to engineer variant TNF proteins that rapidly form heterotrimers with native TNF. These heterotrimers neither bind to TNF receptors nor trigger TNF-mediated responses, so the engineered variant proteins functionally act as dominant-negative agents (Steed et al., 2003). In principle, one could also engineer variants of BLyS that would complex with native BLyS to form heterotrimers that could neither bind nor trigger the BLyS receptors.

3.3. Which Patients Are Candidates for BLyS Antagonist Therapy?

In terms of promoting development of autoimmunity, BLyS may assume at least two distinct roles. On the one hand, BLyS may serve as a *contributor* to development of disease. BLyS per se may not cause loss of tolerance to self-antigens, but once such tolerance is broken, the ever-present nature of the autoantigen permits it to repetitively stimulate the host immune system and elicit a detectable autoimmune response. Autoreactive B cells may have a greater BLyS dependency than do non-autoreactive B cells (Lesley et al., 2004; Thien et al., 2004), so in the presence of increasing amounts of BLyS, the autoimmune response is exaggerated. When coupled to additional permissive genetic and/or environmental factors, this exaggerated autoimmune response can lead to frank clinical disease.

Accordingly, a reduction in BLyS levels to "normal" should ameliorate disease by suppressing the BLyS-driven acceleration or exaggeration of the autoimmune response. Although self-tolerance would still be "broken," the magnitude of the autoimmune response would be now insufficient to drive clinical disease. Thus, patients with the most elevated circulating BLyS levels should be the ones most responsive to BLyS antagonist therapy. Those patients with normal circulating BLyS levels might be relatively insensitive to BLyS antagonist therapy, since "excess" BLyS in these patients would not be driving the clinical autoimmunity.

On the other hand, BLyS may serve as a passive *facilitator* in development of disease. In this model, development of the pathologic anti-self response is inherently BLyS-independent. The magnitude of the autoimmune response is similar regardless of whether BLyS levels are normal or elevated. That is, the autoimmune response is so robust that it is not further amplified by increased amounts of BLyS. Indeed, the fact that a considerable number of patients with autoimmune rheumatic diseases do not overtly overexpress BLyS strongly suggests that BLyS overexpression is not absolutely essential to development of disease. Nevertheless, given the crucial role of BLyS in B cell development, a certain threshold level of BLyS is required to permit meaningful B cell responses (including autoantibody responses). When BLyS levels are reduced below this critical threshold level, the ability to fully mount an autoimmune response (along with other B cell and humoral responses) is impaired. Accordingly, the patients most responsive to BLyS antagonist therapy might be those with normal, rather than elevated, circulating BLyS levels, since in such patients, less neutralization of BLyS would be required to reach the critical threshold level.

These two models are not mutually exclusive. There likely are individuals in whom BLyS plays a contributor role, and there likely are others in whom BLyS plays a facilitator role. From a therapeutic perspective, the models may operationally be viewed as a continuum, with some patients requiring relatively more BLyS neutralization and other patients requiring relatively less BLyS neutralization before clinical benefits are achieved.

3.4. Concluding Comments

Based on an increasing body of *in vivo* and *ex vivo* evidence in mice and humans, BLyS likely plays a key role in the pathogenesis of several systemic autoimmune rheumatic diseases, including SLE, RA, and SS. Clinical efficacy of BLyS antagonists has been decisively demonstrated in multiple mouse models, and safety of at least one BLyS antagonist has already been documented in humans. Although additional investigation is yet needed, one can be cautiously optimistic that BLyS antagonism will secure a prominent place in the therapeutic management of patients with autoimmune rheumatic diseases.

References

Avery, D.T., Kalled, S.L., Ellyard, J.I., Ambrose, C., Bixler, S.A., Thien, M., Brink, R., Mackay, F., Hodgkin, P.D., and Tangye, S.G. (2003). BAFF selectively enhances the survival of plasmablasts generated from human memory B cells. *J. Clin. Invest.*, **112**, 286–297.

Baker, K.P., Edwards, B.M., Main, S.H., Choi, G.H., Wager, R.E., Halpern, W.G., Lappin, P.B., Riccobene, T., Abramian, D., Sekut, L., Sturm, B., Poortman, C., Minter, R.R., Dobson, C.L., Williams, E., Carmen, S., Smith, R., Roschke, V., Hilbert, D.M., Vaughan, T.J., and Albert, V.R. (2003). Generation and characterization of LymphoStat-B, a human monoclonal antibody that antagonizes the bioactivities of B lymphocyte stimulator. *Arthritis Rheum.*, **48**, 3253–3265.

Balázs, M., Martin, F., Zhou, T., and Kearney, J.F. (2002). Blood dendritic cells interact with splenic marginal zone B cells to initiate T-independent immune responses. *Immunity*, **17**, 341–352.

Batten, M., Groom, J., Cachero, T.G., Qian, F., Schneider, P., Tschopp, J., Browning, J.L., and Mackay, F. (2000). BAFF mediates survival of peripheral immature B lymphocytes. *J. Exp. Med.*, **192**, 1453–1465.

Briones, J., Timmerman, J.M., Hilbert, D.M., and Levy, R. (2002). BLyS and BLyS receptor expression in non-Hodgkin's lymphoma. *Exp. Hematol.*, **30**, 135–141.
Castigli, E., Scott, S., Dedeoglu, F., Bryce, P., Jabara, H., Bhan, A.K., Migozuchi, E., and Geha, R.S. (2004). Impaired IgA class switching in APRIL-deficient mice. *Proc. Natl. Acad. Sci.*, **101**, 3903–3908.
Cheema, G.S., Roschke, V., Hilbert, D.M., and Stohl, W. (2001). Elevated serum B lymphocyte stimulator levels in patients with systemic immune-based rheumatic diseases. *Arthritis Rheum.*, **44**, 1313–1319.
Claudio, E., Brown, K., Park, S., Wang, H., and Siebenlist, U. (2002). BAFF-induced NEMO-independent processing of NK-κB2 in maturing B cells. *Nat. Immunol.*, **3**, 958–965.
Craxton, A., Magaletti, D., Ryan, E.J., and Clark, E.A. (2003). Macrophage- and dendritic cell-dependent regulation of human B-cell proliferation requires the TNF family ligand BAFF. *Blood*, **101**, 4464–4471.
Do, R.K.G., Hatada, E., Lee, H., Tourigny, M.R., Hilbert, D., and Chen-Kiang, S. (2000). Attenuation of apoptosis underlies B lymphocyte stimulator enhancement of humoral immune response. *J. Exp. Med.*, **192**, 953–964.
Furie, R., Stohl, W., Ginzler, E., Becker, M., Mishra, N., Chatham, W., Merrill, J.T., Weinstein, A., McCune, W.J., Zhong, J., Freimuth, W., and Lymphostat-B Study Group (2003). Safety, pharmacokinetic and pharmacodynamic results of a phase 1 single and double dose-escalation study of Lymphostat-B (human monoclonal antibody to BLyS) in SLE patients. *Arthritis Rheum.*, **48**, S377.
Gavin, A.L., Aït-Azzouzene, D., Ware, C.F., and Nemazee, D. (2003). ΔBAFF, an alternate splice isoform that regulates receptor binding and biopresentation of the B cell survival cytokine, BAFF. *J. Biol. Chem.*, **278**, 38220–38228.
Gorelik, L., Cutler, A.H., Thill, G., Miklasz, S.D., Shea, D.E., Ambrose, C., Bixler, S.A., Su, L., Scott, M.L., and Kalled, S.L. (2004). Cutting edge: BAFF regulates CD21/35 and CD23 expression independent of its B cell survival function. *J. Immunol.*, **172**, 762–766.
Groom, J., Kalled, S.L., Cutler, A.H., Olson, C., Woodcock, S.A., Schneider, P., Tschopp, J., Cachero, T.G., Batten, M., Wheway, J., Mauri, D., Cavill, D., Gordon, T.P., Mackay, C.R., and Mackay, F. (2002). Association of BAFF/BLyS overexpression and altered B cell differentiation with Sjögren's syndrome. *J. Clin. Invest.*, **109**, 59–68.
Gross, J.A., Dillon, S.R., Mudri, S., Johnston, J., Littau, A., Roque, R., Rixon, M., Schou, O., Foley, K.P., Haugen, H., McMillen, S., Waggie, K., Schreckhise, R.W., Shoemaker, K., Vu, T., Moore, M., Grossman, A., and Clegg, C.H. (2001). TACI-Ig neutralizes molecules critical for B cell development and autoimmune disease: Impaired B cell maturation in mice lacking BLyS. *Immunity*, **15**, 289–302.
Gross, J.A., Johnston, J., Mudri, S., Enselman, R., Dillon, S.R., Madden, K., Xu, W., Parrish-Novak, J., Foster, D., Lofton-Day, C., Moore, M., Littau, A., Grossman, A., Haugen, H., Foley, K., Blumberg, H., Harrison, K., Kindsvogel, W., and Clegg, C.H. (2000). TACI and BCMA are receptors for a TNF homologue implicated in B-cell autoimmune disease. *Nature*, **404**, 995–999.
Hahne, M., Kataoka, T., Schröter, M., Hofmann, K., Irmler, M., Bodmer, J.-L., Schneider, P., Bornand, T., Holler, N., French, L.E., Sordat, B., Rimoldi, D., and Tschopp, J. (1998). APRIL, a new ligand of the tumor necrosis factor family, stimulates tumor cell growth. *J. Exp. Med.*, **188**, 1185–1190.
Harless, S.M., Lentz, V.M., Sah, A.P., Hsu, B.L., Clise-Dwyer, K., Hilbert, D.M., Hayes, C.E., and Cancro, M.P. (2001). Competition for BLyS-mediated signaling through Bcmd/BR3 regulates peripheral B lymphocyte numbers. *Curr. Biol.*, **11**, 1986–1989.
Hatada, E.N., Do, R.K.G., Orlofsky, A., Liou, H.-C., Prystowsky, M., MacLennan, I.C.M., Caamano, J., and Chen-Kiang, S. (2003). NF-κB1 p50 is required for BLyS attenuation of apoptosis but dispensable for processing of NF-κB2 p100 to p52 in quiescent mature B cells. *J. Immunol.*, **171**, 761–768.
Hatzoglou, A., Roussel, J., Bourgeade, M.-F., Rogier, E., Madry, C., Inoue, J., Devergne, O., and Tsapis, A. (2000). TNF receptor family member BCMA (B cell maturation) associates with TNF receptor-associated factor (TRAF) 1, TRAF2, and TRAF3 and activates NF-κB, Elk-1, c-Jun N-terminal kinase, and p38 mitogen-activated protein kinase. *J. Immunol.*, **165**, 1322–1330.

He, B., Chadburn, A., Jou, E., Schattner, E.J., Knowles, D.M., and Cerutti, A. (2004). Lymphoma B cells evade apoptosis through the TNF family members BAFF/BLyS and APRIL. *J. Immunol.*, **172**, 3268–3279.

Hsu, B.L., Harless, S.M., Lindsley, R.C., Hilbert, D.M., and Cancro, M.P. (2002). Cutting edge: BLyS enables survival of transitional and mature B cells through distinct mediators. *J. Immunol.*, **168**, 5993–5996.

Huard, B., Arlettaz, L., Ambrose, C., Kindler, V., Mauri, D., Roosnek, E., Tschopp, J., Schneider, P., and French, L.E. (2004). BAFF production by antigen-presenting cells provides T cell co-stimulation. *Int. Immunol.*, **16**, 467–475.

Huard, B., Schneider, P., Mauri, D., Tschopp, J., and French, L.E. (2001). T cell costimulation by the TNF ligand BAFF. *J. Immunol.*, **167**, 6225–6231.

Kanakaraj, P., Migone, T.-S., Nardelli, B., Ullrich, S., Li, Y., Olsen, H.S., Salcedo, T.W., Kaufman, T., Cochrane, E., Gan, Y., Hilbert, D.M., and Giri, J. (2001). BLyS binds to B cells with high affinity and induces activation of the transcription factors NF-κB and ELF-1. *Cytokine*, **13**, 25–31.

Karpusas, M., Cachero, T.G., Qian, F., Boriack-Sjodin, A., Mullen, C., Strauch, K., Hsu, Y.-M., and Kalled, S.L. (2002). Crystal structure of extracellular human BAFF, a TNF family member that stimulates B lymphocytes. *J. Mol. Biol.*, **315**, 1145–1154.

Kawasaki, A., Tsuchiya, N., Fukazawa, T., Hashimoto, H., and Tokunaga, K. (2001). Presence of four major haplotypes in human BCMA gene: Lack of association with systemic lupus erythematosus and rheumatoid arthritis. *Genes Immun.*, **2**, 276–279.

Kawasaki, A., Tsuchiya, N., Fukazawa, T., Hashimoto, H., and Tokunaga, K. (2002). Analysis on the association of human BLYS (BAFF, TNFSF13B) polymorphisms with systemic lupus erythematosus and rheumatoid arthritis. *Genes Immun.*, **3**, 424–429.

Kayagaki, N., Yan, M., Seshasayee, D., Wang, H., Lee, W., French, D.M., Grewal, I.S., Cochran, A.G., Gordon, N.C., Yin, J., Starovasnik, M.A., and Dixit, V.M. (2002). BAFF/BLyS receptor 3 binds the B cell survival factor BAFF ligand through a discrete surface loop and promotes processing of NF-κB2. *Immunity*, **10**, 515–524.

Kelly, K., Manos, E., Jensen, G., Nadauld, L., and Jones, D.A. (2000). APRIL/TRDL-1, a tumor necrosis factor-like ligand, stimulates cell death. *Cancer Res.*, **60**, 1021–1027.

Kern, C., Cornuel, J.-F., Billard, C., Tang, R., Rouillard, D., Steunou, V., Defrance, T., Ajchenbaum-Cymbalista, F., Simonin, P.-Y., Feldbaum, S., and Kolb, J.-P. (2004). Involvement of BAFF and APRIL in the resistance to apoptosis of B-CLL through an autocrine pathway. *Blood*, **103**, 679–688.

Khare, S.D., Sarosi, I., Xia, X.-Z., McCabe, S., Miner, K., Solovyev, I., Hawkins, N., Kelley, M., Chang, D., Van, G., Ross, L., Delaney, J., Wang, L., Lacey, D., Boyle, W.J., and Hsu, H. (2000). Severe B cell hyperplasia and autoimmune disease in TALL-1 transgenic mice. *Proc. Natl. Acad. Sci.*, **97**, 3370–3375.

Kim, H.M., Yu, K.S., Lee, M.E., Shin, D.R., Kim, Y.S., Paik, S.-G., Yoo, O.J., Lee, H., and Lee, J.-O. (2003). Crystal structure of the BAFF-BAFF-R complex and its implications for receptor activation. *Nat. Struct. Biol.*, **10**, 342–348.

Koyama, T., Tsukamoto, H., Masumoto, K., Himeji, D., Hayashi, K., Harada, M., and Horiuchi, T. (2003). A novel polymorphism of the human *APRIL* gene is associated with systemic lupus erythematosus. *Rheumatology*, **42**, 980–985.

Laabi, Y., Gras, M.-P., Brouet, J.-C., Berger, R., Larsen, C.-J., and Tsapis, A. (1994). The BCMA gene, preferentially expressed during B lymphoid maturation, is bidirectionally transcribed. *Nucleic Acids Res.*, **22**, 1147–1154.

Lavie, F., Miceli-Richard, C., Quillard, J., Roux, S., Leclerc, P., and Mariette, X. (2004). Expression of BAFF (BLyS) in T cells infiltrating labial salivary glands from patients with Sjögren's syndrome. *J. Pathol.*, 202, 496–502.

Lesley, R., Xu, Y., Kalled, S.L., Hess, D.M., Schwab, S.R., Shu, H.-B., and Cyster, J.G. (2004). Reduced competitiveness of autoantigen-engaged B cells due to increased dependence on BAFF. *Immunity*, **20**, 441–453.

Litinskiy, M.B., Nardelli, B., Hilbert, D.M., He, B., Schaffer, A., Casali, P., and Cerutti, A. (2002). DCs induce CD40-independent immunoglobulin class switching through BLyS and APRIL. *Nat. Immunol.*, **3**, 822–829.

Liu, W., Szalai, A., Zhao, L., Liu, D., Martin, F., Kimberly, R.P., Zhou, T., and Carter, R.H. (2004). Control of spontaneous B lymphocyte autoimmunity with adenovirus-encoded soluble TACI. *Arthritis Rheum.*, **50**, 1884–1896.

Liu, Y., Hong, X., Kappler, J., Jiang, L., Zhang, R., Xu, L., Pan, C.-H., Martin, W.E., Murphy, R.C., Shu, H.-B., Dai, S., and Zhang, G. (2003). Ligand-receptor binding revealed by the TNF family member TALL-1. *Nature*, **423**, 49–56.

Liu, Y., Xu, L., Opalka, N., Kappler, M., Shu, H.-B., and Zhang, G. (2002). Crystal structure of sTALL-1 reveals a virus-like assembly of TNF family ligands. *Cell*, **108**, 383–394.

Mackay, F., Woodcock, S.A., Lawton, P., Ambrose, C., Baetscher, M., Schneider, P., Tschopp, J., and Browning, J.L. (1999). Mice transgenic for BAFF develop lymphocytic disorders along with autoimmune manifestations. *J. Exp. Med.*, **190**, 1697–1710.

Mariette, X., Roux, S., Zhang, J., Bengoufa, D., Lavie, F., Zhou, T., and Kimberly, R. (2003). The level of BLyS (BAFF) correlates with the titre of autoantibodies in human Sjögren's syndrome. *Ann. Rheum. Dis.*, **62**, 168–171.

Marsters, S.A., Yan, M., Pitti, R.M., Haas, P.E., Dixit, V.M., and Ashkenazi, A. (2000). Interaction of the TNF homologues BLyS and APRIL with the receptor homologues BCMA and TACI. *Curr. Biol.*, **10**, 785–788.

Moore, P.A., Belvedere, O., Orr, A., Pieri, K., LaFleur, D.W., Feng, P., Soppet, D., Charters, M., Gentz, R., Parmelee, D., Li, Y., Galperina, O., Giri, J., Roschke, V., Nardelli, B., Carrell, J., Sosnovtseva, S., Greenfield, W., Ruben, S.M., Olsen, H.S., Fikes, J., and Hilbert, D.M. (1999). BLyS: Member of the tumor necrosis factor family and B lymphocyte stimulator. *Science*, **285**, 260–263.

Moreaux, J., Legouffe, E., Jourdan, E., Quittet, P., Rème, T., Lugagne, C., Moine, P., Rossi, J.-F., Klein, B., and Tarte, K. (2004). BAFF and APRIL protect myeloma cells from apoptosis induced by IL-6 deprivation and dexamethasone. *Blood*, **103**, 3148–3157.

Mukhopadhyay, A., Ni, J., Zhai, Y., Yu, G.-L., and Aggarwal, B.B. (1999). Identification and characterization of a novel cytokine, THANK, a TNF homologue that activates apoptosis, nuclear factor-κB, and c-Jun NH$_2$-terminal kinase. *J. Biol. Chem.*, **274**, 15978–15981.

Nardelli, B., Belvedere, O., Roschke, V., Moore, P.A., Olsen, H.S., Migone, T.S., Sosnovtseva, S., Carrell, J.A., Feng, P., Giri, J.G., and Hilbert, D.M. (2001). Synthesis and release of B-lymphocyte stimulator from myeloid cells. *Blood*, **97**, 198–204.

Ng, L.G., Sutherland, A.P.R., Newton, R., Qian, F., Cachero, T.G., Scott, M.L., Thompson, J.S., Wheway, J., Chtanova, T., Groom, J., Sutton, I.J., Xin, C., Tangye, S.G., Kalled, S.L., Mackay, F., and Mackay, C.R. (2004). B cell-activating factor belonging to the TNF family (BAFF)-R is the principal BAFF receptor facilitating BAFF costimulation of circulating T and B cells. *J. Immunol.*, **173**, 807–817.

Novak, A.J., Bram, R.J., Kay, N.E., and Jelinek, D.F. (2002). Aberrant expression of B-lymphocyte stimulator by B chronic lymphocytic leukemia cells: A mechanism for survival. *Blood*, **100**, 2973–2979.

Novak, A.J., Darce, J.R., Arendt, B.K., Harder, B., Henderson, K., Kindsvogel, W., Gross, J.A., Greipp, P.R., and Jelinek, D.F. (2004). Expression of BCMA, TACI, and BAFF-R in multiple myeloma: A mechanism for growth and survival. *Blood*, **103**, 689–694.

O'Connor, B.P., Raman, V.S., Erickson, L.D., Cook, W.J., Weaver, L.K., Ahonen, C., Lin, L.-L., Mantchev, G.T., Bram, R.J., and Noelle, R.J. (2004). BCMA is essential for the survival of long-lived bone marrow plasma cells. *J. Exp. Med.*, **199**, 91–97.

Oren, D.A., Li, Y., Volovik, Y., Morris, T.S., Dharia, C., Das, K., Galperina, O., Gentz, R., and Arnold, E. (2002). Structural basis of BLyS receptor recognition. *Nat. Struct. Biol.*, **9**, 288–292.

Petri, M., Stohl, W., Chatham, W., McCune, W.J., Butler, T., Ryel, J., Zhong, J., Recta, J., and Freimuth, W. (2003). BLyS plasma concentrations correlate with disease activity and levels of anti-dsDNA autoantibodies and immunoglobulins (Ig) in a SLE patient observational study. *Arthritis Rheum.*, **48**, S655.

Rennert, P., Schneider, P., Cachero, T.G., Thompson, J., Trabach, L., Hertig, S., Holler, N., Qian, F., Mullen, C., Strauch, K., Browning, J.L., Ambrose, C., and Tschopp, J. (2000). A soluble form of B cell maturation antigen, a receptor for the tumor necrosis factor family member APRIL, inhibits tumor cell growth. *J. Exp. Med.*, **192**, 1677–1683.

Rodriguez, B., Valdez, H., Freimuth, W., Butler, T., Asaad, R., and Lederman, M.M. (2003). Plasma levels of B-lymphocyte stimulator increase with HIV disease progression. *AIDS*, **17**, 1983–1985.

Roschke, V., Sosnovtseva, S., Ward, C.D., Hong, J.S., Smith, R., Albert, V., Stohl, W., Baker, K.P., Ullrich, S., Nardelli, B., Hilbert, D.M., and Migone, T.-S. (2002). BLyS and APRIL form biologically active heterotrimers that are expressed in patients with systemic immune-based rheumatic diseases. *J. Immunol.*, **169**, 4314–4321.

Scapini, P., Nardelli, B., Nadali, G., Calzetti, F., Pizzolo, G., Montecucco, C., and Cassatella, M.A. (2003). G-CSF-stimulated neutrophils are a prominent source of functional BLyS. *J. Exp. Med.*, **197**, 297–302.

Schiemann, B., Gommerman, J.L., Vora, K., Cachero, T.G., Shulga-Morskaya, S., Dobles, M., Frew, E., and Scott, M.L. (2001). An essential role for BAFF in the normal development of B cells through a BCMA-independent pathway. *Science*, **293**, 2111–2114.

Schneider, P., MacKay, F., Steiner, V., Hofmann, K., Bodmer, J.-L., Holler, N., Ambrose, C., Lawton, P., Bixler, S., Acha-Orbea, H., Valmori, D., Romero, P., Werner-Favre, C., Zubler, R.H., Browning, J.L., and Tschopp, J. (1999). BAFF, a novel ligand of the tumor necrosis factor family, stimulates B cell growth. *J. Exp. Med.*, **189**, 1747–1756.

Seshasayee, D., Valdez, P., Yan, M., Dixit, V.M., Tumas, D., and Grewal, I.S. (2003). Loss of TACI causes fatal lymphoproliferation and autoimmunity, establishing TACI as an inhibitory BLyS receptor. *Immunity*, **18**, 279–288.

Shu, H.-B., Hu, W.-H., and Johnson, H. (1999). TALL-1 is a novel member of the TNF family that is down-regulated by mitogens. *J. Leukoc. Biol.*, **65**, 680–683.

Shu, H.-B. and Johnson, H. (2000). B cell maturation protein is a receptor for the tumor necrosis factor family member TALL-1. *Proc. Natl. Acad. Sci.*, **97**, 9156–9161.

Steed, P.M., Tansey, M.G., Zalevsky, J., Zhukovsky, E.A., Desjarlais, J.R., Szymkowski, D.E., Abbott, C., Carmichael, D., Chan, C., Cherry, L., Cheung, P., Chirino, A.J., Chung, H.H., Doberstein, S.K., Eivazi, A., Filikov, A.V., Gao, S.X., Hubert, R.S., Hwang, M., Hyun, L., Kashi, S., Kim, A., Kim, E., Kung, J., Martinez, S.P., Muchhal, U.S., Nguyen, D.-H.T., O'Brien, C., O'Keefe, D., Singer, K., Vafa, O., Vielmetter, J., Yoder, S.C., and Dahiyat, B.I. (2003). Inactivation of TNF signaling by rationally designed dominant-negative TNF variants. *Science*, **301**, 1895–1898.

Stein, J.V., López-Fraga, M., Elustondo, F.A., Carvalho-Pinto, C.E., Rodríguez, D., Gómez-Caro, R., de Jong, J., Martínez-A., C., Medema, J.P., and Hahne, M. (2002). APRIL modulates B and T cell immunity. *J. Clin. Invest.*, **109**, 1587–1598.

Stohl, W., Cheema, G.S., Briggs, W., Xu, D., Sosnovtseva, S., Roschke, V., Ferrara, D.E., Labat, K., Sattler, F.R., Pierangeli, S.S., and Hilbert, D.M. (2002). B lymphocyte stimulator protein-associated increase in circulating autoantibody levels may require CD4[+] T cells: Lessons from HIV-infected patients. *Clin. Immunol.*, **104**, 115–122.

Stohl, W., Metyas, S., Tan, S.-M., Cheema, G.S., Oamar, B., Xu, D., Roschke, V., Wu, Y., Baker, K.P., and Hilbert, D.M. (2003). B lymphocyte stimulator overexpression in patients with systemic lupus erythematosus: Longitudinal observations. *Arthritis Rheum.*, **48**, 3475–3486.

Stohl, W., Metyas, S., Tan, S.-M., Cheema, G.S., Oamar, B., Roschke, V., Wu, Y., Baker, K.P., and Hilbert, D.M. (2004). Inverse association between circulating APRIL levels and serologic and clinical disease activity in patients with systemic lupus erythematosus. *Ann. Rheum. Dis.*, **63**, 1096–1103.

Tan, S.-M., Xu, D., Roschke, V., Perry, J.W., Arkfeld, D.G., Ehresmann, G.R., Migone, T.-S., Hilbert, D.M., and Stohl, W. (2003). Local production of B lymphocyte stimulator protein and APRIL in arthritic joints of patients with inflammatory arthritis. *Arthritis Rheum.*, **48**, 982–992.

Tardivel, A., Tinel, A., Lens, S., Steiner, Q.-G., Sauberli, E., Wilson, A., Mackay, F., Rolink, A.G., Beermann, F., Tschopp, J., and Schneider, P. (2004). The anti-apoptotic factor Bcl-2 can functionally substitute for the B cell survival but not for the marginal zone B cell differentiation activity of BAFF. *Eur. J. Immunol.*, **34**, 509–518.

Thien, M., Phan, T.G., Gardam, S., Amesbury, M., Basten, A., Mackay, F., and Brink, R. (2004). Excess BAFF rescues self-reactive B cells from peripheral deletion and allows them to enter forbidden follicular and marginal zone niches. *Immunity*, **20**, 785–798.

Thompson, J.S., Bixler, S.A., Qian, F., Vora, K., Scott, M.L., Cachero, T.G., Hession, C., Schneider, P., Sizing, I.D., Mullen, C., Strauch, K., Zafari, M., Benjamin, C.D., Tschopp, J., Browning, J.L., and Ambrose, C. (2001). BAFF-R, a novel TNF receptor that specifically interacts with BAFF. *Science*, **293**, 2108–2111.

Thompson, J.S., Schneider, P., Kalled, S.L., Wang, L., Lefevre, E.A., Cachero, T.G., MacKay, F., Bixler, S.A., Zafari, M., Liu, Z.-Y., Woodcock, S.A., Qian, F., Batten, M., Madry, C., Richard, Y., Benjamin, C.D., Browning, J.L., Tsapis, A., Tschopp, J., and Ambrose, C. (2000). BAFF binds to the tumor necrosis factor receptor-like molecule B cell maturation antigen and is important for maintaining the peripheral B cell population. *J. Exp. Med.*, **192**, 129–135.

Tribouley, C., Wallroth, M., Chan, V., Paliard, X., Fang, E., Lamson, G., Pot, D., Escobedo, J., and Williams, L.T. (1999). Characterization of a new member of the TNF family expressed on antigen presenting cells. *Biol. Chem.*, **380**, 1443–1447.

Varfolomeev, E., Kischkel, F., Martin, F., Seshasayee, D., Wang, H., Lawrence, D., Olsson, C., Tom, L., Erickson, S., French, D., Schow, P., Grewal, I.S., and Ashkenazi, A. (2004). APRIL-deficient mice have normal immune system development. *Mol. Cell. Biol.*, **24**, 997–1006.

von Bülow, G.-U. and Bran, R.J. (1997). NF-AT activation induced by a CAML-interacting member of the tumor necrosis factor receptor superfamily. *Science*, **278**, 138–141.

von Bülow, G.-U., van Deursen, J.M., and Bram, R.J. (2001). Regulation of the T-independent humoral response by TACI. *Immunity*, **14**, 573–582.

Wang, H., Marsters, S.A., Baker, T., Chan, B., Lee, W.P., Fu, L., Tumas, D., Yan, M., Dixit, V.M., Ashkenazi, A., and Grewal, I.S. (2001). TACI-ligand interactions are required for T cell activation and collagen-induced arthritis in mice. *Nat. Immunol.*, **2**, 632–637.

Williams-Blangero, S., VandeBerg, J.L., Subedi, J., Aivaliotis, M.J., Rai, D.R., Upadhayay, R.P., Jha, B., and Blangero, J. (2003). Genes of chromosomes 1 and 13 have significant effects on *Ascaris* infection. *Proc. Natl. Acad. Sci.*, **99**, 5533–5538.

Wu, Y., Bressette, D., Carrell, J.A., Kaufman, T., Feng, P., Taylor, K., Gan, Y., Cho, Y.H., Garcia, A.D., Gollatz, E., Dimke, D., LaFleur, D., Migone, T.S., Nardelli, B., Wei, P., Ruben, S.M., Ullrich, S.J., Olsen, H.S., Kanakaraj, P., Moore, P.A., and Baker, K.P. (2000). Tumor necrosis factor (TNF) receptor superfamily member TACI is a high affinity receptor for TNF family members APRIL and BLyS. *J. Biol. Chem.*, **275**, 35478–35485.

Xia, X.-Z., Treanor, J., Senaldi, G., Khare, S.D., Boone, T., Kelley, M., Theill, L.E., Colombero, A., Solovyev, I., Lee, F., McCabe, S., Elliott, R., Miner, K., Hawkins, N., Guo, J., Stolina, M., Yu, G., Wang, J., Delaney, J., Meng, S.-Y., Boyle, W.J., and Hsu, H. (2000). TACI is a TRAF-interacting receptor for TALL-1, a tumor necrosis factor family member involved in B cell regulation. *J. Exp. Med.*, **192**, 137–143.

Xu, L.-G. and Shu, H.-B. (2002). TNFR-associated factor-3 is associated with BAFF-R and negatively regulates BAFF-R-mediated NF-κB activation and IL-10 production. *J. Immunol.*, **169**, 6883–6889.

Xu, S. and Lam, D.-P. (2001). B-cell maturation protein, which binds the tumor necrosis factor family members BAFF and APRIL, is dispensable for humoral immune responses. *Mol. Cell. Biol.*, **21**, 4067–4074.

Yan, M., Brady, J.R., Chan, B., Lee, W.P., Hsu, B., Harless, S., Cancro, M., Grewal, I.S., and Dixit, V.M. (2001a). Identification of a novel receptor for B lymphocyte stimulator that is mutated in a mouse strain with severe B cell deficiency. *Curr. Biol.*, **11**, 1547–1552.

Yan, M., Wang, H., Chan, B., Roose-Girma, M., Erickson, S., Baker, T., Tumas, D., Grewal, I.S., and Dixit, V.M. (2001b). Activation and accumulation of B cells in TACI-deficient mice. *Nat. Immunol.*, **2**, 638–643.

Yu, G., Boone, T., Delaney, J., Hawkins, N., Kelley, M., Ramakrishnan, M., McCabe, S., Qiu, W.-r, Kornuc, M., Xia, X.-Z., Guo, J., Stolina, M., Boyle, W.J., Sarosi, K., Hsu, H., Senaldi, G., and Theill, L.E. (2000). APRIL and TALL-1 and receptors BCMA and TACI: System for regulating humoral immunity. *Nat. Immunol.*, **1**, 252–256.

Zarnegar, B., He, J.Q., Oganesyan, G., Hoffmann, A., Baltimore, D., and Cheng, G. (2004). Unique CD40-mediated biological program in B cell activation requires both type 1 and type 2 NF-κB activation pathways. *Proc. Natl. Acad. Sci.*, **101**, 8108–8113.

Zhang, J., Roschke, V., Baker, K.P., Wang, Z., Alarcón, G.S., Fessler, B.J., Bastian, H., Kimberly, R.P., and Zhou, T. (2001). Cutting edge: A role for B lymphocyte stimulator in systemic lupus erythematosus. *J. Immunol.*, **166**, 6–10.

Zhukovsky, E.A., Lee, J.-O., Villegas, M., Chan, C., Chu, S., and Mroske, C. (2004). Is TALL-1 a trimer of a virus-like cluster? *Nature*, **427**, 413–414.

23

Control and Induction of Autoimmunity by Cytokine and Anti-cytokine Treatments

Pierre Miossec

1. Introduction

Cytokines represent a growing list of mediators associated with a large number of mechanisms. They are involved as a first line of innate immunity and also contribute to the amplification of adaptive immunity. Their multiple functions are the key factors for their contribution to the expression of various diseases. Cytokines themselves have been used for treatment, starting with interferon α (IFNα) for viral hepatitis. Later, the inhibition of cytokines has been a key step forward for the control of the most severe inflammatory diseases. At the same time, administration of cytokines or of their inhibitors has been associated with side effects. Some of these manifestations are indicators of the contribution of cytokines and/or regulatory pathways involving cytokines.

Treatment with TNFα inhibitors has revealed the critical role of a single key cytokine first in the pathogenesis of rheumatoid arthritis (RA) and of Crohn's disease (CD) and, then, of many other inflammatory diseases (Feldmann and Maini, 2001). Although their inducing mechanisms are far from being clear, these diseases have been classified as autoimmune. The failure of identification of a causal agent or a specific trigger mechanism was for some time the major limitation for the improvement of current treatments. These new results have clearly demonstrated the role of these soluble factors, which are nonspecific mediators, in the clinical manifestations of diseases, dominated by inflammation and matrix destruction. The nonspecific effect of cytokines is shown when the same clinical results were also observed in CD for which the clinical expression, anatomical distribution, and underlying mechanisms are very different from those of RA (Van Assche and Rutgeerts, 2000). In turn, it indicates that numerous pathways

Pierre Miossec • Department of Immunology and Rheumatology, Hôpital Edouard Herriot, 69437 Lyon Cedex 03, France.

Molecular Autoimmunity: In commemoration of the 100th anniversary of the first description of human autoimmune disease, edited by Moncef Zouali. Springer Science+Business Media, Inc., New York, 2005.

responsible for clinical expression of these chronic inflammatory diseases are in fact often nonspecific.

In this chapter, we focus first on the inhibition of TNFα (mode of action, results, and limitations). Then, we consider the inhibition of other cytokines. Finally we review the autoimmune manifestations induced by cytokine treatment, focusing on the use of IFNα.

2. TNFα and Its Receptors

It is important to realize that the clinical inhibition of TNFα was made possible through the progress in biotechnologies to obtain first the structure of the protein and its receptors, and then the high-scale production of its therapeutic inhibitors, antibodies, or soluble receptors. TNFα is the founding father of a growing family, which includes many critical factors of the immune system. Many important ligand-receptor pairs have been identified. TNFα and lymphotoxin α (LTα) act as trimers on two receptors: the type I TNF receptor (p55-TNF-R) and the type II TNF receptor (p75-TNF-R, Figure 23.1). These two cytokines control inflammation and apoptosis (Figure 23.2). Lymphotoxin β (LTβ) binds to a specific receptor (LTβR). LTα and LTβ have a critical role in the formation and

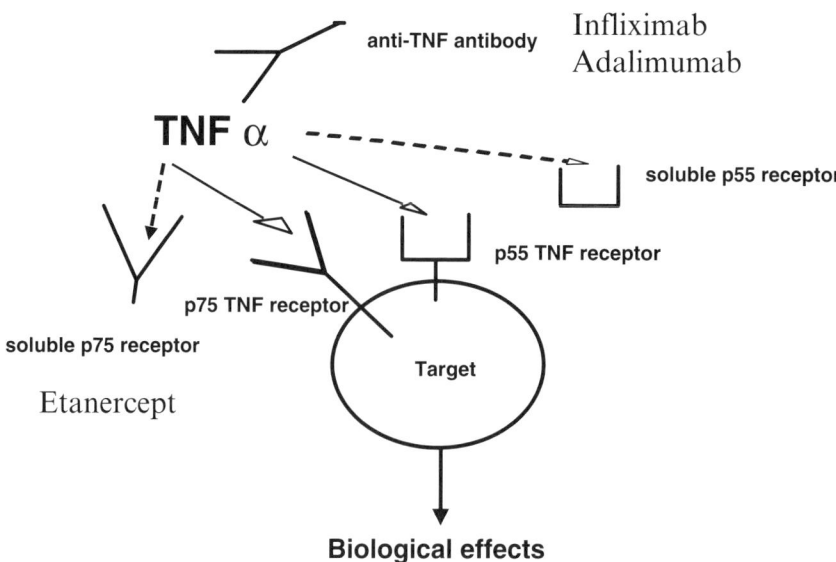

Figure 23.1. Interactions between TNFα, its receptors, and current specific inhibitors. TNFα acts on two p55 and p75 membrane receptors. Such effect is downregulated by the release of the soluble forms of these receptors. Therapeutic inhibitors are either a modified p75-soluble receptor (etanercept) or monoclonal antibodies (remicade, adalimumab).

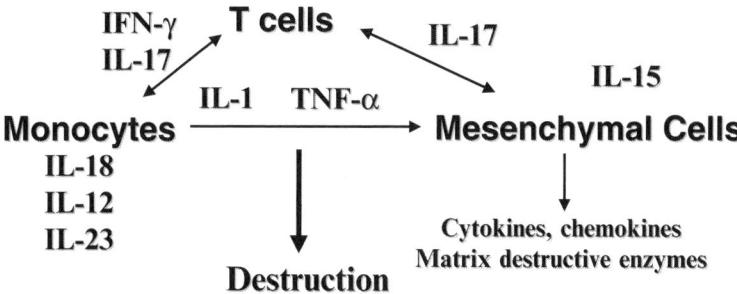

Figure 23.2. Interactions between the major proinflammatory cytokines. Monocyte-derived cytokines, such as IL-1 and TNFα, act on mesenchymal cells, such as synoviocytes, and trigger the release of destructive enzymes. This activity is regulated at the level of T cells through the production of Th1 cytokines, such as IFNγ and IL-17. The Th1 profile is further enhanced by other cytokines, such as IL-12 and IL-18.

function of lymphoid organs (Ruddle, 1999). Fas ligand (CD95 ligand) acts on the Fas receptor (CD95), and is involved in the control of the cellular proliferation and apoptosis. RANK ligand (RANKL) or osteoprotegerin ligand (OPGL) binds to RANK and controls the activation of osteoclasts and bone degradation. OPG acts as a soluble receptor with an inhibitory effect. CD40 ligand expressed by T cells binds to CD40 on B cells and other antigen-presenting cells and controls cellular interactions, in particular between T and B lymphocytes.

TNFα is secreted as a trimer. However, it also exists as a transmembrane biologically active monomer, which is important for local cell–cell interactions (Burger and Dayer, 2002). This molecule is released as a soluble form under the effect of a membrane metallo-proteinase, the TNF-converting enzyme (TACE).

The fixation of TNFα to its receptors leads to activation of signal transduction pathways including MAP kinases and NF-κB. The p55, but not the p75, TNF receptor has a death domain involved in apoptosis controlled by Fas, following activation by TNFα. The other pathway is involved in the inflammatory reaction with the synthesis of cytokines, chemokines, and proteases. For a given activated cell, there is a selective choice between the inflammatory or the apoptotic pathways, so that only one of them prevails.

3. Mode of Action of the Specific TNFα Inhibitors

In reference to the natural control of action and production of a cytokine, therapeutic control can be specific of a given cytokine (antibody, soluble receptor) or more global, acting on a group of proinflammatory cytokines (methotrexate, leflunomide, inhibitors of common intracellular pathways). When considering the specific inhibitors of TNFα, inhibition with an anti-TNFα antibody represents the simplest strategy (Figure 23.1). There are several types of anti-TNFα antibodies, either fully human (adalimumab) or chimeric, keeping a more or less important part of the antibody binding site of murine origin (infliximab). These agents are not natural molecules, since natural autoantibodies against TNFα have not been described. Monoclonal antibodies specific for TNFα bind specifically to epitopes of the TNFα trimer, which interact with the membrane receptors. This results in inhibition of the TNFα capacity to bind the membrane receptors, resulting in functional inhibition. In addition, these antibodies will also recognize the membrane TNFα, thus inhibiting cell interactions.

On the contrary, the use of the soluble receptors (p55 or p75) represents the enhancement of a natural regulation, since the same molecules already control the action of endogenous TNFα. Etanercept is a fusion protein with two p75 receptors, connected to an Fc fragment of an IgG1. This Fc fragment improves the pharmacokinetics of the complex. TNFα and LTα use the same two p55 and p75 TNF receptors to control inflammation and apoptosis. Advantages and limitations of the exclusive inhibition of TNFα (with a specific anti-TNFα antibody) and those associated to that of the LTα (with a soluble receptor) have not been clarified.

4. The Local and Systematic Effects of TNFα Inhibition

The initial clinical results obtained with the administration of anti-TNFα antibodies and TNF-soluble receptors have justified their development on a large scale. Table 23.1 shows a summary of the current use of TNF inhibitors. Other diseases will be added in the near future. The rapid effect on systemic manifestations and on the levels of acute phase proteins confirms the importance of TNFα in systemic inflammation (Figure 23.3). The feeling of well-being reported rapidly by these patients is the confirmation of the effect of TNFα on the brain, in particular on the hypothalamus.

Table 23.1. Major Recognized Indications of TNFα Inhibitors

Infliximab (Remicade)
Rheumatoid arthritis
Spondylarthropathies
Chronic juvenile arthritis
Crohn's disease

Etanercept (Enbrel)
Rheumatoid arthritis
Chronic juvenile arthritis
Spondylarthropathies
Psoriatic arthritis
Psoriasis

Adalimumab (Humira)
Rheumatoid arthritis

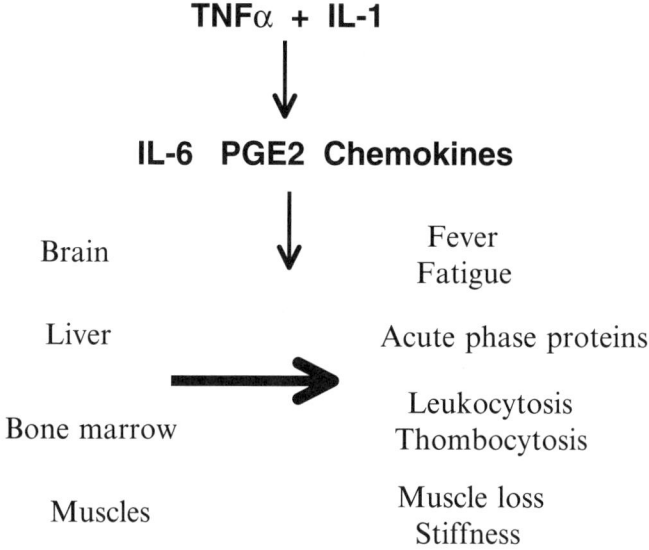

Figure 23.3. Contribution of TNFα and IL-1 to the systemic manifestations of chronic inflammation. These cytokines act with others on the brain leading to fever and fatigue, on the liver leading to the release of acute phase proteins, on the bone marrow leading to precursor activation (leukocytosis and thrombocytosis), and on muscles leading to muscle loss and stiffness.

The anti-inflammatory effect results from local actions on the inflammatory reaction such as the synovitis (Figure 23.4). The migration of inflammatory cells contributes to the initiation and the chronicity of the inflammatory process, leading to matrix destruction. The formation of any inflammatory reaction relies on new blood vessel formation, which is critical for the migration of these inflammatory cells. TNFα induces the expression of adhesion molecules on endothelial cells. These effects favor the migration of T cells with a memory phenotype, expressing preferentially the chemokine receptors CCR1 and CCR5. These selective mechanisms direct cells toward skin, joint, gut, and eye sites, resulting in clinical pictures specific to the respective diseases (Hjelmstrom et al., 2000).

The increase of angiogenesis is an important characteristic of rheumatoid synovitis. Local concentrations of vascular endothelium growth factor were found to decrease in response to infliximab. The sequential biopsies of synovial membrane of treated patients showed a reduction of the cell infiltrate and angiogenesis (Paleolog, 1997). This reduction of inflammatory cells and their interactions contributes directly to a protective effect by reducing the local production of enzymes involved in destruction. The rapid reduction of joint swelling is an impressive effect of anti-TNFα treatment. *In vitro*, infliximab, but not etanercept, can induce death of cells expressing membrane TNFα in the presence of complement through a mechanism of antibody-dependent cell-mediated cytoxicity. However, no increase in cell death by apoptosis has been found in synovial biopsies after treatment with infliximab (Tak et al., 1996). The absence of induction

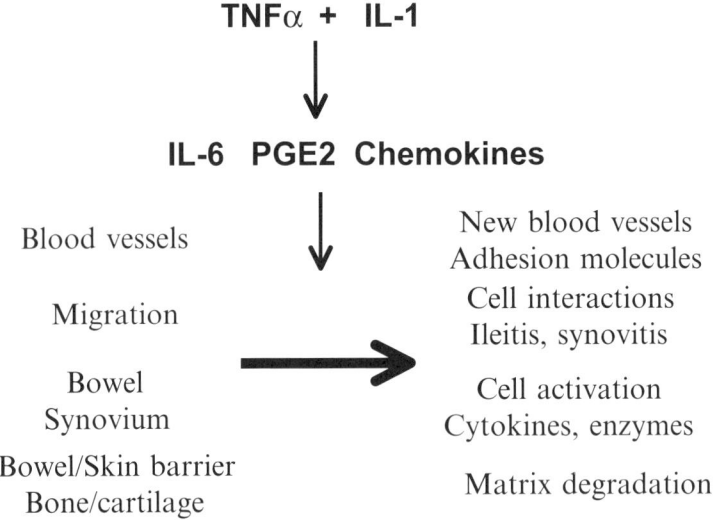

Figure 23.4. Contribution of TNFα and IL-1 to the local manifestations of chronic inflammation. These cytokines act with others on vessels leading to new blood vessel formation and on target tissues leading to increased cell–cell interactions involved in extracellular matrix destruction.

of apoptosis with etanercept could explain its lack of efficacy in CD, where infliximab is effective (Van den Brande *et al.*, 2003).

Clinical results showed a reduction of the rate of articular destruction measured by the absence of new radiological joint damage. This effect is critical because of a major depression of repair capacities. In CD, such effect results in the closing of fistulas. An action on proteases involved in the destruction of bone and cartilage, and of gut and skin matrix, represents the main mode of action of these inhibitors, as has been reported in studies *in vitro* and in animal models (Van den Berg, 2002). Bone destruction in RA results from an activation of osteoclasts combined with the recruitment of osteoclast precursors, which is amplified by TNFα. This inhibitory effect can influence the formation of osteoclasts by reducing the recruitment of precursors, which are common to monocytes and dendritic cells, probably involving the RANK/RANK-L pathway. RANK-L is expressed by osteoblasts, T lymphocytes, and synoviocytes and activates the RANK receptor on osteoclasts and plays a critical role in bone destruction. RA blood concentrations of soluble RANK-L were normalized with infliximab (Ziolkowska *et al.*, 2002).

Induction of repair remains the final achievement to protect joint. In transgenic mice expressing human TNFα, neutralization of this cytokine inhibits bone and cartilage degradation (Shealy *et al.*, 2002). An effect on repair is fully obtained by the coadministration of anti-TNFα and OPG, blocking at the same time the TNFα-TNF receptor and RANK–RANK-L interactions (Zwerina *et al.*, 2004).

5. Understanding the Side Effects of TNFα Inhibitors

The inhibition of a central molecule, such as TNFα, has been associated with side effects, allowing a better understanding of the physiological role of TNFα. In addition, the heterogeneity of the response, and the risk of disease reactivation when stopping the inhibition, indicates that these inflammatory diseases cannot be simplified as diseases of the TNFα pathway.

TNFα is a critical molecule in the control of acute and chronic infections. Not surprisingly, a greater mortality was observed during the treatment of toxic shock with TNFα inhibitors. For long-term treatment, the major complication has been the appearance of opportunistic infections. If all types of opportunistic infections were observed, tuberculosis has been by far the most frequent cause (Keane et al., 2001). Its severity associated to an unusual mortality rate quickly drew attention. Epidemiological studies showed that it was essentially a reactivation of known or undiagnosed tuberculosis, suggesting an acquired defect of cell-mediated immunity. Primary immune defects have been described in association with mutations of genes coding for interferon γ and IL-12 receptor (Casanova and Abel, 2002). These defects are responsible for severe mycobacterial infections usually secondary to BCG vaccination.

In the context of chronic inflammation, such as occurs in RA, there is already a systemic response defect to IL-12 and IL-18, key cytokines for the production of interferon γ (Kawashima and Miossec, 2004). The synergy between these two cytokines is related to the effect of IL-12 on the induction of a functional IL-18 receptor (Kawashima and Miossec, 2003). Such defect in RA results in a lower production of interferon γ by blood cells in response to IL-12 and IL-18. This defect is proportional to the level of systemic inflammation, as measured by C-reactive protein levels. This could explain the increased frequency of tuberculosis in RA, even in the absence of anti-TNFα treatment (Carmona et al., 2003). At initiation of a treatment with a TNFα inhibitor, the disease is usually still very active. The additive effect of the inhibitor explains the rapid appearance and the severity of these reactivations. Later, the risk is reduced because the improvement of the clinical situation has been able to correct the systemic immune defect related to inflammation, reducing the risk of reactivation. The specific anti-TNFα effect results from an inhibition of cell–cell interactions and has a positive impact on local inflammation, resulting in the beneficial clinical effect. Conversely, inhibition of these interactions also results in granuloma disintegration, leading to the diffusion of mycobacteria that were kept under control in these granulomas.

The induction of antinuclear antibodies is common, but rarely associated with clinical manifestations of lupus (Shakoor et al., 2002). A positive connection between TNFα and lupus has been established in lupus mice where the inhibition of TNFα increases incidence and mortality from renal disease (Kontoyiannis and Kollias, 2000). Conversely, administration of TNFα has a protective effect on mouse lupus. The anti-inflammatory cytokine IL-10 is directly involved in the production of autoantibodies and IgG (Llorente et al., 1995). These properties indicate a mutual inhibition between TNFα and IL-10 actions. Furthermore, the inhibition of TNFα favors the production of IL-10, leading to the production of autoantibodies and the orientation of the Th1–Th2 cytokine balance toward Th2 (Miossec and van den Berg, 1997).

These notions also must be taken into account to estimate the possible effect of TNFα inhibitors on the incidence of lymphomas (Brown *et al.*, 2002), but their interpretation is extremely difficult because of a greater frequency of lymphomas in the general population, a frequency further increased in association with RA and even more, with Sjögren's syndrome. The contribution of TNFα to apoptosis and the amplified immunosuppressive effect of IL-10 must be considered. Conversely, TNFα is involved in lymph node hypertrophy, a common sign of activity of inflammatory diseases (McLachlan *et al.*, 2003).

The beneficial effect obtained in the treatment of RA and CD would suggest that all inflammatory diseases, and in particular those associated with a Th1 profile, could be controlled with the same TNFα inhibitors. However, TNFα inhibition in multiple sclerosis was associated to an increase in clinical and radiological signs (Mohan *et al.*, 2001). The mechanism has not been clarified, but could be related to the anatomical site and to the contribution of the blood brain barrier (Robinson *et al.*, 2001). The importance of TNFα inhibition also could be demonstrated with the reintroduction of infliximab in a case of aseptic meningitis (Marotte *et al.*, 2001).

6. Other Cytokine Inhibitors

More recently additional cytokines have been the targets of treatment, but at this early stage, it is still difficult to compare the results with those obtained with TNFα inhibition. IL-1 has been categorized as a critical cytokine in chronic inflammation. Its mode of action is similar to that of TNFα, but with important differences (Arend and Dayer, 1995). The two IL-1α and IL-1β act through two receptors, but the critical receptor for biological responses is the membrane type I IL-1 receptor, which, when combined with an accessory protein, forms the fully functional IL-1 receptor and transduces a signal. The membrane type II IL-1 receptor does not transduce a signal. It is rather an endogenous regulator released as a soluble form that traps soluble IL-1. In addition to the two IL-1 molecules with an agonistic effect, IL-1 receptor antagonist (IL-1 RA) is a receptor antagonist, which can bind to the membrane type I receptor, without inducing a signal (the accessory chain is not recruited).

IL-1 RA has been used in the treatment of RA with beneficial effects. Although significant enough to get marketing approval from health authorities for the treatment of RA, the efficacy appears lower than that of TNFα inhibition (Bresnihan, 2002). A major limitation with IL-1 RA is the need for a continuous high level of the compound since at least a 1:100 ratio has to be obtained between the IL-1 and IL-1 RA levels. At present, other means of blocking IL-1 are in progress, such as anti-IL-1 antibodies, IL-1-soluble receptors used alone or in complex structures in an IL-1 trap.

As for IL-1 and TNF, IL-6 is highly present in the context of chronic inflammation. There has been some debate on the classification of IL-6 as a proinflammatory or an anti-inflammatory cytokine, and controlling IL-6 as a therapeutic approach was the best way to clarify this issue. For that purpose an anti-IL-6 receptor antibody named MRA has been developed and has shown efficacy for the treatment of RA and CD. The molecule is now in phase III trials (Ito et al., 2004; Nishimoto et al., 2004).

IL-15 is a cytokine involved in chronic inflammation, especially in the activation of T cells. Recent positive results have been obtained with a monoclonal anti-IL-15 antibody for the treatment of RA (McInnes and Gracie, 2004).

7. Other Cytokines as Treatment Targets

The favorable clinical results obtained with TNFα and IL-1 inhibitors may suggest that the study of the contribution of cytokines to arthritis is almost over. However, additional cytokines also contributing to joint inflammation have been considered as possible targets (Figure 23.2). IL-17 is a T cell–derived cytokine, which could be classified as a Th1 cytokine. It often acts in synergy with TNF and IL-1 (Miossec, 2003). In addition, IL-17 increases IL-1 and TNF production by monocytes. *In vitro* studies and animal models have strongly indicated the interest in blocking IL-17 for the treatment of chronic inflammation.

IL-18, IL-12, and IL-23 are cytokines produced by monocytes and other antigen-presenting cells. They interact through synergistic interactions, favoring a Th1 cytokine profile (Trinchieri et al., 2003). IL-12 targeting is ongoing with an anti-IL-12 antibody. IL-18 action is regulated by its endogenous inhibitor IL-18 binding protein (IL-18 BP). Preclinical results have been obtained to suggest the use of IL-18 BP as a treatment agent (Kawashima et al., 2004). IL-23 has been more recently described as a cytokine responsible for some of the proinflammatory effects first associated with IL-12. In addition, IL-23 increases IL-17 production by T cells (Aggarwal et al., 2003).

8. Targeting One or More than One Cytokine

It remains to be understood how blocking a single cytokine such as TNFα can still be effective. Currently, the list of cytokines, chemokines, and growth factors involved in inflammation is up to 100. It was first thought that TNFα could be located upstream of a cytokine cascade. However, a list of cytokines, such as IL-12, IL-18 and IL-17, has been shown to act on the production of TNFα itself. It should be recognized that the concentrations expected *in vivo* may be much lower that those used in a cell culture system with isolated cells stimulated with a single cytokine. In addition, interactions between cytokines can be synergistic through common intracellular pathways located downstream of the receptors that confer specificity. Transcription factors, such as NFκB, are shared for the activation of IL-1, TNF, and IL-17. These common pathways are therapeutic targets of small chemical molecules now in clinical development.

Combination of low concentrations of cytokines was used to dissect these synergistic interactions. For example, combination of low concentrations of TNFα, IL-1, and IL-17, with no effect when used alone, was able to induce transcription factor activation with an effect higher than that observed with high concentrations of a single cytokine (Granet *et al.*, 2004). More importantly, this combination was able to increase the recruitment of transcription factors, some not activated by high concentrations of a single cytokine (Granet and Miossec, 2004). Such experiments mimic the *in vivo* situation with local interactions between cytokines produced by monocytes, such as TNFα and IL-1, and by T cells, such as IL-17.

Similar conclusions were obtained when low doses of soluble receptors for IL-1, TNFα, and IL-17 were combined with samples of synovium and juxtarticular bone (Chabaud and Miossec, 2001). As observed in patients, two thirds of the RA synovium samples responded to etanercept. The rate of response was increased up to 90% when the three receptors were combined. However, such combinations have not yet been tested in the clinic. One trial tested the combination of IL-1Ra and TNF sR in RA. No improvement of efficacy was observed despite an increase in incidence of infections (Genovese *et al.*, 2004).

9. Understanding the Heterogeneity of the Response to TNFα Inhibitors

In clinical practice, it appears that about one third of RA patients do not improve. This heterogeneity could be, at least partially, explained by the absence of a TNFα contribution in some patients (Ulfgren *et al.*, 2000). As indicated above, the same rate of response is observed when incubating fragments of RA synovial membrane with etanercept (Chabaud and Miossec, 2001). Although not directly proven, such differences could be related to cytokine gene polymorphisms affecting TNFα or other cytokines (Mugnier *et al.*, 2003). It is also common to observe a progressive loss of clinical response to inhibitors previously effective. For compounds such as infliximab, induction of an antibody response directed against the mouse part of the antibody has been demonstrated in patients with CD (Baert *et al.*, 2003). Progressive induction of a TNFα–independent inflammatory pathway is another mechanism further supporting the downstream role of TNFα in the pathogenesis of these diseases.

Finally, these treatments have a suspensive effect, with reappearance of symptoms after prolonged treatment discontinuation, reflecting an effect on the action and not so much on the production of TNFα. Stopping the inhibition is then followed by a rebound effect. Furthermore, TNFα inhibition decreases the production of soluble TNFα receptors, thus reducing the endogenous anti-inflammatory regulation. These data also suggest the contribution of other factors and other cell types. It remains to be demonstrated if a combined or sequential control of T cells, B cells, dendritic cells, and of their interactions would be necessary for the induction of a remission.

10. Autoimmune Manifestations with Cytokine Administration

The list of cytokines with possible therapeutic use has been growing. As for any adverse reaction with a new compound, the interpretation of the underlying mechanisms is often difficult. The incidence is obviously related to the properties of the molecule itself, its dose regimen, and mode of administration, but also to the underlying disease as well as to the size of the exposed population. Among the cytokines already in clinical application, IFNα has been the most commonly used, particularly in patients with chronic viral hepatitis. For these reasons, both this cytokine and this disease have been associated with the largest list of side effects. Thus, we review the side effects observed with this cytokine first.

The major indications of INFα are type B and C chronic viral hepatitis, and cancer (mostly renal cell carcinoma and melanoma, and hematological malignancies). The large number of treated patients, the largest for any cytokine, allows a good assessment of its safety (Miossec, 1997).

Induction of autoimmune events appears to be a frequent feature (Fattovich et al., 1996; Wilson et al., 2002). This includes an exhaustive list of manifestations of autoimmunity and associated diseases (Miossec, 1997). This has led to the recommendation to exclude patients with concomitant clinically overt autoimmune disease from the use of IFNα for the treatment of viral hepatitis. The range goes from the mere presence or induction of autoantibodies with no clinical consequence to the most severe autoimmune disease. The pathogenicity of autoantibodies is often unclear and far from being always associated with disease manifestations. Their targets include blood cells (red cells, leukocytes, platelets), coagulation factors (factor VIII, lupus anticoagulant), immunoglobulins (rheumatoid factor with or without cryoglobulin activity), intracellular components (nucleus, enzymes), hormones (thyroid hormones, insulin), and the skin (epidermis) (Fattovich et al., 1991). In particular, exacerbation of hepatitis C–related cryoglobulinemia has been described leading to severe clinical consequences, including polyneuritis and even fatal cases (Batisse et al., 2004). This has to be taken into account since an association with hepatitis C is found in almost 50% of all cases of mixed cryoglobulinemia (Agnello et al., 1992), although treatment with IFNα is usually helpful, particularly in combination with ribavirin (Ferri et al., 1993; Mazzaro et al., 2003).

Blood cells and coagulation factors are frequent targets. These manifestations include idiopathic or autoimmune hemolytic anemia and thrombocytopenia (Murakami et al., 1994). Regarding acquired factor VIII inhibitor, some patients with hemophilia A and hepatitis C developed antibodies to factor VIII (Stricker et al., 1994). Induction of lupus anticoagulant has been implicated in the development of thrombotic events.

Regarding the list of organ-specific diseases, thyroid abnormalities appear to be the most common manifestations (Lisker-Melman et al., 1992), but the exact incidence remains unclear. In a survey including 11,241 patients with hepatitis, 71 developed autoimmune thyroid disease during IFNα treatment (Fattovich et al., 1996) and various thyroid abnormalities have been observed. Half of them had hypothyroidism, 30% hyperthyroidism, and 20% a biphasic (hyperthyroidism followed by hypothyroidism) pattern. This includes the two ends of the spectrum, ranging from patients clinically and biochemically hypothyroid, but negative for thyroid autoantibodies, to patients remaining euthyroid, but with thyroid autoantibodies. Idiopathic thyroiditis is a very common autoimmune disease, often associated with Sjögren's syndrome, resulting in lymphocytic infiltration of the exocrine glands. It is important to keep in mind that such features are commonly found in patients with hepatitis C in the absence of any cytokine treatment (Haddad et al., 1992). Along the same line, induction of insulin antibodies and onset of insulin-dependent diabetes with increased antiglutamic acid decarboxylase antibody levels have been observed in 10 out of the 11,241 patients in the survey described above (Fattovich et al., 1996).

Other manifestations include almost the entire list of autoimmune diseases (Miossec, 1997). Regarding arthritis manifestations, IFNα was responsible for the induction or flare of various types of inflammatory arthritis, either associated to RA, psoriasis, lupus, spondyloarthropathy, or as yet unclassified (Conlon et al., 1990; Kiely and Bruckner, 1994). Regarding muscular manifestations, cases of dermatomyositis and polymyositis have been described mostly in patients with chronic hepatitis C (Conlon et al., 1990). Similarly, myasthenia gravis with anti-acetylcholine receptor antibodies and Guillain-Barré syndrome have been observed (Batocchi et al., 1995).

Features of systemic lupus erythematosus including severe cases with nephropathy, cerebral vasculitis, and chorea have been reported (Wilson et al., 2002). More recently, the role of IFNα in the pathogenesis of lupus has been demonstrated, suggesting that IFNα may represent a target for treatment (Pascual et al., 2003; Schmidt and Ouyang, 2004). However, induction of lupus remains a rare event in patients treated with IFNα.

Interference with graft survival also has been a major consequence of such treatment, including graft versus host disease following therapy for relapsed leukemia after allogeneic bone marrow transplantation, as well as allograft rejection following treatment of hepatitis C after liver and renal transplantation (Saab et al., 2004). Treated recipients may also develop progressive cirrhosis despite achieving a sustained virological response.

Other cytokine side effects also are of interest. IL-2 was the first cytokine used as a recombinant product. In early studies, it was used for the *ex vivo* culture of autologous peripheral blood lymphocytes before reinjection and production of lymphokine-activated killer (LAK) cells. Similarly tumor-infiltrating lymphocytes (TILs) have been cultured the same way. Reduction of metastases was observed in patients with extensive melanoma and renal cell carcinoma (Rosenberg et al., 1988). Most recent studies have used IL-2 either alone or in combination with other cytokines, mainly IFNα. One of the major adverse reactions with IL-2 was the acute accumulation of body fluid related to a capillary leak syndrome (Ballmer-Weber et al., 1995). *In vivo* studies have shown the activation of vascular endothelial cells by proinflammatory cytokines, the production of which was stimulated by IL-2. When used on a more chronic basis, such effect may contribute to the migration of inflammatory cells, mostly lymphocytes, to the perivascular site. A number of side effects observed with IFNα have been described with IL-2 (Gaspari, 1994), including the induction of chronic arthritis, myositis, and thyroid manifestations.

Interferon β (IFNβ) is currently used for the treatment of multiple sclerosis (Francis, 2004). Although the clinical experience is not as large as with IFNα, autoimmune manifestations do not appear to be a frequent concern, although thyroiditis is also the most common autoimmune reported event.

Interferon γ (IFNγ) is a NK/Th1-derived cytokine and some of the comments regarding IFNα apply to IFNγ. The experience with the latter cytokine, although limited, indicates its autoimmune potential. In patients with hepatitis C, thyroid dysfunction was uncommon in contrast to the induction of antinuclear antibodies. In patients with psoriasis arthritis or spondyloarthropathy, increased arthritis activity was observed. In patients with RA, no benefit was demonstrated, but in some cases, induction of antinuclear antibodies was observed (Cannon *et al.*, 1993). In some patients with multiple sclerosis, treatment with IFNγ led to increased disease activity (Panitch *et al.*, 1987). More recently, however, autoimmune manifestations were not observed in patients with cancer or opportunistic infections related to HIV infection (Riddell *et al.*, 2001; Stuart *et al.*, 2004).

Interleukin 12 (IL-12) is a cytokine produced by monocytes and antigen-presenting cells with a major effect on cell-mediated immunity in part through the production of IFNγ. IL-12 has been recently used as protein and through gene therapy for cancer treatment (Cebon *et al.*, 2003; Sangro *et al.*, 2004). Autoimmune manifestations have not been observed but further data with prolonged exposure are needed.

The family of colony-stimulating factors (CSFs) includes IL-3 or multi-CSF, granulocyte-CSF (G-CSF), and granulocyte-monocyte-CSF (GM-CSF). Such cytokines are used for the stimulation of bone marrow precursors after spontaneous or postchemotherapy bone marrow depression. Most autoimmune manifestations have been observed with G-CSF or GM-CSF. They have been used in patients with RA, particularly those with Felty's syndrome, defined as the combination of rheumatoid factor–positive destructive RA, severe neutropenia, and splenomegaly. Improvement of the neutropenia has been observed, sometimes with increased thrombocytopenia and anemia (Hoshina *et al.*, 1994). Some patients also showed increased arthritis activity (Yasuda *et al.*, 1994). However, such flare-up was not a constant observation.

11. Conclusions

The use of TNFα inhibitors has provided clear evidence for the direct role of cytokines in complex inflammatory diseases. The simplest approach, already in practice, is the specific inhibition of their action. The stimulation of endogenous regulatory mechanisms can represent a more physiological way to restore an adequate balance. The lack of response and the occurrence of adverse reactions observed in some patients exposed to cytokines and their inhibitors imply taking into account the level of production and regulation of the target cytokines. Part of such heterogeneity is genetically determined. Considering these issues will allow a better risk–benefit assessment of the treatment for each patient.

References

Aggarwal, S., Ghilardi, N., Xie, M.H., de Sauvage, F.J., and Gurney, A.L. (2003). Interleukin-23 promotes a distinct CD4 T cell activation state characterized by the production of interleukin-17. *J. Biol. Chem.*, **278**, 1910–1914.

Agnello, V., Chung, R.T., and Kaplan, L.M. (1992). A role for hepatitis C virus infection in type II cryoglobulinemia. *N. Engl. J. Med.*, **327**, 1490–1495.

Arend, W.P. and Dayer, J.M. (1995). Inhibition of the production and effects of interleukin-1 and tumor necrosis factor alpha in rheumatoid arthritis. *Arthritis Rheum.*, **38**, 151–160.

Baert, F., Noman, M., Vermeire, S., Van Assche, G., D'Haens, G. Carbonez, A., and Rutgeerts, P. (2003). Influence of immunogenicity on the long-term efficacy of infliximab in Crohn's disease. *N. Engl. J. Med.*, **348**, 601–608.

Ballmer-Weber, B.K., Dummer, R., Kung, E., Burg, G., and Ballmer, P.E. (1995). Interleukin 2-induced increase of vascular permeability without decrease of the intravascular albumin pool. *Br. J. Cancer*, **71**, 78–82.

Batisse, D., Karmochkine, M., Jacquot, C., Kazatchkine, M.D., and Weiss, L. (2004). Sustained exacerbation of cryoglobulinaemia-related vasculitis following treatment of hepatitis C with peginterferon alfa. *Eur. J. Gastroenterol. Hepatol.*, **16**, 701–703.

Batocchi, A.P., Evoli, A., Servidei, S., Palmisani, M.T., Apollo, F., and Tonali, P. (1995). Myasthenia gravis during interferon alfa therapy. *Neurology*, **45**, 382–383.

Bresnihan, B. (2002). Anakinra as a new therapeutic option in rheumatoid arthritis: Clinical results and perspectives. *Clin. Exp. Rheumatol.*, **20**, S32–S34.

Brown, S.L., Greene, M.H., Gershon, S.K., Edwards, E.T., and Braun, M.M. (2002). Tumor necrosis factor antagonist therapy and lymphoma development: Twenty-six cases reported to the Food and Drug Administration. *Arthritis Rheum.*, **46**, 3151–3158.

Burger, D. and Dayer, J.M. (2002). Cytokines, acute-phase proteins, and hormones: IL-1 and TNF-alpha production in contact-mediated activation of monocytes by T lymphocytes. *Ann. NY Acad. Sci.*, **966**, 464–473.

Cannon, G.W., Emkey, R.D., Denes, A., Cohen, S.A., Saway, P.A., Wolfe, F., Jaffer, A.M., Weaver, A.L., Manaster, B.J., and McCarthy, K.A. (1993). Prospective 5-year followup of recombinant interferon-gamma in rheumatoid arthritis. *J. Rheumatol.*, **20**, 1867–1873.

Carmona, L., Hernandez-Garcia, C., Vadillo, C., Pato, E., Balsa, A., Gonzalez-Alvaro, I., Belmonte, M.A., Tena, X., and Sanmarti, R. (2003). Increased risk of tuberculosis in patients with rheumatoid arthritis. *J. Rheumatol.*, **30**, 1436–1439.

Casanova, J.L. and Abel, L. (2002). Genetic dissection of immunity to mycobacteria: The human model. *Annu. Rev. Immunol.*, **20**, 581–620.

Cebon, J., Jager, E., Shackleton, M.J., Gibbs, P., Davis, I.D., Hopkins, W., Gibbs, S., Chen, Q., Karbach, J., Jackson, H., MacGregor, D.P., Sturrock, S., Vaughan, H., Maraskovsky, E., Neumann, A., Hoffman, E., Sherman, M.L., and Knuth, A. (2003). Two phase I studies of low dose recombinant human IL-12 with Melan-A and influenza peptides in subjects with advanced malignant melanoma. *Cancer Immun.*, **3**, 7.

Chabaud, M. and Miossec, P. (2001). The combination of tumor necrosis factor alpha blockade with interleukin-1 and interleukin-17 blockade is more effective for controlling synovial inflammation and bone resorption in an ex vivo model. *Arthritis Rheum.*, **44**, 1293–1303.

Conlon, K.C., Urba, W.J., Smith, J.W., 2nd, Steis, R.G., Longo, D.L., and Clark, J.W. (1990). Exacerbation of symptoms of autoimmune disease in patients receiving alpha-interferon therapy. *Cancer*, **65**, 2237–2242.

Fattovich, G., Betterle, C., Brollo, L., Pedini, B., Giustina, G., Realdi, G., Alberti, A., and Ruol, A. (1991). Autoantibodies during alpha-interferon therapy for chronic hepatitis B. *J. Med. Virol.*, **34**, 132–135.

Fattovich, G., Giustina, G., Favarato, S., and Ruol, A. (1996). A survey of adverse events in 11,241 patients with chronic viral hepatitis treated with alfa interferon. *J. Hepatol.*, **24**, 38–47.

Feldmann, M. and Maini, R.N. (2001). Anti-TNF alpha therapy of rheumatoid arthritis: What have we learned? *Annu. Rev. Immunol.*, **19**, 163–196.

Ferri, C., Marzo, E., Longombardo, G., Lombardini, F., La Civita, L., Vanacore, R., Liberati, A.M., Gerli, R., Greco, F., Moretti, A. (1993). Interferon-alpha in mixed cryoglobulinemia patients: A randomized, crossover-controlled trial. *Blood*, **81**, 1132–1136.

Francis, G. (2004). Benefit-risk assessment of interferon-beta therapy for relapsing multiple sclerosis. *Expert Opin. Drug Saf.*, **3**, 289–303.

Gaspari, A.A. (1994). Autoimmunity as a complication of interleukin 2 immunotherapy. Many unanswered questions. *Arch. Dermatol.*, **130**, 894–898.

Genovese, M.C., Cohen, S., Moreland, L., Lium, D., Robbins, S., Newmark, R., and Bekker, P. (2004). Combination therapy with etanercept and anakinra in the treatment of patients with rheumatoid arthritis who have been treated unsuccessfully with methotrexate. *Arthritis Rheum.*, **50**, 1412–1419.

Granet, C., Maslinski, W., and Miossec, P. (2004). Increased AP-1 and NF-kappaB activation and recruitment with the combination of the proinflammatory cytokines IL-1beta, tumor necrosis factor alpha and IL-17 in rheumatoid synoviocytes. *Arthritis Res. Ther.*, **6**, R190–R198.

Granet, C. and Miossec, P. (2004). Combination of the pro-inflammatory cytokines IL-1, TNF-alpha and IL-17 leads to enhanced expression and additional recruitment of AP-1 family members, Egr-1 and NF-kappaB in osteoblast-like cells. *Cytokine*, **26**, 169–177.

Haddad, J., Deny, P., Munz-Gotheil, C., Ambrosini, J.C., Trinchet, J.C., Pateron, D., Mal, F., Callard, P., and Beaugrand, M. (1992). Lymphocytic sialadenitis of Sjögren's syndrome associated with chronic hepatitis C virus liver disease. *Lancet*, **339**, 321–323.

Hjelmstrom, P., Fjell, J., Nakagawa, T., Sacca, R., Cuff, C.A., and Ruddle, N.H. (2000). Lymphoid tissue homing chemokines are expressed in chronic inflammation. *Am. J. Pathol.*, **156**, 1133–1138.

Hoshina, Y., Moriuchi, J., Nakamura, Y., Arimori, S., and Ichikawa, Y. (1994). CD4+ T cell-mediated leukopenia of Felty's syndrome successfully treated with granulocyte-colony-stimulating factor and methotrexate. *Arthritis Rheum.*, **37**, 298–299.

Ito, H., Takazoe, M., Fukuda, Y., Hibi, T., Kusugami, K., Andoh, A., Matsumoto, T., Yamamura, T., Azuma, J., Nishimoto, N., Yoshizaki, K., Shimoyama, T., and Kishimoto, T. (2004). A pilot randomized trial of a human anti-interleukin-6 receptor monoclonal antibody in active Crohn's disease. *Gastroenterology*, **126**, 989–996.

Kawashima, M. and Miossec, P. (2003). Heterogeneity of response of rheumatoid synovium cell subsets to interleukin-18 in relation to differential interleukin-18 receptor expression. *Arthritis Rheum.*, **48**, 631–637.

Kawashima, M. and Miossec, P. (2004). Decreased response to IL-12 and IL-18 of blood cells in rheumatoid arthritis. *Arthritis Res. Ther.*, **6**, R39–R45.

Kawashima, M., Novick, D., Rubinstein, M., and Miossec, P. (2004). Regulation of interleukin-18 binding protein production by blood and synovial cells from patients with rheumatoid arthritis. *Arthritis Rheum.*, **50**, 1800–1805.

Keane, J., Gershon, S., Wise, R.P., Mirabile-Levens, E., Kasznica, J., Schwieterman, W.D., Siegel, J.N., and Braun, M.M. (2001). Tuberculosis associated with infliximab, a tumor necrosis factor alpha-neutralizing agent. *N. Engl. J. Med.*, **345**, 1098–1104.

Kiely, P.D. and Bruckner, F.E. (1994). Acute arthritis following interferon-alpha therapy. *Br. J. Rheumatol.*, **33**, 502–503.

Kontoyiannis, D. and Kollias, G. (2000). Accelerated autoimmunity and lupus nephritis in NZB mice with an engineered heterozygous deficiency in tumor necrosis factor. *Eur. J. Immunol.*, **30**, 2038–2047.

Lisker-Melman, M., Di Bisceglie, A.M., Usala, S.J., Weintraub, B., Murray, L.M., and Hoofnagle, J.H. (1992). Development of thyroid disease during therapy of chronic viral hepatitis with interferon alfa. *Gastroenterology*, **102**, 2155–2160.

Llorente, L., Zou, W., Levy, Y., Richaud-Patin, Y., Wijdenes, J., Alcocer-Varela, J., Morel-Fourrier, B., Brouet, J.C., Alarcon-Segovia, D., Galanaud, P. (1995). Role of interleukin 10 in the B lymphocyte hyperactivity and autoantibody production of human systemic lupus erythematosus. *J. Exp. Med.*, **181**, 839–844.

Marotte, H., Charrin, J.E., and Miossec, P. (2001). Infliximab-induced aseptic meningitis. *Lancet*, **358**, 1784.

Mazzaro, C., Zorat, F., Comar, C., Nascimben, F., Bianchini, D., Baracetti, S., Donada, C., Donadon, V., and Pozzato, G. (2003). Interferon plus ribavirin in patients with hepatitis C virus positive mixed cryoglobulinemia resistant to interferon. *J. Rheumatol.*, **30**, 1775–1781.

McInnes, I.B. and Gracie, J.A. (2004). Interleukin-15: A new cytokine target for the treatment of inflammatory diseases. *Curr. Opin. Pharmacol.*, **4**, 392–397.

McLachlan, J.B., Hart, J.P., Pizzo, S.V., Shelburne, C.P., Staats, H.F., Gunn, M.D., and Abraham, S.N. (2003). Mast cell-derived tumor necrosis factor induces hypertrophy of draining lymph nodes during infection. *Nat. Immunol.*, **4**, 1199–1205.

Miossec, P. (1997). Cytokine-induced autoimmune disorders. *Drug Saf.*, **17**, 93–104.

Miossec, P. (2003). Interleukin-17 in rheumatoid arthritis: If T cells were to contribute to inflammation and destruction through synergy. *Arthritis Rheum.*, **48**, 594–601.

Miossec, P. and van den Berg, W. (1997). Th1/Th2 cytokine balance in arthritis. *Arthritis Rheum.*, **40**, 2105–2115.

Mohan, N., Edwards, E.T., Cupps, T.R., Oliverio, P.J., Sandberg, G., Crayton, H., Richert, J.R., and Siegel, J.N. (2001). Demyelination occurring during anti-tumor necrosis factor alpha therapy for inflammatory arthritides. *Arthritis Rheum.*, **44**, 2862–2869.

Mugnier, B., Balandraud, N., Darque, A., Roudier, C., Roudier, J., and Reviron, D. (2003). Polymorphism at position −308 of the tumor necrosis factor alpha gene influences outcome of infliximab therapy in rheumatoid arthritis. *Arthritis Rheum.*, **48**, 1849–1852.

Murakami, C.S., Zeller, K., Bodenheimer, H.C., Jr., and Lee, W.M. (1994). Idiopathic thrombocytopenic purpura during interferon-alpha 2B treatment for chronic hepatitis. *Am. J. Gastroenterol.*, **89**, 2244–2245.

Nishimoto, N., Yoshizaki, K., Miyasaka, N., Yamamoto, K., Kawai, S., Takeuchi, T., Hashimoto, J., Azuma, J., Kishimoto, T., Ito, H., Takazoe, M., Fukuda, Y., Hibi, T., Kusugami, K., Andoh, A., Matsumoto, T., Yamamura, T., and Shimoyama, T. (2004). Treatment of rheumatoid arthritis with humanized anti-interleukin-6 receptor antibody: A multicenter, double-blind, placebo-controlled trial. *Arthritis Rheum.*, **50**, 1761–1769.

Paleolog, E. (1997). Target effector role of vascular endothelium in the inflammatory response: Insights from the clinical trial of anti-TNF alpha antibody in rheumatoid arthritis. *Mol. Pathol.*, **50**, 225–233.

Panitch, H.S., Hirsch, R.L., Haley, A.S., and Johnson, K.P. (1987). Exacerbations of multiple sclerosis in patients treated with gamma interferon. *Lancet*, **1**, 893–895.

Pascual, V., Banchereau, J., and Palucka, A.K. (2003). The central role of dendritic cells and interferon-alpha in SLE. *Curr. Opin. Rheumatol.*, **15**, 548–556.

Riddell, L.A., Pinching, A.J., Hill, S., Ng, T.T., Arbe, E., Lapham, G.P., Ash, S., Hillman, R., Tchamouroff, S., Denning, D.W., and Parkin, J.M. (2001). A phase III study of recombinant human interferon gamma to prevent opportunistic infections in advanced HIV disease. *AIDS Res. Hum. Retroviruses*, **17**, 789–797.

Robinson, W.H., Genovese, M.C., and Moreland, L.W. (2001). Demyelinating and neurologic events reported in association with tumor necrosis factor alpha antagonism: By what mechanisms could tumor necrosis factor alpha antagonists improve rheumatoid arthritis but exacerbate multiple sclerosis? *Arthritis Rheum.*, **44**, 1977–1983.

Rosenberg, S.A., Packard, B.S., Aebersold, P.M., Solomon, D., Topalian, S.L., Toy, S.T., Simon, P., Lotze, M.T., Yang, J.C., Seipp, C.A. (1988). Use of tumor-infiltrating lymphocytes and interleukin-2 in the immunotherapy of patients with metastatic melanoma. A preliminary report. *N. Engl. J. Med.*, **319**, 1676–1680.

Ruddle, N.H. (1999). Lymphoid neo-organogenesis: Lymphotoxin's role in inflammation and development. *Immunol. Res.*, **19**, 119–125.

Saab, S., Kalmaz, D., Gajjar, N.A., Hiatt, J., Durazo, F., Han, S., Farmer, D.G., Ghobrial, R.M., Yersiz, H., Goldstein, L.I., Lassman, C.R., and Busuttil, R.W. (2004). Outcomes of acute rejection after interferon therapy in liver transplant recipients. *Liver Transplant.*, **10**, 859–867.

Sangro, B., Mazzolini, G., Ruiz, J., Herraiz, M., Quiroga, J., Herrero, I., Benito, A., Larrache, J., Pueyo, J., Subtil, J.C., Olague, C., Sola, J., Sadaba, B., Lacasa, C., Melero, I., Qian, C., and Prieto, J. (2004). Phase I trial of intratumoral injection of an adenovirus encoding interleukin-12 for advanced digestive tumors. *J. Clin. Oncol.*, **22**, 1389–1397.

Schmidt, K.N. and Ouyang, W. (2004). Targeting interferon-alpha: A promising approach for systemic lupus erythematosus therapy. *Lupus*, **13**, 348–352.

Shakoor, N., Michalska, M., Harris, C.A., and Block, J.A. (2002). Drug-induced systemic lupus erythematosus associated with etanercept therapy. *Lancet*, **359**, 579–580.

Shealy, D.J., Wooley, P.H., Emmell, E., Volk, A., Rosenberg, A., Treacy, G., Wagner, C.L., Mayton, L., Griswold, D.E., and Song, X.Y. (2002). Anti-TNF-alpha antibody allows healing of joint damage in polyarthritic transgenic mice. *Arthritis Res.*, **4**, R7.

Stricker, R.B., Barlogie, B., and Kiprov, D.D. (1994). Acquired factor VIII inhibitor associated with chronic interferon-alpha therapy. *J. Rheumatol.*, **21**, 350–352.

Stuart, K., Levy, D.E., Anderson, T., Axiotis, C.A., Dutcher, J.P., Eisenberg, A., Erban, J.K., and Benson, I.A. (2004). Phase II study of interferon gamma in malignant carcinoid tumors (E9292): A trial of the Eastern Cooperative Oncology Group. *Invest. New Drugs*, **22**, 75–81.

Tak, P.P., Taylor, P.C., Breedveld, F.C., Smeets, T.J., Daha, M.R., Kluin, P.M., Meinders, A.E., and Maini, R.N. (1996). Decrease in cellularity and expression of adhesion molecules by anti-tumor necrosis factor alpha monoclonal antibody treatment in patients with rheumatoid arthritis. *Arthritis Rheum.*, **39**, 1077–1081.

Trinchieri, G., Pflanz, S., and Kastelein, R.A. (2003). The IL-12 family of heterodimeric cytokines: New players in the regulation of T cell responses. *Immunity*, **19**, 641–644.

Ulfgren, A.K., Andersson, U., Engstrom, M., Klareskog, L., Maini, R.N., and Taylor, P.C. (2000). Systemic anti-tumor necrosis factor alpha therapy in rheumatoid arthritis down-regulates synovial tumor necrosis factor alpha synthesis. *Arthritis Rheum.*, **43**, 2391–2396.

Van Assche, G. and Rutgeerts, P. (2000). Anti-TNF agents in Crohn's disease. *Expert Opin. Invest. Drugs*, **9**, 103–111.

Van den Berg, W.B. (2002). Lessons from animal models of arthritis. *Curr. Rheumatol. Rep.*, **4**, 232–239.

Van den Brande, J.M., Braat, H., van den Brink, G.R., Versteeg, H.H., Bauer, C.A., Hoedemaeker, I., van Montfrans, C., Hommes, D.W., Peppelenbosch, M.P., and van Deventer, S.J. (2003). Infliximab but not etanercept induces apoptosis in lamina propria T-lymphocytes from patients with Crohn's disease. *Gastroenterology*, **124**, 1774–1785.

Wilson, L.E., Widman, D., Dikman, S.H., and Gorevic, P.D. (2002). Autoimmune disease complicating antiviral therapy for hepatitis C virus infection. *Semin. Arthritis Rheum.*, **32**, 163–173.

Yasuda, M., Kihara, T., Wada, T., Shiokawa, S., Furuta, E., Suenagu, Y., Nonaka, S., Nobunaga, M., Yoshiok, K., and Isayama, T. (1994). Granulocyte colony-stimulating factor induction of improved leukocytopenia with inflammatory flare in a Felty's syndrome patient. *Arthritis Rheum.*, **37**, 145–146.

Ziolkowska, M., Kurowska, M., Radzikowska, A., Luszczykiewicz, G., Wiland, P., Dziewczopolski, W., Filipowicz-Sosnowska, A., Pazdur, J., Szechinski, J., Kowalczewski, J., Rell-Bakalarska, M., and Maslinski, W. (2002). High levels of osteoprotegerin and soluble receptor activator of nuclear factor kappa B ligand in serum of rheumatoid arthritis patients and their normalization after anti-tumor necrosis factor alpha treatment. *Arthritis Rheum.*, **46**, 1744–1753.

Zwerina, J., Hayer, S., Tohidast-Akrad, M., Bergmeister, H., Redlich, K., Feige, U., Dunstan, C., Kollias, G., Steiner, G., Smolen, J., and Schett, G. (2004). Single and combined inhibition of tumor necrosis factor, interleukin-1, and RANKL pathways in tumor necrosis factor-induced arthritis: Effects on synovial inflammation, bone erosion, and cartilage destruction. *Arthritis Rheum.*, **50**, 277–290.

24

Hematopoietic Stem Cell Transplantation for the Treatment of Severe Autoimmune Diseases

Alan Tyndall and Paul Hasler

1. Introduction

The concept of autoimmunity encompasses a complex network of humoral and cellular events resulting in unwanted tissue responses, called autoimmune disease (AD). Genetic and environmental factors are involved, and there is no unifying pathophysiological concept, just as there is no single successful treatment strategy. Recent thinking has focused on the innate immune system as the critical component in the pathway leading to autoinflammatory events, but despite effective targeted therapies such as anti-tumor necrosis factor-alpha (TNF-α) strategies, the successful eradication of AD remains elusive.

Most patients with severe AD are treated with a combination of glucocorticosteroids and immunosuppressive agents, but some either do not respond or require more toxic drugs to achieve or maintain clinical remission, and this subgroup poses a serious treatment dilemma.

The dose of cytotoxic drugs such as cyclophosphamide (Cy) has been limited by bone marrow toxicity, but improving hematopoietic stem cell transplantation (HSCT) techniques allows one to exceed these limits, then "rescue" the patient with autologous hematopoietic stem cells (HSCs). The observation in some patients receiving HSCT for conventional indications that a coexisting AD also improved suggested that HSCT could be a viable option for selected AD patients. The concept was also supported by animal model data. This led to an international collaboration, and, currently worldwide around 700 patients have received an HSCT as treatment of an AD. The experience gained from the phase I and II studies so far has been exploited in designing the running phase III randomized comparative trials and at the same time the study of basic biological consequences such as immune reconstitution.

Alan Tyndall • Department of Rheumatology, University Hospital, Basle, Switzerland.
Paul Hasler • Department of Rheumatology, Kantonsspital, Aarau, Switzerland.

Molecular Autoimmunity: In commemoration of the 100th anniversary of the first description of human autoimmune disease, edited by Moncef Zouali. Springer Science+Business Media, Inc., New York, 2005.

The first consensus statement concerning the use of HSCT in the treatment of severe AD was published in April 1995 (Marmont *et al.*, 1995) and the first case report in October 1996 (Tamm *et al.*, 1996). Results of the autologous HSCT programs have suggested that in favorable outcomes, a resetting of a dysregulated autoaggressive immune system may be occurring, rather than total ablation of autoimmune-inducing cells. This chapter reviews the animal model and coincidental human data that led to the prospective clinical trials and summarizes the results of the phase I and II trials from which the phase II trials were developed. In addition, data concerning immune reconstitution and possible future directions are presented.

2. Autoimmune Disease Mechanisms

AD is thought to result from the aberrant activation of the immune system together with the failure of immunoregulatory mechanisms that normally maintain immune tolerance. Despite the heterogeneous clinical expression of AD in humans, it seems clear that most ADs share several or all of the following features. They are polyclonal, with rarely a defined inciting single antigenic epitope and by the time of clinical disease expression, there has been extensive epitope spreading and effector cell recruitment (Davidson and Diamond, 2001). The innate immune system and tissue environment probably play a vital role in determining whether an antigen will evoke an immune reaction or anergy/tolerance (Medzhitov and Janeway, 2002) and a genetic component is present, but not sufficient. Some genes encode for major histocompatibility complex (MHC) molecules, but multiple other genes on different chromosomes play a role. In insulin-dependent diabetes mellitus for example, at least 19 such regions have been mapped (Tisch and McDevitt, 1996); disease initiation and perpetuation probably involve activation and disturbance of specific subsets of regulatory T cells. The recent reevaluation of a subset of $CD4^+CD25^+$ T cells with suppressor activity (Shevach, 2002) supports this concept of dysregulation, rather than "all or nothing" events, as in malignant clonal disease. The complexity of these diseases is further illustrated by the observation that clinical expression is often dependent on a mixture of inflammatory and scarring processes.

Presentation of self-antigens probably occurs continuously, but under normal circumstances produces either apoptosis, anergy, or tolerance if presented without costimulatory molecules. Typically, such peripheral tolerance occurs when nonprofessional antigen-presenting cells (APCs) lacking B7 (CD80) (Kamradt and Mitchison, 2001) are involved. Which T cells are needed for this autoaggressive reaction? It is known that autoreactive T cells escape thymic deletion and remain in the periphery, but with low affinity. Under the circumstances described above, these lymphocytes may be activated and induce an autoimmune process. This reaction is probably, in turn, controlled by regulatory T cell subsets, especially early in the process.

Breakdown of this regulatory network over time allows clinical expression and the development of chronic AD. Reversal of this vicious circle and reinstitution of the normal regulatory network, but not eradication of the last single autoreactive cell, is one of the postulated mechanisms behind the concept of HSCT for treatment of AD.

3. Coincidental AD in Patients Receiving HSCT for Another Indication

A number of case reports have been published over the past 20 years describing patients receiving HSCT for a conventional indication (e.g., aplastic anemia or malignancy) in which a coincidental AD improved or even fully remitted. In the majority of initial reports allogeneic HSCT was employed (Table 24.1). Many of these patients remained free of both diseases, the hematological and the AD. In some patients relapse occurred, and in one such patient full engraftment with donor-type lymphocytes (McKendry *et al.*, 1996) was observed. More recent reports have included response following autologous HSCT (Table 24.2),

Table 24.1. Coincidental Autoimmune Disease and Allogeneic HSC Transplantation

Disease for which transplant is performed	AD present	Outcome of AD	Patient outcome	Reference
SAA	RA	Remission	Died	(Baldwin *et al.*, 1977)
SAA	RA	Remission	Died	(Baldwin *et al.*, 1977)
SAA	RA	Remission	Died	(Baldwin *et al.*, 1977)
SAA	RA	Remission	Well	(Baldwin *et al.*, 1977)
SAA	RA	Partial remission	Well	(Jacobs *et al.*, 1986)
SAA	RA	Remission	Well	(Lowenthal *et al.*, 1993)
SAA	RA	Remission	Well	(Lowenthal *et al.*, 1993)
AML	Psoriasis	Remission	Well	(Eedy *et al.*, 1990)
CML	Psoriasis	Remission	Well	(Yin and Jowitt, 1992)
AML	Ulcerative colitis	Remission	Well	(Yin and Jowitt, 1992)
ALL	Autoimmune hepatitis	Remission	Well	(Vento *et al.*, 1996)
CML	Multiple sclerosis		Well	(McAllister *et al.*, 1997)
Various	Hyperthyroidism IDDM SLE, RA Crohn's disease Vasculitis Dermatitis herpetiformis	No recurrence		(Nelson *et al.*, 1997)
MALT lymphoma	Sjögren's syndrome	No effect	Alive	(Ferraccioli *et al.*, 2000)

SAA: severe aplastic anemia; AML: acute myeloid leukemia; CML: chronic myeloid leukemia; ALL: acute lymphoblastic leukemia; RA: rheumatoid arthritis; IDDM: insulin-dependent diabetes mellitus; SLE: systemic lupus erythematosus.

Table 24.2. Coincidental Autoimmune Disease and Autologous HSC Transplantation

Disease for which transplant is performed	AD present	Outcome of AD	Patient outcome	Reference
NHL	Myasthenia gravis	Remission	Well	(Salzman and Jackson, 1994)
Ovarian cancer	Thyroiditis	Relapse	Alive	(Euler et al., 1996)
NHL	Myasthenia	Relapse	Died	(Euler et al., 1996)
NHL	SLE	Relapse	Alive	(Euler et al., 1996)
NHL	Atopic dermatitis	Relapse	Alive	(Euler et al., 1996)
NHL	RA	Remission	Alive	(Snowden et al., 1997)
CML	SLE	Remission	Alive	(Meloni et al., 1997)
NHL	SLE	Remission	Alive	(Jondeau et al., 1997)
NHL	RA	Relapse	Alive	(Cooley et al., 1997)
NHL	RA	Relapse	Alive	(Cooley et al., 1997)
AML	Psoriasis	Relapse	Alive	(Cooley et al., 1997)
Plasma cell leukemia	Psoriasis	Relapse	Alive	(Cooley et al., 1997)
NHL	Crohn's disease	Remission	Alive	(Kashyap and Forman, 1998)
Hodgkin's	Crohn's disease	Remission	Alive	(Musso et al., 2000)

NHL: non-Hodgkin's lymphoma. Others as given in Table 24.1.

emphasizing the fact that genetic predisposition alone is not sufficient for AD expression (Davidson and Diamond, 2001).

There are also case reports of transfer of AD through allogeneic HSCT, including myasthenia gravis, thyroid disease, insulin-dependent diabetes mellitus, celiac disease, and psoriasis with arthritis (Minchinton et al., 1982). In one patient, production of autoantibodies (anti-Clq) was detected in a recipient following HSCT from a donor with known systemic lupus erythematosus (SLE), but clinical disease did not develop (Sturfelt et al., 1996).

In interpreting these case reports, it is important to remember that there is selection bias and details of AD severity or extent are often lacking, making it difficult to determine the clinical relevance of the outcome. Thus, there is sufficient evidence in these reports to assume some modification of the AD process following HSCT, justifying further clinical trials.

4. Animal Models

Support for the concept of HSCT in the treatment of AD is also found in animal models. Since the original observation by Denman in 1996 that SLE could be transferred from a susceptible to a nonsusceptible strain through allogeneic bone marrow transplantation (Denman et al., 1969), and later by Morton (Morton and Siegel, 1974), many proofs of concept observations have been published. This has recently been extensively reviewed (van Bekkum, 2000; Ikehara, 2001). In interpreting these data it is important to distinguish models in which AD is genetically and inevitably programmed, e.g., the MLR/lpr mouse and those in which a genetic component plus a trigger are required, e.g., adjuvant arthritis in the buffalo rat. The latter is more like human AD, as reflected in concordance rates between identical twins, i.e., 15% in SLE, 18% in rheumatoid arthritis (RA), 25% in multiple sclerosis (MS), and 50% in insulin-dependent diabetes mellitus (IDDM). In addition, it is important to distinguish between HSCT performed to prevent AD occurring, or HSCT to treat established AD. As in human AD, the autoimmune process has been active at a cellular level, often long before the clinical features become manifest. The data have been summarized by van Bekkum (2000), whose work in adjuvant arthritis, and later experimental allergic encephalomyelitis (EAE), demonstrated that not just allogeneic but also autologous HSCT could prevent and treat AD (Knaan-Shanzer et al., 1991). In addition, a significant peripheral immunological tolerance was induced, especially in the arthritis model. It is hoped that such immunomodulation will also occur in humans, and that HSCT will induce more than just profound immunosuppression.

5. Treatment of Human Autoimmune Disease with Hematopoietic Stem Cell Transplantation

Currently around 700 patients worldwide have received a blood marrow transplant (BMT) as treatment of an AD alone, 468 of whom are registered in the European Group for Blood and Marrow Transplants (EBMT) and the European League against Rheumatism (EULAR) database (Table 24.3). The majority of patients have had either severe MS or systemic sclerosis (SSc), also called scleroderma. This reflects the fact that there is no reliably effective alternative treatment option in these disorders. However, as the experience grew, other ADs were transplanted, mostly in the context of combined phase I and II trials and following the consensus guidelines developed at international meetings (McSweeney et al., 1997; Tyndall and Gratwohl, 1997) early in the program.

The quintessence of these guidelines was:

1. *HSCT regimes*: A limited number of protocols only should be employed (Table 24.4). This was mostly followed and allowed some comparison of intensity versus toxicity/benefit to be drawn.
2. *Patient selection*: Patients should have had failed conventional therapy and have a poor prognosis concerning life or vital organ function. There should be enough reversible or maintainable vital organ function to ensure a decent quality of life if the immunological/inflammatory process were arrested or reversed. The patient should have sufficient capacity to withstand the HSCT procedure.

As the program proceeded, certain clinical parameters and treatment-related factors emerged as being associated with an unacceptable risk, such as a mean pulmonary artery pressure >50 mmHg in SSc, high disability scores in MS, and total body irradiation (TBI) without lung shielding in SSc. This experience was then exploited in the design of the phase III randomized studies. Following the initial consensus meetings, single case reports and small series of patients transplanted for

Table 24.3. EBMT/EULAR Autoimmune Disease Autologous HSCT Database (Status: August 2002)

Disease and disease category	N
Neurological disorders	
Multiple sclerosis	135
Myasthenia gravis	2
Polyneuropathy	2
Amyotrophic lateral sclerosis	2
Rheumatological disorders	
Systemic sclerosis	72
Rheumatoid arthritis	72
Juvenile idiopathic arthritis	51
Systemic lupus erythematosus	55
Dermatomyositis	7
Mixed connective tissue disease	4
Morbus Behcet	3
Psoriatic arthritis	2
Ankylosing spondylitis	2
Sjögren's syndrome	1
Vasculitides	
Wegener's granulomatosis	3
Cryoglobulinemia	4
Not classified	2
Hematological immuncytopenias	
Immune thrombopenia	12
Pure red cell aplasia	4
Autoimmune hemolytic anemia	4
Thrombotic thrombocytopenic anemia	3
Evans syndrome	2
Gastrointestinal	
Enteropathy	2
Inflammatory bowel disease	1
Others	3
Total	453

Table 24.4. Conditioning Regimens Used with HSCT in Autoimmune Diseases

Conditioning regimen	Total
Cyclophosphamide	115
Cyclophosphamide ± ATG ± other drugs	110
Cyclophosphamide + radiation ± other drugs or ATG	43
Busulfan ± cyclophosphamide ± ATG ± other drugs	25
BEAM ± ATG	80
Others/missing	66
Total	439

TBI: total body irradiation (includes some patients with total lymphoid irradiation); ATG: antithymocyte globulin; BEAM: BCNU, VP16, ara-C, melphalan.

the treatment of severe AD have been published. These reports demonstrate a heterogeneity of patient selection, target AD, and outcome, and have in part formed the basis for further prospective trials. In the EBMT/EULAR database the most commonly transplanted diseases are MS, SSc, RA, juvenile idiopathic arthritis (JIA), and SLE, the data coming from over 100 transplant centers in more than 20 countries. There were long-lasting responses in all disease categories, but they were achieved at a price, the overall actuarially adjusted transplant-related mortality (TRM) being 7% (Gratwohl et al., 2001). This was higher than the predicted 3% for autologous HSCT overall and reflects the general overall level of illness and multiorgan involvement of many AD patients compared with, for instance, breast cancer patients undergoing high-dose chemotherapy and HSCT. In fact, there is a marked difference between AD groups with a TRM of 11% in SLE and only one patient (1.4%) with RA. There are also different response rates and types. In RA, JIA, and SLE more patients responded early but later relapsed than for MS and SSc.

6. Systemic Sclerosis

SSc is a multiorgan AD with immunological, vascular, and collagen overproduction components. In the first 45 patients, an improvement of 25% or more in the skin score (measured by the modified Rodnan method) was seen in 70% of the patients, with a TRM of 17% (Binks et al., 2001). Several protocols were used, mostly either Cy-based (4 g/m^2 Cy mobilization and Cy 200 mg/kg body weight conditioning or radiation 8 Gy/Cy 120 mg/kg body weight). With further patient recruitment and longer-term follow-up, the TRM of the EBMT-registered patients fell, considered to be related to more careful patient selection. Lung function tended to stabilize and some factors were identified as potentially hazardous for HSCT, e.g., pulmonary hypertension >50 mmHg mean pulmonary arterial pressure, severe cardiac involvement, severe pulmonary fibrosis, and uncontrolled systemic hypertension. A long-term follow-up of this cohort showed an overall TRM of 8.5%, no further transplant related deaths, and trend to durable remissions (Farge et al., 2004).

A multicenter US study of 19 SSc patients utilizing a regimen of Cy 120 mg/kg, TBI 8 Gy, and equine ATG 90 mg/kg body weight and a CD34-selected graft product showed a sustained benefit in 12 patients at median follow-up of 14.7 months (McSweeney et al., 2002). Four patients died, three from treatment-related causes and one from disease progression. In two cases a fatal regimen-related pulmonary toxicity occurred, which was not seen in the subsequent 11 patients in whom lung shielding was employed. Twelve patients had a sustained and significant improvement of skin score and functional status to a degree not previously seen with other treatment modalities.

7. Rheumatoid Arthritis

A retrospective analysis of the first 78 registered patients showed significant improvement, with 67% achieving an ACR-50 response at some time after transplant (Snowden et al., 2004). Most of the patients had failed a median of

5 (range 2–9) conventional disease-modifying antirheumatic drugs (DMARDs) before the transplant. Some degree of relapse was seen in 73% of patients after transplant, but was in most cases relatively easy to control with drugs that had proven ineffective before transplant. At 12 months after transplant, more than half of the patients had achieved an ACR-50 or more, and of these, just over 50% had not restarted DMARDs. The median follow-up was 18 (6–40) months, and at this time the majority of patients received a conditioning regimen of Cy 200 mg/m^2 alone and received peripherally harvested stem cells after either granulocyte-colony stimulating factor (G-CSF) or Cy/G-CSF (equal numbers) mobilization. Only one TRM was reported, a patient who, 5 months after transplant (Busulphan/Cy), died of sepsis, with a coincidental non-small-cell lung carcinoma being discovered at autopsy. In the opinion of the investigators, this was not considered to be a transplant-induced tumor. A multicenter trial in Australia failed to show any advantage of CD34 selection of the graft after nonmyeloablative conditioning with Cy (Moore et al., 2002).

8. Juvenile Idiopathic Arthritis

A total of 51 children with JIA, mostly the systemic form called Still's disease, have been registered. Most of these cases were treated in two Dutch centers using a bone marrow–obtained stem cell source and a conditioning protocol of Cy 200 mg/kg body weight, TBI 4 Gy, and ATG (Barron et al., 2001; Wulffraat et al., 2003). In the whole group there were 15 complete remissions and 3 partial remissions reported. In those attaining remission, the corticosteroid dose could be reduced and some patients experienced puberty and catch-up growth. Three patients died from hemophagocytic syndrome, also called the macrophage activation syndrome, thought to be related to intercurrent infection or uncontrolled systemic activity of the disease at the time of transplantation. Protocols were modified accordingly, such that systemic activity is controlled before the transplant with methyl prednisolone intravenously. Since this modification, no such deaths have occurred.

9. Systemic Lupus Erythematosus

Of the 55 registrations in the EBMT/EULAR database, most had either renal and/or central nervous system (CNS) involvement, and 21 had failed conventional Cy treatment. A peripheral stem cell source after mobilization with Cy and G-CSF was used in the majority. Twenty-three patients received a conditioning with Cy and ATG, 11 Cy plus TBI and 4 other regimens were employed. An unselected graft was used in 29, with CD34 selection in 19. There were five deaths due to treatment and one from progressive disease, resulting in an actuarially adjusted TRM of 10% (2–20). In those 53 patients with sufficient data for analysis, 66% achieved a "remission," defined as an SLE Disease Activity Index (SLEDAI) of ≤3 and steroid reduction to <10 mg/day. Of those achieving remission, 32% subsequently relapsed to some degree and were mostly easily controlled on standard agents that had previously been ineffective. There were 12 deaths after 1.5 (0–48) months, of which seven (12%) were related to the procedure (Jayne and Tyndall, 2004; Jayne et al., 2004).

Traynor et al. (2002) reported on nine patients with severe SLE who were mobilized in a transplant protocol. One died as a result of infection following mobilization and another 3 months later from active CNS SLE, having not proceeded to transplant. The seven remaining were free of signs of active SLE at a median follow-up of 25 months after transplant. The high-dose chemotherapy consisted of Cy 200 mg/kg, methylprednisolone 1 g, and equine ATG 90 mg/kg. The numbers of cases with vasculitis, Behçets disease, relapsing polychondritis, and other ADs are too small to draw meaningful conclusions, with further phase I and II standardized protocol pilot studies proceeding.

10. Prospective Randomized Controlled Clinical Trials

Criteria for moving to phase III randomized controlled trials (RCTs) are: enough information is available from phase I/II trials; inherent mortality of the disease justifies the risk of the procedure; prognostic factors of the disease are known to define patients at high risk for disease progression; HSCT morbidity and mortality is acceptably low; risk of disease progression after HSCT is low; little or no alternative conventional therapy is available. Such criteria are currently sufficiently met for SSc, MS, and RA.

In the Autologous Stem Cell Transplantation International Scleroderma (ASTIS) Trial, patients who have less than 4 years of diffuse skin involvement and evidence of progressive and organ- or life-threatening disease are selected. The primary end point on which the trial is powered is event-free survival at 2 years, events being arbitrarily but precisely defined to capture irreversible and severe end-organ failure or death. Exclusion criteria are based on the phase I and II data to avoid an unacceptably high TRM risk together with a minimal chance of clinically significant improvement. The treatment arm is mobilization with Cy $4g/m^2$ and G-CSF, followed by Cy 200 mg/kg body weight conditioning plus ATG and a CD34-selected graft. The control arm is monthly IV pulse Cy 750 mg/m^2 for 12 months. The ASTIS trial is running, and further details are available on the website www.astistrial.com. So far, 41 patients have been randomized and there has been no treatment-related mortality.

A similar study is being planned by a US consortium (P. McSweeney, personal communication). In the Autologous Stem Cell Transplantation International Rheumatoid Arthritis (ASTIRA) Trial, active RA patients who have failed at least four DMARDs, including methotrexate and TNF-α blocking agents with a disease duration between 2 and 15 years, will all receive stem cell mobilization with Cy 4 g/m^2 and G-CSF. Randomization will then occur to either continued conventional therapy with either methotrexate or leflunomide or conditioning with Cy 200 mg/m^2 and ATG. The graft will not be manipulated. The primary end point is the number of patients reaching a good or moderate EULAR response and/or an ACR 20 at 6 months. Sixteen patients in each arm are required, calculated on a >50% difference in the two groups and the trial is running. In SLE, a phase II study is being planned to assess the role if any of posttransplant maintenance (e.g., mycophenolate mofetil) therapy to retain remission. The results of phase I/II trials in JIA using Cy alone versus Cy and TBI suggested no advantage of the TBI (N. Wulffraat, personal communication). Further phase II studies will be performed to assess the optimal regimen for a phase III study.

11. Open Issues

11.1. Allogeneic HSCT

Several factors have been limiting for autologous transplants. First, regimens have been less than maximally intensive and thus have not achieved eradication of immunologic memory. Second, reinfusion of potentially pathogenic T and B cells is a potential cause of relapse. The latter has been thought to justify allogeneic transplantation, with its attendant higher risks (10–30% mortality), mainly relating to graft versus host disease (GVHD). Several case reports and series support this notion. A child with autoimmune hemolytic anemia that was refractory to immunosuppression and splenectomy had only a 7-week remission after autologous transplantation, but was still in remission 18 months after an HLA-identical unrelated donor transplant (De Stefano *et al.*, 1999). A graft-versus-autoimmunity effect has been proposed in a patient given an allogeneic stem cell transplant for chronic myeloid leukemia, who also had severe psoriasis (Slavin *et al.*, 2000). This is also compatible with long-term control of RA in a small number of allografted patients. These patients had received conditioning regimens very similar to that given in trials of autografting in RA (i.e., Cy 200 mg/kg), and the longer remission in the allografted patients suggest that the type of graft rather than the conditioning regimen determined outcome. However, both long-term remissions with autologous HSCT and relapses following allogeneic HSCT (with full donor chimerism) have been observed in AD after transplant, as well as in newly occurring AD (Rouquette-Gally *et al.*, 1987, 1988; Snowden and Heaton, 1997; Baron *et al.*, 1998).

Novel concepts with nonmyeloablative conditioning regimens may reduce early transplant-related mortality to less than 10%, making allogeneic HSCT for AD more acceptable. Still, the risk of GVHD will remain (Binks *et al.*, 2001), and it is unclear whether the target for HSCT in AD can be defined as clearly as in malignant and inherited disorders. Carefully selected cases with early but high risk disease and low risk for transplant-related mortality (young age, HLA-identical siblings) should provide an answer. For these reasons, allogeneic HSCT for the treatment of severe AD must await further refinement of the transplant procedure and, in particular, the prevention of GVHD. The international guidelines stipulated that autologous HSCT should be the preferred approach (Gratwohl and Tyndall, 1997). So far, this has been mostly adhered to with allogeneic HSCT for AD alone having been performed mainly in refractory cytopenias (Gratwohl *et al.*, 2001).

Despite new developments, arguments not to use allogeneic HSCT in the first instance remain the same. Treatment-related toxicity is high, GVHD cannot yet be avoided and might interfere with the preexisting disease without the potential additional benefit of "graft-versus-autoimmunity." Unlike malignancy, there is no definable clone of autoaggressive cells to be eradicated. Furthermore, incomplete or slowed immune reconstitution after allogeneic HSCT might lead to late development of a donor-type AD, even more so in predisposed patients.

It remains open to debate whether reduced intensity conditioning regimens might alter these perspectives, since they have been shown to reduce early mortality. So far, they have not reduced risk of GVHD and long-term follow-up is required. Still, there is a consensus that it might be appropriate under carefully selected conditions to begin the planning of phase I/II studies to evaluate the role of allogeneic HSCT. Conditioning with Cy ± ATG as used for aplastic anemia for many years might be the most appropriate choice.

11.2. Immune Reconstitution

In general, immune reconstitution must be considered separately for autologous and allogeneic transplantation. According to the many variations of each procedure (see also Table 24.4), the impact of individual regimens on immune reconstitution may differ, even before considering the effect of different diseases and age groups. Among the possible measures for immune reconstitution are cell surface markers to determine the appearance and development of different cell lines, the response to infectious agents and vaccination, and the repertoire of adaptive immunity to non-self and self-antigens. In the autologous setting, the method used for mobilization, the extent of the immunoablative regimen, and negative and positive selection of the autograft may influence reconstitution of the hemopoietic and immune systems. In children, $CD4^+CD45RO^+$ peripheral T cell expansion occurred within 16 days after transplant. Natural killer (NK) cell ($CD16^+/CD56^+$) counts normalized rapidly. After day 30 an inverted CD4/CD8 ratio was still present, and T cell ($CD3^+$, $CD3^+CD4^+$, $CD4^+CD45RA^+$) recovery was delayed until 24 months after the autograft (Kalwak *et al.*, 2002). In adults, lymphocyte subset recovery and T cell receptor (TCR) beta-chain variable region did not differ between patients receiving unmanipulated or CD34-selected autologous HSCT (Peggs *et al.*, 2003). A study of dendritic cell (DC) subset reconstitution after autologous HSCT involving 58 patients showed that peripheral blood $CD11c^+CD123^{low}$ type 1 DC (DC1) and $CD11c^-CD123^+$ type 2 DC (DC2) counts reached preconditioning levels 20 days after unmanipulated autologous HSCT. When CD34 selection was performed, recovery to levels after mobilization was delayed until day 60. Levels of DC1 and DC2 approached normal from day 180 in the group receiving unmanipulated grafts, while in those receiving CD34-selected HSCT, DC1 and DC2 counts remained persistently lower than normal (Damiani *et al.*, 2002).

The reconstitution of the immune system after allogeneic HSCT is complex, and may be significantly impaired by chronic GVHD. After allogeneic HSCT with full donor chimerism, recovery of TCR rearrangement diversity is more rapid in younger individuals. Recovery in older patients was slower, but little difference remained at 9 months. In this series, low levels of TCR rearrangement correlated with severe opportunistic infections and with GVHD (Lewin et al., 2002). The reconstitution of recipient NK cell repertoire after HLA-matched HSCT followed the killer immunoglobulin-like receptor (KIR) pattern of the donor in the majority of cases, whereas in the others no uniform pattern was evident, and severe clinical complications occurred (Shilling et al., 2003). DC recovery after allo-HSCT was rapid, DC1 exceeding DC2 (Chklovskaia et al., 2004), in contrast to the levels after autologous HSCT (Damiani et al., 2002). NK cell reconstitution was also rapid, with a high proportion of interferon-gamma-producing $CD56^{high}CD16^{-/low}$ NK cells and reduced $CD56^{low}CD16^{high}$ cells (Chklovskaia et al., 2004). This differs from the data for autologous HSCT (Damiani et al., 2002).

The finding of T cell receptor excision circles (TRECs) in T cells recently exiting the thymus (Douek et al., 2000) has allowed a more detailed analysis of normal and autoaggressive T cell reactions following HSCT for AD. Following HSCT for AD, some adult patients have shown an increase in the number of lymphocytes bearing TRECs, indicating that the thymus may become reactivated and theoretically capable of inducing central tolerance.

Recipients of autologous HSCT with CD34 selection had higher rates of infection with agents other than cytomegalovirus (CMV) than those of unselected autografts. The principal cause of the difference were varicella zoster virus, parainfluenza virus 3, and bacterial infections (Crippa et al., 2002). Fungal infections showed no significant differences between CD34-selected and -non-selected recipients (Crippa et al., 2002). Donor T cell immunity against CMV, as measured by HLA-A2 tetrameric complexes targeting viral phosphoprotein UL83, was established in most recipients unless prolonged immunosuppression was required for GVHD (Aubert et al., 2001). CMV infection was associated with higher $CD8^+$ counts and decreased TCR beta-chain variable region diversity (Peggs et al., 2003).

Patients treated with allogeneic HSCT have a higher rate of infections compared with those receiving autologous HSCT (Einsele et al., 2003). EBV activation after allogeneic HSCT was associated with low $CD8^+$ levels, and a high cellular viral load preceded reactivation (Clave et al., 2004). The rate of severe infections after allogeneic HSCT appears to depend on the source of the graft, since it was significantly higher in marrow recipients than in patients receiving filgrastim-mobilized peripheral blood stem cell grafts (Storek et al., 2001).

11.3. Ablative Therapy without HSCT

HSCs resist the cytotoxic effects of cyclophosphamide, and therefore theoretically, an HSCT is not needed following aplasia induction and G-CSF-supported reconstitution. Such a strategy has been successfully employed in aplastic anemia and applied to SLE (Petri et al., 2003). Early results are encouraging, but a significant number of patients had not had conventional pulse cyclophosphamide therapy and the reconstitution times, especially for platelets, were prolonged compared to rescue with HSCT. Both procedures remain research-based rather than standard therapy.

12. Conclusions

The role of stem cell transplantation in the treatment of severe, therapy-refractory AD remains experimental, with data on around 700 patients being sufficiently encouraging to proceed to randomized prospective trials in the major diseases: SSc, RA, MS, and soon JIA and SLE. An impressive international collaboration has and is reducing duplication of efforts with shared databases, protocols, patient selection, and end points. The concept of resetting a dysbalance in the complex immune network, rather than total eradication of clonal autoimmunity, is emerging. Further clinical trials are required to establish the place, if any, HSCT has in such treatment, and a basic science program continues to explain the pathophysiological mechanisms of these immune-modulating strategies.

References

Aubert, G., Hassan-Walker, A.F., Madrigal, J.A., Emery, V.C., Morte, C., Grace, S., Koh, M.B., Potter, M., Prentice, H.G., Dodi, I.A., and Travers, P.J. (2001). Cytomegalovirus-specific cellular immune responses and viremia in recipients of allogeneic stem cell transplants. *J. Infect. Dis.*, **184**, 955–963.

Baldwin, J.L., Storb, R., Thomas, E.D., and Mannik, M. (1977). Bone marrow transplantation in patients with gold-induced marrow aplasia. *Arthritis Rheum.*, **20**, 1043–1048.

Baron, F.A., Hermanne, J.P., Dowlati, A., Weber, T., Thiry, A., Fassotte, M.F., Fillet, G., and Beguin, Y. (1998). Bronchiolitis obliterans organizing pneumonia and ulcerative colitis after allogeneic bone marrow transplantation. *Bone Marrow Transplant.*, **21**, 951–954.

Barron, K.S., Wallace, C., Woolfrey, C.E.A., Laxer, R.M., Hirsch, R., Horwitz, M., Siegel, J., Filipovich, L., Wulffraat, N., Passo, M., and Rider, L.G. (2001). Autologous stem cell transplantation for pediatric rheumatic diseases. *J. Rheumatol.*, **28**, 2337–2358.

Binks, M., Passweg, J.R., Furst, D., McSweeney, P., Sullivan, K., Besenthal, C., Finke, J., Peter, H.H., van Laar, J., Breedveld, F.C., Fibbe, W.E., Farge, D., Gluckman, E., Locatelli, F., Martini, A., van den Hoogen, F., van de Putte, L., Schattenberg, A.V., Arnold, R., Bacon, P.A., Emery, P., Espigado, I., Hertenstein, B., Hiepe, F., Kashyap, A., Kotter, I., Marmont, A., Martinez, A., Pascual, M.J., Gratwohl, A., Prentice, H.G., Black, C., and Tyndall, A. (2001). Phase I/II trial of autologous stem cell transplantation in systemic sclerosis: Procedure related mortality and impact on skin disease. *Ann. Rheum. Dis.*, **60**, 577–584.

Chklovskaia, E., Nowbakht, P., Nissen, C., Gratwohl, A., Bargetzi, M., and Wodnar-Filipowicz, A. (2004). Reconstitution of dendritic and natural killer-cell subsets after allogeneic stem cell transplantation: Effects of endogenous flt3 ligand. *Blood*, **103**, 3860–3868.

Clave, E., Agbalika, F., Bajzik, V., de Latour, R.P., Trillard, M., Rabian, C., Scieux, C., Devergie, A., Socie, G., Ribaud, P., Ades, L., Ferry, C., Gluckman, E., Charron, D., Esperou, H., Toubert, A., and Moins-Teisserenc, H. (2004). Epstein-Barr virus (EBV) reactivation in allogeneic stem-cell transplantation: Relationship between viral load, EBV-specific T-cell reconstitution and rituximab therapy. *Transplantation*, **77**, 76–84.

Cooley, H.M., Snowden, J.A., Grigg, A.P., and Wicks, I.P. (1997). Outcome of rheumatoid arthritis and psoriasis following autologous stem cell transplantation for hematologic malignancy. *Arthritis Rheum.*, **40**, 1712–1715.

Crippa, F., Holmberg, L., Carter, R.A., Hooper, H., Marr, K.A., Bensinger, W., Chauncey, T., Corey, L., and Boeckh, M. (2002). Infectious complications after autologous CD34-selected peripheral blood stem cell transplantation. *Biol. Blood Marrow Transplant.*, **8**, 281–289.

Damiani, D., Stocchi, R., Masolini, P., Michelutti, A., Sperotto, A., Geromin, A., Skert, C., Cerno, M., Michieli, M., Baccarani, M., and Fanin, R. (2002). Dendritic cell recovery after autologous stem cell transplantation. *Bone Marrow Transplant.*, **30**, 261–266.

Davidson, A. and Diamond, B. (2001). Autoimmune diseases. *N. Engl. J. Med.*, **345**, 340–350.

De Stefano, P., Zecca, M., Giorgiani, G., Perotti, C., Giraldi, E., and Locatelli, F. (1999). Resolution of immune haemolytic anaemia with allogeneic bone marrow transplantation after an unsuccessful autograft. *Br. J. Haematol.*, **106**, 1063–1064.

Denman, A.M., Russell, A.S., and Denman, E.J. (1969). Adoptive transfer of the diseases of New Zealand black mice to normal mouse strains. *Clin. Exp. Immunol.*, **5**, 567–595.

Douek, D.C., Vescio, R.A., Betts, M.R., Brenchley, J.M., Hill, B.J., Zhang, L., Berenson, J.R., Collins, R.H., and Koup, R.A. (2000). Assessment of thymic output in adults after haematopoietic stem-cell transplantation and prediction of T-cell reconstitution. *Lancet*, **355**, 1875–1881.

Eedy, D.J., Burrows, D., Bridges, J.M., and Jones, F.G. (1990). Clearance of severe psoriasis after allogenic bone marrow transplantation. *Br. Med. J.*, **300**, 908.

Einsele, H., Bertz, H., Beyer, J., Kiehl, M.G., Runde, V., Kolb, H.J., Holler, E., Beck, R., Schwerdfeger, R., Schumacher, U., Hebart, H., Martin, H., Kienast, J., Ullmann, A.J., Maschmeyer, G., Kruger, W., Niederwieser, D., Link, H., Schmidt, C.A., Oettle, H., and Klingebiel, T. (2003). Infectious complications after allogeneic stem cell transplantation: Epidemiology and interventional therapy strategies—Guidelines of the Infectious Diseases Working Party (AGIHO) of the German Society of Hematology and Oncology (DGHO). *Ann. Hematol.*, **82**(Suppl. 2), S175–S185.

Euler, H.H., Marmont, A.M., Bacigalupo, A., Fastenrath, S., Dreger, P., Hoffknecht, M., Zander, A.R., Schalke, B., Hahn, U., Haas, R., and Schmitz, N. (1996). Early recurrence or persistence of autoimmune diseases after unmanipulated autologous stem cell transplantation. *Blood*, **88**, 3621–3625.

Farge, D., Passweg, J., van Laar, J.M., Marjanovic, Z., Besenthal, C., Finke, J., Peter, H.H., Breedveld, F.C., Fibbe, W.E., Black, C., Denton, C., Koetter, I., Locatelli, F., Martini, A., Schattenberg, A.V., van den Hoogen, F., van de Putte, L., Lanza, F., Arnold, R., Bacon, P.A., Bingham, S., Ciceri, F., Didier, B., Diez-Martin, J.L., Emery, P., Feremans, W., Hertenstein, B., Hiepe, F., Luosujarvi, R., Leon Lara, A., Marmont, A., Martinez, A.M., Pascual Cascon, H., Bocelli-Tyndall, C., Gluckman, E., Gratwohl, A., and Tyndall, A. (2004). Autologous stem cell transplantation in the treatment of systemic sclerosis: Report from the EBMT/EULAR Registry. *Ann. Rheum. Dis.*, **63**, 974–981.

Ferraccioli, G., Damato, R., De Vita, S., Fanin, R., Damiani, D., and Baccarani, M. (2000). Haematopoietic stem cell transplantation (HSCT) in a patient with Sjögren's syndrome and lung malt lymphoma cured lymphoma not the autoimmune disease. *Ann. Rheum. Dis.*, **60**, 174–176.

Gratwohl, A., Passweg, J., Gerber, I., and Tyndall, A. (2001). Stem cell transplantation for autoimmune diseases. *Best Pract. Res. Clin. Haematol.*, **14**, 755–776.

Gratwohl, A. and Tyndall, A. (1997). Hematopoietic stem cell transplantations in treatment of autoimmune diseases. *Z. Rheumatol.*, **56**, 173–177.

Ikehara, S. (2001). Treatment of autoimmune diseases by hematopoietic stem cell transplantation. *Exp. Hematol.*, **29**, 661–669.

Jacobs, P., Vincent, M.D., and Martell, R.W. (1986). Prolonged remission of severe refractory rheumatoid arthritis following allogeneic bone marrow transplantation for drug-induced aplastic anaemia. *Bone Marrow Transplant.*, **1**, 237–239.

Jayne, D., Passweg, J., Marmont, A., Farge, D., Zhao, X., Arnold, R., Hiepe, F., Lisukov, I., Musso, M., Ou-Yang J., Marsh, J., Wulffraat, N., Besalduch, J., Bingham, S.J., Emery, P., Brune, M., Fassas, A., Faulkner, L., Ferster, A., Fiehn, C., Fouillard, L., Geromin, A., Greinix, H., Rabusin, M., Saccardi, R., Schneider, P., Zintl, F., Gratwohl, A., and Tyndall, A. (2004). Autologous stem cell transplantation for systemic lupus erythematosus. *Lupus*, **13**, 168–176.

Jayne, D. and Tyndall, A. (2004). Autologous stem cell transplantation for systemic lupus erythematosus. *Lupus*, **13**, 359–365.

Jondeau, K., Job-Deslandre, C., Bouscary, D., Khanlou, N., Menkes, C.J., and Dreyfus, F. (1997). Remission of nonerosive polyarthritis associated with Sjögren's syndrome after autologous hematopoietic stem cell transplantation for lymphoma. *J. Rheumatol.*, **24**, 2466–2468.

Kalwak, K., Gorczynska, E., Toporski, J., Turkiewicz, D., Slociak, M., Ussowicz, M., Latos-Grazynska, E., Krol, M., Boguslawska-Jaworska, J., and Chybicka, A. (2002). Immune reconstitution after haematopoietic cell transplantation in children: Immunophenotype analysis with regard to factors affecting the speed of recovery. *Br. J. Haematol.*, **118**, 74–89.

Kamradt, T. and Mitchison, N.A. (2001).Tolerance and autoimmunity. *N. Engl. J. Med.*, **344**, 655–664.

Kashyap, A. and Forman, S.J. (1998). Autologous bone marrow transplantation for non-Hodgkin's lymphoma resulting in long-term remission of coincidental Crohn's disease. *Br. J. Haematol.*, **103**, 651–652.

Knaan-Shanzer, S., Houben, P., Kinwel-Bohre, E.P., and van Bekkum, D.W. (1991). Remission induction of adjuvant arthritis in rats by total body irradiation and autologous bone marrow transplantation. *Bone Marrow Transplant.*, **8**, 333–338.

Lewin, S.R., Heller, G., Zhang, L., Rodrigues, E., Skulsky, E., van den Brink, M.R, Small, T.N., Kernan, N.A., O'Reilly, R.J., Ho, D.D., and Young, J.W. (2002). Direct evidence for new T-cell generation by patients after either T-cell-depleted or unmodified allogeneic hematopoietic stem cell transplantations. *Blood*, **100**, 2235–2242.

Lowenthal, R.M., Cohen, M.L., Atkinson, K., and Biggs, J.C. (1993). Apparent cure of rheumatoid arthritis by bone marrow transplantation. *J. Rheumatol.*, **20**, 137–140.

Marmont, A., Tyndall, A., Gratwohl, A., and Vischer, T. (1995). Haemopoietic precursor-cell transplants for autoimmune diseases. *Lancet*, **345**, 978.

McAllister, L.D., Beatty, P.G., and Rose, J. (1997). Allogeneic bone marrow transplant for chronic myelogenous leukemia in a patient with multiple sclerosis. *Bone Marrow Transplant.*, **19**, 395–397.

McKendry, R.J., Huebsch, L., and Leclair, B. (1996). Progression of rheumatoid arthritis following bone marrow transplantation. A case report with a 13-year followup. *Arthritis Rheum.*, **39**, 1246–1253.

McSweeney, P.A., Nash, R.A., Storb, R., Furst, D.E., Gauthier, J., and Sullivan, K.M. (1997). Autologous stem cell transplantation for autoimmune diseases: Issues in protocol development. *J. Rheumatol.*, **24**(Suppl. 48), 79–84.

McSweeney, P.A., Nash, R.A., Sullivan, K.M., Storek, J., Crofford, L.J., Dansey, R., Mayes, M.D., McDonagh, K.T., Nelson, J.L., Gooley, T.A., Holmberg, L.A., Chen, C.S., Wener, M.H., Ryan, K., Sunderhaus, J., Russell, K., Rambharose, J., Storb, R., and Furst, D.E. (2002). High-dose immunosuppressive therapy for severe systemic sclerosis: Initial outcomes. *Blood*, **100**, 1602–1610.

Medzhitov, R. and Janeway, C.A., Jr. (2002). Decoding the patterns of self and nonself by the innate immune system. *Science*, **296**, 298–300.

Meloni, G., Capria, S., Vignetti, M., Mandelli, F., and Modena, V. (1997). Blast crisis of chronic myelogenous leukemia in long-lasting systemic lupus erythematosus: Regression of both diseases after autologous bone marrow transplantation. *Blood*, **89**, 4659.

Minchinton, R.M., Waters, A.H., Kendra, J., and Barrett, A.J. (1982). Autoimmune thrombocytopenia acquired from an allogeneic bone-marrow graft. *Lancet*, **2**, 627–629.

Moore, J., Brooks, P., Milliken, S., Biggs, J., Ma, D., Handel, M., Cannell, P., Will, R., Rule, S., Joske, D., Langlands, B., Taylor, K., O'Callaghan, J., Szer, J., Wicks, I., McColl, G., Passeullo, F., and Snowden, J. (2002). A pilot randomized trial comparing CD34-selected versus unmanipulated hemopoietic stem cell transplantation for severe, refractory rheumatoid arthritis. *Arthritis Rheum.*, **46**, 2301–2309.

Morton, J.I. and Siegel, B.V. (1974). Transplantation of autoimmune potential. I. Development of antinuclear antibodies in H-2 histocompatible recipients of bone marrow from New Zealand black mice. *Proc. Natl. Acad. Sci. USA*, **71**, 2162–2165.

Musso, M., Porretto, F., Crescimanno, A., Bondi, F., Polizzi, V., and Scalone, R. (2000).Crohn's disease complicated by relapsed extranodal Hodgkin's lymphoma: Prolonged complete remission after unmanipulated PBPC autotransplant. *Bone Marrow Transplant.*, **26**, 921–923.

Nelson, J.L., Torrez, R., Louie, F.M., Choe, O.S., Storb, R., and Sullivan, K.M. (1997). Pre-existing autoimmune disease in patients with long-term survival after allogeneic bone marrow transplantation. *J. Rheumatol.*, **24**(Suppl. 48), 23–29.

Peggs, K.S., Verfuerth, S., Pizzey, A., Khan, N., Moss, P., Goldstone, A.H., Yong, K., and Mackinnon, S. (2003). Reconstitution of T-cell repertoire after autologous stem cell transplantation: Influence of CD34 selection and cytomegalovirus infection. *Biol. Blood Marrow Transplant.*, **9**, 198–205.

Petri, M., Jones, R.J., and Brodsky, R.A. (2003). High-dose cyclophosphamide without stem cell transplantation in systemic lupus erythematosus. *Arthritis Rheum.*, **48**, 166–173.

Rouquette-Gally, A.M., Boyeldieu, D., Gluckman, E., Abuaf, N., and Combrisson, A. (1987). Autoimmunity in 28 patients after allogeneic bone marrow transplantation: Comparison with Sjögren's syndrome and scleroderma. *Br. J. Haematol.*, **66**, 45–47.

Rouquette-Gally, A.M., Boyeldieu, D., Prost, A.C., and Gluckman, E. (1988). Autoimmunity after allogeneic bone marrow transplantation. A study of 53 long-term-surviving patients. *Transplantation*, **46**, 238–240.

Shevach, E.M. (2002). CD4+CD25+ suppressor T cells: More questions than answers. *Nat. Rev. Immunol.*, **2**, 389–400.

Shilling, H.G., McQueen, K.L., Cheng, N.W., Shizuru, J.A., Negrin, R.S., and Parham, P. (2003). Reconstitution of NK cell receptor repertoire following HLA-matched hematopoietic cell transplantation. *Blood*, **101**, 3730–3740.

Slavin, S., Nagler, A., Varadi, G., and Or, R. (2000). Graft vs autoimmunity following allogeneic non-myeloablative blood stem cell transplantation in a patient with chronic myelogenous leukemia and severe systemic psoriasis and psoriatic polyarthritis. *Exp. Hematol.*, **28**, 853–857.

Snowden, J.A. and Heaton, D.C. (1997). Development of psoriasis after syngeneic bone marrow transplant from psoriatic donor: Further evidence for adoptive autoimmunity. *Br. J. Dermatol.*, **137**, 130–132.

Snowden, J.A., Passweg, J., Moore, J.J., Milliken, S., Cannell, P., Van Laar, J., Verburg, R., Szer, J., Taylor, K., Joske, D., Rule, S., Bingham, S.J., Emery, P., Burt, R.K., Lowenthal, R.M., Durez, P., McKendry, R.J., Pavletic, S.Z., Espigado, I., Jantunen, E., Kashyap, A., Rabusin, M., Brooks, P., Bredeson, C., and Tyndall, A. (2004). Autologous hemopoietic stem cell transplantation in severe rheumatoid arthritis: A report from the EBMT and ABMTR. *J. Rheumatol.*, **31**, 482–488.

Snowden, J.A., Patton, W.N., O'Donnell, J.L., Hannah, E.E., and Hart, D.N. (1997). Prolonged remission of longstanding systemic lupus erythematosus after autologous bone marrow transplant for non-Hodgkin's lymphoma. *Bone Marrow Transplant.*, **19**, 1247–1250.

Storek, J., Dawson, M.A., Storer, B., Stevens-Ayers, T., Maloney, D.G., Marr, K.A., Witherspoon, R.P., Bensinger, W., Flowers, M.E., Martin, P., Storb, R., Appelbaum, F.R., and Boeckh, M. (2001). Immune reconstitution after allogeneic marrow transplantation compared with blood stem cell transplantation. *Blood*, **97**, 3380–3389.

Sturfelt, G., Lenhoff, S., Sallerfors, B., Nived, O., Truedsson, L., and Sjoholm, A.G. (1996). Transplantation with allogenic bone marrow from a donor with systemic lupus erythematosus (SLE): Successful outcome in the recipient and induction of an SLE flare in the donor. *Ann. Rheum. Dis.*, **55**, 638–641.

Tamm, M., Gratwohl, A., Tichelli, A., Perruchoud, A.P., and Tyndall, A. (1996). Autologous haemopoietic stem cell transplantation in a patient with severe pulmonary hypertension complicating connective tissue disease. *Ann. Rheum. Dis.*, **55**, 779–780.

Tisch, R. and McDevitt, H. (1996). Insulin-dependent diabetes mellitus. *Cell*, **85**, 291–297.

Traynor, A.E., Barr, W.G., Rosa, R.M., Rodriguez, J., Oyama, Y., Baker, S., Brush, M., and Burt, R.K. (2002). Hematopoietic stem cell transplantation for severe and refractory lupus. Analysis after five years and fifteen patients. *Arthritis Rheum.*, **46**, 2917–2923.

Tyndall, A. and Gratwohl, A. (1997). Hemopoietic blood and marrow transplants in the treatment of severe autoimmune disease. *Curr. Opin. Hematol.*, **4**, 390–394.

van Bekkum, D.W. (2000). Stem cell transplantation in experimental models of autoimmune disease. *J. Clin. Immunol.*, **20**, 10–16.

Vento, S., Cainelli, F., Renzini, C., Ghironzi, G., and Concia, E. (1996). Resolution of autoimmune hepatitis after bone-marrow transplantation. *Lancet*, **348**, 544–545.

Wulffraat, N.M., Brinkman, D., Ferster, A., Opperman, J., ten Cat, R., Wedderburn, L., Foster, H., Abinun, M., Prieur, A.M., Horneff, G., Zintl, F., de Kleer, I., and Kuis, W. (2003). Long-term follow-up of autologous stem cell transplantation for refractory juvenile idiopathic arthritis. *Bone Marrow Transplant.*, **32**(Suppl. 1), S61–S64.

Yin, J.A. and Jowitt, S.N. (1992). Resolution of immune-mediated diseases following allogeneic bone marrow transplantation for leukaemia. *Bone Marrow Transplant.*, **9**, 31–33.

25

Molecular Mimicry in Autoimmune Uveitis: From Pathogenesis to Therapy

Gerhild Wildner, Maria Diedrichs-Moehring, and Stephan R. Thurau

1. Introduction

Autoimmune uveitis is an inflammatory disease affecting the inner eye of about 2‰ of the western population. The disease is mediated by CD4$^+$ Th1 cells (Calder et al., 1999), which recruit macrophages and granulocytes after recognition of ocular autoantigens. Those inflammatory cells can irreversibly destroy photoreceptors and neuronal tissue within the eye, leading to decreased vision or even to blindness (Figure 25.1, see color insert).

The mechanisms of the primary induction of the autoimmune response directed to intraocular autoantigens are still unclear. The eye as an immune-privileged organ can only be entered by already activated T cells, for the blood–retina barrier prevents invasion of naive lymphocytes. T cells specific for ocular autoantigen will subsequently get reactivated by local antigens to escape apoptosis and to induce uveitis, whereas activated cells not specific for ocular antigens cannot survive the immunosuppressive environment within the eye (Thurau et al., 2004).

Therefore, the autoimmune response leading to uveitis must be initiated outside the eye. Considering the sequestered expression of retinal autoantigens the extraocular antigens eliciting the immune response must mimic retinal epitopes. Only in the case of sympathetic ophthalmia the autoaggressive T cells are activated within an injured eye by local antigen after ocular trauma, and subsequently invade the noninjured eye to cause uveitis. We have described several peptides that mimic a retinal autoantigen peptide with respect to amino acid similarities. These peptides are derived from (a) another autoantigen, namely HLA-B, which is useful as an oral

---

Gerhild Wildner, Maria Diedrichs-Moehring, and Stephan R. Thurau • Section of Immunobiology, Department of Ophthalmology, Ludwig-Maximilians-University, Mathildenstr. 8, 80336 Munich, Germany.

Molecular Autoimmunity: In commemoration of the 100th anniversary of the first description of human autoimmune disease, edited by Moncef Zouali. Springer Science+Business Media, Inc., New York, 2005.

tolerogen for the treatment of autoimmune uveitis, (b) from a gastrointestinal pathogen (a rotavirus), and (c) from bovine milk casein, a common nutritional antigen. Even though the latter two peptides are pathogenic in the rat model of experimental uveitis, they have not been capable of inducing oral tolerance. Thus, our data show multiple cases of antigenic mimicry to cause uveitis. They also provide evidence that pathogenic antigens are not obligatory oral tolerogens.

2. Retinal Autoantigens and Mimicry Peptides

The major retinal autoantigens defined by their ability to induce experimental uveitis in animals, especially in mice and Lewis rats (Figure 25.1), are retinal soluble antigen (S-antigen, S-Ag) (de Kozak et al., 1981) and interphotoreceptor retinoid-binding protein (IRBP) (Broekhuyse et al., 1986), both expressed by photoreceptor cells. The immunogenic and pathogenic epitopes are well characterized (Broekhuyse et al., 1986; Donoso et al., 1988; De Smet et al., 1993; Rai et al., 2001).

Some years ago, we found peripheral blood lymphocytes of uveitis patients proliferating in response to a peptide from S-Ag (PDSAg, aa 342–355) as well as a peptide from HLA-B (B27PD, aa 125–138) (Wildner and Thurau, 1994). The peptides share discontinuous amino acid sequence homologies (Figure 25.2), which are thought to represent anchor motifs for major histocompatibility complex (MHC) binding and/or T cell receptor (TCR) contact residues.

Mimicry of a sequestered, tissue-specific autoantigen and a ubiquitously expressed MHC antigen (which is an autoantigen as well) is a rather unconventional idea for the cellular pathogenesis of an autoimmune disease. However, it has been proposed that for their promotion of homeostasis and survival naive peripheral T cells in lymphoid follicles need to be "tickled" with low-affinity self-peptides (e.g., MHC-derived) presented on MHC molecules, but without getting activated (Goldrath and Bevan, 1999). Thus, T cells escaping thymic selection despite recognizing an equivalent of peptide B27PD might have a potential to become autoaggressive lymphocytes that cause uveitis by cross-reacting with retinal autoantigen peptide. We, and others, postulate that these HLA peptide–specific T cells could be activated peripherally, either by bystander activation or by a third cross-reactive environmental antigen (Wildner et al., 2002; Choi and Craft, 2004). Evidence for the latter possibility will be shown later in

B27PD	A	L	N	E	D	L	S	S	W	T	A	A	D	T	
PDSAg	F	L	G	E	L	T	S	S	E	V	A	T	E	V	
Rota				W	T	E	V	S	E	V	A	T	E	V	
Cas				S	E	E	S	A	E	V	A	T	E	E	V

Figure 25.2. Mimicry peptides. Amino acid sequences (one letter code) of peptides mimicking retinal peptide PDSAg: S-antigen, aa 342–355, B27PD: HLA-B, aa 125–138, Rota: rotavirus outer capsid protein vp4 (aa 591–601), Cas: bovine alpha s2-casein (aa 73–84). Identical or homologous amino acids are marked with gray boxes.

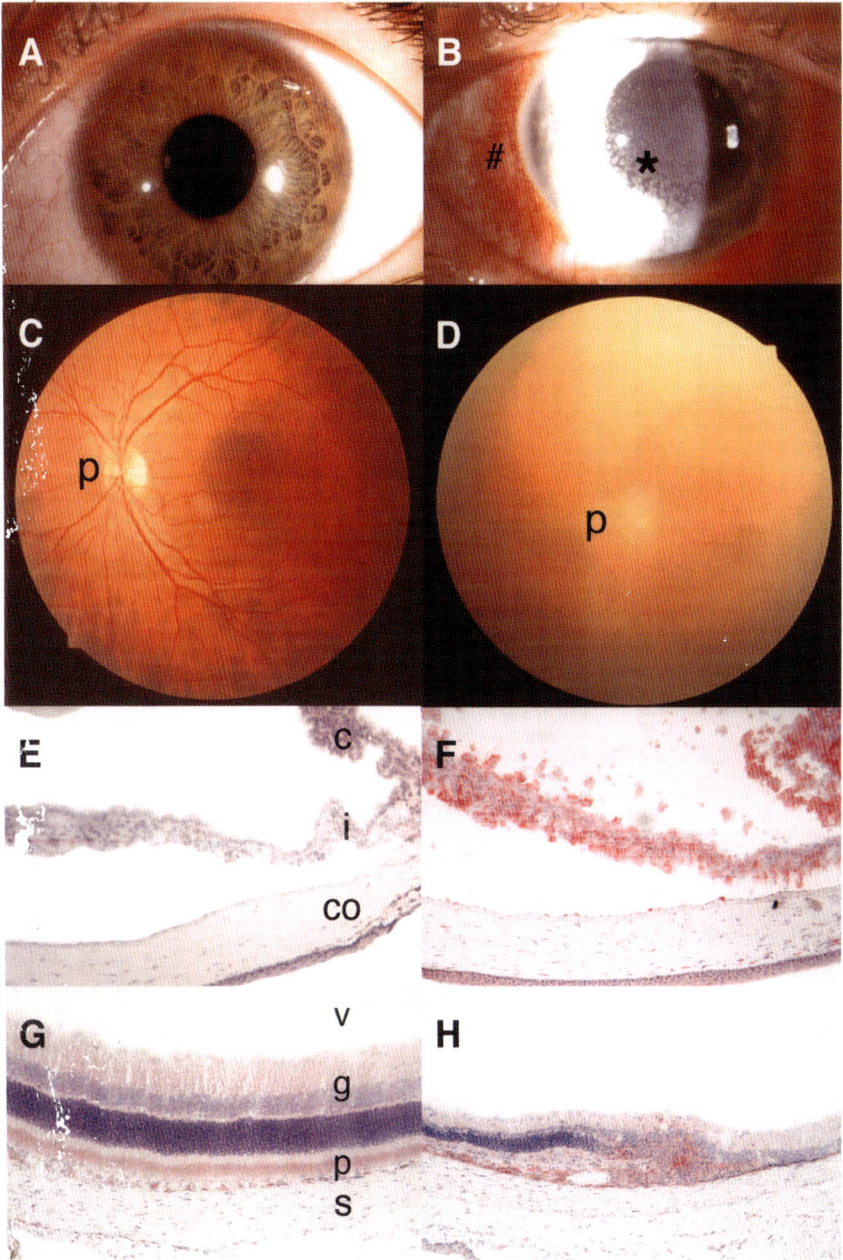

Figure 25.1. Uveitis in human and rat eyes. (A–D) Human eyes. (A) Healthy human anterior part of the eye. (B) Anterior uveitis (iritis), *leucocyte precipitates at the corneal endothelium; # conjunctivitis. (C) Fundoscopy of a healthy posterior part of a human eye, p: papilla (optic nerve head). (D) Fundoscopy, posterior uveitis with severe vitritis, p: papilla. (E–H) Histology of rat eyes (cryosections). (E) Normal anterior part of a rat eye, c: ciliary body, i: iris, co: cornea. (F) Iritis, red staining: CD4+ cells (T cells and macrophages). (G) Normal rat retina, v: vitreous, g: ganglion cells, p: photoreceptors, s: sclera. (H) Posterior uveitis: destruction of the retinal architecture, infiltration of CD4+ cells (red).

this chapter, describing the immunogenic and uveitogenic capacity of two environmental antigens.

3. HLA Peptide B27PD in EAU

The immunological properties of peptide B27PD were tested *in vivo* in the Lewis rat model of EAU (Figure 25.3). Following immunization with HLA peptide B27PD most rats develop a very mild uveitis, often only characterized by infiltrating lymphocytes without retinal destruction. In contrast, immunization with the retinal peptide PDSAg, which is highly uveitogenic with an incidence of more than 80%, gave rise to a severe, destructive form of uveitis. However, oral tolerance induction by feeding B27PD or PDSAg strongly suppresses uveitis caused by immunization with S-Ag or PDSAg (Wildner and Thurau, 1994).

The mechanism of oral tolerance (Mowat, 1987) is usually effective for nutritional proteins, preventing adverse reactions that potentially lead to food allergies. When autoantigens are fed, autoimmune reactions can be suppressed as well, indicating a potential for oral tolerance as a therapeutic approach for autoimmune diseases (Weiner, 1997). Tolerance is mediated by suppressor cells specific for the respective antigen; however, the exact mechanisms underlying orally induced suppression are not yet fully elucidated. It is assumed that suppressor T cells recognize the respective antigen and secrete suppressive cytokines, such as TGF-β (Miller *et al.*, 1992), IL-10 (Rizzo *et al.*, 1999; Slavin *et al.*, 2001) (Th3, Tr type), or cytokines belonging to the respective antagonistic Th type of the pathogenic immune response (Weiner, 2001).

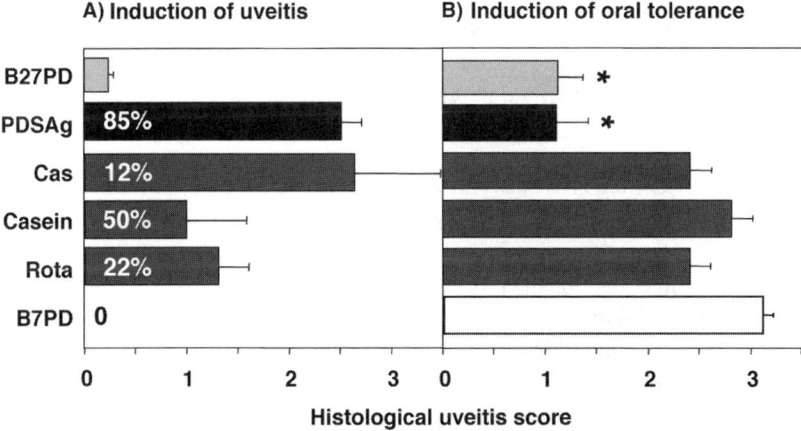

Figure 25.3. Immunological activities of mimicry peptides. (A) Pathogenicity was tested by immunization of rats with the respective peptides. B7PD (ALNEDLRSWTAADT), derived from HLA-B7, differs from B27PD in only one amino acid (131 S ⇒ R). Mean histological uveitis score + SE is shown. Numbers in percent indicate the incidence (positive eyes). (B) Rats were fed with the respective peptides to show suppression of PDSAg-induced uveitis. Asterisks indicate significant reduction of disease.

Even IRBP-induced uveitis can be significantly downregulated by oral application of peptide B27PD, but not by S-Ag or peptide PDSAg, which points to a not yet identified mimotope in the sequence of IRBP. This is supported by the findings of Deeg *et al.* (2002), who have established a new model of experimentally induced uveitis in horses. Following immunization with IRBP all horses developed uveitis. Peripheral blood lymphocytes proliferated in response to IRBP peptide R14 (bovine IRBP, aa 1169–1191) as well as to HLA peptide B27PD, but not to S-Ag peptide PDSAg.

The antigen specificity was confirmed by the failure of recognizing control peptide B7PD, which differs from the sequence of B27PD by only a single amino acid. B7PD is not uveitogenic and is unable to induce oral tolerance and to prevent uveitis (Wildner and Thurau, 1994) (Figure 25.3).

Gamma-delta T cells isolated from spleens of mucosally tolerized rats can transfer tolerance to naive animals (McMenamin *et al.*, 1995). Rats can be protected from PDSAg-induced uveitis by preinoculation of γ/δ TCR$^+$ splenocytes from donor rats either fed with PDSAg itself or with B27PD, but not with B7PD (Wildner *et al.*, 1996). These regulatory γ/δ T cells are CD8$^+$. They proliferate *in vitro* in response to the tolerizing antigen (PDSAg or B27PD) or its respective mimicry peptides, but not to control peptide B7PD, in coculture with antigen-specific activated α/β^+ effector cells (Wildner *et al.*, 2004). The respective mode of antigen recognition/presentation of γ/δ and α/β T cells is still unknown, but our data suggest that there is a direct cell–cell contact between effector and suppressor cells. Although in our rat model effector and regulatory T cells are different with respect to type (CD4 vs. CD8) and TCR (α/β vs. γ/δ), they seem to similarly recognize the antigen peptides.

Nevertheless, retinal peptide PDSAg and HLA peptide B27PD differ with respect to pathogenicity and tolerogenicity: while PDSAg is both uveitogenic and orally tolerogenic, B27PD is predominantly tolerogenic. We thus attempted to define those amino acid residues that stimulate pathogenic effector cells versus those that confer tolerance (Wildner and Diedrichs-Möhring, 2003b).

4. Pathogenic and Tolerogenic Epitopes of the Retinal Peptide PDSAg and Its Mimotope B27PD

To determine the localization of the pathogenic and/or the tolerogenic epitope within the 14-mer peptides PDSAg and B27PD, we analyzed N- and C-terminally truncated and chimeric variants (Figure 25.4). Truncation of up to 3 N-terminal amino acids of PDSAg resulted in a peptide that still retains the full capability of inducing uveitis and oral tolerance, but its ability to stimulate proliferation of PDSAg-specific T cell lines was decreased. In contrast, the C-terminus of PDSAg was highly sensitive to alterations: truncation of one amino acid drastically reduced the pathogenicity and oral tolerogenicity, whereas stimulation of PDSAg-specific T cell lines *in vitro* was still obtained. Truncation of two C-terminal amino acids (peptide PDSAgV12) completely destroyed the immunologic activity of the peptide.

Molecular Mimicry in Autoimmune Uveitis: From Pathogenesis to Therapy

Amino acid sequence	Code	T cell proliferation	Uveitis	Oral tolerance
FLGELTSSEVATEV	PDSAg	++++	+++	+++
.LGELTSSEVATEV	PDSAgV2	++	+++	+++
..GELTSSEVATEV	PDSAgV3	++	+++	+++
...ELTSSEVATEV	PDSAgV4	++	+++	+++
....LTSSEVATEV	PDSAgV5	++	–	+
.....TSSEVATEV	PDSAgV6	+	–	–
FLGELTSSEVAT..	PDSAgV12	–	–	+
FLGELTSSEVATE.	PDSAgV13	++	+	+
ALNEDLSSWTAADT	B27PD	–	–	+++
...EDLSSWTAADT	B27V4	–	–	–
....DLSSWTAADT	B27V5	–	–	+
.....LSSWTAADT	B27V6	–	–	–
FLGELTSSWTAADT	S-B	–	–	–
FLGELTSWTAADT.	S-1-B	–	–	–
ALNEDLSSEVATEV	B-S	++	–	–
....DLSSSTVADE	Chi1(10)	–	–	–
...ELTSWEAATT.	Chi2(10)	–	±	–

Figure 25.4. Epitope mapping of pathogenic and tolerogenic amino acids. Variants of retinal autoantigen peptide PDSAg and mimicry peptide B27PD were tested in the rat model *in vitro* for induction of proliferation of a PDSAg-specific T cell line, and *in vivo* for uveitogenicity and oral tolerogenicity. The latter was tested by preventing PDSAg-induced uveitis.

Peptide PDSAg as well as peptide B27PD are presented by RT1.B^l (rat MHC class II, equivalent to I-A and DQ, respectively), as proven by inhibition of T cell proliferation by MHC class II–specific antibodies (for PDSAg) or competition of B27PD with PDSAg for MHC binding (Wildner and Diedrichs-Möhring, 2003b). Neither the 14-mer B27PD nor its truncated variants stimulated proliferation of highly PDSAg-specific T cell lines *in vitro* or generated uveitogenic T cell responses *in vivo*. Despite its lack of pathogenicity, HLA peptide B27PD is a potent inducer of oral tolerance to PDSAg-mediated uveitis (Figure 25.4). N-terminal-truncated variants of B27PD still bound to RT1.B^l, but did neither induce disease nor oral tolerance.

Whereas only the full-length 14-mer B27PD binds efficiently to RT1.B^l and is highly tolerogenic, the *in vivo* activity of PDSAg is even retained when three N-terminal amino acids are lacking. The pathogenic as well as the tolerogenic epitope of PDSAg seems thus to be located at the C-terminus, because the absence of only one C-terminal amino acid altered its uveitogenicity and pathogenicity more dramatically than the truncation of three N-terminal amino acids. Therefore, chimeric peptides were synthesized, which consisted of combinations of the N- and the C-termini of PDSAg and B27PD (peptides S-B, S-1-B, B-S), respectively, or had single amino acid exchanges (peptides Chi1(10) and Chi2(10)) (Figure 25.4).

Only the chimeric peptide B-S, containing the N-terminal part of B27PD and the C-terminus of PDSAg, strongly stimulated proliferation of PDSAg-specific T cell lines *in vitro*. *In vivo* this peptide was neither pathogenic nor tolerogenic. Its counterpart, S-B, consisting of the N-terminus of PDSAg and the C-terminus of B27PD, as well a as a variant (S-1-B), lacking the Ser at position 8, were ineffective *in vitro* both in stimulating a PDSAg-specific T cell line as well as binding to the restriction element RT1.B^l. *In vivo*, S-B and S-1-B were neither pathogenic nor tolerogenic (Figure 25.4).

The chimeric peptide Chi1(10), composed of amino acids from both founder peptides and including a N-terminal stretch of amino acids that correspond to the sequence of B27PD, was ineffective *in vivo* and did not stimulate PDSAg-specific T cells *in vitro*, but bound to RT1.B^l. Peptide Chi2(10), inducing uveitis in 1 of 10 immunized Lewis rats but no oral tolerance, did not compete with PDSAg for RT1.B^l-binding or T cell stimulation (Figure 25.4). This peptide lacks the postulated favored RT1.B anchors (Reizis *et al.*, 1996) at amino acid positions 4 and 9, however, positions 2 (Thr, T), 5 (Glu, E), and 8 (Thr, T), which are thought to be recognized by the TCR (Hemmer *et al.*, 2000), are identical with those of peptide PDSAg. Reizis *et al.* (1996) described Phe (F) as a preferred amino acid at position 4. Here Trp (W) is located at position 4, indicating that Trp might function as an anchor as well. The lack of an aliphatic or negatively charged anchor at position 9, favored for RT1.B^l binding, might explain the low incidence of uveitis by a rare TCR with a strong avidity for Chi2(10). Nevertheless, in our model we could not exclude the presentation of peptide Chi2(10) by RT1.D, the other rat MHC-class II molecule, for which the binding motifs are still unknown, thus stimulating a nondominant RT1.D-restricted T cell response.

In conclusion, we could not clearly define MHC-binding motifs or amino acids responsible for recognition by uveitogenic α/β^+ cells, nor for the activation of orally induced γ/δ^+ T cells. Amino acids flanking the core region or certain combinations of amino acids can influence the structure of a peptide in an unpredictable way (Conant and Swanborg, 2003) and thus favor or abrogate MHC presentation and/or activation of T cells. For the orally induced γ/δ T lymphocytes, neither restriction elements nor motifs for binding to the restriction element or the TCR are known. Nevertheless, our data reveal differences between *in vitro* and *in vivo* T cell responses. *In vitro*, specific T cell lines can recognize peptides that highly differ from their original antigen, whereas such cross-reactive peptides are inactive *in vivo* (Anderson *et al.*, 2000). So far, in humans, where only *in vitro* assays are available for the definition of potentially pathogenic or tolerogenic epitopes, the results from cell culture assays might sometimes be misleading and not always representative for the *in vivo* situation. On the other hand, our data show that mimicry peptides, such as B27PD, might be suitable for therapeutic application, e.g., oral tolerance induction, for their lack of pathogenicity and high capacity of specific tolerance induction. The opposite example is shown by mimotopes with high pathogenicity, as given in the next paragraph, but with no capacity of inducing therapeutic oral tolerance.

5. Antigenic Mimicry of Retinal Autoantigen and Environmental Antigens

Normally, autoaggressive immune responses are suppressed. It is possible that their initiation follows antigenic mimicry of self-antigens and an antigen provided by an infectious agent, e.g., bacterium or virus (Singh *et al.*, 1989;

Wucherpfennig and Strominger, 1995; Lawson, 2000). The initiating event of uveitis by HLA peptide B27PD mimicking retinal peptide PDSAg was thought to be "bystander activation," caused by any immunological stress like an infection, for an appropriate microbial or viral mimicry antigen was not known (Wildner and Thurau, 1994). Recently we described two potential environmental mimicry partners for PDSAg: a peptide (Rota) derived from rotavirus, a gastrointestinal pathogen, and a second peptide (Cas) from bovine milk casein, a common nutritional antigen. Both peptides share the C-terminal amino acid sequence with PDSAg, which we suspect to be the major epitope for uveitogenic T cells. (Figures 25.2 and 25.4) (Wildner and Diedrichs-Möhring, 2003a). Peptides Rota and Cas, or casein protein in complete Freund's adjuvant, cause uveitis after subcutaneous immunization of Lewis rats (Figure 25.3). The disease is similar to that induced with S-Ag peptide PDSAg with respect to clinic and histology, but has a lower incidence (Figure 25.3). However, T cell lines specific for Rota or Cas remain cross-reactive with PDSAg, but not with the other mimotope Rota or Cas, respectively. Although all three peptides sequence homologies are thought to be responsible for TCR cross-reactivity there seems to be no strict trimolecular mimicry, but rather an oligoclonal T cell pool with restricted bimolecular cross-reactivity.

Since nutritional antigens like milk or enteroviruses, such as Rotaviruses, contact the immune system via the gastrointestinal tract, we decided to immunize rats via the oral route. We used native cholera toxin as an adjuvant, which induces a mucosal Th1 response in rats (Wilson *et al.*, 1991). None of the peptides or retinal S-antigen protein causes uveitis in rats, whereas a single application of 1 mg of casein protein or 500 µl of bovine milk in addition to cholera toxin is uveitogenic. Disease can be adoptively transferred to naive animals with spleen cells of rats orally immunized with casein or milk (unpublished data). The "casein protein" preparation is a mixture of αs1- and αs2-casein, of which only the αs2-casein includes the mimicry peptide Cas. The content of αs2-casein in this preparation, and thus the dose of uveitogenic antigen used for oral immunization, is 180 µg. Highly purified αs2-casein is not available. Moreover, 500 µl bovine milk contains an equivalent of only 14 µg αs2-casein, which is obviously sufficient to induce an autoimmune response. Immunostimulatory properties of milk proteins have already been described (Bhattacharyya and Das, 1999; Gill *et al.*, 2000). We propose that α-caseins (or other proteins contained in bovine milk) might support mimotope-specific Th1 responses, independent of their own antigenicity.

This observation suggests that milk consumption during a gastrointestinal Th1 response, like a viral infection, might be able to cause uveitis in humans as well. This conclusion is supported by the enhanced serological and cellular immune responses to peptide Rota and casein antigens found in uveitis patients (Figure 25.5). Bovine milk protein is a major component of food in industrialized countries, where the frequency of autoimmune uveitis is 2‰. We therefore speculate that under certain conditions the gastrointestinal immune response to bovine milk could also target the retinal S-Ag-mimotope Cas and therefore activate cross-reactive T cells that elicit ocular inflammation. On the other hand, it is expected that under normal circumstances consumption of bovine αs2-casein

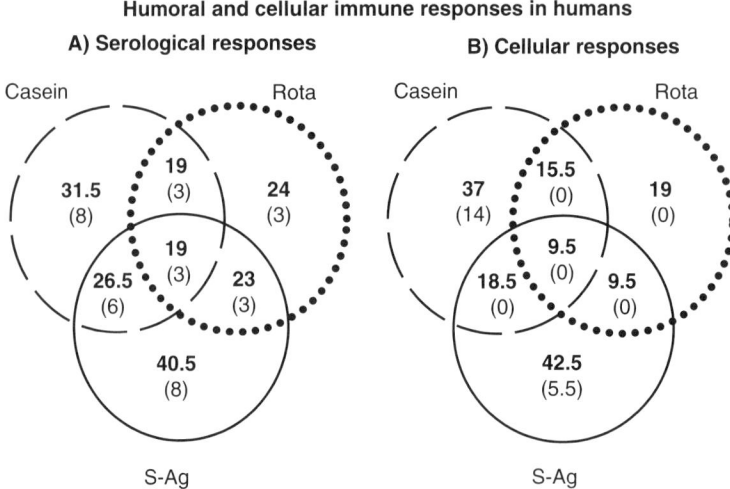

Figure 25.5. Humoral and cellular immune responses in humans. (A) Serological responses of uveitis patients ($n = 98$) and healthy blood donors ($n = 36$) were tested in ELISA. Numbers in bold indicate % positive sera of uveitis patients, numbers in brackets are those of healthy donors. A value was defined as positive when exceeding the average O.D. of healthy donors + 2 SD. (B) *In vitro* proliferation of peripheral blood lymphocytes of uveitis patients (bold numbers, $n = 32$) and healthy donors (in brackets, $n = 18$). Overlapping circles include those sera or peripheral blood lymphocytes that react with more than one antigen. Responses to S-Ag or PDSAg or to casein protein or peptide Cas, respectively, are combined.

would cause oral tolerance rather than oral immunization, thus preventing uveitis. However, casein has an immunogenic potential, as it is the major antigen of infantile food allergies (Schwartz, 1991). Furthermore, in the rat model it was ineffective in inducing oral tolerance to prevent uveitis. This was also observed with the peptides Cas and Rota (Wildner and Diedrichs-Möhring, 2003a).

As already observed by testing peptide variants of PDSAg and HLA peptide B27PD, we once more demonstrated that pathogenic proteins/peptides must not always be effective as oral tolerogens (and vice versa), a fact that complicates the selection of oral tolerogens for clinical trials.

6. Treatment of Uveitis Patients with Oral Peptide B27PD

The fact that in rats the HLA peptide B27PD was at least as effective in inducing oral tolerance but much less uveitogenic than the retinal autoantigen peptide PDSAg (Figure 25.3) rendered it attractive as a potential oral tolerogen for patients with autoimmune uveitis. The therapeutic potential of oral tolerance in uveitis patients has been demonstrated by Nussenblatt *et al.* (1997), who conducted a controlled, double-blinded phase I/II trial. The antigens used were purified retinal S-Ag, a crude retinal extract consisting of many different retinal antigens and a preparation of retinal extracts enriched with S-Ag. These antigen preparations were tested against placebo and therapeutic effects, as defined by the possibility to replace conventional immunosuppressive therapy and the prevention of relapses, were shown for purified S-Ag, although this study did not reach statistical significance, probably due to small sample size.

Our experimental animal results as well as the patient's lymphocyte responses (Wildner and Thurau, 1994; Thurau *et al.*, 1997) prompted us to test the peptide B27PD as an oral tolerogen in uveitis patients. After approval of the local ethical committee, patients with severe and chronic uveitis resistant to conventional immunosuppressive therapy or suffering from severe side effects were selected for oral tolerance induction with peptide B27PD and followed for up to 5 years or longer (Thurau *et al.*, 1999). Eight patients with average disease duration of 13 years were enrolled in our study. The clinical diagnosis included anterior uveitis (without or with underlying disease, like juvenile idiopathic arthritis or ankylosing spondylitis), intermediate uveitis and posterior uveitis with retinal vasculitis and cystoid macular edema, or panuveitis associated with systemic sarcoidosis. The patients received 4 mg of encapsulated peptide 3 times a week for 12 weeks (oral tolerance induction period). As concomitant immunosuppressive or anti-inflammatory medication only 20 mg of prednisolone or equivalent were allowed, which was adjusted to visual acuity and intraocular inflammatory activity. Since the active mechanism of oral tolerance may be blocked by immunosuppressive therapy, two patients had to stop azathioprine treatment before oral peptide application.

During the 12 weeks of tolerance induction all patients improved in visual acuity of intraocular inflammation. Therefore systemic steroid medication could be reduced significantly. Within the 1st year after initiation of oral tolerance intraocular inflammation decreased in seven of the eight patients and visual acuity increased or stabilized in 14 of 16 eyes.

Four of the patients were retreated with oral peptide due to recurrences of uveitis 1 year after the first course of treatment, and all responded with a decrease of inflammation, suggesting that oral tolerance can be reinduced.

During the extended follow-up of 4 years after the 1st year post study entry the average daily steroid dose was further decreased, either because of a therapeutic long-term effect of oral tolerance induction or the retreatment of four patients with peptide (Figure 25.6). Within the subsequent follow-up two of the patients had progressive disease activity. In these patients conventional immunosuppression was reinstituted during the 3rd and 4th year.

During the 5-year study period we did not see any side effects that could be correlated with the peptide medication, not even in those patients who were retreated with oral B27PD. Three of the patients had previously suffered severely from side effects of their conventional immunosuppressive therapy, which was one of the reasons to offer the oral peptide to these patients. Due to the reduction or cessation of conventional immunosuppression the patients recovered from side effects such as signs of Cushing's syndrome in all patients receiving corticosteroids, steroid-induced depression, aseptic hip necrosis, osteomyelitis, leukopenia as a sign of bone marrow suppression, and hair loss in patients with azathioprine.

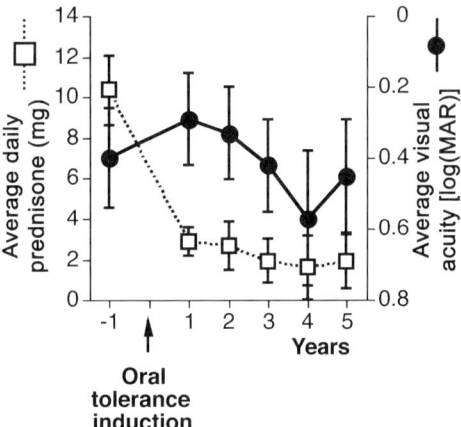

Figure 25.6. Oral tolerance induction in uveitis patients. Eight patients were treated during the first therapeutic trial. Open squares: average daily prednisone during 1 year before oral tolerance induction and in the 5 following years. Visual acuity (closed circles) is shown as log(MAR) (logarithm of minimal angle of resolution); lower values indicate better visual acuity. Compared to the pretreatment period (year 1) visual acuity improved and the average daily prednisolone decreased after oral tolerance induction.

This study did not include a control group since this was the first time that humans received the peptide. Nevertheless, a therapeutic as well as a steroid-sparing effect was demonstrated, which warrants further investigation of this new therapeutic approach. The idea that a mimotope-peptide as oral tolerogen might be superior to the respective autoantigen protein was supported by the fact that the peptides were not pathogenic (at least in the rat model), even when applied orally with a Th1-promoting adjuvant (Wildner and Diedrichs-Möhring, 2003a). Furthermore, using a single epitope represented by a peptide might focus the tolerogenic immune response to the respective determinant, rather than offering the whole peptide repertoire of a certain protein. A peptide could also be directly presented by MHC molecules without the need for prior internalization and processing. Finally, the tolerogenic effect of peptide B27PD on uveitis induced with IRBP in the rat model might represent a broader therapeutic effect not observed with S-Antigen protein or its peptide (Wildner and Thurau, 1994), and a small, well-defined synthetic peptide might be safer than a whole protein, generated recombinantly or purified from animal tissue.

References

Anderson, A.C., Waldner, H., Turchin, V., Jabs, C., Das, M.P., Kuchroo, V.K., and Nicholson, L.B. (2000). Autoantigen-responsive T cell clones demonstrate unfocused TCR cross-reactivity toward multiple related ligands: Implications for autoimmunity. *Cell. Immunol.*, **202**, 88–96.

Bhattacharyya, J. and Das, K.P. (1999). Molecular chaperone-like properties of an unfolded protein, alpha(s)-casein. *J. Biol. Chem.*, **274**, 15505–15509.

Broekhuyse, R.M., Winkens, H.J., and Kuhlmann, E.D. (1986). Induction of experimental autoimmune uveoretinitis and pinealitis by IRBP. Comparison to uveoretinitis induced by S-antigen and opsin. *Curr. Eye Res.*, **5**, 231–240.

Calder, V.L., Shaer, B., Muhaya, M., McLauchlan, M., Pearson, R.V., Jolly, G., Towler, H.M., and Lightman, S. (1999). Increased CD4+ expression and decreased IL-10 in the anterior chamber in idiopathic uveitis. *Invest. Ophthalmol. Vis. Sci.*, **40**, 2019–2024.

Choi, J.Y. and Craft, J. (2004). Activation of naive CD4+ T cells in vivo by a self-peptide mimic: Mechanism of tolerance maintenance and preservation of immunity. *J. Immunol.*, **172**, 7399–7407.

Conant, S.B. and Swanborg, R.H. (2003). MHC class II peptide flanking residues of exogenous antigens influence recognition by autoreactive T cells. *Autoimmun. Rev.*, **2**, 8–12.

Deeg, C.A., Thurau, S.R., Gerhards, H., Ehrenhofer, M., Wildner, G., and Kaspers, B. (2002). Uveitis in horses induced by interphotoreceptor retinoid-binding protein is similar to the spontaneous disease. *Eur. J. Immunol.*, **32**, 2598–2606.

de Kozak, Y., Sakai, J., Thillaye, B., and Faure, J.P. (1981). S antigen-induced experimental autoimmune uveo-retinitis in rats. *Curr. Eye Res.*, **1**, 327–337.

De Smet, M.D., Bitar, G., Roberge, F.G., Gery, I., and Nussenblatt, R.B. (1993). Human S-antigen: Presence of multiple immunogenic and immunopathogenic sites in the Lewis rat. *J. Autoimmun.*, **6**, 587–599.

Donoso, L.A., Merryman, C.F., Sery, T.W., Vrabec, T., Arbizo, V., and Fong, S.L. (1988). Human IRBP: Characterization of uveitopathogenic sites. *Curr. Eye Res.*, **7**, 1087–1095.

Gill, H.S., Doull, F., Rutherfurd, K.J., and Cross, M.L. (2000). Immunoregulatory peptides in bovine milk. *Br. J. Nutr.*, **84**, S111–S117.

Goldrath, A.W. and Bevan, M.J. (1999). Selecting and maintaining a diverse T-cell repertoire. *Nature*, **402**, 255–262.

Hemmer, B., Pinilla, C., Gran, B., Vergelli, M., Ling, N., Conlon, P., McFarland, H.F., Houghten, R., and Martin, R. (2000). Contribution of individual amino acids within MHC molecule or antigenic peptide to TCR ligand potency. *J. Immunol.*, **164**, 861–871.

Lawson, C.M. (2000). Evidence for mimicry by viral antigens in animal models of autoimmune disease including myocarditis. *Cell. Mol. Life Sci.*, **57**, 552–560.

McMenamin, C., McKersey, M., Kuhnlein, P., Hunig, T., and Holt, P.G. (1995). Gammadelta T cells down-regulate primary IgE responses in rats to inhaled soluble protein antigens. *J. Immunol.*, **154**, 4390–4394.

Miller, A., Lider, O., Roberts, A.B., Sporn, M.B., and Weiner, H.L. (1992). Suppressor T cells generated by oral tolerization to myelin basic protein suppress both in vitro and in vivo immune responses by the release of transforming growth factor beta after antigen-specific triggering. *Proc. Natl. Acad. Sci. USA*, **89**, 421–425.

Mowat, A.M. (1987). The regulation of immune responses to dietary protein antigens. *Immunol. Today*, **8**, 93–98.

Nussenblatt, R.B., Gery, I., Weiner, H.L., Ferris, F.L., Shiloach, J., Remaley, N., Perry, C., Caspi, R.R., Hafler, D.A., Foster, C.S., and Whitcup, S.M. (1997). Treatment of uveitis by oral administration of retinal antigens: Results of a phase I/II randomized masked trial. *Am. J. Ophthalmol.*, **123**, 583–592.

Rai, G., Saxena, S., Kumar, H., and Singh, V.K. (2001). Human retinal S-antigen: T cell epitope mapping in posterior uveitis patients. *Exp. Mol. Pathol.*, **70**, 140–145.

Reizis, B., Mor, F., Eisenstein, M., Schild, H., Stefanovic, S., Rammensee, H.G., and Cohen, I.R. (1996). The peptide binding specificity of the MHC class II I-A molecule of the Lewis-rat, RT1.B(l). *Int. Immunol.*, **8**, 1825–1832.

Rizzo, L.V., Morawetz, R.A., Miller Rivero, N.E., Choi, R., Wiggert, B., Chan, C.C., Morse, H.C., Nussenblatt, R.B., and Caspi, R.R. (1999). IL-4 and IL-10 are both required for the induction of oral tolerance. *J. Immunol.*, **162**, 2613–2622.

Schwartz, R.H. (1991). IgE-mediated allergic reactions to cow's milk. *Immunol. Allergy Clin. North Am.*, **11**, 717–741.

Singh, V.K., Yamaki, K., Abe, T., and Shinohara, T. (1989). Molecular mimicry between uveitopathogenic site of retinal S-antigen and *Escherichia coli* protein: Induction of experimental autoimmune uveitis and lymphocyte cross-reaction. *Cell. Immunol.*, **122**, 262–273.

Slavin, A.J., Maron, R., and Weiner, H.L. (2001). Mucosal administration of IL-10 enhances oral tolerance in autoimmune encephalomyelitis and diabetes. *Int. Immunol.*, **13**, 825–833.

Thurau, S.R., Diedrichs-Möhring, M., Fricke, H., Arbogast, S., and Wildner, G. (1997). Molecular mimicry as a therapeutic approach for an autoimmune disease: Oral treatment of uveitis-patients with an MHC-peptide crossreactive with autoantigen—First results. *Immunol. Lett.*, **57**, 193–201.

Thurau, S.R., Diedrichs-Möhring, M., Fricke, H., Burchardi, C., and Wildner, G. (1999). Oral tolerance with an HLA-peptide mimicking retinal autoantigen as a treatment of autoimmune uveitis. *Immunol. Lett.*, **68**, 205–212.

Thurau, S.R., Mempel, T., Flügel, A., Diedrichs-Möhring, M., Krombach, F., Kawakami, N. and Wildner, G. (2004). The fate of autoreactive, GFP+ T cells in rat models of uveitis analyzed by intravital fluorescence microscopy and FACS. *Int. Immunol.*, **16**, 1573–1582.

Weiner, H.L. (1997). Oral tolerance: Immune mechanisms and treatment of autoimmune diseases. *Immunol. Today*, **18**, 335–343.

Weiner, H.L. (2001). Oral tolerance: Immune mechanisms and the generation of Th3-type TGF-beta-secreting regulatory cells. *Microbes Infect.*, **3**, 947–954.

Wildner, G. and Diedrichs-Möhring, M. (2003a). Autoimmune uveitis induced by molecular mimicry of peptides from rotavirus, bovine casein and retinal S-antigen. *Eur. J. Immunol.*, **33**, 2577–2587.

Wildner, G. and Diedrichs-Möhring, M. (2003b). Differential recognition of a retinal autoantigen peptide and its variants by rat T cells in vitro and in vivo. *Int. Immunol.*, **15**, 927–935.

Wildner, G., Diedrichs-Möhring, M., and Thurau, S.R. (2002). Arthritis and uveitis by antigenic mimicry of HLA-B27 and cytokeratin peptides. *Eur. J. Immunol.*, **32**, 299–306.

Wildner, G., Hunig, T., and Thurau, S.R. (1996). Orally induced, peptide-specific gamma/delta TCR+ cells suppress experimental autoimmune uveitis. *Eur. J. Immunol.*, **26**, 2140–2148.

Wildner, G. and Thurau, S.R. (1994). Cross-reactivity between an HLA-B27-derived peptide and a retinal autoantigen peptide: A clue to major histocompatibility complex association with autoimmune disease. *Eur. J. Immunol.*, **24**, 2579–2585.

Wildner, G., Thurau, S.R., and Diedrichs-Möhring, M. (2004). Gamma-delta T cells as orally induced suppressor cells in rats: In vitro characterization. *Ann. NY Acad. Sci.*, **1029**, 416–422.

Wilson, A.D., Bailey, M., Williams, N.A., and Stokes, C.R. (1991). The in vitro production of cytokines by mucosal lymphocytes immunized by oral administration of keyhole limpet hemocyanin using cholera toxin as an adjuvant. *Eur. J. Immunol.*, **21**, 2333–2339.

Wucherpfennig, K.W. and Strominger, J.L. (1995). Molecular mimicry in T cell-mediated autoimmunity: Viral peptides activate human T cell clones specific for myelin basic protein. *Cell*, **80**, 695–705.

26

Molecular Pathogenesis of the Antiphospholipid Syndrome: Toward Novel Therapeutic Targets

Silvia S. Pierangeli, Mariano Vega-Ostertag, Azzudin E. Gharavi, E. Nigel Harris

1. Introduction

Antiphospholipid syndrome (APS) is a disorder of recurrent thrombosis, pregnancy loss, and thrombocytopenia associated with persistently positive anticardiolipin (aCL) antibodies and/or lupus anticoagulant (LA) tests (Harris, 1987; Wilson et al., 1999). APS is a multisystem disorder that may be associated with systemic lupus erythematosus (SLE).

The mechanisms by which antiphospholipid (aPL) antibodies mediate disease are only partially understood and our knowledge is limited by the apparent polyreactivity of the antibodies, the multiple potential end-organ targets, and the variability of clinical context that disease may present. aPL antibodies are heterogenous and it is likely that more than one mechanism may be involved in causing thrombosis (Roubey, 1998; Meroni and Riboldi, 2001). In fact, aPL may cause thrombosis *in vitro* by interfering with activation of protein C, or inactivation of factor V by activated protein C, inhibition of endothelial prostacyclin production, impairment of fibrinolysis, and exert a stimulatory effect on platelet function (Campbell et al., 1995; Roubey, 1998; Esmon et al., 2000; Meroni and Riboldi, 2001). From studies that utilized animal models of thrombosis, endothelial cell (EC) activation, and pregnancy loss, there is also evidence that aPL antibodies are pathogenic *in vivo* (Branch et al., 1990; Pierangeli et al., 1995, 2001). Recently there has also been convincing evidence that activation of complement mediates thrombogenic effects of aPL and fetal loss in APS (Holers et al., 2002).

Silvia S. Pierangeli and Mariano Vega-Ostertag • Department of Microbiology, Biochemistry and Immunology, Morehouse School of Medicine, Atlanta, Georgia. **Azzudin E. Gharavi** • Department of Medicine, Morehouse School of Medicine, Atlanta, Georgia, Dr Azzudin Gharavi passed away on Oct 13th, 2004. **E. Nigel Harris** • Vice-chancellor, Univ of the West Indies, Mona 7, Kingston, Jamaica.

Molecular Autoimmunity: In commemoration of the 100th anniversary of the first description of human autoimmune disease, edited by Moncef Zouali. Springer Science+Business Media, Inc., New York, 2005.

The treatment of the APS can be directed toward preventing thromboembolic events by using antithrombotic medications or by modulating the immune response with immunotherapy. In the case of thrombotic manifestations, both approaches have been used with considerable side effects (Khamashta et al., 1995; Krnic-Barrie et al., 1997; Crowther et al., 2003). On the one hand, current treatment recommendations are clear. In patients with a history of thrombosis, there is a high risk of recurrence and oral anticoagulation at a relative high international normalized ration (INR) is frequently used for a long period of time. In patients with aPL antibodies without a previous thrombotic event, most clinicians would recommend prophylaxis with low-dose aspirin (Khamashta et al., 1995; Krnic-Barrie et al., 1997; Crowther et al., 2003). On the other hand, the risk of bleeding with oral anticoagulation is high. Furthermore, there is a need for frequent monitoring and patient compliance with diet and lifestyle to minimize the risk of thrombosis recurrence. Hence, there is need for new, more efficient, and less harmful modalities of treatment for APS.

This chapter discusses new findings with respect to molecular and intracellular pathways mediated by aPL antibodies in platelets and ECs that lead to the thrombotic event. A more precise definition of the nature of the aPL–target tissue interaction and the mechanism(s) by which these antibodies cause thrombosis may help in devising new targeted modalities for the treatment of APS patients.

2. Antiphospholipid Antibodies and Platelets

2.1. Effects of aPL on Platelets *In Vitro* and *In Vivo*

Studies from our group have shown that affinity-purified aCL antibodies from patients with APS enhanced activation of platelets treated with suboptimal doses of ADP, thrombin, or collagen (Campbell et al., 1995). The platelet GPIIb/IIIa receptor mediates platelet aggregation induced by all physiologic agonists and is the receptor for fibrinogen and a marker of platelet activation. We have shown that aPL antibodies enhance the expression of platelet membrane glycoproteins, particularly GPIIb/IIIa and GPIIIa, when platelets are pretreated with suboptimal doses of a thrombin receptor agonist peptide (TRAP) (Espinola et al., 2002). In another study, rabbit aCL antibodies were shown to enhance collagen-induced platelet activation (Ling and Wang, 1992). However, no studies have shown whether aPL antibodies contribute to thrombosis by activating platelets *in vivo*. We addressed that question in a recent study by examining how infusions of an anti-GPIIb/IIIa monoclonal antibody (1B5) affect aPL-mediated enhanced thrombus formation in a mouse model of thrombosis. Furthermore, aPL-mediated thrombosis *in vivo* was evaluated in GPIIb/IIIa-deficient mice (β_3-null mice). In brief, CD1 mice in groups of 18 were either injected i.p. twice with 500 µg of human affinity-purified aPL or control IgG (IgG-NHS). Seventy-two hours after the first injection, the dynamics of thrombus formation was examined. Then, mice treated with aPL and IgG-NHS were i.v. infused with 0.1 ml of 1B5 (50 µg/ml) or with saline solution. Thirty minutes later, a second thrombus was induced in each animal and its area (in µm^2) was determined and compared to values obtained before the i.v. infusions. Mean thrombus size in aPL-treated mice was significantly larger when compared to IgG-NHS-treated mice (4935 ± 1609 vs. 1138 ± 97 μm^2). Infusions of 1B5 decreased significantly the thrombus size of aPL-treated mice (from 4935 ± 1609 to 1833 ± 878 μm^2 (65% decrease)). Neither thrombus size nor aCL titers were affected by saline infusions in aPL- or IgG-NHS-treated mice. In another set of experiments and in order to confirm the effects of 1B5 on aPL-thrombus formation, β_3-null mice ($n = 5$) or wild-type (WT) mice ($n = 5$) (kindly provided by Dr. B. Coller, Rockefeller University, New York) were infused i.p. with 1 ml of aPL on two occasions. Seventy-two hours after the first injection, the aCL titer of all animals was >80 GPL units and thrombus dynamics was examined, as described (Pierangeli et al., 1995). A significant decrease in thrombus size was observed in β_3-null mice (52%) when compared with WT mice treated with aPL. The data clearly indicate that aPL antibodies contribute to thrombosis by activating platelets *in vivo* and this may explain, at least in part, the procoagulant and thrombogenic properties of aPL. These observations were confirmed recently by Jankowski et al., who demonstrated that one monoclonal antibody with anti-β_2GPI activity and LA activity produced a platelet-rich thrombus in an animal model of thrombosis, concomitant with an increase in platelet aggregation, when platelets were treated with low concentrations of ADP (Jankowski et al., 2003).

GPIIb/IIIa antagonists such as the Fab fragment of the mouse/chimeric antibody 7E3 (abciximab, a platelet membrane glycoprotein IIb/IIIa receptor inhibitor and a GPIIb/IIIa antagonist) have been used successfully in the treatment of myocardial infarction and stroke (Ferguson et al., 1998). The data indicate that GPIIb/IIIa antagonists such as abciximab may prove to be useful in the treatment of an acute thrombotic event—particularly arterial—in patients with APS.

2.2. Hydroxychloroquine in aPL-Mediated Thrombosis

Hydroxychloroquine (HQ) is also gaining increased attention as a therapeutic agent in APS. HQ is currently used in the treatment of patients with SLE and has been associated with a decreased risk of thrombosis in SLE patients with aPL (Petri, 1996). HQ is known to have an anticoagulant effect and was, in fact, used for postoperative prophylaxis of deep vein thrombosis and pulmonary embolism following hip arthroplasty by orthopedic surgeons (Johnson and Charnley, 1979). In a previous study, our group demonstrated that HQ significantly diminished thrombus size and time of thrombus persistence in mice injected with aPL (Edwards et al., 1997). We then speculated that the effect of the HQ might be due to its antiplatelet activity. We examined the effects of HQ on activation of platelets *in vitro* by aPL in the presence of TRAP. aPL antibodies enhanced significantly the expression of GPIIb/IIIa and GPIIIa on platelets and these effects were completely abrogated by HQ in a dose-dependent fashion (Espinola et al., 2002). How this agent exerts its effect at the molecular level is not completely understood and will be the subject of further studies.

It is conceivable that HQ may be of benefit in APS patients who are unable to tolerate high levels of oral anticoagulation due to hemorrhagic side effects, or in those who continue to experience thrombotic events despite oral anticoagulation. Alternatively, HQ may also be useful in patients found to have significant titers of aCL antibodies or a positive LA test and in those who have not had any previous thromboembolic events. Clinical trials are needed to establish the efficacy of HQ in long-term prospective studies for the treatment of aPL-mediated thrombosis.

2.3. Intracellular Events in aPL-Mediated Platelet Activation

Thromboxane A2 (TXA2) is the major eicosanoid produced in platelets and has a potent aggregatory and vasoconstrictor activity. Robbins *et al.* (1998) showed a significant increase in the urinary excretion of 11-dehydro-thromboxane B2 (11-DH-TXB2) in seven patients with aPL compared to controls. The same group recently demonstrated that an increase of cAMP through agonists, such as PGI2, or inhibitors of the phosphodiesterase, such as theophylline, abrogated aPL-mediated thromboxane production in platelets (Opara *et al.*, 2003). In another study, Forastiero *et al.* (1998) showed a significant correlation between the levels of anti-β_2GPI-IgG and urinary excretion of 11-DH-TXB2 IgG, in a well-characterized group of 34 patients with aPL. Martinuzzo *et al.* (1993) showed a significantly increased urinary excretion of the platelet-derived thromboxane metabolite 11-DH-TXB2 in a group of six patients with LA and aPL antibodies compared with healthy controls. They also demonstrated an increase in the TXB2 production when normal platelets were incubated with low doses of thrombin and F(ab')$_2$ fragments isolated from patients with aPL antibodies (Wang *et al.*, 2002). Lin and Wang (1992) showed that aCL antibodies raised in rabbits produced platelet activation, involving a change in the shape of the platelets, phosphorylation of a 20-kDa protein, TXA2 production, and later events such as phosphorylation of a 47-kDa protein and increased $(Ca^{2+})_i$ mobilization. More recently, the same group showed a synergistic activation of human platelets by the combination of low concentrations of collagen and rabbit aPL antibodies, demonstrating that indomethacin (an inhibitor of TXA2) inhibited Ca^{2+} mobilization and increased phospholipase C activity and aggregation of platelets (Lin and Wang, 1992). Interestingly, in an *in vivo* study, investigators showed that a thromboxane receptor antagonist (BMS) induced a significant reduction in fetal resorption and an increase in the platelet count in a mouse model of APS (Shoenfeld and Blank, 1994). Hence, there are data to support that aPL antibodies can activate platelets, providing an explanation for aPL-mediated thrombosis (particularly arterial). However, the intracellular events that aPL trigger in platelets and the target receptor(s) are not completely known.

Platelets contain family members of mitogen-activated protein kinases (MAPKs): ERK1 (p44 MAPK) and ERK2 (p42 MAPK) and p38 MAPK. p38 MAPK is a member of a family of proline-directed serine/threonine kinases that is dual-phosphorylated on a threonine and tyrosine residue, separated by a single amino acid (Buschbeck *et al.*, 1999). In platelets, p38 MAPK is activated by stress such as heat and osmotic shock, arsenite, H_2O_2, thrombin, collagen, and a thromboxane analog (Borsch-Haubold *et al.*, 1999) and is involved in the cytosolic phospholipase A2 (cPLA2) phosphorylation, with subsequent production of TXB2. Thrombin has also been shown to induce phosphorylation of ERK1/ERK2, involving protein kinase C (PKC), phospholipase Cβ (PLCβ), and the intracellular mobilization of Ca^{2+} (Nadal-Wollbold *et al.*, 2002).

In order to study intracellular pathways activated by aPL in platelets, we examined the effects of aPL on phosphorylation of p38 MAPK, ERK1/ERK2, and cPLA2, on intracellular Ca^{2+} mobilization, and on TXB2 production. The effects of the specific inhibitor for p38 MAPK SB203580 (4-(4-fluorophenyl)-2-(4-methylsulfinylphenyl)-5-(4-pyridyl)-1-imidazole), on aPL-mediated enhancement of platelet aggregation and on TXB2 production were also determined. Treatment of the platelets with IgG-aPL antibodies or with their F(ab')$_2$ fragments resulted in a significant increase in phosphorylation of p38 MAPK (Vega-Ostertag *et al.*, 2004a). Neither IgG-aPL nor their F(ab')$_2$ fragments increased significantly phosphorylation of ERK1/ERK2 (Vega-Ostertag *et al.*, 2004a). Furthermore, pretreatment of the platelets with SB203580 (a p38 MAPK inhibitor) completely abrogated aPL-mediated enhanced platelet aggregation. Platelets treated with aPL-F(ab')$_2$ produced significantly larger amounts of TXB2 when compared to controls, and this effect was completely abrogated by treatment with SB203580. cPLA2 was also significantly phosphorylated in platelets treated with thrombin and aPL-F(ab')$_2$. There were no significant changes in intracellular Ca^{2+} mobilization when platelets were treated with aPL antibodies and low doses of thrombin. The data strongly indicate that aPL antibodies induce TXB2 production mainly through the activation of p38 MAPK and subsequent phosphorylation of cPLA2, and that the ERK1/ERK2 pathway does not seem to be involved, at least in early stages of aPL-mediated platelet activation.

In summary, our studies show that aPL-mediated platelet activation occurs selectively through the p38 MAPK pathway. These findings may be important in designing new modalities of targeted therapies for treatment of thrombosis in APS patients. Because of the broad proinflammatory role of p38 MAPK in several *in vitro* systems, inhibition of this pathway has been advocated as a novel therapeutic strategy for inflammatory diseases (Lee *et al.*, 2000). More *in vivo* studies in animal models need to be carried out in order to evaluate the effects of p38 MAPK-specific inhibitors on proinflammatory and prothrombotic effects of aPL antibodies.

3. Antiphospholipid Antibodies and Endothelial Cells

3.1. Effects of aPL on Endothelial Cells

Recently, interest has focused on the role of aPL in the induction of procoagulant activity (PCA) in ECs and monocytes. Studies have shown that ECs expressed significantly higher amounts of adhesion molecules (ICAM-1, VCAM-1, and E-selectin) when incubated with aPL antibodies and ß$_2$GP1 *in vitro* (Simantov *et al.*, 1996). Similarly, the incubation of ECs with antibodies reacting with ß$_2$GP1 has been shown to induce EC activation with upregulation of adhesion molecules, IL-6 production, and alteration in prostaglandin metabolism (Meroni *et al.*, 2000).

Utilizing mouse models, our group has shown that human polyclonal and monoclonal aPL antibodies activate endothelium and enhance thrombus formation *in vivo* (Gharavi *et al.*, 1999, Pierangeli *et al.*, 1999). Utilizing mice deficient in ICAM-1, E-selectin, P-selectin, and specific anti-VCAM-1 monoclonal antibodies, our group demonstrated that these EC-activating properties of aPL are mediated by ICAM-1, E-selectin, P-selectin, and VCAM-1 (Pierangeli *et al.*, 2001).

Among the mechanisms suggested to explain the prothrombotic and proinflammatory activities of aPL is the upregulation of tissue factor (TF) (Amengual *et al.*, 1998), a transmembrane protein present on the surface of activated cells and a member of the class II cytokine and hematopoietic growth factor receptor family (Nemerson, 1988). It is located on the surface of a number of cell types, primarily monocytes, vascular ECs, and smooth muscle cells (SMCs). The extracellular domain of TF serves as a receptor for activated factor VII (VIIa) (Nemerson, 1988). In the absence of TF, factor VIIa exhibits poor activity toward its substrates, factor IX and X. Once bound to TF, however, this complex rapidly initiates the coagulation cascade by catalyzing the conversion of factors IX and X to factor IXa and Xa, respectively. Under normal circumstances, circulating factor VII is not exposed to TF. When the integrity of the vasculature is breached, ECs are induced to express cell surface TF, and TF may then interact with factor VIIa and initiate blood coagulation (Nemerson, 1988).

TF is normally undetectable on ECs *in vivo*. However, inflammatory mediators and antigenic factors, such as vascular injury and repair, induce its cell surface expression (Semerato and Colucci, 1997). Proinflammatory cytokines, such as TNF-α and also bacterial lipopolysaccharide (LPS), induce activation of ECs involving translocation of nuclear factor κB (NFκB) to the nucleus of the cell and upregulation of adhesion molecules and TF expression (Osterud and Bjorklid, 2001). In addition, in ECs, TF expression has been reported *in vivo* in association with neoplastic disease and cytokine activation in association with sepsis (Osterud and Bjorklid, 2001). This inappropriate expression of TF may be responsible for thrombotic disorders and fibrin deposition, as seen in disseminated intravascular coagulation and thromboembolic disease. TF induction has also been shown in patients with APS (Amengual *et al.*, 1998).

p38 MAPK is an important component of intracellular signaling cascades that initiate various inflammatory cellular responses. The p38 MAPK has been implicated as an important regulator of the coordinated release of cytokines by immunocompetent cells and the functional response of neutrophils to inflammatory stimuli (Cohen, 1997). Different stimuli can activate p38 MAPK, including LPS and other bacterial products, cytokines such as TNF-α and IL-1, growth factors, and stresses such as heat shock, hypoxia, and ischemia/reperfusion. In addition, p38 MAPK positively regulates a variety of genes involved in inflammation, such as TNF-α, IL-1, IL-6, IL-8, cyclooxygenase-2, and collagenase (Cohen, 1997).

As indicated, studies have also shown that aPL upregulate expression of adhesion molecules and TF on ECs (Amengual *et al.*, 1998) (Figure 26.1), but the intracellular mechanisms involved are not completely understood. We therefore examined the effects of aPL on transcription, expression, and function of tissue factor (TF) on ECs and on phosphorylation of p38 MAPK (Vega-Ostertag *et al.*,

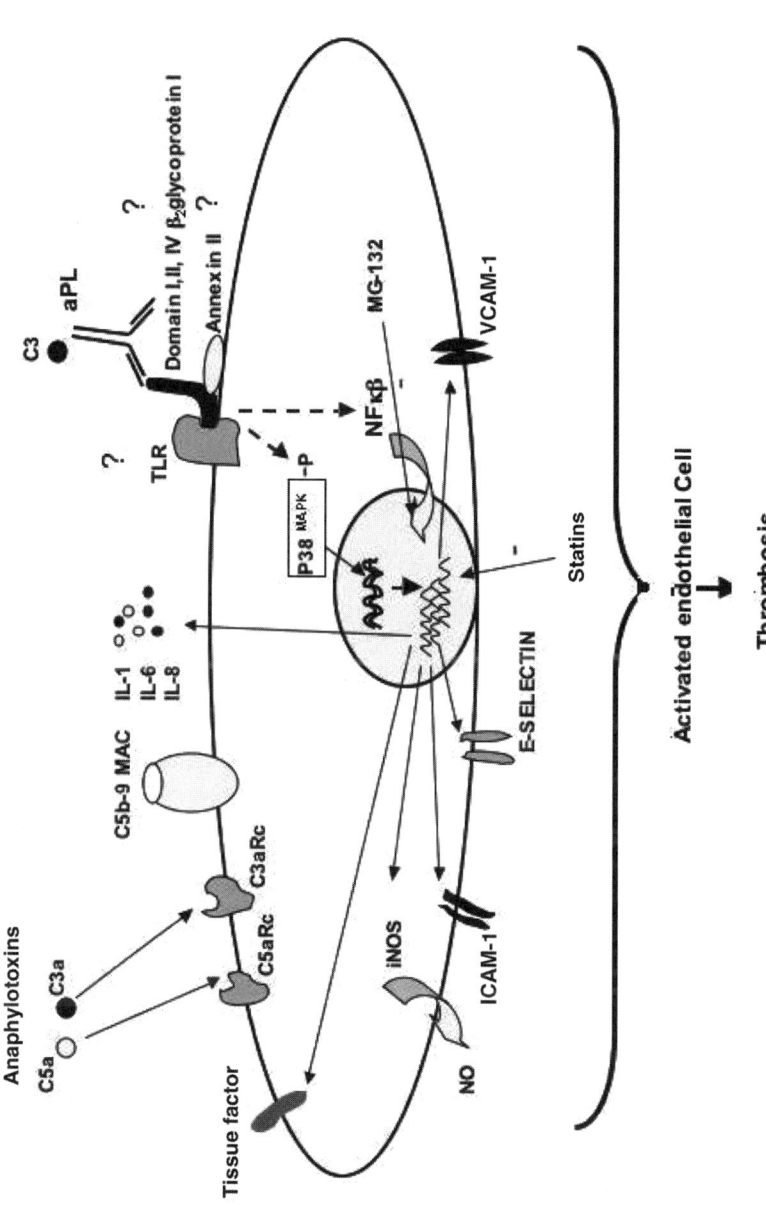

Figure 26.1. Intracellular and molecular events in endothelial cells (ECs) activated by aPL. The figure shows some intracellular pathways affected by aPL antibodies in ECs and the effects of specific inhibitors: (+) activation; (−) inhibitory effects; INOS, inducible nitric oxide synthase; NO, nitric oxide. MG132 is a specific inhibitor of NFκB activation. SB203580 is a specific p38 MAPK inhibitor.

2005). The effects of the specific p38 MAPK inhibitor SB203580 were also evaluated to confirm p38 MAPK involvement. We also examined the effects of aPL on activation of NFκB and production of two inflammatory cytokines, IL-6 and IL-8, on ECs. IgG-aPL were purified from patients with APS that had high titers of aCL antibodies and LA activity. Human umbilical vein endothelial cells (HUVECs) were grown to confluency in medium and treated for 3–4 hr with 100 µg/ml IgG-aPL, with control IgG (IgG-NHS), with medium alone or with LPS (3 µg/ml), as a positive control. Cytokine levels (IL-6 and IL-8) were measured in the supernatants of the tissue culture by ELISA. LPS and IgG-aPL ($n = 2$) increased the production of IL-6 by ECs significantly, when compared to control. LPS and the two IgG-aPL also increased significantly the production of IL-8 by HUVECs over the control values. TF expression was determined by ELISA on cultured HUVECs treated for 4 hr with IgG-aPL antibodies ($n = 4$) or with IgG-NHS. Phorbol myristate acetate (PMA) was used as a positive control. PMA and the four IgG-aPL increased expression of TF on ECs significantly in a dose-dependent fashion. IgG-NHS had no effect on expression of TF on ECs (Pierangeli et al., 2004; Vega-Ostertag et al., 2005). TF expression upregulation by aPL was inhibited when HUVECs were preincubated with SB203580 (Figure 26.1). TF function on lysates of HUVECs was determined using a commercial chromogenic assay that measures factor Xa after activation by TF-FVII complex. TF function was significantly increased when HUVECs were treated with the four aPL preparations. IgG-NHS had no effect on TF function. aPL-induced TF function was inhibited by 34–54% when cells were pretreated with SB203580. p38 MAPK phosphorylation was determined by Western blot in lysates of HUVECs treated with aPL or with IgG-NHS, utilizing a specific monoclonal antibody (28B10) (Figure 26.1). aPL induced a significant phosphorylation of p38 MAPK in HUVECs at 30 min (that increased even more with 60 min incubation). TF mRNA was determined by real time PCR (RT-PCR). Incubation of HUVECs with the four aPL preparations increased TF mRNA by 2.0-, 2.4-, 2.8-, and 1.8-fold (means of three separate experiments). These effects were blocked when cells were pretreated with SB203580. NFκB activation was significantly increased in cultured ECs when treated with aPL or with TNF-α (Vega Ostertag et al., 2005). These effects were inhibited by MG132 (a specific NFκB inhibitor) (Figure 26.1).

These results show that aPL induce TF transcription, and of functional TF expression on ECs, and that p38 MAPK activation and NFκB translocation to the nucleus are required in aPL-mediated signaling pathway that leads to a proadhesive and prothrombotic phenotype. The results also indicate that proinflammatory cytokines are produced by ECs when treated with aPL (Figure 26.1). These cytokines may play a role as regulatory factors in aPL-mediated EC activation and may also contribute to the proadhesive and proinflammatory phenotype observed in ECs of APS patients contributing to a hypercoagulable state.

Some studies have shown that aPL antibodies bind to ß$_2$GPI bound to EC (through domain V) and then EC activation is induced (Del Papa et al., 1998). Another recent study has indicated that aPL/anti-ß$_2$GPI antibodies bind ß$_2$GPI that in turn binds annexin II on the surface of ECs and then intracellular signaling is induced (Ma et al., 2000). A recent publication has suggested that aPL antibodies may bind toll-like receptor-2 (TLR-2), a receptor for bacterial endotoxin, present in ECs and induce intracellular signal transduction (Raschi et al., 2003) (Figure 26.1). The exact nature of the receptor(s) for aPL in ECs is not known and is the subject of current investigations.

3.2. The Statins and Antiphospholipid Antibodies

The 3-hydroxy-3-methylglutaryl-coenzyme A (HMG-CoA) reductase inhibitors, or statins, are potent inhibitors of cholesterol synthesis in the mevalonate pathway. Clinical trials of statin therapy have demonstrated beneficial effects in primary and secondary prevention of coronary heart disease as well as ischemic stroke. Their beneficial effects are only partially explained by their ability to lower cholesterol levels. Statins have also been shown to modify the function of ECs, SMCs, platelets, and monocytes/macrophages (Niwa et al., 1996). Their effects include decreasing the expression of adhesion molecules in monocytes and leukocyte–endothelial interactions, inhibiting platelet function, and inhibiting TF expression by mononuclear cells, downregulating inflammatory cytokines in ECs, increasing fibrinolytic activity and immunomodulation by decreasing the expression of class II major histocompatibility (MHC) complex antigens (Kwak et al., 2000; Eto and Luscher, 2003). Recent studies have suggested that fluvastatin has beneficial effects on aPL-mediated pathogenic effects. Meroni et al. (2001) recently showed that fluvastatin prevented the expression of adhesion molecules and IL-6 in ECs treated with aPL antibodies and these effects seemed to be mediated by NFκB. Our group showed that the thrombogenic and proinflammatory effects of aPL antibodies in vivo could be abrogated in mice fed with fluvastatin for 15 days (Ferrara et al., 2003) and this effect was independent of the cholesterol-lowering effects of the drug. Fluvastatin inhibited the effects of aPL on TF expression on ECs in vitro at doses utilized to reduce cholesterol levels in patients, and these effects were reversed by mevalonate, an intermediate metabolite in the HMG-CoA pathway (Ferrara et al., 2004). Because of the suggested pathogenic role of aPL on induction of TF on ECs, the data presented provide a rationale for using statins as a therapeutic tool in treatment of thrombosis in APS.

3.3. Activation of the Complement Cascade and Antiphospholipid Antibodies

Recent studies have suggested that activation of the complement cascade is necessary for aPL-mediated thrombophilia and fetal loss. First, using the C3 convertase inhibitor complement receptor 1–related gene protein y (Crry)-Ig, we found that inhibition of the complement cascade in vivo blocks aPL-induced fetal loss and growth retardation and inhibit aPL-mediated thrombosis (Holers et al., 2002). Furthermore, mice deficient in complement C3 and C5 were resistant to fetal injury induced by aPL antibodies (Salmon et al., 2002). It was also shown that the interaction of complement component 5a (C5a) with its receptor (C5aR) is necessary for thrombosis of placental vasculature (Girardi et al., 2003). Similarly aPL thrombogenic effects were abrogated in C3- and in C5-deficient mice. Furthermore, a monoclonal anti-C5 antibody reversed thrrombogenic properties of aPL *in vivo*, confirming that complement activation is involved in aPL-induced thrombosis (Pierangeli *et al.*, 2005).

We conclude that complement activation is a necessary intermediary event in the pathogenesis of thrombosis and fetal loss associated with aPL antibodies. We postulate that activated complement fragments themselves have the capacity to bind and activate inflammatory and ECs and to induce a prothrombotic phenotype, either directly through the membrane attack complex (MAC) or through C5a receptor (CD88)-mediated effects (Figure 26.1) (Salmon *et al.*, 2002). We propose the following mechanism for the pathogenic effects of aPL antibodies on thrombosis. First, aPL antibodies bind to ECs, induce their activation and a procoagulant state, as demonstrated in *in vivo* and *in vitro* studies. This activity, however, does not seem to be sufficient to cause thrombosis. Activation of the complement cascade by aPL may amplify these effects by stimulation of the generation of potent mediators of platelet and EC activation, including C3a and C5a and the C5b-MAC (Hattori *et al.*, 1989; Wetsel, 1995; Shin *et al.*, 1996; Munakata *et al.*, 2000; Girardi *et al.*, 2003). The addition of these complement activation products has been shown to cause thrombosis, tissue hypoxia, and inflammation (Davis and Brey, 1992; Munakata *et al.*, 2000).

C5a is a potent soluble inflammatory anaphylotoxic and chemotactic molecule that promotes recruitment and activation of neutrophils (PMN) and monocytes and mediates EC activation through its receptor (C5R, CD88) (Salmon *et al.*, 2002). Binding of C5b to the target initiates the nonenzymatic assembly of the C5b-9MAC. C5b-9MAC activates proinflammatory signaling pathways through the interaction of membrane-associated MAC proteins with heterotrimeric G proteins (Hattori *et al.*, 1989; Davis and Brey, 1992; Wetsel, 1995; Shin *et al.*, 1996; Munakata *et al.*, 2000; Girardi *et al.*, 2003).

In summary, these pathogenic antibodies bind to target cells and, then to complement (Figure 26.1). The complement system in turn damages the ECs, leading to a procoagulant state and undesirable thrombosis (Figure 26.1). We hypothesize that in APS patients, due to aPL deposition targeted to the endothelium, complement activation may be increased locally and may overwhelm normally adequate inhibitory mechanisms. Hence activation of complement may be a critical proximal effector mechanism in aPL-associated thrombosis. We speculate that inhibition of complement activation should ameliorate vascular thrombosis. Therefore, complement may be an important new target for therapy in patients with thrombotic and vascular complications in APS (Quigg, 2002).

4. Conclusions

In the last few years, significant knowledge has been gained on the effects of aPL on platelets and ECs, and the molecular events that these antibodies trigger. The data strongly suggest activation of p38 MAPK in platelets and ECs by aPL and may justify the use of specific inhibitors as a new approach to treat and prevent thrombosis (Table 26.1, Figure 26.1). Most importantly, data indicate that statins (currently used to treat hypercholesterolemia) and the antimalarial drug HQ are also good candidates to be used in treatment and prevention of thrombosis in APS. Furthermore, antagonists of GPIIb/IIIa may be useful, particularly in treatment of acute thrombotic events. Recent findings provide convincing evidence that activation of complement contributes to aPL-mediated thrombosis and pregnancy loss, and suggest that specific inhibitors of complement activation may be used to ameliorate clinical manifestations of APS.

In most autoimmune conditions including APS, therapeutic modalities include steroids and immunosuppressive cytotoxic agents that are counterbalanced by the toxicity and side effects of these medications. In APS, long-term anticoagulation (frequently at high doses) is used to prevent recurrences and to treat thrombotic complications, with the risk of serious complications. Unraveling the molecular mechanisms induced by aPL and the epitope specificity of aPL antibodies may allow the design of new targeted therapeutic agents that would provide a more specific and less harmful approach to treatment than the anticoagulation or general immunosuppression currently used in APS. Based on the new information discussed above, several possibilities (Table 26.1) may become available in the near future.

Acknowledgments

This study was partially funded by a Research Center in Minority Institution Grant (NIH grant number G12-RR03034) and a Minority Biomedical Research Support grant from the NIH grant number S02GMM08248.

Table 26.1. Potential New Agents for Treatment of Thrombosis in APS

Agent	Demonstrated effects
GPIIbIIIa antagonists	Inhibition of aPL-mediated thrombosis *in vivo* in a mouse model (Vega-Ostertag *et al.*, 2004b)
Statins	Abrogation of aPL-induced EC activation and upregulation of tissue factor *in vitro* (Meroni *et al.*, 2000; Ferrara *et al.*, 2004)
	Inhibition of aPL-mediated thrombosis *in vivo* (Ferrara *et al.*, 2003)
Hydroxychloroquine	Inhibition of aPL-mediated platelet activation *in vitro* (Espinola *et al.*, 2002)
	Inhibition of aPL-mediated thrombosis in a mouse model (Edwards *et al.*, 1997)
Specific complement inhibitors (Crry and anti-C5 monoclonal antibody)	Inhibition of aPL-mediated pregnancy loss in mice (Holers *et al.*, 2002; Girardi *et al.*, 2003)
	Inhibition of aPL-mediated thrombosis and EC activation *in vivo* in a mouse model (Holers *et al.*, 2002; Pierangeli *et al.*, 2005)
p38 MAPK inhibitors	Abrogation of aPL-induced upregulation of tissue factor in ECs *in vitro*. (Vega-Ostertag *et al.*, 2005)
	Abrogation of aPL-induced platelet aggregation and activation *in vitro* (Vega-Ostertag *et al.*, 2004a)

References

Amengual, O., Atsumi, T., Khamashta, M.A., and Hughes, G.R.V. (1998). The role of the tissue factor pathway in the hypercoagulable state in patients with the antiphospholipid syndrome. *Thromb. Haemost.*, **79**, 276–281.

Borsch-Haubold, A.G., Ghomashchi, F., Pasquet, S., Goedert, M., Cohen, P., Gelb, M.H., and Watson, S.P. (1999). Phosphorylation of cytosolic phospholipase A2 in platelets is mediated by multiple stress-activated protein kinase pathways. *Eur. J. Biochem.*, **265**, 195–203.

Branch, D.W., Dudley, D.J., Mitchell, M.D., Creighton, K.A., Abbott, T.M., Hammond, E.H., and Daynes, R.A. (1990). Immunoglobulin G fractions from patients with antiphospholipid antibodies cause fetal death in BALB/c mice: A model for autoimmune fetal loss. *Am. J. Obstet. Gynecol.*, **163**, 210–216.

Buschbeck, M., Ghomashchi, F., Gelb, M.H., Watson, S.P., and Borsch-Haubold, A.G. (1999). Stress stimuli increase calcium-induced arachidonic acid release through phosphorylation of cytosolic phospholipase A2. *Biochem. J.*, **344**, 359–366.

Campbell, A.L., Pierangeli, S.S., Wellhausen, S., and Harris, E.N. (1995). Comparison of the effect of anticardiolipin antibodies from patients with the antiphospholipid syndrome and with syphilis on platelet activation and aggregation. *Thromb. Haemost.*, **73**, 529–534.

Cohen, P. (1997). The search for physiological substrates of MAP and SAP kinases in mammalian cells. *Trends Cell. Biol.*, **7**, 353–358.

Crowther, M.A., Ginsberg, J.S., Julian, J., Denburg, J., Hirsch, J., Douketis, J., Laskin, C., Fortin, P., Anderson, D., Kearon, C., Clarke, A., Geerts, W., Forgie, M., Green, D., Costantini, L., Yacura, W., Wilson, S., Gent, M., and Kovacs, M.J. (2003). A comparison of two intensities of warfarin for the prevention of recurrent thrombosis in patients with the antiphospholipid antibody syndrome. *N. Engl. J. Med.*, **349**, 1133–1138.

Davis, W.D. and Brey, R.L. (1992). Antiphospholipid antibodies and complement activation in patients with cerebral ischemia. *Clin. Exp. Immunol.*, **10**, 455–460.

Del Papa, N., Sheng, Y.H., Raschi, E., Kandiah, D.A., Tincani, A., Khamashta, M.A., Atsumi, T., Hughes, G.R., Ichikawa, K., Koike, T., Balestrieri, G., Krilis, S.A., and Meroni, P.L. (1998). Human β_2-glycoprotein I binds to endothelial cells through a cluster of lysine residues that are critical for anionic phospholipids binding and offers epitopes for anti-$\beta_2$glycoprotein I antibodies. *J. Immunol.*, **160**, 5572–5578.

Edwards, M.H., Pierangeli, S., Liu, X.W., Barker, J.H., Anderson, G., and Harris, E.N. (1997). Hydroxychloroquine reverses thrombogenic properties of antiphospholipid antibodies in mice. *Circulation*, **96**, 4380–4384.

Esmon, N.L., Safa, O., Smirnov, M., and Esmon, C.T. (2000). Antiphospholipid antibodies and the protein C pathway. *J. Autoimmun.*, **15**, 221–225.

Espinola, R.G., Pierangeli, S.S., Gharavi, A.E., and Harris, E.N. (2002). Hydroxychloroquine reverses platelet activation induced by human IgG antiphospholipid antibodies. *Thromb. Haemost.*, **87**, 518–522.

Eto, M. and Luscher, T.F. (2003). Modulation of coagulation and fibrinolytic pathways by statins. *Endothelium*, **10**(1), 35–41.

Ferguson, J.J., Kereiakes, D.J., Adgey, A.A., Fox, K.A., Hillegass, W.B., Jr., Pfisterer, M., Vassanelli, C. (1998). Safe use of GPIIb/IIIa inhibitors. *Eur. Heart J.*, Suppl. D40–51. Review.

Ferrara, D.E., Liu, X., Espinola, R.G., Meroni, P.L., Abukhalaf, I., Harris, E.N., and Pierangeli, S.S. (2003). Inhibition of the thrombogenic and inflammatory properties of antiphospholipid antibodies by fluvastatin in an in vivo animal model. *Arthritis Rheum.*, **48**(11), 3272–3279.

Ferrara, D.E., Swerlick, R., Casper, K., Meroni, P.L., Vega-Ostertag, M., Harris, E.N., and Pierangeli, S.S. (2004). Fluvastatin inhibits upregulation of tissue factor expression by antiphospholipid antibodies on endothelial cells. *J. Thromb. Haemost.*, **2**, 1558–1563.

Forastiero, R., Martinuzzo, M., Carreras, L.O., and Maclouf, J. (1998). Anti-β_2 glycoprotein I antibodies and platelet activation in patients with antiphospholipid antibodies: Association with increased excretion of platelet-derived thromboxane urinary metabolites. *Thromb. Haemost.*, **79**, 42–45.

Molecular Pathogenesis of the Antiphospholipid Syndrome 389

Gharavi, A.E., Pierangeli, S.S., Colden-Stanfield, M., Liu, X.W., Espinola, R.G., and Harris, E.N. (1999). GDKV-induced antiphospholipid antibodies enhance thrombosis and activate endothelial cells *in vivo* and *in vitro. J. Immunol.*, **163**, 2922–2927.

Girardi, G., Berman, J., Redecha, P., Spruce, L., Thruman, J.M., Kraus, D., Hollmann, T., Casali, P., Caroll, M.C., Wetsel, R.A., Lambris, J.D., Holers, M., and Salmon, J.E. (2003). Complement C5a receptors and neutrophils mediate fetal injury in the antiphospholipid syndrome. *J. Clin. Invest.*, **112**, 1644–1654.

Harris, E.N. (1987). Syndrome of the black swan. *Br. J. Rheumatol.*, **26**, 324–326.

Hattori, E., Hamilton, K.K., McEver, R.P., and Sims, P.J. (1989). Complement proteins C5b-9 induce secretion of high molecular weight multimers of endothelial von Willebrand factor and translocation of granule membrane protein GMP-140 to the cell surface. *J. Biol. Chem.*, **264**, 9053–9060.

Holers, V.M., Girardi, G., Mo, L., Guthridge, J.M., Molina, H., Pierangeli, S.S., Espinola, R.G., Liu, X., Mao, D., Vialpando, C.G., and Salmon, J.E. (2002). Complement C3 activation is required for antiphospholipid antibody-induced fetal loss. *J. Exp. Med.*, **195**, 211–220.

Jankowski, M., Vreys, I., Wittevrongel, C., Boon, D., Vermylen, J., Hoyaerts, M.F., and Arnout, J. (2003). Thrombogenicity of β_2-glycoprotein I-dependent antiphospholipid antibodies in a photochemically induced thrombosis model in the hamster. *Blood*, **101**, 157–162.

Johnson, R. and Charnley, J. (1979). Hydroxychloroquine in prophylaxis of pulmonary embolism following hip arthroplasty. *Orthop. Clin.*, **144**, 174–177.

Khamashta, M.A., Cuadrado, M.J., Mujic, T., Taub, N.A., Hunt, B.M., and Hughes, G.R.V. (1995). The management of thrombosis in the antiphospholipid antibody syndrome. *N. Engl. J. Med.*, **332**, 993–997.

Krnic-Barrie, S., O'Connor, C.R., Looney, S.W., Pierangeli, S.S., and Harris, E.N. (1997). A retrospective review of 61 patients with antiphospholipid syndrome. *Arch. Intern. Med.*, **157**, 2101–2108.

Kwak, B., Mulhaupt, F., Myit, S., and Mach, F. (2000). Statins as a newly recognized type of immunomodulator. *Nat. Med.*, **6**, 1399–1402.

Lee, J.C., Kumar, S., Griswold, D.E., Underwood, D.C., Votta, B.J., and Adamas, J.L. (2000). Inhibition of p38 MAPK as a therapeutic strategy. *Immunopharmacology*, **47**, 185.

Lin, Y.L. and Wang, C.T. (1992). Activation of human platelets by the rabbit anticardiolipin antibodies. *Blood*, **80**, 3135–3143.

Ma, K., Simantov, R., Zhang, J.C., Silverstein, R., Hajjar, K.A., and McCrae, K.R. (2000). High-affinity binding of β_2 glycoprotein I to human endothelial cells is mediated by annexin II. *J. Biol. Chem.*, **275**(20), 15541–15548.

Martinuzzo, M.E., Maclouf, J., Carreras, L.O., and Levy-Toledano, S. (1993). Antiphospholipid antibodies enhance thrombin-induced platelet activation and thromboxane formation. *Thromb. Haemost.*, **70**, 667–671.

Meroni, P.L., Raschi, E., Camera, M., Testoni, C., Nicoletti, F., Tincani, A., Khamashta, M.A., Balestrieri, G., Tremoli, E., and Hess, D.C. (2000). Endothelial activation by aPL: A Potential pathogenetic mechanism for the clinical manifestations of the syndrome. *J. Autoimmun.*, **30**, 237–240.

Meroni, P.L., Raschi, E., Testoni, C., Tincani, A., Balestrieri, G., Molteni, R., Khamashta, M.A., Tremoli, E., and Camera, M. (2001). Statins prevent endothelial cell activation induced by antiphospholipid (anti-β_2-glycoprotein I) antibodies: Effect on the proadhesive and proinflammatory phenotype. *Arthritis Rheum.*, **44**, 2870–2878.

Meroni, P.L. and Riboldi, P. (2001). Pathogenic mechanisms mediating antiphospholipid syndrome. *Curr. Opin. Rheumatol.*, **13**, 377–382.

Munakata, Y., Saito, T., Matsuda, K., Serina, J., Shibata, S., and Sasaki, T. (2000). Detection of complement-fixing antiphospholipid antibodies in association with thrombosis. *Thromb. Haemost.*, **83**, 728–731.

Nadal-Wollbold, F., Pawlowski, M., Levy-Toledano, S., Berrou, E., Rosa, J.P., and Bryckaert, M. (2002). Platelet ERK2 activation by thrombin is dependent on calcium and conventional protein kinases C but not Raf-1 or B-Raf, *FEBS Lett.*, **531**(3), 475–482.

Nemerson, Y. (1988). Tissue factor and hemostasis. *Blood*, **71**, 1–8.
Niwa, S., Totsuka, T., and Hayashi, S. (1996). Inhibitory effect of fluvastatin, an HMG-CoA reductase inhibitor on the expression of adhesion molecules on human monocyte cell line. *Int. J. Immunopharmacol.*, **18**, 669–675.
Opara, R., Robbins, D.L., and Ziboh, V.A. (2003). Cyclic-AMP agonists inhibit antiphospholipid/β_2 glycoprotein I induced synthesis of human platelet thromboxane A2 *in vitro*. *J. Rheumatol.*, **30**, 55–58.
Osterud, B. and Bjorklid, E. (2001). The tissue factor pathway in disseminated intravascular coagulation. *Semin. Thromb. Haemost.*, **27**, 605–617.
Petri, M. (1996). Thrombosis and systemic lupus erythematosus: The Hopkins lupus cohort perspective. *Scand. J. Rheumatol.*, **25**, 91–193.
Pierangeli, S.S., Colden-Stanfield, M., Liu, X., Barker, J.H., Anderson, G.H., and Harris, E.N. (1999). Antiphospholipid antibodies from antiphospholipid syndrome patients activate endothelial cells *in vitro* and *in vivo*. *Circulation*, **99**, 1997–2000.
Pierangeli, S.S., Espinola, R.G., Liu, X., and Harris, E.N. (2001). Thrombogenic effects of antiphospholipid antibodies are mediated by intercellular cell adhesion molecule-1, vascular cell adhesion molecule-1, and P-selectin. *Circ. Res.*, **88**, 245–250.
Pierangeli, S.S., Girardi, G., Vega-Ostertag, M.E., Liu, X., Espinola, R.G., Salmon, J.E. (2005). Requirement of activation of complement C3 and C5 for antiphospholipid-antibody-mediated thrombophilia. *Arthritis Rheum.* **52**, 1545–1554.
Pierangeli, S.S., Liu, X.W., Anderson, G.H., Barker, J.H., and Harris, E.N. (1995). Induction of thrombosis in a mouse model by IgG, IgM and IgA immunoglobulins from patients with the antiphospholipid syndrome. *Thromb. Haemost.*, **74**, 1361–1367.
Pierangeli, S.S., Vega-Ostertag, M., and Harris, E.N. (2004). Intracellular signaling triggered by antiphospholipid antibodies in platelets and endothelial cells: A pathway to targeted therapies. *Thromb. Res.*, **114**, 467–476.
Quigg, R.J. (2002). Use of complement inhibitors in tissue injury. *Trends Mol. Med.*, **8**, 430–436.
Raschi, E., Testoni, C., Bosisio, D., Borghi, M.O., Koike, T., Mantovani, A., and Meroni, P.L. (2003). The role of MYD88 transduction signaling pathway in endothelial activation by antiphospholipid antibodies. *Blood*, **101**(9), 3495–3500.
Robbins, D.L., Leung, S., Miller-Blair, D.J., and Ziboh, V. (1998). Effect of anticardiolipin/β_2 glycoprotein I complexes on production of thromboxane A2 by platelets from patients with the antiphospholipid syndrome. *J. Rheumatol.*, **25**(1), 51–56.
Roubey, R.A.S. (1998). Mechanisms of autoantibody-mediated thrombosis. *Lupus*, **7**, S114–S119.
Salmon, J.E., Girardi, G., and Holers, V.M. (2002). Complement activation as a mediator of antiphospholipid antibody induced pregnancy loss and thrombosis. *Ann. Rheum. Dis.*, **61**, 46–50.
Semerato, N. and Colucci, M. (1997). Tissue factor in health and disease. *Thromb. Haemost.*, **78**, 759–764.
Shin, M.L., Ros, H.G., and Nicolescu, F.I. (1996). Membrane attack by complement assembly and biology of terminal complement complexes. *Biomembranes*, **4**, 123–149.
Shoenfeld, Y. and Blank, M. (1994). Effect of long-acting thromboxane receptor antagonist (BMS180,291) on experimental antiphospholipid syndrome. *Lupus*, **3**, 397–400.
Simantov, R., Lo, S.K., Gharavi, A., Sammaritano, L.R., Salmon, J.E., and Silverstein, R.L. (1996). Antiphospholipid antibodies activate vascular endothelial cells. *Lupus*, **5**, 440–441.
Vega-Ostertag, M.E., Harris, E.N., and Pierangeli, S.S. (2004). Inracellular events in platelet activation induced by antiphospholipid antibodies in the presence of low doses of thrombis. *Arthritis Rheum.*, **50**, 2911–2919.
Vega-Ostertag, M., Kasper, K., Swerlick, R., Ferrara, D.E., Harris, E.N., and Pierangeli, S.S. (2005). P38 mitogen activated protein kinase is involved in upregulation of tissue factor by antiphospholipid antibodies, *Arthritis Rheum.* **52**, 1545–1554.
Wang, L., Su, C.-Y., Chou, K.-Y., and Wang, C.T. (2002). Enhancement of human platelet activation by the combination of low concentrations of collagen and rabbit anticardiolipin antibodies. *Br. J. Haematol.*, **118**, 1152–1162.

Wetsel, R.A. (1995). Structure, function and cellular expression of complement anaphylotoxin receptors. *Curr. Opin. Immunol.*, **7**, 48–53.

Wilson, W.A., Gharavi, A.E., Koike, T., Lockshin, M.D., Branch, D.W., Piette, J.C., Brey, R., Derksen, R., Triplett, D.A., Harris, E.N., Hughes, G.R., and Khamashta, M.A. (1999). International consensus statement on preliminary classification criteria for definite antiphospholipid syndrome: Report of an international workshop. *Arthritis Rheum.*, **42**, 1309–1311.

27

A Novel Approach to the Prevention of Atherosclerosis

Sun-Ah Kang and Marc Monestier

1. Introduction

Although atherosclerosis is one of the leading causes of death in industrialized societies, our understanding of its pathogenesis and prevention remains limited. Human studies have revealed several important risk factors, such as genetic and environmental elements, and the generation of genetically modified animal models of atherosclerosis has improved our understanding of the cellular and molecular events in this disease. These studies suggest that atherosclerosis is a chronic inflammatory disease and that the immune system is involved in disease progression. Autoantibodies are present during atherosclerosis and several recent studies have demonstrated a paradoxically protective role for some of these autoantibodies. Although the exact function of autoantibodies in atherosclerosis is still not understood, these studies suggest that manipulating humoral autoimmunity may represent a novel therapeutic approach in this disease. In this chapter, we review the pathogenesis of atherosclerosis, with a particular emphasis on the role of the immune system. We also discuss studies that have addressed the importance of autoantibodies in this disease and examine the possible use of antibody-based clinical interventions for the treatment or prophylaxis of atherosclerosis.

2. Atherosclerosis

Atherosclerosis is a disease characterized by lipid accumulation and deposition of extracellular matrix components in the intima of mid- to large-size arterial walls (Wick et al., 1995; Rader and FitzGerald, 1998; Lusis, 2000). In addition to several genetic and environmental factors (Table 27.1), elevated levels of plasma lipoproteins such as low-density lipoproteins (LDLs) are a

---

Sun-Ah Kang and Marc Monestier • Department of Microbiology and Immunology, Temple University School of Medicine, Philadelphia, Pennsylvania 19140.

Molecular Autoimmunity: In commemoration of the 100th anniversary of the first description of human autoimmune disease, edited by Moncef Zouali. Springer Science+Business Media, Inc., New York, 2005.

Table 27.1. Possible Risk Factors Associated with Atherosclerosis Development

Genetic factors	Environmental factors
High VLDL/LDL level	Smoking
Low HDL level	High-fat diet
High blood pressure	Infectious agents
Elevated lipoprotein levels	Stress
Diabetes	Lack of exercise
Family history	
High homocysteine levels	
Gender	

primary risk factor since increased levels of cholesterol can be sufficient by themselves to stimulate atherosclerotic plaque development in humans and in experimental animals (Krieger, 1998; Glass and Witztum, 2001). Disease development is illustrated by three steps: fatty lesion formation, intermediate fibrous fatty plaque formation, and advanced lesion formation and thrombosis (Lusis, 2000; Ross, 1995). At the early stage of atherosclerosis, fatty streak lesions are formed by the subendothelial accumulation of lipid-laden macrophages (foam cells) and by recruited monocytes. Such fatty streaks are not clinically significant, but they are the precursors of advanced atherosclerotic plaques that eventually cause myocardial infarction, stroke, or gangrene through plaque rupture and thrombosis (Lusis, 2000; Rader and FitzGerald, 1998).

2.1. Lesion Initiation

Early lesions of atherosclerosis, called fatty streaks, are initiated dominantly by elevated levels of plasma cholesterol. In humans, LDLs containing apolipoprotein B-100 (apoB) are the main sources of plasma cholesterol delivery (Glass and Witztum, 2001). These LDLs are taken up by LDL receptors (LDLRs) that recognize the amino-terminus of apoB, and the surface expression of LDLRs is subject to a negative feedback mechanism regulated by intracellular cholesterol level. Cholesterol removal is mediated by very low-density lipoproteins (VLDLs), of which the major protein component is apolipoprotein E (apoE). A disruption of the rate of LDL uptake affects circulating LDL levels, as observed in several studies in genetically modified mouse models of atherosclerosis such as LDLR knockout or apoE knockout strains (Breslow *et al.*, 1996; Sanan *et al.*, 1998). In these animals, increased levels of circulating LDLs result in increased transportation and subendothelial accumulation of LDLs. Passively transported LDLs through endothelial cell junctions can be retained in the intima via interactions between LDLs and negatively charged extracellular matrix proteoglycans (Skalen *et al.*,. 2002; Boren *et al.*, 1998). This accumulation of LDL particles is a preliminary to the formation of fatty streaks.

2.2. Fatty Streak Formation

LDL particles that have accumulated in the intima become susceptible to either enzymatic or nonenzymatic modifications. These various modification processes include oxidation, lipolysis, and proteolysis, but oxidation plays a pivotal role in atherosclerosis, as attested by the fact that antioxidant treatment reduces atherosclerotic lesion development in animal models (Steinberg, 1991). The oxidation process is a continuum that can result in LDL particles undergoing varying levels of oxidative modification ranging from minimally modified LDL (mmLDL) to oxidized LDL (oxLDL). mmLDLs can promote the expression of cell adhesion molecules such as P-, E-selectins, VLA-4, and ICAMs, and thus stimulate the entry of circulating cells into the intima (Allen et al., 1998; Dong et al., 1998; Shih et al., 1999). Accumulated mmLDL particles also stimulate endothelial cells (ECs) and smooth muscle cells (SMCs) from the arterial wall to release monocyte chemotactic protein (MCP-1), resulting in additional monocyte adhesion and migration to the subendothelial area (Liao et al., 1991; Shih et al., 1999).

Only strong oxidative modifications of LDL can influence the uptake rate of lipid particles by macrophages. Because extreme oxidation results in the fragmentation of apoB, oxLDLs are not recognized by LDLRs on macrophages or SMCs (Steinberg, 1996; Heinecke, 1997; Horkko et al., 2000). Inhibition experiments using specific antibodies or the use of knockout mice indicate that several surface molecules, such as scavenger (CD36, SR-A, CD68) and Fc receptors, are involved in macrophage uptake of modified LDLs (Ling et al., 1997; Terpstra et al., 1997; Febbraio et al., 2000; Zibara et al., 2002). Uptake of oxLDL mostly induces foam cell formation due to the lack of negative feedback control after reaching elevated intracellular cholesterol levels. Activated foam cells may contribute further to the process by increasing the rate of LDL oxidation, which results in additional foam cell formation and accumulation. In addition, several investigations implicate a T cell involvement in this step via the secretion of proinflammatory cytokines such as IFN-γ and TNF-α (Ross, 1995; Wick et al., 1995; Lusis, 2000). Therefore, multiple processes in the arterial wall involving modified lipoproteins, macrophages, lymphocytes, ECs, and SMCs conspire to create fatty streaks.

2.3. Fibrous Plaques

Fibrous plaques are characterized by a growing necrotic core composed of extracellular lipids and cell debris, proliferating SMCs, and densely accumulated fibrous connective tissue matrix (Ross, 1995; Libby, 2000; Lusis, 2000). Cytokines secreted by macrophages and T cells are also important for SMC migration and proliferation at this stage of disease. Furthermore, the balance between inhibitory and stimulatory cytokines involved in SMC growth and extracellular matrix production greatly influences the stability of the fibrous cap (Libby, 2000). Studies have demonstrated the importance of CD40–CD40L interactions among T cells, macrophages, SMCs, and ECs in maintaining the stability of fibrous plaque, further emphasizing the importance of the relationship between immune response and disease development (Laman *et al.*, 1997). In addition to immune cell interactions, risk factors such as homocysteine, hormones, and infections are also crucial in plaque maintenance and development (Lusis, 2000).

2.4. Plaque Rupture and Thrombosis

An imbalance in the production of inhibitory and stimulatory cytokines can also result in plaque rupture. For example, IFN-γ from activated T cells can inhibit collagen production by SMCs and lead to increased expression of major histocompatibility complex (MHC) class II molecules resulting in the formation of unstable plaques that may rupture, leading to exposure of tissue factor in the necrotic core. This is indirectly confirmed by the elevated expression of MHC class II molecules and surface markers of activated T cells at the site of plaque rupture (Laman *et al.*, 1997). Also, various proteolytic enzymes such as matrix metalloproteinase or elastase are released from macrophages upon T cell stimulation and are then involved in plaque rupture by degrading structurally important constituents of the endothelial cell matrix. Frequent apoptotic events around the necrotic core, in response to inflammatory cytokines from immune cells, also contribute to the progression to thrombosis (Sata and Walsh, 1998). Rupture may thus lead to platelet aggregation, followed by thrombosis, resulting in disastrous clinical consequences such as myocardial infarction or stroke (Libby, 2000).

3. Immune Cells in Atherosclerosis

In recent years, evidence has accumulated that atherosclerosis is a chronic inflammatory disease. The localization of both macrophages and lymphocytes to human atherosclerotic lesions suggests that the immune response contributes to disease progression. It has been known for many years that monocyte-derived macrophages localize in atherosclerotic lesions and most lipid-laden foam cells bear macrophage differentiation markers. This has been confirmed by histochemical studies of atherosclerotic lesions using macrophage-specific monoclonal antibodies (reviewed in Libby and Hansson, 1991; Roselaar *et al.*, 1996; Hansson, 1997, 2001). The possible involvement of immune cells in atherosclerotic lesions is also suggested by the increased expression of MHC class II molecules on the SMC surfaces in the lesions. Elevated expression of cell adhesion molecules and secreted cytokines found in the lesions also suggest an active monocyte response. LDLR$^{-/-}$ mice that are also deficient in RAG expression (and therefore do not have any B or T cells) exhibit reduced and delayed lesion progression, thus indicating the importance of lymphocytes in disease development (Song *et al.*, 2001). The involvement of dendritic cells (DCs) and mast cells in atherosclerosis has also been reported (Hansson, 2001; Perrin-Cocon *et al.*, 2001; Alderman *et al.*, 2002; Coutant *et al.*, 2004). All of these immune cells may potentially contribute their unique roles to different stages of the disease.

The recruitment of monocytes to the lesion, followed by infiltration of these cells into the subendothelial space of the arterial wall, is a crucial stage in atherosclerosis. Yet this step is not clearly understood, although evidence suggests that lipoprotein modifications, especially oxidation, greatly increase monocyte adhesion and infiltration in atherosclerotic lesions. At the initial stage, elevated levels of LDLs stimulate ECs to express monocyte-specific cell adhesion molecules, such as P- and E-selectins (Dong et al., 1998; Allen et al., 1998). MCP-1 released from ECs after stimulation by mmLDLs further increases the expression of integrins (VLA-4, VLA-5) and cell adhesion molecules (VCAM-1, ICAM-1) to enhance monocyte adherence and migration (Liao et al., 1991; Shih et al., 1999). These various adhesion molecules are proatherogenic, as indicated by blockage experiments performed in atherosclerosis-prone animal models (Allen et al., 1998; Dong et al., 1998; Shih et al., 1999). The expression of these adhesion molecules is further induced by accumulated oxLDL, foam cells, and activated immune cells in the intima, hence allowing additional monocyte recruitment and infiltration.

4. Cellular Immunity in Atherosclerosis

Macrophages are a major regulator of immune responses in both the innate and the adaptive immune systems. During atherosclerosis, the importance of macrophages depends not only upon their role in immunological regulation, but also upon their participation as foam cells, one of the actual constituents of atherosclerotic plaques. Monocyte differentiation into macrophages in atherosclerotic intima is induced by macrophage colony-stimulating factor (M-CSF) secreted by macrophages as well as vascular and stromal cells. Lack of M-CSF inhibits macrophage proliferation and differentiation, as observed in the M-CSF-deficient osteopetrotic (op/op) mouse strain. Studies of op/op mice crossed with LDLR-deficient animals demonstrate that M-CSF participates critically in fatty streak formation and progression to advanced fibrous plaques (Rajavashisth et al., 1998). This cross exhibits dramatically reduced lesion sizes even when the op mutation is heterozygous (op$^{+/-}$/LDLR$^{-/-}$) (Rajavashisth et al., 1998).

Monocyte-derived macrophages may interact with various forms of lipoproteins in the arterial intima. Accumulated lipoproteins are taken up by macrophages via LDLRs, scavenger receptors, or the lectin-like oxLDL receptor (LOX-1), for which surface expression is tightly regulated by various cytokines. Among these different receptors, class B scavenger receptor CD36 may be particularly proatherogenic because of its elevated level of expression in atherosclerotic lesions and because its targeted genetic deficiency reduces lesion sizes in atherosclerosis-prone mice (Kim et al., 1997; Febbraio et al., 2000; Zibara et al., 2002).

Cell surface receptor–mediated uptake of various forms of LDLs modulates immune responses via cytokine production (Table 27.2). LDL binding to LDLR on macrophages leads to the production of IL-10, a cytokine known to promote Th2 polarization and humoral immune responses, whereas oxLDL can stimulate the production of IL-12, a proinflammatory Th1 cytokine (Varadhachary et al., 2001). Surface expression of LDLR, however, is negatively regulated by intracellular cholesterol levels. Hence, during atherogenic conditions with increased levels of circulating LDLs, LDLR expression is downregulated once intracellular cholesterol level reaches its threshold. This also decreases LDLR ligation and results in reduced production of IL-10 and a reciprocal increase in IL-12 secretion.

Table 27.2. Some Immune Response Mediators Involved in Atherosclerosis

Cytokines/chemokines	Origin	Functional properties	Atherosclerotic events	Effect on atherosclerosis
MCP-1	SMC, EC	Monocyte chemoattractant	↑ Monocyte infiltration	←
TNF-α	Th1, MØ	↑ Adhesion molecules ↑ Thrombosis coagulation	↑ Monocyte infiltration	←
IFN-γ	Th1	↑ Adhesion molecules ↓ SMC proliferation ↓ Collagen synthesis ↑MHCII expression on MØ and EC	↑ Monocyte infiltration ↓ Plaque formation ↓ Plaque stability, ↑ rupture ↑ Inflammation in lesion	←
PDGF	MØ, EC, SMC	Leukocyte migration	↑ Monocyte infiltration	←
IL-1	MØ, EC, SMC	↑ Adhesion molecules ↑ Thrombosis coagulation ↑ SMC proliferation	↑ Monocyte infiltration	←
M-CSF	MØ, EC	↑ Monocyte differentiation	↑ Plaque formation ↑ Foam cell formation	←
IL-4	Th2	↑ CD36 ↑ NOS ↑ PPAR-γ dependent genes	↑ Foam cell formation	←
IL-10	Th2	↑ Th2 type cytokine production ↓ Th1 type responses	↓ Disease progression	→
IL-12	Th1	↑ IFN-γ ↓ Th2 type responses	↑ Immune responses in intima ↑ Disease progression	←

This table (based upon both human and murine studies) summarizes the role of some of the immune mediators responsible for the pathogenesis of atherosclerosis.
EC, endothelial cell; IFN-γ, interferon-γ; IL-4, interleukin-4; IL-10, interleukin-10; IL-12, interleukin-12; MØ, macrophage; MCP-1, monocyte chemotactic protein-1; M-CSF, macrophage colony-stimulating factor; MHC, major histocompatibility complex; NOS, nitric oxide synthase; PDGF, platelet-derived growth factor; PPAR-γ, peroxisome proliferator–activated receptor-γ; SMC, smooth muscle cell; TNF-α, tumor necrosis factor-α.

An increase in IL-12 production may also result from macrophage uptake of oxLDLs via a scavenger receptor such as CD36, and this elevated IL-12 production may worsen atherosclerosis by promoting inflammation (Table 27.2) (Varadhachary et al., 2001). Therefore, macrophages are critically involved in atherosclerosis not only via their transformation into lipid-laden foam cells but also by producing different cytokines engaged in immune responses.

The detection of T cells in atherosclerotic lesions suggests their involvement in disease progression. Early studies with RAG-deficient mice, which showed reduced atherosclerosis in the absence of lymphocytes, support the importance of T cells in disease progression (Dansky et al., 1997; Song et al., 2001). The majority of T cells present in atherosclerotic lesions are $CD3^+CD4^+$, T cell receptor $\alpha\beta^+$ T cells (Hansson, 1997; Libby, 2000). $CD4^+$ T cells can differentiate into Th1 or Th2 subsets in response to various cytokine cocktails. For instance, IL-10 is a Th2 inducer and IL-12 favors Th1 differentiation. The reciprocal roles of these two cytokines in T cell polarization are also well established. These two different subsets of T cells can also be distinguished by their cytokine profiles. For example, Th1 cells produce IFN-γ, a potent proinflammatory cytokine, while anti-inflammatory cytokines such as IL-4 or IL-10 are secreted by the Th2 subset.

Most studies have focused on the role of Th1 cells in atherosclerosis because of their abundance among T cells in lesions. In apoE-deficient mice, inhibition of Th1 polarization using pentoxifylline, a phosphodiesterase inhibitor, resulted in a 60% reduction in lesion size (Laurat et al., 2001). The colocalization of T cells and IFN-γ to atherosclerotic lesions also implicates an involvement of Th1 cells (Hansson, 1997; Mallat et al., 1999). IFN-γ triggers macrophage activation, increased expression of MHC molecules, and inhibition of SMC proliferation. Mice deficient for both apoE and IFN-γ exhibit reduced plaque size, as well as decreased lesional lipid accumulation and cellularity, but a remarkable increase in collagen level when compared to $apoE^{-/-}$ IFN-γ-sufficient mice (Gupta et al., 1997). IFN-γ is thus proatherogenic by virtue of its potent proinflammatory roles in lesions and its involvement in plaque destabilization. In addition to IFN-γ, the lesional localization of other proinflammatory cytokines such as IL-1 and TNF-α has been observed, but their importance in atherosclerosis is not fully elucidated (Table 27.2). They may represent powerful inducers of inflammation in vessel walls because of their ability to promote macrophage activation and to increase the expression of cell adhesion molecules (Hansson, 1997).

In contrast to Th1 cells, Th2 cells and cytokines are less abundant in atherosclerotic lesions. This is possibly due to the elevated expression of Th1 type cytokines, especially IL-12, since the two subsets are reciprocally controlled by their major cytokines, IL-10 and IL-12. For these reasons, it has been suggested that Th2 cells can potentially prevent atherosclerosis via the production of IL-10, which can indeed be detected in atherosclerotic plaques at various stages of development (Hansson, 2001). Its role in atherosclerosis has been suggested by studies in IL-10-deficient mice that showed raised cell infiltration, low percentages of collagen, and increased level of IFN-γ in atherosclerotic lesions (Mallat et al., 1999). The protective role of IL-10 in atherosclerosis was confirmed by transfer of IL-10 cDNA into IL-10-deficient mice and by bone marrow transplantation experiments. Introduction of IL-10 cDNA into IL-10-deficient mice dramatically reduces the size of fatty streak (Mallat *et al.*, 1999), and transplantation of bone marrow from IL-10 transgenic mice into LDLR$^{-/-}$ recipients reduces lesion size and necrotic core, and induces IgG1 production, indicating a Th2 profile of B cell-activation (Pinderski *et al.*, 2002).

In contrast with IL-10, another Th2 cytokine, IL-4, may be proatherogenic. Indeed, its deficiency in LDLR$^{-/-}$ or apoE$^{-/-}$ mice results in decreased atherosclerotic lesions (King *et al.*, 2002; Davenport and Tipping, 2003). These results are surprising at first glance since IL-4 can downregulate IFN-γ production, Th1 responses, and macrophage activation. The proatherogenic effect of IL-4 may be due to its capacity to increase CD36 expression on macrophages, a scavenger receptor important for oxLDL uptake and therefore foam cell formation (Huang *et al.*, 1999). Thus, the full contribution of Th1 and Th2 helper cells in atherosclerosis progression remains to be elucidated.

5. Humoral Immunity in Atherosclerosis

The role of B cells in atherosclerosis has received less attention than that of T cells and macrophages. Overall, the experimental evidence suggests that B cells may be protective. LDLR-deficient mice have normal number of B cells, but bone marrow transplantation from B cell–deficient mice into LDLR-deficient recipients results in complete lack of B cells and antibodies, as well as increased atherosclerotic lesion size (Major *et al.*, 2002). Splenectomy in apoE$^{-/-}$ mice further promotes atherosclerosis, but this progression can be prevented by the transfer of B cells from apoE-deficient mice (Caligiuri *et al.*, 2002). These data suggest that B cells may prevent disease progression in atherosclerosis possibly via antibody production and immune regulation via antigen presentation.

The oxidation process creates multiple chemical modifications on both the protein and the lipid components of the LDL particles (Palinski *et al.*, 1990, 1995b, 1996), resulting in the formation of neo-self-antigens that can elicit the production of anti-oxLDL antibodies. The presence of anti-oxLDL antibodies has been documented in humans and in mouse or rabbit strains prone to atherosclerosis (Palinski and Witztum, 2000; Vaarala, 2000; Thiagarajan, 2001). Although the role of these antibodies in atherosclerosis is still controversial, the correlation of anti-oxLDL titer and disease progression suggests that anti-oxLDL antibody titers are a marker of atherosclerosis progression (Palinski *et al.*, 1990, 1995b; Romero *et al.*, 2000; Thiagarajan, 2001).

Patients with systemic lupus erythematosus (SLE) or antiphospholipid antibody syndrome (APS) are prone to develop arterial complications including arterial thrombosis and atherosclerosis (Roubey, 1996; Haviv, 2000; Hughes, 2000; Urowitz and Gladman, 2000). This fact has attracted attention on the possible role of the autoimmune process in atherosclerosis development. Antibodies to oxLDLs are frequently detected in SLE patients, and several reports have suggested that they also cross-react with phospholipids, suggesting the existence of a significant overlap between anti-oxLDL and antiphospholipid autoantibody populations (Romero *et al.*, 2000; Vaarala, 2000; Thiagarajan, 2001; Nicolo and Monestier, 2004).

The specificities of anti-oxLDL antibodies are quite diverse. Some of them bind to oxidized lipid moieties in the LDL particles, whereas others may recognize the oxidative modifications of apoB (Thiagarajan, 2001). Cross-reactivities toward oxLDLs have been suggested for various autoantibody populations associated with systemic autoimmunity. APS patients have autoantibodies to a phospholipid-binding protein, ß2-glycoprotein I, which can in turn bind to LDL (Horkko *et al.*, 1997; Reddel *et al.*, 2000). Overall, antibody specificities such as anticardiolipin, anti-ß2 glycoprotein I, or anti-ß2 glycoprotein I-cardiolipin complexes have been reported to cross-react with various epitopes of oxLDL (Mizutani *et al.*, 1995; Horkko *et al.*, 2001; Matsuura *et al.*, 2002). It has also been suggested that the clinical associations of these antibodies may be different. For instance, anti-ß2-glycoprotein I-cardiolipin-complex antibodies are associated with both arterial and venous thrombosis in SLE, whereas anti-oxLDL antibodies are only associated with arterial thrombosis in APS patients (Vaarala, 2000).

With respect to atherosclerosis, an important group of anti-oxLDL antibodies are naturally occurring autoantibodies directed against oxidized phospholipids. In the atherosclerosis-prone apoE-deficient mouse strain, oxLDL-specific B cells produce IgM antibodies that recognize oxidized phospholipids (Shaw *et al.*, 2000) and share structural and functional properties with naturally occurring T15 idiotype-positive antiphosphatidylcholine antibodies (Rose and Afanasyeva, 2003; Shaw *et al.*, 2000, 2003). These T15 antibodies are of B-1 cell origin, can be protective against certain microbial infections, and can cross-react with apoptotic cells and mediate their clearance (Shaw *et al.*, 2000, 2003; Cocca *et al.*, 2001; Rose and Afanasyeva, 2003). Some anti-oxLDL antibodies indeed recognize epitopes common to both oxLDL and apoptotic cells (Chang *et al.*, 1999).

The possible role of these autoantibodies in atherosclerosis remains unclear. The initial assumption was that they were probably pathogenic, possibly via the formation of immune complexes. For instance, *in vitro* studies showed that the presence of LDL-immune complexes promotes macrophage uptake via FcγRI and formation of foam cells (Lopes-Virella *et al.*, 1997). In contrast, more recent studies have suggested a potential protective role for autoantibodies in atherosclerosis. For example, some antiphospholipid antibodies including anti-oxLDL antibodies could interrupt oxLDL uptake of macrophage by blocking their binding to scavenger receptors (Chang et al., 1999; Shaw *et al.*, 2000).

6. Vaccination or Immunoglobulin Administration in Atherosclerosis

The large body of evidence showing that the immune response plays a role in the progression of atherosclerosis suggests that manipulating the immune system represents a possible prophylactic or therapeutic avenue for this disease. In view of the pathogenic role of the Th1 response, altering the Th1/Th2 balance is an attractive idea, but the proatherogenic effects of IL-4 suggest that this approach is also risky (King et al., 2002; Davenport and Tipping, 2003). In recent years, the most encouraging data for the immunological manipulation of atherosclerosis stem from studies involving immunization with atherosclerosis-associated autoantigens or immunoglobulin treatment. As indicated above, several different antigens engaged in atherosclerosis have been reported. Modified LDL molecules, usually via oxidation, are formed throughout disease progression and elicit autoimmune responses. Interestingly, immunization with one of the antigens that result from the oxidation process, malondialdehyde-modified LDL, reduces atherosclerosis in several animal models (Palinski et al., 1995a; Ameli et al., 1996; Freigang et al., 1998; George et al., 1998; Zhou et al., 2001). Some of these studies suggest that protection depends upon the induction of high titer of antibodies that may interrupt oxLDL recognition by macrophages or increase the clearance of oxLDL (Freigang et al., 1998; Zhou et al., 2001). Because elevated antibodies in immunized animals are of the IgG isotype, T cells could also be involved.

Microbial antigens may also play a role in the pathogenesis of atherosclerosis. For instance, certain microbial infections such as *Chlamydia pneumoniae* have been frequently reported in patients with atherosclerosis and it has been suggested that molecular mimicry may exist between these bacteria and modified LDLs (Hansson, 2001; Binder et al., 2003). As described above, strong similarities exist between anti-oxLDL antibodies isolated from apoE-deficient mice and naturally occurring T15 antibodies that are protective against common pathogens such as pneumococci (Shaw et al., 2000). Immunization of LDLR$^{-/-}$ mice with *Streptococcus pneumoniae* resulted in the production of elevated levels of anti-oxLDL, decreased atherosclerotic lesions, and plasma antibodies from these mice prevented the binding of oxLDL to macrophages (Binder et al., 2003).

In addition to active immunization, passive administration of antiphospholipid antibodies can also improve disease in atherosclerosis-prone mice (Nicolo et al., 2003). We obtained a panel of antiphospholipid IgG monoclonal antibodies from (NZW × BXSB) F_1 mice, a cross that spontaneously develops systemic autoimmunity reminiscent of lupus and APS (Monestier et al., 1996). We observed that biweekly administration of one such monoclonal antibody, FB1, prevents plaque formation in LDLR$^{-/-}$ mice when compared to PBS- or control antibody-treated animals (Nicolo et al., 2003).

7. Conclusions

Studies in recent years have shown that the immune system plays a greater role in the pathogenesis of atherosclerosis than previously recognized. These findings open new possibilities of treatment or prevention of this disease, and approaches using antibodies are especially promising. Future studies should aim at further characterizing the fine specificities of beneficial antibodies and at optimizing the means of their delivery (active immunization versus passive administration).

References

Alderman, C.J.J., Bunyard, P.R., Chain, B.M., Foreman, J.C., Leake, D.S., and Katz, D.R. (2002). Effects of oxidized low density lipoprotein on dendritic cells: A possible immunoregulatory component of the atherogenic micro-environment? *Cardiovasc. Res.*, **55**, 806–819.

Allen, S., Khan, S., Futwan, A.-M., Batten, P., and Yacoub, M. (1998). Native low density lipoprotein induced calcium transients trigger VCAM-1 and E selectin expression in cultured human vascular endothelial cells. *J. Clin. Invest.*, **101**, 1064–1075.

Ameli, S., Hultgardh-Nilsson, A., Regnstrom, J., Calara, F., Yano, J., Cercek, B., Shah, P.K., and Nilsson, J. (1996). Effect of immunization with homologous LDL and oxidized LDL on early atherosclerosis in hypercholesterolemic rabbits. *Arterioscler. Thromb. Vasc. Biol.*, **16**, 1074–1079.

Binder, C.J., Horkko, S., Dewan, A., Chang, M.K., Kieu, E.P., Goodyear, C.S., Shaw, P.X., Palinski, W., Witztum, J.L., and Silverman, G.J. (2003). Pneumococcal vaccination decreases atherosclerotic lesion formation: Molecular mimicry between *Streptococcus pneumoniae* and oxidized LDL. *Nat. Med.*, **9**, 736–743.

Boren, J., Olin, K., Lee, I., Chait, A., Wight, T.N., and Innerarity, T.L. (1998). Identification of the principal proteoglycan-binding site in LDL. *J. Clin. Invest.*, **101**, 2658–2664.

Breslow, J.L., Plump, A., and Dammerman, M. (1996). New mouse models of lipoprotein disorders and atherosclerosis. In V. Fuster, R. Ross, and E.J. Topol (eds) *Atherosclerosis and Coronary Artery Disease*. Lippincott-Raven, Philadelphia. pp. 363–378.

Caligiuri, G., Nicoletti, A., Poirier, B., and Hansson, G.K. (2002). Protective immunity against atherosclerosis carried by B cells of hypercholesterolemic mice. *J. Clin. Invest.*, **109**, 745–753.

Chang, M.K., Bergmark, C., Laurila, A., Horkko, S., Han, K.H., Friedman, P., Dennis, E.A., and Witztum, J.L. (1999). Monoclonal antibodies against oxidized low-density lipoprotein bind to apoptotic cells and inhibit their phagocytosis by elicited macrophages: Evidence that oxidation-specific epitopes mediate macrophage recognition. *Proc. Natl. Acad. Sci. USA*, **96**, 6353–6358.

Cocca, B.A., Seal, S.N., D'Agnillo, P., Mueller, Y.M., Katsikis, P.D., Rauch, J., Weigert, M., and Radic, M.Z. (2001). Structural basis for autoantibody recognition of phosphatidylserine-beta 2 glycoprotein I and apoptotic cells. *Proc. Natl. Acad. Sci. USA*, **98**, 13826–13831.

Coutant, F., Agauge, S., Perrin-Cocon, L., Andre, P., and Lotteau, V. (2004). Sensing environmental lipids by dendritic cell modulated its function. *J. Immunol.*, **172**, 54–60.

Dansky, H.M., Charlton, S.A., Harper, M.M., and Smith, J.D. (1997). T and B lymphocytes play a minor role in atherosclerotic plaque formation in the apolipoprotein E-deficient mouse. *Proc. Natl. Acad. Sci. USA*, **94**, 4642–4646.

Davenport, P.T. and Tipping, P.G. (2003). The role of interleukin-4 and interleukin-12 in the progression of atherosclerosis in apolipoprotein E-deficient mice. *Am. J. Pathol.*, **163**, 1117–1125.

Dong, M., Chapman, M., Brown, A.A., Frenette, P.S., Hynes, R.O., and Wagner, D.D. (1998). The combined role of P and E selectin in atherosclerosis. *J. Clin. Invest.*, **102**, 145–152.

Febbraio, M., Podrez, E.A., Smith, J.D., Hajjar, D.P., Hazen, S.L., Hoff, H.F., Sharma, K., and Silverstein, R.L. (2000). Targeted disruption of the class B scavenger receptor CD36 protects against atherosclerotic lesion development in mice. *J. Clin. Invest.*, **105**, 1049–1056.

Freigang, S., Horkko, S., Miller, E., Witztum, J.L., and Palinski, W. (1998). Immunization of LDL receptor-deficient mice with homologous malondialdehyde-modified and native LDL reduces progression of atherosclerosis by mechanisms other than induction of high titers of antibodies to oxidative neoepitopes. *Arterioscler. Thromb. Vasc. Biol.*, **18**, 1972–1982.

George, J., Afek, A., Gilburd, B., Levkovitz, H., Shaish, A., Goldberg, I., Kopolovic, Y., Wick, G., Shoenfeld, Y., and Harats, D. (1998). Hyperimmunization of apo-E-deficient mice with homologous malondialdehyde low-density lipoprotein suppresses early atherogenesis. *Atherosclerosis*, **138**, 147–152.

Glass, C.K. and Witztum, J.L. (2001). Atherosclerosis. The road ahead. *Cell*, **104**, 503–516.

Gupta, S., Pablo, A.M., Jiang, X.C, Wang, N., Tall, A.R., and Schindler, C. (1997). IFN-gamma potentiates atherosclerosis in ApoE knock-out mice. *J. Clin. Invest.*, **99**, 2752–2761.

Hansson, G.K. (1997). Cell-mediated immunity in atherosclerosis. *Curr. Opin. Lipidol.*, **8**, 301–311.

Hansson, G.K. (2001). Immune mechanisms in atherosclerosis. *Arterioscler. Thromb. Vasc. Biol.*, **21**, 1876–1890.

Haviv, Y.S. (2000). Association of anticardiolipin antibodies with vascular injury: Possible mechanisms. *Postgrad. Med. J.*, **76**, 625–628.

Heinecke, J.W. (1997). Mechanisms of oxidative damage of low density lipoprotein in human atherosclerosis. *Curr. Opin. Lipidol.*, **8**, 268–274.

Horkko, S., Binder, C.J., Shaw, P.X., Chang, M.K., Silverman, G., Palinski, W., and Witztum, J.L. (2000). Immunological responses to oxidized LDL. *Free Radic. Biol. Med.*, **28**, 1771–1779.

Horkko, S., Miller, E., Branch, D.W., Palinski, W., and Witztum, J.L. (1997). The epitopes for some antiphospholipid antibodies are adducts of oxidized phospholipid and beta2 glycoprotein 1 (and other proteins). *Proc. Natl. Acad. Sci. USA*, **94**, 10356–10361.

Horkko, S., Olee, T., Mo, L., Branch, D.W., Woods, V.L.J., Palinski, W., Chen, P.P., and Witztum, J.L. (2001). Anticardiolipin antibodies from patients with the antiphospholipid antibody syndrome recognize epitopes in both beta(2)-glycoprotein 1 and oxidized low-density lipoprotein. *Circulation*, **103**, 941–946.

Huang, J.T., Welch, J.S., Ricote, M., Binder, C.J., Willson, T.M., Kelly, C., Witztum, J.L., Funk, C.D., Conrad, D., and Glass, C.K. (1999). Interleukin-4-dependent production of PPAR-gamma ligands in macrophages by 12/15-lipoxygenase. *Nature*, **400**, 378–382.

Hughes, G.R.V. (2000). Immunology, lupus and atheroma. *Lupus*, **9**, 159–160.

Kim, J.G., Keshava, C., Murphy, A.A., Pitas, R.E., and Parthasarathy, S. (1997). Fresh mouse peritoneal macrophages have low scavenger receptor activity. *J. Lipid Res.*, **38**, 2207–2215.

King, V.L., Szilvassy, S.J., and Daugherty, A. (2002). Interleukin-4 deficiency decreases atherosclerotic lesion formation in a site-specific manner in female LDL receptor–/– mice. *Arterioscler. Thromb. Vasc. Biol.*, **22**, 456–461.

Krieger, M. (1998). The best of cholesterols, the worst of cholesterols: A tale of two receptors. *Proc. Natl. Acad. Sci. USA*, **95**, 4077–4080.

Laman, J.D., de Smet, B.J.G.L., Schoneveld, A., and Meurs, M.V. (1997). CD40-CD40L interactions in atherosclerosis. *Immunol. Today*, **18**, 272–277.

Laurat, E., Poirier, B., Tupin, E., Caligiuri, G., Hansson, G.K., Bariety, J., and Nicoletti, A. (2001). In vivo downregulation of T helper cell 1 immune responses reduces atherogenesis in apolipoprotein E knockout mice. *Circ. Res.*, **104**, 197–202.

Liao, F., Berliner, J.A., Mehrabian, M., Navab, M., Demer, L.L., Lusis, A.J., and Fogelman, A.M. (1991). Minimally modified low density lipoprotein is biologically active *in vivo* in mice. *J. Clin. Invest.*, **87**, 2253–2257.

Libby, P. (2000). Changing concepts of atherogenesis. *J. Intern. Med.*, **247**, 349–358.

Libby, P. and Hansson, G.K. (1991). Biology of disease—Involvement of the immune system in human atherogenesis: Current knowledge and unanswered questions. *Lab. Invest.*, **64**, 5–15.

Ling, W., Lougheed, M., Suzuki, H., Buchan, A., Kodama, T., and Steinbrecher, U.P. (1997). Oxidized or acetylated low density lipoproteins are rapidly cleared by the liver in mice with disruption of the scavenger receptor class A type I/II gene. *J. Clin. Invest.*, **100**, 244–252.

Lopes-Virella, M.F., Binzafar, N., Rackley, S., Takei, A., La Via, M., and Virella, G. (1997). The uptake of LDL-IC by human macrophages: Predominant involvement of the FcγRI receptor. *Atherosclerosis*, **135**, 161–170.

Lusis, A.J. (2000). Atherosclerosis. *Nature*, **407**, 233–241.

Major, A.S., Fazio, S., and Linton, M.F. (2002). B lymphocyte deficiency increases atherosclerosis in LDL receptor null mice. *Arterioscler. Thromb. Vasc. Biol.*, **22**, 1892–1898.

Mallat, Z., Besnard, S., Duriez, M., Deleuze, V., Emmanuel, F., Bureau, M.F., Soubrier, F., Esposito, B., Duez, H., Fievet, C., Staels, B., Duverger, N., Scherman, D., and Tedgui, A. (1999). Protective role of interleukin-10 in atherosclerosis. *Circ. Res.*, **85**, e17–e24.

Matsuura, E., Kobayashi, K., Kasahara, J., Yasuda, T., and Makino, H. (2002). Anti-beta2 glycoprotein 1 autoantibodies and atherosclerosis. *Int. Rev. Immunol.*, **21**, 41–66.

Mizutani, H., Kurata, Y., Kosugi, S., Shiraga, M., Kashiwagi, H., Tomiyama, Y., Kanakura, Y., Good, R.A., and Matsuzawa, Y. (1995). Monoclonal anticardiolipin autoantibodies established from the

(New Zealand white × BXSB) F1 mouse model of antiphospholipid syndrome cross-react with oxidized low-density lipoprotein. *Arthritis Rheum.*, **38**, 1382–1388.

Monestier, M., Kandiah, D.A., Kouts, S., Novick, K.E., Ong, G.L., Radic, M.Z., and Krilis, S.A. (1996). Monoclonal antibodies from NZW × BXSB F_1 mice to b_2-glycoprotein I and cardiolipin. Species specificity and charge-dependent binding. *J. Immunol.*, **156**, 2631–2641.

Nicolo, D., Goldman, B.I., and Monestier, M. (2003). Passive administration of antiphospholipid antibody reduces atherosclerosis in LDLR–/– mice. *Arthritis Rheum.*, **48**, 2974–2978.

Nicolo, D. and Monestier, M. (2004). Antiphospholipid antibodies and atherosclerosis. *Clin. Immunol.*, **112**, 183–189.

Palinski, W., Horkko, S., Miller, E., Steinbrecher, U.P., Powell, H.C., Curtiss, L.K., and Witztum, J. L. (1996). Cloning of monoclonal autoantibodies to epitopes of oxidized lipoproteins from apolipoprotein E-deficient mice. Demonstration of epitopes of oxidized low density lipoprotein in human plasma. *J. Clin. Invest.*, **98**, 800–814.

Palinski, W., Miller, E., and Witztum, J.L. (1995a). Immunization of low density lipoprotein (LDL) receptor-deficient rabbits with homologous malondialdehyde-modified LDL reduces atherogenesis. *Proc. Natl. Acad. Sci. USA*, **92**, 821–825.

Palinski, W., Tangirala, R.K., Miller, E., Young, S.G., and Witztum, J.L. (1995b). Increased autoantibody titers against epitopes of oxidized LDL in LDL receptor-deficient mice with increased atherosclerosis. *Arterioscler. Thromb. Vasc. Biol.*, **15**, 1569–1576.

Palinski, W. and Witztum, J.L. (2000). Immune responses to oxidative neoepitopes on LDL and phospholipids modulate the development of atherosclerosis. *J. Intern. Med.*, **247**, 371–380.

Palinski, W., Yla-Herttuala, S., Rosenfeld, M.E., Butler, S.W., Socher, S.A., Parthasarathy, S., Curtiss, L.K., and Witztum, J.L. (1990). Antisera and monoclonal antibodies specific for epitopes generated during oxidative modification of low density lipoprotein. *Arteriosclerosis*, **10**, 325–335.

Perrin-Cocon, L., Coutant, F., Agaugue, S., Deforges, S., Andre, P., and Lotteau, V. (2001). Oxidized low-density lipoprotein promotes mature dendritic cell transition from differentiating monocyte. *J. Immunol.*, **167**, 3785–3791.

Pinderski, L.J., Fischbein, M.P., Subbanagounder, G., Fishbein, M.C., Kubo, N., Cheroutre, H., Curtiss, L.K., Berliner, J.A., and Boisvert, W.A. (2002). Overexpression of interleukin10 by activated T lymphocytes inhibits atherosclerosis in LDL receptor deficient mice by altering lymphocyte and macrophage phenotype. *Circ. Res.*, **90**, 1064–1071.

Rader, D.J. and FitzGerald, G.A. (1998). State of the art: Atherosclerosis in a limited edition. *Nat. Med.*, **4**, 899–900.

Rajavashisth, T., Qiao, J., Tripathi, S., Tripathi, J., Mishra, N., Hua, M., Wang, X.P., Loussararian, A., Clinton, S., Libby, P., and Lusis A.J. (1998). Heterozygous osteopetrotic(op) mutation reduces atherosclerosis in LDL receptor deficient mice. *J. Clin. Invest.*, **101**, 2702–2710.

Reddel, S.W., Wang, Y.X., Sheng, Y.H., and Krilis, S.A. (2000). Epitope studies with anti-beta 2-glycoprotein I antibodies from autoantibody and immunized sources. *J. Autoimmun.*, **15**, 91–96.

Romero, F.I., Khamashta, M.A., and Hughes, G.R.V. (2000). Lipoprotein(a) oxidation and autoantibodies; a new path in atherothrombosis. *Lupus*, **9**, 206–209.

Rose, N. and Afanasyeva, M. (2003). Autoimmunity: Busting the atherosclerotic plaque. *Nat. Med.*, **9**, 641–642.

Roselaar, S.E., Kakkanathu, P.X., and Daugherty, A. (1996). Lymphocyte populations in atherosclerotic lesions of apoE–/– and LDL receptor–/– mice. Decreasing density with disease progression. *Arterioscler. Thromb. Vasc. Biol.*, **16**, 1013–1018.

Ross, R. (1995). Cell biology of atherosclerosis. *Annu. Rev. Physiol.*, **57**, 791–804.

Roubey, R.A.S. (1996). Immunology of the antiphospholipid antibody syndrome. *Arthritis Rheum.*, **39**, 1444–1454.

Sanan, D.A., Newland, D.L., Tao, R., Marcovina, S., Wang, J., Mooser, V., Hammer, R.E., and Hobbs, H.H. (1998). Low density lipoprotein receptor-negative mice expressing human apolipoprotein B-100 develop complex atherosclerotic lesions on a chow diet: No accentuation by apolipoprotein(a). *Proc. Natl. Acad. Sci. USA*, **95**, 4544–4549.

Sata, M. and Walsh, K. (1998). Oxidized LDL activates fas-mediated endothelial cell apoptosis. *J. Clin. Invest.*, **102**, 1682–1689.

Shaw, P.X., Goodyear, C.S., Chang, M.K., Witztum, J.L., and Silverman, G.J. (2003). The autoreactivity of antiphosphorylcholine antibodies for atherosclerosis associated neo-antigens and apoptotic cells. *J. Immunol.*, **170**, 6151–6157.

Shaw, P.X., Horkko, S., Chang, M.K., Curtiss, L.K., Palinski, W., Silverman, G.J., and Witztum, J.L. (2000). Natural antibodies with the T15 idiotype may act in atherosclerosis, apoptotic clearance, and protective immunity. *J. Clin. Invest.*, **105**, 1731–1740.

Shih, P.T., Elices, M.J., Fang, Z.T., Ugarova, T.P., Strahl, D., Territo, M.C., Frank, J.S., Kovach, N.L., Cabanas, C., Berliner, J.A., and Vora, D.K. (1999). Minimally modified low density lipoprotein induces monocyte adhesion to endothelial connecting segment-1 by activating alpha1 integrin. *J. Clin. Invest.*, **103**, 613–625.

Skalen, K., Gustaffson, M., Rydberg, E.K., Hulten, L.M., Wiklund, O., Innerarity, T.L., and Boren, J. (2002). Subendothelial retention of atherogenic lipoproteins in early atherosclerosis. *Nature*, **417**, 750–754.

Song, L., Leung, C., and Schindler, C. (2001). Lymphocytes are important in early atherosclerosis. *J. Clin. Invest.*, **108**, 251–259.

Steinberg, D. (1991). Antioxidants and atherosclerosis. A current assessment. *Circulation*, **84**, 1420–1425.

Steinberg, D. (1996). Oxidized low density lipoprotein – An extreme example of lipoprotein heterogeneity. *Isr. J. Med. Sci.*, 32, 469–472.

Terpstra, V., Kondratenko, N., and Steinberg, D. (1997). Macrophages lacking scavenger receptor A show a decrease in binding and uptake of acetylated low-density lipoprotein and of apoptotic thymocytes, but not of oxidatively damaged red blood cells. *Proc. Natl. Acad. Sci. USA*, **94**, 8127–8131.

Thiagarajan, P. (2001). Atherosclerosis, autoimmunity, and systemic lupus erythematosus. *Circulation*, **104**, 1876–1877.

Urowitz, M.B. and Gladman, D.D. (2000). Accelerated atheroma in lupus-background. *Lupus*, **9**, 161–165.

Vaarala, O. (2000). Autoantibodies to modified LDLs and other phospholipid-protein complexes as markers of cardiovascular diseases. *J. Intern. Med.*, **247**, 381–384.

Varadhachary, A.S., Monestier, M., and Salgame, P. (2001). Reciprocal induction of IL-10 and IL-12 from macrophages by low density lipoproteins and its oxidized forms. *Cell. Immunol.*, **213**, 45–51.

Wick, G., Schett, G., Amberger, A., Kleindienst, R., and Xu, Q. (1995). Is atherosclerosis an immunologically mediated disease? *Immunol. Today*, **16**, 27–33.

Zhou, X., Caligiuri, G., Hamsten, A., Lefvert, A.K., and Hansson, G.K. (2001). LDL immunization induces T-cell-dependent antibody formation and protection against atherosclerosis. *Arterioscler. Thromb. Vasc. Biol.*, **21**, 108–114.

Zibara, K., Malaud, E., and McGregor, J.L. (2002). CD36 mRNA and protein expression levels are significantly increased in the heart and testis of apoE deficient mice in comparison to wild type(C57BL/6). *J. Biomed. Biotechnol.*, **2**, 14–21.

28
Antigen-Specific Regulation of Autoimmunity

Amy E. Juedes and Matthias G. von Herrath

1. Introduction

Autoreactive T cells that escape negative selection exist in the periphery of normal individuals. Activation of these T cells can result in autoimmunity. Organ-specific autoimmune diseases, including multiple sclerosis (MS), and its animal model experimental autoimmune encephalomyelitis (EAE), type 1 diabetes (T1D), and rheumatoid arthritis (RA), are thought to be caused by autoaggressive T cell responses against self-antigens. Autoaggressive CD4 and CD8 T cells that mediate destruction during autoimmunity are usually of the Th1/Tc1 type, producing cytokines like interferon (IFN)-γ and tumor necrosis factor (TNF)-α. In contrast, regulation and sometimes protection from autoimmunity is associated with Th2/Th3 cytokines like IL-4, IL-10, and TGF-β. One of the promising methods developed recently for treating autoimmune disease involves vaccination with autoantigens in a tolerogenic fashion (nasal administration, oral feeding, and DNA vaccination). This is thought to be protective via induction of regulatory T cells (Tregs) that produce anti-inflammatory Th2/Th3 cytokines. When Tregs are activated by local antigen-presenting cells (APCs), they can suppress autoaggressive T cells specific for other antigens, a process known as bystander suppression. The mechanism of suppression may involve a combination of anti-inflammatory cytokine production or cell–cell contact that downmodulates APC function, or possibly acts directly on autoaggressive T cells themselves. Thus, the induction of autoantigen-specific Tregs can enable the local dampening of autoimmune processes, even if the antigen specificities of the autoaggressive T cells are not known. This chapter focuses on factors that influence the induction of autoantigen-specific regulatory T cells as well as potential mechanisms involved in the protection by different Treg populations.

Amy E. Juedes and Matthias G. von Herrath • Division of Developmental Immunology, La Jolla Institute for Allergy and Immunology, San Diego, California 92121.

Molecular Autoimmunity: In commemoration of the 100th anniversary of the first description of human autoimmune disease, edited by Moncef Zouali. Springer Science+Business Media, Inc., New York, 2005.

2. Antigen-Specific Therapy

The concept of antigen-specific therapy for treating autoimmune disease involves administration of autoantigens via a tolerogenic route. For example, it is well known that delivery of antigens to mucosal surfaces can lead to systemic tolerance. Tolerance or suppression can be induced not only to the fed antigen itself, but also to newly encountered antigens through bystander suppression. This strategy has been exploited as a potential therapy to prevent autoimmunity by generating antigen-specific regulatory T cells that will traffic to the target organ and subsequently inhibit autoreactive T cell responses. The efficacy of such an approach has been demonstrated in animal models of T1D. Nasal or oral administration of insulin or glutamic acid decarboxylase (GAD) can protect mice from T1D in the nonobese diabetic (NOD) mouse model (Tian et al., 1996; Polanski et al., 1997; Ploix et al., 1998; Bergerot et al., 1999; Maron et al., 1999; Aspord and Thivolet, 2002), and also using a model triggered by lymphocytic choriomeningitis virus (LCMV) infection of mice expressing an LCMV antigen transgenically in the beta cells of the pancreas (RIP-LCMV model) (Homann et al., 1999). Likewise, in experimental EAE, oral or nasal administration of myelin basic protein (MBP) can protect from disease (Higgins and Weiner, 1988; Bai et al., 1997; Slavin et al., 2001). Finally, feeding with type II collagen can protect mice from collagen-induced arthritis (Nagler-Anderson et al., 1986; Myers et al., 2002).

Another promising avenue for autoantigen therapy involves the use of naked plasmid DNA immunization. This approach requires relatively small doses of plasmid-expressing autoantigens that are administered intramuscularly. This type of therapy has been successful in preventing T1D (Coon et al., 1999; Tisch et al., 2001; Urbanek-Ruiz et al., 2001) after vaccination with plasmids encoding insulin or GAD. Similarly, treatment with plasmids encoding myelin oligodendrocyte glycoprotein (MOG) can protect from EAE (Garren et al., 2001). Importantly, DNA vaccination is even successful in reversing ongoing disease (Garren et al., 2001).

3. Antigen-Induced Regulatory T cells

Vaccination with autoantigens via a tolerogenic route is thought to protect from autoimmune diseases through the induction of Tregs. Depending on the mode of vaccination, the antigen used, the timing and dosage, these Tregs variously produce regulatory cytokines such as IL-4, IL-10, and transforming growth factor (TGF)-β, and suppress autoaggressive T cells via bystander suppression. It is important to note that antigen-induced Tregs are probably distinct from another much-investigated subset of Tregs, the naturally occurring $CD4^+CD25^+$ T cells. $CD4^+CD25^+$ Tregs are generated in the thymus possibly by active positive selection on autoantigens (Shevach et al., 2001; Apostolou et al., 2002). It is clear that $CD4^+CD25^+$ T cells play an important role in normal immune homeostasis, as depletion of $CD4^+CD25^+$ T cells by thymectomy results in the spontaneous development of various autoimmune diseases (Sakaguchi et al., 1985). *In vitro*, $CD4^+CD25^+$ T cells can suppress the proliferation and cytokine production of naive T cells (Thornton and Shevach, 1998; Jonuleit et al., 2001; Levings et al., 2001; Piccirillo and Shevach, 2001). While the mechanism of suppression is not known, it requires cell–cell contact.

One distinction that can be drawn between CD4$^+$CD25$^+$ Tregs and antigen-induced Tregs is the baseline level that they are present at. CD4$^+$CD25$^+$ Tregs exist in unmanipulated hosts (i.e., "naturally" occurring), while autoantigen-specific Tregs will arise only after immunization with antigens. Some studies have reported an increase in CD4$^+$CD25$^+$ T cells after immunization with autoantigens, and correlated the presence of these cells with protection from autoimmune disease (Mukherjee et al., 2003). However, these cells are likely still distinct from naturally occurring CD4$^+$CD25$^+$ Tregs, as CD25 (the IL-2R) can also be expressed on activated cells. More recently, very interesting studies have provided strong evidence that the forkhead transcription factor FoxP3 might be a better phenotypic and functional marker than CD25 for naturally occurring Tregs (Fontenot et al., 2003; Khattri et al., 2003).

It is not presently known whether autoantigen-induced Tregs represent a distinct subset of T cells, or whether they are simply T cells with specific effector functions. Currently, these cells are defined primarily according to their protective functions. For example, there is no evidence for the expression of unique genes by antigen-induced Tregs such as FoxP3 in the case of the naturally occurring Tregs. In addition, unlike CD4$^+$CD25$^+$ naturally occurring Tregs, which seem to suppress many immune response independently of antigen specificity, antigen-induced Tregs only suppress certain immune responses via cytokine-mediated bystander suppression. Thus, it seems that antigen-induced Tregs may be defined as T cells that can be induced after immunization with autoantigens via a tolerogenic route, and protect against autoaggressive or inflammatory immune responses via their ability to produce anti-inflammatory mediators such as IL-4, IL-10, and TGF-β. At this point, it is unclear how many different subspecies of autoantigen-induced Tregs we have to expect. However, the major players are Th2-like cells (producing IL-4), Th3 cells (producing TGF-β), Tr1 cells (producing IL-10), or even cells producing combinations of these cytokines (Table 28.1).

For example, in T1D, oral insulin feeding results in a population of Tregs that produce predominantly IL-4 (Th2-like) (Ploix et al., 1998; Bergerot et al., 1999; Homann et al., 1999). The critical role of IL-4 in this system is demonstrated by the fact that mice deficient in IL-4 or STAT6 cannot be protected by oral feeding (Homann et al., 1999). In contrast, during EAE, oral administration

Table 28.1. Examples of Autoantigen-Specific Regulatory T Cells

Predominant cytokine	T cell type	Disease	Route of immunization
TGF-β	Th3	EAE	Oral[a]
IL-4	Th2	IDDM, EAE	Oral, DNA vaccine[b]
IL-10	Tr1	IDDM, EAE	Intranasal[c]

[a](Chen et al., 1994); [b](Homann et al., 1999; Garren et al., 2001; Tisch et al., 2001); [c](Burkhart et al., 1999; Slavin et al., 2001). EAE, experimental autoimmune encephalomyelitis; IDDM, insulin-dependent diabetes mellitus.

of MBP protects mice by inducing regulatory T cells that produce predominantly TGF-β (Chen et al., 1994; Slavin et al., 2001). Nasal administration of antigens, on the other hand, is typically associated with predominantly IL-10-producing regulatory T cells (Burkhart et al., 1999; Slavin et al., 2001; Aspord and Thivolet, 2002). These studies demonstrate that there are several factors that may influence the cytokine production by Tregs. In addition, depending on the particular disease, certain cytokines may be more or less effective in terms of protection.

4. Factors Involved in Treg Induction

The mechanism by which regulatory T cells are induced will likely have an impact on the phenotype of these cells. Regulatory T cells are thought to be primed by antigen-presenting dendritic cells (DCs). Recent evidence has suggested that many subsets of DCs exist that vary in their surface molecule expression, cytokine production, as well as the class of T cell response they are capable of inducing (Moser and Murphy, 2000). Thus, one factor that may influence the outcome of therapeutic vaccination with autoantigens is the local subsets of DCs present at the site of immunization. For example, subsets of DCs present in the Peyer's patches can preferentially induce IL-4- and IL-10-secreting T cells (Iwasaki and Kelsall, 1999), as compared to DCs isolated from the spleen. Further, studies by Akbari et al. (2001) demonstrate that pulmonary DCs preferentially produce IL-10, and induce IL-4- and IL-10-producing T cells after intranasal immunization. In contrast, DCs isolated from the gut produce TGF-β and induce IL-4-, IL-10-, and TGF-β-producing T cells after oral feeding of antigen.

It may also be possible to influence the efficacy of therapeutic vaccination with autoantigens by using cytokines as adjuvants. Administration of IL-4 or IL-10 along with oral MBP enhances the differentiation of regulatory T cells and results in better protection from EAE (Inobe et al., 1998; Slavin et al., 2001). Likewise, DNA vaccine studies have demonstrated that the inclusion of plasmids expressing IL-4 or IL-10 can enhance protection (Weaver et al., 2001; Wolfe et al., 2002; Seifarth et al., 2003). Presumably, administration of cytokines acts at the site of priming of Treg and may influence characteristics of local DCs. Along this line, studies with *in vitro* cultured DCs have demonstrated that cytokine-modulated DCs may be able to preferentially induce Tregs. One recent study demonstrated that culture with IL-10 induces the differentiation of a distinct subset of DCs that can induce tolerance through the differentiation of IL-10-secreting regulatory T cells *in vitro* and *in vivo* (Wakkach et al., 2003).

Finally, another general consideration when performing antigen-specific therapy to treat autoimmune disease is that vaccination with autoantigens is not always protective, and in fact can sometimes exacerbate disease (Weaver et al., 2001; Wolfe et al., 2002). Disease exacerbation is usually associated with the induction of a Th1 response to the autoantigen, as opposed to a regulatory Th2/Th3 response. There are likely to be multiple factors involved in determining whether vaccination results in a protective or exacerbating response. For example, the timing of vaccination is likely to be important, and in fact most therapies are started very early or even before initiation of disease. The timing for prevention of autoimmune disease is probably linked to the status of the autoimmune repertoire at the time of therapeutic intervention. Studies in the RIP-LCMV mouse model of diabetes have demonstrated that it is very difficult to "switch" the class of the response if a very strong autoaggressive T cell for a particular autoantigen already exists in the host (Coon et al., 1999). The induction of Tregs might be further influenced by the expression level of a particular antigen in the host. This is supported by studies demonstrating that transgenic overexpression of GAD in islets resulted in exacerbation after GAD-specific therapy, as opposed to the protection observed in mice that only expressed endogenous levels of GAD (Wolfe et al., 2002). Again, cytokine adjuvants may also be effective in preventing some of these deleterious effects, as inclusion of IL-4 along with GAD could mediate protection (Wolfe et al., 2002).

4.1. Mechanisms of Protection

Treating autoimmune disease by autoantigen immunization in a tolerogenic fashion involves many variables that can be optimized to generate a population of Tregs with protective capacity. However, a distinct issue is the mechanism by which such regulatory T cells, once induced, mediate their protective effect. This is thought to occur via bystander suppression, a process mediated by the production of anti-inflammatory cytokines such as IL-4, IL-10, and TGF-β.

While regulatory T cells are most likely primed in lymphoid structures draining the site of vaccination, they must traffic to the site of inflammation to exert their protective effects. This presumably occurs in the target organ itself or its draining lymph node, where the autoantigen in question is being presented. Evidence from the RIP-LCMV model of T1D suggests that regulation occurs in draining lymph nodes of the target organ as opposed to the target organ. This was demonstrated after transfer of purified CFSE-labeled Tregs that proliferate specifically in the pancreatic draining lymph node, and not in the pancreas (Homann et al., 1999). Thus, the level of a particular autoantigen in the draining lymph node will critically influence the effectiveness of protection via bystander suppression. In addition, the timing of presentation of the "regulatory" autoantigen will have to coincide with the presence of autoaggressive T cells in the same location. This is particularly relevant because responses to autoantigens evolve during the course of autoimmune disease. Numerous studies have demonstrated the existance of epitope spreading, a process whereby the specificity of the immune response spreads to include self-epitopes other than those used to initiate the inflammatory process (Vanderlugt and Miller, 2002). The development of the autoreactive T cell repertoire correlates with the presentation of autoantigens by local APCs in a temporal fashion (Katz-Levy et al., 2000). Thus, it is likely that different autoantigens may be presented at different times in the draining lymph node, and this fact may at least partially explain why certain autoantigens are much more effective in terms of protection than others.

Bystander suppression is thought to be mediated by the production of anti-inflammatory cytokines. As discussed above, depending on the immunization regimen, regulatory T cells populations have been variously reported to produce cytokines like IL-4, IL-10, and TGF-β. It is not clear at this point exactly which cytokines are required to be produced by the Tregs in order to mediate suppression. This might be highly dependent on the antigen and disease model under investigation, as well as on the timing of therapeutic intervention. However, many studies have investigated the role of particular cytokines in various models of autoimmune disease.

IL-4 has been demonstrated to be essential for the protection mediated by DNA vaccination with GAD (Tisch et al., 2001) or by oral insulin feeding (Homann et al., 1999), as mice deficient in IL-4 are not protected. While it is not clear from these studies whether IL-4 is necessary as an effector cytokine or simply is required for full Th2 differentiation, other studies have demonstrated that the presence of IL-4 in the target organ can be very effective in preventing T1D. Transgenic mice that express IL-4 in the beta cells are protected from LCMV-induced diabetes or spontaneous diabetes on the NOD background (Mueller et al., 1996; King et al., 2001).

Another potent anti-inflammatory cytokine, IL-10, can also suppress autoimmune disease. Transgenic IL-10 protects from EAE (Cua et al., 1999), and has also been demonstrated to be important in protection mediated by regulatory T cells induced after intranasal or IV administration of MBP (Burkhart et al., 1999; Wildbaum et al., 2002). However, the role of IL-10 may be dependant on the disease. Surprisingly, NOD mice that transgenically express IL-10 in their beta cells develop enhanced diabetes (Wogensen et al., 1994). TGF-β is another cytokine that has been reported to be induced after oral feeding with islet antigens. While TGF-β has well-known immune suppressor functions (Gorelik and Flavell, 2002), its role in protection mediated by antigen-induced Tregs remains unknown.

Finally, in addition to cytokines, a recent study has also demonstrated the importance of cell–cell contact at the level of the APC in mediating bystander suppression (Alpan et al., 2004). This study was performed using a model of oral tolerance in which mice are fed one antigen, and later simultaneously immunized with the oral antigen and another unrelated antigen in complete Freund adjuvant (CFA). Regulatory T cells specific for the fed antigen are then able to suppress the response to the unrelated antigen via bystander suppression. This study demonstrated that the regulatory T cells specific for the fed antigen require the production of IL-4 and IL-10 to mediate suppression, but interestingly also require cell–cell contact. Specifically, the regulatory T cell must have contact with the same APC that also presents the other unrelated antigen (Alpan et al., 2004). Thus, cytokines by themselves are not sufficient to replicate the *in vivo* effect of regulatory T cells, which also require cell–cell contact to mediate their protective effects (Figure 28.1).

4.2. Application to Human Disease

While promising results have so far been obtained in animal models using antigen-specific therapies, the results in human clinical trials have been somewhat disappointing. Oral administration of insulin to treat T1D patients (Chaillous et al., 2000) or MBP to MS patients (Weiner et al., 1993; Fukaura

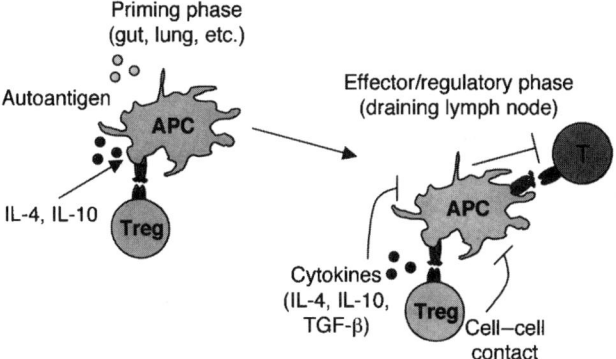

Figure 28.1. The mechanism of protection by regulatory T cells involves both cytokines and cell–cell contact. Protective regulatory T cells can be induced after oral feeding or nasal immunization with autoantigens. Administration of cytokines like IL-4 and IL-10 enhances protection, presumably by acting on antigen-presenting DCs that prime Tregs. After being primed, Tregs come to the target organ draining lymph node, and encounter their antigen presented by a local DC. Bystander suppression by Tregs requires both the action of cytokines and cell–cell contact with the same antigen-presenting cell (APC) that encounters the autoaggressive T cells.

et al., 1996) has not resulted in any significant clinical benefit. However, oral tolerance can be induced in humans, as has been demonstrated after subjects were fed keyhole limpet hemacyanin (KLH), resulting in inhibition of T cell recall responses *in vitro* (Husby *et al.*, 1994). In addition, some clinical benefit has been obtained in another autoimmune disease, RA, after feeding patients type II collagen (Trentham *et al.*, 1993; Barnett *et al.*, 1998). As discussed above, the potential of autoantigen-specific therapy is likely to be influenced by several factors including the dose, route of administration, timing of vaccination, and antigen selection, and these factors may need to be optimized for successful treatment of human disease.

One issue that might be very important, especially for human disease, is the status of the autoaggressive repertoire at the time of therapy. As discussed above, it may be very difficult to induce a regulatory response if autoaggressive T cells are already at a high level. This may be the case in humans in particular, where therapy often cannot be initiated until diagnosis of an already ongoing disease. Therefore, one avenue for achieving success with autoantigen-specific therapy in humans may involve the use of more systemic modulators, such as cytokine adjuvants. Additionally, systemic modulators may be useful in "resetting" the aggressive repertoire. For example, one promising new therapy for treating T1D involves the use of nonmitogenic anti-CD3 antibody. Anti-CD3 antibodies can reverse diabetes in animal models (Chatenoud *et al.*, 1997; von Herrath *et al.*, 2002). This treatment also has clinical benefit in humans as well, improving insulin production and metabolic control during the first year (Herold *et al.*, 2002). It may be possible to use therapies like anti-CD3 antibodies that potentially downregulate autoimmune aggressors, combined with antigen-specific therapies to induce a regulatory population that ensures continued suppression of disease. In this regard, it will also be critical to develop tools that enable the tracking of both autoaggressive and T regulatory populations in the blood to evaluate the effectiveness of the therapeutic intervention. So far, this has been difficult, even in animal models, where regulatory T cells are frequently difficult to identify, even in the peripheral lymphoid organs.

5. Conclusions

Antigen-specific therapy is a promising avenue for treating autoimmune diseases. The protection after vaccination with autoantigens in a tolerogenic manner is mediated by the induction of regulatory T cell populations. The exact nature of these Tregs is variable, depending on the type of therapy, but in general they mediate their protective effect by bystander suppression. This process involves the production of anti-inflammatory cytokines such as IL-4, IL-10, and TGF-β, as well as cell–cell contact with the same APC that encounters the autoaggressive T cells. The effectiveness of antigen-specific therapy is influenced by many factors such as antigen selection, and the route and timing of immunization. In addition, the exact dynamics of the T cell repertoire in the host, as well as the particular disease in question, needs to be considered when applying such a therapy. So far, in animal models, success has been achieved in the prevention of autoimmune diseases like T1D and EAE after vaccination with autoantigens via a mucosal or nasal route, and also after immunization with naked plasmid DNA expressing autoantigens. In contrast, it has, as of yet, been more difficult to see an immediate benefit in human clinical trials. However, by applying strategies such as systemic therapies aimed at reducing the autoaggressive T cells with antigen-specific therapy, the results may be improved. This type of strategy, along with the development of techniques to track autoaggressive and regulatory T cells in the blood, will likely make antigen-specific therapy an effective means of treating autoimmune diseases.

References

Akbari, O., Dekruyff, R.H., and Umetsu, D.T. (2001). Pulmunary dendritic cells producing IL-10 mediate tolerance induced by respiratory exposure to antigen. *Nat. Immunol.*, **2**, 725–731.

Alpan, O., Bachelder, E., Isil, E., Arnheiter, H., and Matzinger, P. (2004). 'Educated' dendritic cells act as messengers from memory to naive T helper cells. *Nat. Immunol.*, **5**, 615–622.

Apostolou, I., Sarukhan, A., Klein, L., and von Boehmer, H. (2002). Origin of regulatory T cells with known specificity for antigen. *Nat. Immunol.*, **3**, 756–763.

Aspord, C. and Thivolet, C. (2002). Nasal administration of CTB-insulin induces active tolerance against autoimmune diabetes in non-obese diabetic (NOD) mice. *Clin. Exp. Immunol.*, **130**, 204–211.

Bai, X.F., Shi, F.D., Xiao, B.G., Li, H.L., van der Meide, P.H., and Link, H. (1997). Nasal administration of myelin basic protein prevents relapsing experimental autoimmune encephalomyelitis in DA rats by activating regulatory cells expressing IL-4 and TGF-beta mRNA. *J. Neuroimmunol.*, **80**, 65–75.

Barnett, M.L., Kremer, J.M., St Clair, E.W., Clegg, D.O., Furst, D., Weisman, M., Fletcher, M.J., Chasan-Taber, S., Finger, E., Morales, A., Le, C.H., and Trentham, D.E. (1998). Treatment of rheumatoid arthritis with oral type II collagen. Results of a multicenter, double-blind, placebo-controlled trial. *Arthritis Rheum.*, **41**, 290–297.

Bergerot, I., Arreaza, G.A., Cameron, M.J., Burdick, M.D., Strieter, R.M., Chensue, S.W., Chakrabarti, S., and Delovitch, T.L. (1999). Insulin B-chain reactive CD4+ regulatory T-cells induced by oral insulin treatment protect from type 1 diabetes by blocking the cytokine secretion and pancreatic infiltration of diabetogenic effector T-cells. *Diabetes*, **48**, 1720–1729.

Burkhart, C., Liu, G.Y., Anderton, S.M., Metzler, B., and Wraith, D.C. (1999). Peptide-induced T cell regulation of experimental autoimmune encephalomyelitis: A role for IL-10. *Int. Immunol.*, **11**, 1625–1634.

Chaillous, L., Lefevre, H., Thivolet, C., Boitard, C., Lahlou, N., Atlan-Gepner, C., Bouhanick, B., Mogenet, A., Nicolino, M., Carel, J.C., Lecomte, P., Marechaud, R., Bougneres, P., Charbonnel, B., and Sai, P. (2000). Oral insulin administration and residual beta-cell function in recent-onset type 1 diabetes: A multicentre randomised controlled trial. Diabete Insuline Orale group. *Lancet*, **356**, 545–549.

Chatenoud, L., Primo, J., and Bach, J.F. (1997). CD3 antibody-induced dominant self tolerance in overtly diabetic NOD mice. *J. Immunol.*, **158**, 2947–2954.

Chen, Y., Kuchroo, V.K., Inobe, J., Hafler, D.A., and Weiner, H.L. (1994). Regulatory T cell clones induced by oral tolerance: Suppression of autoimmune encephalomyelitis. *Science*, **265**, 1237–1240.

Coon, B., An, L.L., Whitton, J.L., and von Herrath, M.G. (1999). DNA immunization to prevent autoimmune diabetes. *J. Clin. Invest.*, **104**, 189–194.

Cua, D.J., Groux, H., Hinton, D.R., Stohlman, S.A., and Coffman, R.L. (1999). Transgenic interleukin 10 prevents induction of experimental autoimmune encephalomyelitis. *J. Exp. Med.*, **189**, 1005–1010.

Fontenot, J.D., Gavin, M.A., and Rudensky, A.Y. (2003). Foxp3 programs the development and function of CD4+CD25+ regulatory T cells. *Nat. Immunol.*, **4**, 330–336.

Fukaura, H., Kent, S.C., Pietrusewicz, M.J., Khoury, S.J., Weiner, H.L., and Hafler, D.A. (1996). Induction of circulating myelin basic protein and proteolipid protein-specific transforming growth factor-beta1-secreting Th3 T cells by oral administration of myelin in multiple sclerosis patients. *J. Clin. Invest.*, **98**, 70–77.

Garren, H., Ruiz, P.J., Watkins, T.A., Fontoura, P., Nguyen, L.T., Estline, E.R., Hirschberg, D.L., and Steinman, L. (2001). Combination of gene delivery and DNA vaccination to protect from and reverse Th1 autoimmune disease via deviation to the Th2 pathway. *Immunity*, **15**, 15–22.

Gorelik, L. and Flavell, R.A. (2002). Transforming growth factor-beta in T-cell biology. *Nat. Rev. Immunol.*, **2**, 46–53.

Herold, K.C., Hagopian, W., Auger, J.A., Poumian-Ruiz, E., Taylor, L., Donaldson, D., Gitelman, S.E., Harlan, D.M., Xu, D., Zivin, R.A., and Bluestone, J.A. (2002). Anti-CD3 monoclonal antibody in new-onset type 1 diabetes mellitus. *N. Engl. J. Med.*, **346**, 1692–1698.

Higgins, P.J. and Weiner, H.L. (1988). Suppression of experimental autoimmune encephalomyelitis by oral administration of myelin basic protein and its fragments. *J. Immunol.*, **140**, 440–445.

Homann, D., Holz, A., Bot, A., Coon, B., Wolfe, T., Petersen, J., Dyrberg, T.P., Grusby, M.J., and von Herrath, M.G. (1999). Autoreactive CD4+ T cells protect from autoimmune diabetes via bystander suppression using the IL-4/Stat6 pathway. *Immunity*, **11**, 463–472.

Husby, S., Mestecky, J., Moldoveanu, Z., Holland, S., and Elson, C.O. (1994). Oral tolerance in humans. T cell but not B cell tolerance after antigen feeding. *J. Immunol.*, **152**, 4663–4670.

Inobe, J., Slavin, A.J., Komagata, Y., Chen, Y., Liu, L., and Weiner, H.L. (1998). IL-4 is a differentiation factor for transforming growth factor-beta secreting Th3 cells and oral administration of IL-4 enhances oral tolerance in experimental allergic encephalomyelitis. *Eur. J. Immunol.*, **28**, 2780–2790.

Iwasaki, A. and Kelsall, B.L. (1999). Freshly isolated Peyer's patch, but not spleen, dendritic cells produce interleukin 10 and induce the differentiation of T helper type 2 cells. *J. Exp. Med.*, **190**, 229–239.

Jonuleit, H., Schmitt, E., Stassen, M., Tuettenberg, A., Knop, J., and Enk, A.H. (2001). Identification and functional characterization of human CD4(+)CD25(+) T cells with regulatory properties isolated from peripheral blood. *J. Exp. Med.*, **193**, 1285–1294.

Katz-Levy, Y., Neville, K.L., Padilla, J., Rahbe, S., Begolka, W.S., Girvin, A.M., Olson, J.K., Vanderlugt, C.L., and Miller, S.D. (2000). Temporal development of autoreactive Th1 responses and endogenous presentation of self myelin epitopes by central nervous system-resident APCs in Theiler's virus-infected mice. *J. Immunol.*, **165**, 5304–5314.

Khattri, R., Cox, T., Yasayko, S.A., and Ramsdell, F. (2003). An essential role for Scurfin in CD4+CD25+ T regulatory cells. *Nat. Immunol.*, **4**, 337–342.

King, C., Mueller Hoenger, R., Malo Cleary, M., Murali-Krishna, K., Ahmed, R., King, E., and Sarvetnick, N. (2001). Interleukin-4 acts at the locus of the antigen-presenting dendritic cell to counter-regulate cytotoxic CD8+ T-cell responses. *Nat. Med.*, **7**, 206–214.

Levings, M.K., Sangregorio, R., and Roncarolo, M.G. (2001). Human CD25(+)CD4(+) T regulatory cells suppress naive and memory T cell proliferation and can be expanded in vitro without loss of function. *J. Exp. Med.*, **193**, 1295–1302.

Maron, R., Melican, N.S., and Weiner, H.L. (1999). Regulatory Th2-type T cell lines against insulin and GAD peptides derived from orally- and nasally-treated NOD mice suppress diabetes. *J. Autoimmun.*, **12**, 251–258.

Moser, M. and Murphy, K.M. (2000). Dendritic cell regulation of TH1-TH2 development. *Nat. Immunol.*, **1**, 199–205.

Mueller, R., Krahl, T., and Sarvetnick, N. (1996). Pancreatic expression of interleukin-4 abrogates insulitis and autoimmune diabetes in nonobese diabetic (NOD) mice. *J. Exp. Med.*, **184**, 1093–1099.

Mukherjee, R., Chaturvedi, P., Qin, H.Y., and Singh, B. (2003). CD4(+)CD25(+) regulatory T cells generated in response to insulin B:9–23 peptide prevent adoptive transfer of diabetes by diabetogenic T cells. *J. Autoimmun.*, **21**, 221–237.

Myers, L.K., Sakurai, Y., Tang, B., He, X., Rosloniec, E.F., Stuart, J.M., and Kang, A.H. (2002). Peptide-induced suppression of collagen-induced arthritis in HLA-DR1 transgenic mice. *Arthritis Rheum.*, **46**, 3369–3377.

Nagler-Anderson, C., Bober, L.A., Robinson, M.E., Siskind, G.W., and Thorbecke, G.J. (1986). Suppression of type II collagen-induced arthritis by intragastric administration of soluble type II collagen. *Proc. Natl. Acad. Sci. USA*, **83**, 7443–7446.

Piccirillo, C.A. and Shevach, E.M. (2001). Cutting edge: Control of CD8+ T cell activation by CD4+CD25+ immunoregulatory cells. *J. Immunol.*, **167**, 1137–1140.

Ploix, C., Bergerot, I., Fabien, N., Perche, S., Moulin, V., and Thivolet, C. (1998). Protection against autoimmune diabetes with oral insulin is associated with the presence of IL-4 type 2 T-cells in the pancreas and pancreatic lymph nodes. *Diabetes*, **47**, 39–44.

Polanski, M., Melican, N.S., Zhang, J., and Weiner, H.L. (1997). Oral administration of the immunodominant B-chain of insulin reduces diabetes in a co-transfer model of diabetes in the NOD mouse and is associated with a switch from Th1 to Th2 cytokines. *J. Autoimmun.*, **10**, 339–346.

Sakaguchi, S., Fukuma, K., Kuribayashi, K., and Masuda, T. (1985). Organ-specific autoimmune diseases induced in mice by elimination of T cell subset. I. Evidence for the active participation of T cells in natural self-tolerance; deficit of a T cell subset as a possible cause of autoimmune disease. *J. Exp. Med.*, **161**, 72–87.

Seifarth, C., Pop, S., Liu, B., Wong, C.P., and Tisch, R. (2003). More stringent conditions of plasmid DNA vaccination are required to protect grafted versus endogenous islets in nonobese diabetic mice. *J. Immunol.*, **171**, 469–476.

Shevach, E.M., McHugh, R.S., Piccirillo, C.A., and Thornton, A.M. (2001). Control of T-cell activation by CD4+CD25+ suppressor T cells. *Immunol. Rev.*, **182**, 58–67.

Slavin, A.J., Maron, R., and Weiner, H.L. (2001). Mucosal administration of IL-10 enhances oral tolerance in autoimmune encephalomyelitis and diabetes. *Int. Immunol.*, **13**, 825–833.

Thornton, A.M. and Shevach, E.M. (1998). CD4+CD25+ immunoregulatory T cells suppress polyclonal T cell activation in vitro by inhibiting interleukin 2 production. *J. Exp. Med.*, **188**, 287–296.

Tian, J., Atkinson, M.A., Clare-Salzler, M., Herschenfeld, A., Forsthuber, T., Lehmann, P.V., and Kaufman, D.L. (1996). Nasal administration of glutamate decarboxylase (GAD65) peptides induces Th2 responses and prevents murine insulin-dependent diabetes. *J. Exp. Med.*, **183**, 1561–1567.

Tisch, R., Wang, B., Weaver, D.J., Liu, B., Bui, T., Arthos, J., and Serreze, D.V. (2001). Antigen-specific mediated suppression of beta cell autoimmunity by plasmid DNA vaccination. *J. Immunol.*, **166**, 2122–2132.

Trentham, D.E., Dynesius-Trentham, R.A., Orav, E.J., Combitchi, D., Lorenzo, C., Sewell, K.L., Hafler, D.A., and Weiner, H.L. (1993). Effects of oral administration of type II collagen on rheumatoid arthritis. *Science*, **261**, 1727–1730.

Urbanek-Ruiz, I., Ruiz, P.J., Paragas, V., Garren, H., Steinman, L., and Fathman, C.G. (2001). Immunization with DNA encoding an immunodominant peptide of insulin prevents diabetes in NOD mice. *Clin. Immunol.*, **100**, 164–171.

Vanderlugt, C.L. and Miller, S.D. (2002). Epitope spreading in immune-mediated diseases: Implications for immunotherapy. *Nat. Rev. Immunol.*, **2**, 85–95.

von Herrath, M.G., Coon, B., Wolfe, T., and Chatenoud, L. (2002). Nonmitogenic CD3 antibody reverses virally induced (rat insulin promoter-lymphocytic choriomeningitis virus) autoimmune diabetes without impeding viral clearance. *J. Immunol.*, **168**, 933–941.

Wakkach, A., Fournier, N., Brun, V., Breittmayer, J.P., Cottrez, F., and Groux, H. (2003). Characterization of dendritic cells that induce tolerance and T regulatory 1 cell differentiation in vivo. *Immunity*, **18**, 605–617.

Weaver, D.J., Jr., Liu, B., and Tisch, R. (2001). Plasmid DNAs encoding insulin and glutamic acid decarboxylase 65 have distinct effects on the progression of autoimmune diabetes in nonobese diabetic mice. *J. Immunol.*, **167**, 586–592.

Weiner, H.L., Mackin, G.A., Matsui, M., Orav, E.J., Khoury, S.J., Dawson, D.M., and Hafler, D.A. (1993). Double-blind pilot trial of oral tolerization with myelin antigens in multiple sclerosis. *Science*, **259**, 1321–1324.

Wildbaum, G., Netzer, N., and Karin, N. (2002). Plasmid DNA encoding IFN-gamma-inducible protein 10 redirects antigen-specific T cell polarization and suppresses experimental autoimmune encephalomyelitis. *J. Immunol.*, **168**, 5885–5892.

Wogensen, L., Lee, M.S., and Sarvetnick, N. (1994). Production of interleukin 10 by islet cells accelerates immune-mediated destruction of beta cells in nonobese diabetic mice. *J. Exp. Med.*, **179**, 1379–1384.

Wolfe, T., Bot, A., Hughes, A., Mohrle, U., Rodrigo, E., Jaume, J.C., Baekkeskov, S., and von Herrath, M. (2002). Endogenous expression levels of autoantigens influence success or failure of DNA immunizations to prevent type 1 diabetes: Addition of IL-4 increases safety. *Eur. J. Immunol.*, **32**, 113–121.

Index

A
Ablative therapy without HSCT, 358
  for aplastic anemia, 358
  pulse cyclophosphamide therapy, 358
Acantholysis, 128, 130–131, 136
  inducement by plasminogen activator, 13
Acetylcholine (ACh)
  ADCC as addition mechanism in pathogenic action of, 109
  receptor antibodies, 102, 109, 340
  reduction in the number of receptors, 151
Acetylcholine receptor
  anti-acetylcholine receptor (anti-AChR) antibodies, 151, 156
  intrathymic expression and immunopathogenesis of myasthenia gravis (MG), 151–160
    expression of neuromuscular AchRs by thymic cells, 153–157
    model to examine peripheral T cell entry and activation in thymus, 158–160
    role of thymus in MG pathogenesis, 152–153
    thymus and central immune tolerance, 157
    thymus and T cell trafficking, 157–158
  mutations/variations in the genes of ACR subunits, 102
  on keratinocytes, 130
Acquired autoimmune MG, 102
Acute myeloid leukemia (AML), 349
Acute phase response, 37
Acute post-streptococcal glomerulonephritis, 167–169
  Goodpasture's syndrome, 169–170
  lupus nephritis, 170–174
  other nephritogenic autoantibodies, 174–175
Acute post-streptococcal nephritis, 167
Acute rheumatic fever (ARF), 116
Amino acid sequence, 366, 369
Androgens, administration of, 184
Angiogenesis, 333
Antibodies

anti-α-actinin, 172
anti-acetylcholine receptor (anti-AChR), 151
anti-C1q, 171–172
anti-CD4+, 101
anti-DNA, 171–172, 174
anti-dsDNA, 183
anti-enolase, 175
anti-laminin, 172
anti-nuclear, 335
anti-nucleosome, 171, 174
anti-ribosomal P protein, 174
anti-Sm, 171, 270
  associated with arterial and venous thrombosis in SLE, 401
  pathogenic anti-DNA, 184
Antibody-dependent cell-mediated cytotoxicity (ADCC), 109
Anti-dsDNA autoantibody, 189
Antigen presenting cells (APCs), 8, 148, 407
  C3 deposition, 7
  nonprofessional, 348
  thymic, 157
Antigen (s)
  pathogen-related, 227
  planted, 165, 172
  S-, 366, 371, 374
  self-antigens, autoaggressive T cell responses against, 407
Antigen-specific regulation of autoimmunity, 407–414
  antigen-induced regulatory T cells, 408–410
  antigen-specific therapy, 408
  Treg induction, factors involved in, 410–414
    application to human disease, 412–414
    mechanisms of protection, 411–413
Anti-inflammatory cytokines, 36–37
Anti-oxLDL antibodies associated with arterial thrombosis, 401
Antiphospholipid antibody syndrome (APS), 400
Antiphospholipid syndrome, molecular pathogenesis and novel therapeutic targets, 360–370

Antiphospholipid syndrome (*Continued*)
  antiphospholipid antibodies and endothelial cells, 364–369
    activation of complement cascade and antiphospholipid antibodies, 368–369
    effects of aPL on endothelial cells, 364–368
    statins and antiphospholipid antibodies, 368
  antiphospholipid antibodies and platelets, 361–364
    effects of aPL on platelets *in vitro* and *in vivo*, 361–362
    hydroxychloroquine in aPL-mediated thrombosis, 362–363
    intracellular events in aPL-mediated platelet activation, 363–364
Apolipoprotein E (apoE), 394
Apoptosis
  absence of induction with etanercept, 334
  cell death by, 333
  contribution of TNFα, 336
  control of Fas ligand (CD95 ligand), 331
  defective apoptosis in autoimmunity, 189
  elimination of self-reactive B lymphocytes in the bone marrow, 228
  receptors to control, 332
    TNFα and LTAα, 332
Apoptotic cells
  and clearance, 30–31
  in autoimmune diseases, 30
  phagocytosis of, 31
APRIL
  and its relevance to BlyS, 318–319
  longevity factors, 230
Arthritis, 115
  collagen-induced, 201
  manifestations, 340
  rheumatoid, 407
Association analysis
  haplotype relative risk (HRR), 57
  pedigree disequilibrium test (PDT), 57
  transmission disequilibrium test, TDT, 57
ASTIRA, *see* Autologous Stem Cell Transplantation International Rheumatoid Arthritis
ASTIS, *see* Autologous Stem Cell Transplantation International Scleroderma
Atherosclerosis, 393–396
  antibodies function in, 393
  autoantibodies in, 401
  cellular immunity in atherosclerosis, 397–400
  disease development, 394
  humoral immunity in atherosclerosis, 400–401
  immune cells in atherosclerosis, 396–397
  importance of macrophages, 397
  in the absence of lymphocytes, 399
  macrophages, 400
  oxidation, lipolysis, and proteolysis, 395
  pathogenesis of
    microbial antigens, 402
  prevention of, 393–402
    fatty streak formation, 395
    fibrous plaques, 395–396
    lesion initiation, 394
    plaque rupture and thrombosis, 396
  risk factors associated with, 393–394
    genetic and environmental elements, 393
  role of B cells in, 400
  T cells, 400
  vaccination or immunoglobulin administration in, 401–402
Atherosclerosis-associated autoantigens, 402
Atherosclerotic plaques, 399
  maintenance and development, 396
Atopic dermatitis, 350
Autoantibodies
  against dsDNA and phospholipids in lupus, 184
  against glomerulonephritis, 184
  anti-dsDNA and ss-DNA autoantibodies levels in the absence of IFNγ gene, 190
  as Trojan horse immunomodulators, 294–297
  association of SLE and RA in the production of, 17
  cardiac-specific, 2
  cause-and-effect relationship between disease and, 101
  hallmark of lupus, 190
  IL-10 in the production of, 335
  in atherosclerosis, 401
  in CD patients, 143
  nephritogenic, 174–175
  populations associated with systemic autoimmunity, 401
  role of B cells in development of, 229
Autoantibodies and nephritis, 165–176
Autoantibody genes, characterization of, 229
Autoantigen (s), 227
  atherosclerosis-associated, 402
  HLA-B, 365
  ocular, 365

Index

of Pemphigus vulgaris, 130
retinal 365–366, 369–372
thyroglobulin (Tg), 197
Autoimmune blistering disorders, 135
Autoimmune disease (AD)
  acute lymphoblastic leukemia (ALL), 349
  allogeneic HSC transplantation for coincidental autoimmune diseases, 349
  autologous HSC transplantation for coincidental autoimmune diseases, 350
  autologous HSCT database, 352; see EBMT/EULAR AD autologous HSCT database
  CRP as regulator of, 27–38, see C-reactive protein
  diabetes and ulcerative colitis, 182
  features of, 348
  hematopoietic stem cell transplantation (HSCT) techniques, 347
  thyroiditis, 181
  HLA class II genes in human chromosome 6, susceptibility to, 116
  idiopathic thyroiditis, 339
  inflammatory bowel disease, 104
  insulin-dependent diabetes mellitus, 348
  multiple sclerosis (MS), 182
  multiple sclerosis, 104, 182
  myasthenia gravis (MG), 151–182
  Pemphigus vulgaris (PV), 127
  predominant in women
    rheumatoid arthritis, 182
    scleroderma, 181
    Sjögren's syndrome, 181
  SLE, a chronic relapsing AD, 69
  susceptibility of women to different, 181
  susceptibility to, 182
    systemic lupus erythematosus (SLE), 181
  systemic lupus erythematosus, 104
  transfer of AD through allogeneic HSCT, 350
  transplant-related mortality (TRM), 353
  treatment of
    anti-tumor necrosis factor-alpha (TNF-α) strategies, 347
    blood marrow transplant (BMT), 351
    combination of glucocorticosteroids and immunosuppressive agents, 347
    cytotoxic drugs such as cyclophosphamide (CY), 358
Autoimmune hemolytic anemia, 229, 339
Autoimmune hepatitis, 349
Autoimmune hyperthyroidism, 59
Autoimmune hypothesis, 210, 220
Autoimmune myocarditis, 2–3 see Myocarditis, autoimmune
Autoimmune reactions
  triggered by defect in immune regulation, 217–218
  triggered by infections, 218–219
Autoimmune response, 2
Autoimmune thyroid disease, 339
Autoimmune thyroiditis
  Treg cells influence on susceptibility in, 197–206
Autoimmunity
  against desmogleins in Pemphigus vulgaris, 127–136
  and TLR 9, see Toll-like receptor, 9
  antibody-mediated autoimmunity, 215–216
  B-cell mediated autoimmunity, 215–216
  clinical spectrum, 210–211
  human, 263
  immunotherapies, 216
  in mutliple sclerosis, 209
  pathological spectrum, 211
  protective autoimmunity, 220
  T-cell mediated autoimmunity, 212–215
  treatment with
    TNFα inhibitors, 329
  triggers for autoimmune reactions, 216–219
Autologous Stem Cell Transplantation International Rheumatoid Arthritis (ASTIRA), 355
Autologous Stem Cell Transplantation International Scleroderma (ASTIS), 355
Autoreactive T cells
  activation of result in autoimmunity, 407
  multiple sclerosis (MS), 407
  organ-specific autoimmune diseases, 92

B

B cell activation factor (BAFF), 189; see BlyS longevity factors, 230
B cell receptor, 18, 20, 22
  crippled, (BcR), 227
B cells
  autoreactivity in human autoantibody-associated diseases, 292–293
  human autoimmunity, an abnormality of B cell function, 291–297
  in the spontaneous development of autoantibodies, 229
B lymphocyte depletion therapy in autoimmune disorders BLyD, 291–308
  clinical significance of Trojan horse concept, 297–303
  adverse events associated with BLyD, 302–303

B lymphocyte depletion therapy (*Continued*)
  anti-CD20 therapeutic agents, 299–300
  effector mechanisms in RA, 298–299
  efficacy, 301
    failure of seronegative disease to respond, 302
    logistics of B cell depletion, 299
    repeated cycles of B cell depletion, 303
    rituximab, 300
  BLyD data on Trojan horse concept, 303–306
    clinical response, 305
    kinetics of relapse follow autoantibody, 305–306
    selective fall in autoantibody levels, 303–304
    total Ig levels, 304–305
    two patterns of relapse, 306
  human autoimmunity, an abnormality of B cell function, 291–297
    autoantibodies as effector molecules, 294
    autoantibodies as Trojan horse immunomodulators, 294–297
    B and T cell autoreactivity in human autoantibody-associated diseases, 292–293
    generation of autoreactive cells, 293
B lymphocyte stimulator (BlyS) and autoimmune rheumatic diseases, 313–322
  BlyS and its receptors, 313–319
    APRIL and its relevance to BlyS, 318–319
    biology of, 313–315
    in vivo deficiency of BlyS or its receptors, 315–316
    supranormal levels of BlyS *in vivo*, 316–318
  BlyS antagonism as therapeutic modality, 319–322
    candidates for BlyS antagonist therapy, 321–322
    human experience, 320–321
    mouse models, 319–320
Bacterial DNA, TLR 9 and the immunostimulatory effects of, 18
BAFF, *see* B cell activation factor, *see* BlyS
Behcet's disease, 355
Bisphenol-A (BPA), environmental estrogen, 184
Blistering disorders, autoimmune, 135
Blisters of skin and mucosa, 127
Blood brain barrier, 336

Blood cells, 339
Blood marrow transplant (BMT), treatment of AD, 351
Blood–retina barrier, 365
Bone destruction in rheumatoid arthritis (RA), 334
Bone marrow
  B lymphopoesis in, 187
  suppression, 373
  toxicity, 347
  transplantation, 340
Bystander suppression, 407

C
ζ chains, 246, 249, 251, 253
C4 gene, 86, 89
  C4A gene, 86–87, 92, 94
    mutant *C4A* gene, 90
  C4B gene, 85–87, 94
    mutant *C4B* gene, 90
C4 protein, 89–90, 92, 95
  deficiency, 91–92
  dichotomous gene size variation of C4A and C4B, 86–87
  isotypes C4A and C4B, 87
  isotypic residues of C4A and C4B, 87
  polymorphism, 95
C4A and C4B proteins
  C4A deficiency, 91–95
    heterozygous, 86, 92, 94–95
    homozygous, 85, 91–95
  C4B deficiency, 94–95
    heterozygous, 86, 95
    homozygous, 95
  gene dosage(s), 86, 89
    C4A, 85, 95
    C4B, 85, 92, 95
  gene size, 85–86, 89
  genetic risk factor, 85, 93–94
  genotype, 94
  phenotype, 87, 91–94
  polygenic and gene size variations of, 95
  racial diversities, deficiencies in SLE RCCX modules, 86–87
C-reactive protein (CRP) as regulator of autoimmune disease, 27–38
  and autoantibodies, 33
  and chromatin clearance, 30
  and nephritis, 33–34
  as acute-phase reactant, 28–29
  binding to Fcγ receptors, 27–28
  cardiovascular disease and, 29
  CRP, SAP and nuclear antigen clearance, 30–31

Index

essential role of IL–10 in anti-inflammatory activities of CRP, 37
FcγR as CRP receptors, identification of, 34–35
FcγR in CRP effects of inflammation, role of, 35–37
genetics and autoimmunity, 31
in animal models of autoimmunity, 32–33
in autoimmune disease, current perspective, 37–38
in immune complex nephritis, 33–35
in inflammation, 34
interaction with nuclear antigens, 29–30
levels in human SLE, 31
nuclear localization, 30
polymorphisms, 31
protection from endotoxin shock, 34
receptors, 34–35
regulation of synthesis, 28–29
structural features of, 27–28
transgenic models, 32
uptake of apoptotic cells, 31
Calcineuring, overexpression of, 5
Cardiac myosin-induced autoimmune myocarditis, 2–3
Cardiomyocytes, G protein-coupled receptor β2-AR, 105
Cardiomyopathy, hereditary forms of, 5
Carditis, 115
Carnitine transporter involved in carnitine deficiency *OCTN2*, 61
Casein, 366, 371–372
Caucasians, insulin gene VNTR alleles in, 59
CD4, 365, 368
CD4+ T cells, role in immune regulation, 134
Celiac disease, 350
  bone and cartilage, destruction of, 334
  molecular basis of, 141–148
    additional T cell stimulatory peptides in barley, rye, and oats, 145–146
    generation of safer foods for patients, 147
    hypothesis for disease development, 147–148
    specificity of tTG is linked to gluten toxicity, 145
    T cell recognition of gluten peptides, 142–145
    the HLA gene dose effect, 146–147
Cell adhesion molecules, 397
Cell death by apoptosis, 333
Cell signaling, 268–270
  TLR 9 and intracellular signaling, 18–20

Cell–cell interactions
  inhibition of, 335
  TNFα importance in, 331
Cell-mediated cytoxicity, antibody-dependent, 333
Cerebral vasculitis, 340
CFSE, 159
Chemokine receptors
  CCR1 and CCR5, 333
  in pathogenesis of myocarditis, 5, 11
Chemokines, 2
  affected by 17β-estradiol, natural estrogen, 189
  and myocarditis, 10–11
  MCP–1 or MIP–1*alpha*, in pathogenesis of myocarditis, 5, 11
  synthesis of, 331
Chlamydia pneumoniae, infection in atherosclerosis with, 402
Chloroquine, therapeutic effects in SAD, 21
Cholera toxin, 371
Cholestrol
  elevated intracellular levels, 394–395
  increased levels of removal, 394
Chorea, 340
Chronic myeloid leukemia (CML), 349
CNS inflammation, 209, 213–214, 219–220
Collagen IV, 169
Colony-stimulating factors (CSFs), 341
Complement C4 protein, 1, 89–90
Complement receptor I, 165
Complement receptors, 6–7
Complement system, 2
Congenital myasthenic syndromes, 102
Coxsackievirus B3 (CB3)-induced autoimmune myocarditis, 2
Coxsackieviruses, 2
CpG motifs, 18–20
Crohn's disease, 63, 349–350
  genetic analysis in autoimmunity, 61
  identification of mutations in the *CARD15* gene in, 61
CTLA-4, 249, 252
Cushing's syndrome, 373
Cytokine(s), 2, 329–330, 407
  action and production of, 331
  administration of, 329
  affected by 17β-estradiol, natural estrogen, 189
  and myocarditis, 8–10
  anti-inflammatory, 399
  IL–10, 105, 335

Cytokine(s) (*Continued*)
  by monocytes and APCs
    IL–12, IL–18 and IL–23, 337, 341
    TNFα and IL–1, 338
  by T cells
    IL–17, 338
  clinical application, 339
  control of T cells responsive to self-antigens, 9
  IFNα, in chronic viral hepatitis, 339
  IFN-γ and IL-2 direct influence on neuromuscular transmission, 104
  immunosuppressive cytokine TGF-β, 134
  in adaptive immunity, 329
  in innate immunity, 329
  in pathogenesis of rheumatoid arthritis (RA) and Crohn's disease (CD), 329
  in the production of autoantibodies and IgG, 335
  increased expression of AchRs in MG, 156
  inhibitory and stimulatory, 396
  interactions between the major proinflammatory, 331
  interferon (IFN)-γ, 407
  interferon α (IFNα) for viral hepatitis, 329
  interleukin
    IL–1α, 104
    IL-β, 103
    IL–4, IL–10, 105
  lymphotoxin gene, deficiency of, 103
  molecular mimicry, 122
    by CD4+ T cells, 121
  monocyte derived, 331
  proinflammatory, 201, 399
    IL–1, and IL–2, 122
    IL-lα and IL–1β, 105
    IFN-γ and TNF-α, 395
    TNF-α, 105, 122
  synthesis of, 331
  T cell-derived cytokine, IL–17, 337
  Th1 cytokines
    IFNγ, 331
    IL–1, IL–17, 331
  Th2/Th3 cytokines
    IL–4, 407
    IL–10, 407
  tumor necrosis factor (TNF)-α, 407
    inhibition of, 330–331
    proinflammatory and immunomodulatory functions of, 102
Cytokine and anti-cytokine treatments, control and induction of autoimmunity by, 329–341
  autoimmune manifestations with cytokine administration, 338–341
  heterogeneity of the response to TNFα inhibitors, 338
  other cytokine inhibitors, 336–337
  other cytokines as treatment targets, 337
  targeting one or more than one cytokine, 337–338
  TNFα
    and its receptors, 330–331
    local and systemic effects of inhibition, 332–334
    mode of action of specific inhibitors, 331–332
    side effects of inhibitors, 335–336
Cytokine gene polymorphisms, 338

D

Dermatomyositis, 340
Dermtitis herpetiformis, 349
Desmogleins, auotantibody reactivity against, 129–131
Diabetes, 182, 201
Dilated cardiomyopathy (DCM), 1, 5, 7, 10
Disease-modifying antirheumatic drugs (DMARDs), 354
DNA methylation and drug-induced lupus, 73–77
DNA methylation inhibitors, 70
Drug-induced lupus (DIL), 70

E

EAM, *see* Experimental Autoimmune Myocarditis
EAT, *see* Experimental Autoimmune Thyroiditis
EBMT, *see* European Group for Blood and Marrow Transplants
EBMT/EULAR AD autologous HSCT database, 354
  for hematological immuncytopenias, 352–353
  for neurological disorders, 352–353
  for rheumatological disorders, 352–353
Endocarditis, 115
Enolase
  bacterial, 168
  human, 168
  streptococcal, 168
Epitope, 366, 368–371, 374
ERs, *see* Estrogen receptors
Estrogen
  17β-estradiol on cytokines and chemokines, 189

Index

B cell activation and T cell dysregulation, 183
effects on lupus, 182
environmental, 182, 184
natural, 182
oxogenous, 182
transcripts of ER in SLE patients, 187
Estrogen antagonist, tamoxifen, 184
Estrogen receptor (ER), 183, 189
Estrogen response element (ERE), 187
Estrogen, interferon-gamma, and lupus, 191–192
estrogen and lupus, 184–192
IFNγ in SLE and other autoimmune diseases, 189–190
mechanisms of estrogen effects on immune system, 185–189
biological effects on cells by ER-dependent and –independent mechanisms, 185–188
estrogen alterations of B cells, 188–189
estrogen effects on cytokines, 189
Estrogen-mediialed thymic cortex atrophyhy, 187
EULAR, *see* European League against Rheumatism
European Group for Blood and Marrow Transplants (EBMT), 351
European League against Rheumatism (EULAR), 351
response, 355
Experimental allergic encephalomyelitis (EAE), 351
Experimental autoimmune encephalomyelitis (EAE), 32, 200, 407
Experimental autoimmune glomerulonephritis (EAG), 169
Experimental autoimmune myocarditis (EAM), 1–2, 11
complement system in the initiation of, 6
enhancement by IFN-*gamma* deficiency, 9
IL-10 for limiting inflammation in, 7
in A/J mice, Th2-like phenotype, 9
myocardial lesions in, 10
susceptibility to induction of, 3
Experimental autoimmune thyroiditis (EAT), 200
a model in autoimmune diseases studies, 197
Experimental autoimmune uveitis (EAU), 201, 367
Eye
activation of T cells in sympathetic ophthalmia, 365
an immune-privileged organ, 365
immune-privileged organ, 365
immunosuppressive environment within, 365
irreversible destruction of photoreceptors and neuronal tissue, 365

F

Fcγ receptors (FcγR), 62, 168
as CRP receptors, identification of, 34–35
CRP binding to, 36–37
human, 35
in CRP effects of inflammation, role of, 35–37
in nephrotoxic nephritis, 34
mouse, 35–37
regulation of inflammation by, 35–37
role in CRP protection from endotoxin shock, 36
surface plasmon resonance studies, 35
Felty's syndrome, 341
Fibrin, 175
Full-genome scans for T1D, 59
Functional polymorphism, 62
Fyn, 246, 253

G

Gene therapy for cancer treatment
IL-10, 341
Gene transcription, 266
Genes
allelic heterogeneity, 55
genetic heterogeneity, 55
identification of the genes involved in complex diseases, 55
Genetic association analyses of the genes and haplotypes, 58
Genome scans and linkage analysis, 58
Genome scans for RA, 60
Genotyping, 56
of SNPs, 61
Germinal center hyperplasia, 152–154
Gliadin
activation of the innate immune system, 145
T-cell stimulatory peptides, 145, 147
Gliadin peptide, 145
HLA-DQ2-restricted, 144
Glomerular basement membrane, 168
Glomerulonephritis, 167–169
in the absence of IFNγ gene, 190
Glomerulonephritis, acute, antigenic variations associated with, 116
Glomerulonephritis, autoantibodies against, 184

Glomerulonephritis, hallmark of lupus, 190
Gluten
  food industry, use in, 142
  heterogeneity and complexity, 142
  in disease development, 147
  level of, 147
  source of nitrogen and amino acids, 142
Gluten peptide
  secretion of inflammatory cytokines, 144
  stimulation of T cells of CD patients, 144
  T-cell stimulatory peptides, 144
  T cell stimulatory properties
    identification of, 147
Glutenins, LMW- and HMW, 144
  T-cell stimulatory peptides, 145, 147
Gluten-specific T cell, 143
  response
    amplification of, 148
Goodpasture's syndrome, 167, 169–170, 173, 175
Granulocyte-colony stimulating factor, 354–355
Grave's disease, 59, 229–230
Gross murine leukemia virus (G-MLV), thymotrophic, 158
Group A streptococcal (GAS) throat infection, 115
Guillain–Barre syndrome, 340

H

Haplotype relative risk (HRR), 57
Haplotypes, 57–58
Hematopoietic stem cell transplantation for treatment of severe autoimmune diseases, 347–359
  animal models, 350–351
  autoimmune diseases mechanisms, 348–349
  coincidental AD in patients receiving HSCT for another indication, 349–350
  juvenile idiopathic arthritis, 354
  open issues, 356–359
    ablative therapy without HSCT, 358–359
    allogeneic HSCT, 356–357
    immune reconstitution, 357–358
    prospective randomized controlled clinical trials, 355–356
  rheumatoid arthritis, 353–354
  systemic lupus erythematosus, 354–355
  systemic sclerosis, 353
  treatment of human AD with HSCT, 351–353
Hemophagocytic syndrome, 354
Hemophilia A, 339

Hepatitis C, 339–340
Hepatitis C, thyroid dysfunction, 341
Hepatitis C-related cryoglobulinemia, 339
Histocompatibility leucocyte antigen (HLA)
  DQ2 or HLA-DQ8, 141
HLA, 59
HLA class II alleles
  in PV, 128
  PV-associated, 132
HLA class II antigens, 116
HLA class II molecules
  distribution of, associated with development of RF/RHD in different populations, 118
HLA region, 60
HLA-associated disease
  celiac disease (CD), 141
HLA-DQ2 molecules
  peptide-binding properties of, 142
  types of, 146
    DR3, DR7, 146
HLADRpl, gene most commonly shared among autoimmune diseases, 63
Hormones, 339
HSC, autologous hematopoietic stem cells (HSCs), 347
HSC transplantation
  guidelines, 351
HSCT for AD normal and autoaggressive T cell reactions, 358
  autologous HSCT infection with agents other than cytomegalovirus (CMV), 358
HSCT in autoimmune diseases
  conditioning regimens, 352
HSCT morbidity and mortality, 355
HSCT, Allogeneic
  autoimmune hemolytic anemia
  graft versus host disease (GVHD)
  graft-versus-autoimmunity
HSCT, see Hematopoietic Stem Cell Transplantation
Human autoimmune diseases, the genetics of, 55–63
  genetic analysis in autoimmunity, 58–63
    autoimmune diabetes (T1D), 58–59
    Crohn's disease and ulcerative colitis, 61
    genes shared between autoimmune diseases, 63
    genome scans and linkage analysis in autoimmune diseases, 58
    multiple sclerosis, 59–60
    rheumatoid arthritis, 60–61
    systemic lupus erythematosus, 62–63
  genetics of complex diseases, analysis of, 56–58

association analysis, 57–58
combining linkage and association, 58
linkage analysis, 56–57
Human autoimmunity, 263
Human desmosomal cadherin, Dsg4, 129
Human genome
  sequence and scans in unraveling complex disease genetics, 55–63
Human leucocyte antigen (HLA)
  B27PD, 365–366, 368–369
  inducing effective oral tolerance, 372
  mimicking retinal peptide PDSAg, 371
Human lupus, complement components C4A and C4B in, 85–95
  C4A and C4B defeciency in SLE and immune complex diseases, 90–92
    C4A or C4B deficiencies in human SLE, 92–94
    deficiency of C4B in SLE patients from aborigines, 93–94
    impairment of immune response in C4-deficient patients, 91–92
    low complement activity and C4 protein concentrations in SLE, 92
    molecular basis of complete C4 deficiency, 90–91
    partial deficiencies versus polygenic variations of C4Aand C4B, 94
    partial deficiency of C4A in SLE across multiple ethnic groups, 92–93
  diversity of complement components C4A and C4B in human population, 86–90
    dichotomy in gene sizes, polygenes, and RCCX module variants, 86–88
    diversity of human C4A and C4B proteins, 87–89
    genetic determinants of C4 plasma/serum protein levels, 89–90
Human lupus, failure to maintain T cell DNA methylation, 69–76
  aberrant T cell DNA methylation, gene expression, and cellular function in idiopathic lupus, 77–80
  DNA methylation, 77–78
  gene expression and cellular function, 78–80
  DNA methylation and drug-induced lupus, 73–77
  DNA methylation and autoimmunity, 73–75
  DNA methylation and drug-induced lupus, 75–76
  T cell genes affected by DNA methylation inhibitors, 76–77

DNA methylation, chromatin structure, and gene expression, 70–73
Human myasthenia gravis (MG), non-MHC genetic polymorphisms with functional importance for, 101–110
  β2-adrenergic receptor in MG, 105–106
  proinflammatory and anti-inflammatory cytokines in MG, 102–105
    association of MG to alleles of TNF-α, 102–103
    functional implications of association with IL-1β TaqI RFLP A2 allele, 104
    functional implications of association with TNF-α–308 A2 allele, 103–104
    IL-10 is associated to MG with high autoantibody levels, 105
    lack of associations of MG to genetic variants of IL-4 and IL-6, 105
  T-cell receptor cofactor CTLA-4 in MG, 106–109
    association to MG with thymoma and increased activation of the immune system, 106–107
    Ctla-4 (AT)n is associated to ADCC, 109
    CTLA-4 and thymomas, 109
    functional correlates to the genetic variants of *Ctla*-4, 107–108
    promoter SNPs –1772 (C/T) and –1661 (A/G), 108
    the A/G SNP in CDS1, 108
    the C/T SNP at -318, 108
Hyperplasia, hallmark of lupus, 190
Hyperthyroidism, 349
  autoimmune, 59
  treatment by IFNα, 339

I

IFNα treatment
  thyroid abnormalities, 339
IFNβ
  for the treatment of multiple sclerosis, 340
IFN-γ in macrophage activation, 399
Immune complex clearance, 62
Immune complexes, 165
  circulating, 165
  in situ, 165
Immune homeostasis, 181
Immunoglobulins, 62, 339
  enhanced secretion of, 183
Immunosuppressive agents, 127
Immunosuppressive cytokine TGF-β, 134
Immunotherapies of PV
  immunoadsorption, 127
  plasmapheresis, 127

Inflammation, 28–29, 31, 34–37
   receptors to control, 332
      TNFα and LTAα, 332
INFα, inhibitors of, 331
Inhibition of T cell activation, 63
Innate immune response, major agents
   chemokines, 6
   complement system, 6
   cytokines, 6
   NK cells, 6
Innate immune system and myocarditis, 5–11
   Chemokines and myocarditis, 10–11
   Complement and myocarditis, 6–8
   cytokines and myocarditis, 8–10
   NK cells and myocarditis, 8
Innate immunity in EAM, 1–11; *see* Myocarditis, autoimmune
Insulin gene, causative polymorphism within, 58
Insulin-dependent diabetes mellitus (IDDM), 349–351
Integrins, 397
Interferon (IFN)-γ, 156, 160, 407
Interferon γ (IFN γ), 341
Interleukin, 272
   IL-1 receptor antagonist, 36
Interleukin-4 (IL-4), 105
   biallelic polymorphism in *IL-4* gene, 105
   cofactor in the maturation of B cells, 105
   polymorphism, 105
   variable number of tandem repeat (VNTR), 105
Interleukin–10 (IL–10)
   biological activities of, 105
Interleukin 12 (IL–12) is a cytokine produced by monocytes and APCs, 341
IL-1 receptor antagonist (IL-IRa), 104
IL-1 receptor-associated kinase (IRAK) family, 19, 21
IL 2
   defective production, 255
   deficiency, 255
   gene activation, 255
IL-10
   alleles of, 105
   inhibiting the synthesis of IL-lα and IL-1β and TNF-α, 105
IL-10, 134
   CRP induction of, 36
   induction through Fcγ receptors, 36
Interphotoreceptor retinoid-binding protein (IRBP), 366, 368, 374
IRBP, *see* Interphotoreceptor retinoid-binding protein

J
Juvenile idiopathic arthritis (JIA), 353

K
Keratinocytes, 127–128, 131
   acetylcholine receptor, 130
Killer immunoglobulin-like receptor (KIR), 358

L
LAT, 246
Lck, 246, 249, 252–253]
LDL receptors (LDLRs), 394
Leukemia, relapsed, 340
Leukocytosis, 333
Light meromyosin fragment (LMM), 123
Linkage analysis report
   type 1 diabetes (T1D) mellitus, 58
Linkage disequilibrium, 57
Lipid rafts, 249, 252–253
Lipopolysaccharide, cytokines, 35
Lipoproteins
   LDL molecules, modified, 402
   very low-density lipoproteins (VLDLs), 394
Logarithm of the odds (LOD) score, 57
Lupus, 340
   hallmarks of lupus, 190
   role of IFNα in the pathogenesis of, 340
Lupus, human
   agents associated with, 70
   clinical manifestations of, 69
   drug-induced lupus (DIL), 70
   parts affected by, 69
   pregnancy and flares of, 183
Lupus nephritis, 170–174
Lupus-like syndromes, 229
   agents associated with, 70
Lymphocytes, autologous peripheral blood, 340
Lymphocytes T, 334
Lymphocytes, T and B, 331
   B lymphocytes, role in systemic autoimmune disease, 227
Lymphocytes, traffic of, 157
Lymphocytic choriomeningitis virus (LCMV), 158
Lymphokine-activated killer (LAK) cells, 340
Lymphotoxin gene
   deficiency of, 103
Lymphotoxin α and β, 102
Lymphotoxins (LTα and LTβ), 330
   role in formation and function of lymphoid organs, 331

Index

M

Macrophage activation syndrome, 354
Macrophage colony-stimulating factor (M-CSF), 397
MACS cytokine secretion assay, 132
Major histocompatibility antigen (MHC), 197
   class I antigens, expression by Thymic myoid cells and medullary epithelial cells, 156
   class II antigens, increased expression of, 227
   class II molecules, downregulation the expression of, 105
   class III genes, 86
Mapping
   variable number tandem repeat (VNTR) types, 58–59
Membrane attack complex (MAC), 6–7
Mendelian inheritance mode, 56
Meningitis, infliximab in aseptic, 336; *see also* Types of anti-TNFα antibodies
MHC, *see* Major histocompatibility antigen
Microbial antigens
   in the pathogenesis of atherosclerosis, 402
Microsatellites, 58
   useful polymorphic markers, 56
Mitogen-activated protein kinase (MAPK), 19–20, 246, 249, 254
Mitral valve regurgitation (MVR) in RHD patients, 117
Mixed cryoglobulinemia, 169, 174
Molecular autoimmunity
   antigen-specific regulation of autoimmunity, 407–414
      antigen-specific therapy, 408
      antigen-induced regulatory T cells, 408–410
      Treg induction, factors involved in, 410–414
   antiphospholipid syndrome, molecular pathogenesis and novel therapeutic targets, 360–370
      antiphospholipid antibodies and endothelial cells, 364–369
      antiphospholipid antibodies and platelets, 361–364
   atherosclerosis, prevention of, 393–402
      atherosclerosis, 393–396
      cellular immunity in atherosclerosis, 397–400
      humoral immunity in atherosclerosis, 400–401
      immune cells in atherosclerosis, 396–397
      vaccination or immunoglobulin administration in atherosclerosis, 401–402
   autoantibodies and nephritis, 165–176
      acute poststreptococcal glomerulonephritis, 167–169
      Goodpasture's syndrome, 169–170
      lupus nephritis, 170–174
      other nephritogenic autoantibodies, 174–175
   autoimmunity against desmogleins in Pemphigus vulgaris, 127–136
      active animal model of pemphigus vulgaris, 135
      autoantibody reactivity against desmogleins, 129–131
      autoreactive T lymphocytes in pemphigus, 132–134
      clinical phenotype of Pemphigus vulgaris, 128
      epidemiology of pemphigus and association with HLA class II alleles, 128
      passive animal models of pemphigus vulgaris, 135
      pathogenesis of pemphigus, 128–129
      regulatory T lymphocyes in pemphigus, 134–135
   autoimmnity in multiple sclerosis, 209–220
      autoimmune hypothesis of MS, 210, 220
      multiple facets of multiple sclerosis, 210–217
      protective autoimmunity, 220
      triggers for autoimmune reactions in MS patients, 216–219
   B lymphocyte depletion therapy in autoimmune disorders: chasing Trojan horses, 291–308
      BLyD data on Trojan horse concept, 303–306
      clinical significance of Trojan horse concept, 297–303
      human autoimmunity, an abnormality of B cell function, 291–297
   B lymphocyte stimulator (BlyS) and autoimmune rheumatic diseases, 313–322
      BlyS and its receptors, 313–319
      BlyS antagonism as therapeutic modality, 319–322
   celiac disease, molecular basis of, 141–148
      additional T cell stimulatory peptides in barley, rye, and oats, 145–146

Molecular autoimmunity (*Continued*)
  generation of safer foods for patients, 147
  hypothesis for disease development, 147–148
  specificity of tTG is linked to gluten toxicity, 145
  T cell recognition of gluten peptides, 142–145
  the HLA gene dose effect is linked to the level of gluten presentation, 146–147
control and induction of autoimmunity by cytokine and anti-cytokine treatments, 329–341
  autoimmune manifestations with cytokine administration, 338–341
  heterogeneity of the response to TNFα inhibitors, 338
  other cytokine inhibitors, 336–337
  other cytokines as treatment targets, 337
  targeting one or more than one cytokine, 337–338
  TNFα and its receptors, 330–331
  TNFα, local and systemic effects of inhibition, 332–334
  TNFα, mode of action of specific inhibitors, 331–332
  TNFα, side effects of inhibitors, 335–336
C-reactive protein (CRP) as regulator of autoimmune disease, 27–38
  as acute-phase reactant, 28–29
  CRP, SAP and nuclear antigen clearance, 30–31
  essential role of IL–10 in anti-inflammatory activities of CRP, 37
  FcγR as CRP receptors, identification of, 34–35
  FcγR in CRP effects of inflammation, role of, 35–37
  genetics and autoimmunity, 31
  in animal models of autoimmunity, 32–33
  in autoimmune disease, current perspective, 37–38
  in immune complex nephritis, 33–35
  in inflammation, 34
  interaction with nuclear antigens, 29–30
  levels in human SLE, 31
  structural features of, 27–28
estrogen, interferon-gamma, and lupus, 191–192
  estrogen and lupus, human and animal studies, 192–184

  IFNγ in SLE and other autoimmune diseases, 189–190
  mechanisms of estrogen effects on immune system, 185–189
hematopoietic stem cell transplantation, treatment of autoimmune diseases, 347–359
  ablative therapy without HSCT, 358–359
  allogeneic HSCT, 356–357
  animal models, 350–351
  autoimmune diseases mechanisms, 348–349
  coincidental AD in patients receiving HSCT for another indication, 349–350
  immune reconstitution, 357–358
  juvenile idiopathic arthritis, 354
  prospective randomized controlled clinical trials, 355–356
  rheumatoid arthritis, 353–354
  systemic lupus erythematosus, 354–355
  systemic sclerosis, 353
  treatment of human AD with HSCT, 351–353
human autoimmune diseases, the genetics of, 55–63
  aberrant T cell DNA methylation, gene expression, and cellular function in idiopathic lupus, 77–80
  DNA methylation and drug-induced lupus, 73–77
  DNA methylation, chromatin structure, and gene expression, 70–73
  genetic analysis in autoimmunity, 58–63
  genetics of complex diseases, analysis of, 56–58
  human lupus, T cell DNA methylation and chromatin structure, 69–76
human lupus, complement components C4A and C4B in, 85–95
  C4A and C4B deficiency in SLE and immune complex diseases, 90–92
  deficiencies of C4A or C4B in human SLE, 92–94
  diversity of complement components C4A and C4B in human population, 86–90
human myasthenia gravis (MG), non-MHC genetic polymorphisms with functional importance for, 101–110
  β2-adrenergic receptor in MG, 105–106
  proinflammatory and anti-inflammatory cytokines in MG, 102–105
  T-cell receptor cofactor CTLA–4 in MG, 106–109

Index

human systemic lupus erythematosus, immune cell signaling and gene transcription in, 263–275
   altered pattern of tyrosine phosphorylation and calcium responses, 264–265
   mechanisms of increased TCR/CD3-mediated $[Ca^{2+}]$ response in SLE T cells, 268–270
   protein kinase A function, 271–272
   regulation of transcription determines interleukin 2 deficiency in SLE T cells, 272–274
   TCR ζ chain deficiency, 265–268
innate immunity in experimental autoimmune myocarditis, 1–11
   experimental models of myocarditis, 2–3
   innate immune system and myocarditis, 5–11
   mouse genotype, impact on prevalence and severity of myocarditis, 4–5
intrathymic expression of neuromuscular AchRs and immunopathogenesis of myasthenia gravis (MG), 151–160
   expression of neuromuscular AchRs by thymic cells, 153–157
   model to examine peripheral T cell entry and activation in thymus, 158–160
   role of thymus in MG pathogenesis, 152–153
   thymus and central immune tolerance, 157
   thymus and T cell trafficking, 157–158
molecular mimicry in autoimmune uveitis, pathogenesis to therapy, 365–374
   antigenic mimicry of retinal autoantigen and environmental antigens, 370–372
   HLA peptide B27PD in EAU, 367–368
   pathogenic and tolerogenic epitopes of PDSAg and B27PD, 368–370
   retinal autoantigens and mimicry peptides, 366–367
   treatment uveitis patients with oral peptide B27PD, 372–374
NKT cells and autoimmune type 1 diabetes, 43–50
   future directions, 49–50
   iNkt cells in the pathogenesis of T1D, role of, 45–49
   NKT cells, 44–45
   Type 1 diabetes (T1D), 44, 58–59
rheumatic heart disease (RHD), molecular basis of autoimmune reactions leading to valvular lesions, 115–123
   animal models, 122–123
   cytokines, 122
   genetic susceptibility, 116–118
   molecular mimicry and RF/RHD, 118–122
   the etiopathogenic agent, Streptococcus pyogenes, 116–117
susceptibility in autoimmune thyroiditis, Treg cell, MHC class II gene control, 197–206
   CD4$^+$ T cells as mediators of induced resistance, 199–201
   CD4$^+$CD25$^+$ T cell as peripheral barrier to autoimmune thyroiditis, 204
   CD25 expression on CD4$^+$ T cells in induced resistance, 201–203
   MHC class II gene control of susceptibility, 198–199
   T cell regulation and MHC restriction, 204–205
systemic autoimmunity, crippled B lymphocyte signaling checkpoints in, 227–239
   B cell receptor-mediated signaling checkpoints, 230
   B lymphocyte participate in both innate and adaptive immunity, 228
   critical regulators of B cell receptor signaling, 231–233
   critical role of B cells in autoimmunity, 229–230
   disrupted B cell signaling pathways in human autoimmunity, 237–239
   negative regulators of B cell receptor-mediated signal transduction, 234–237
systemic autoimmunity, disrupted T cell receptor signaling pathways in, 245–256
   signaling pathways in T cells, 246–250
   T cell signaling abnormalities in systemic autoimmune disease, 250–256
systemic lupus erythematosus, accumulation of self-antigens in, 279–286
   antigen clearance and autoimmunity in DNASE1-deficient patients, 281–282
   defective clearance of self-antigens in SLE, 282–286
   T cell in human lupus, 280–281
toll-like receptor (TLR) 9 and autoimmunity, 17–22
   CpG sequences in self-DNA trigger autoantibody production, 20–21
   TLR 9 and the immunostimulatory effects of bacterial DNA, 18
   TLR 9 as a target for regulating RF production, 21–22

Molecular autoimmunity (*Continued*)
    TLR 9 and intracellular signaling, 18–20
    TLRs as receptors for pathogen-associated molecules, 17–18
Molecular mimicry, 168, 227
    and RF/RHD, 118–122
    antigenic mimicry of retinal autoantigen and environmental antigens, 370–372
    HLA peptide B27PD in EAU, 367–368
    in autoimmune uveitis, pathogenesis to therapy, 365–374
    pathogenic and tolerogenic epitopes of retinal peptide PDSAg and its mimotope B27PD, 368–370
    retinal autoantigens and mimicry peptides, 366–367
    treatment uveitis patients with oral peptide B27PD, 372–374
Monoclonal antibodies, CD4 and CDS, 197
Monocyte chemptactic protein (MCP–1), 395
Monocyte-derived macrophages, 397
Mouse genotype, impact on prevalence and severity of myocarditis, 4–5
MRA (anti-IL-6 receptor antibody)
    efficacy for the treatment of RA and CD, 337
Multiple sclerosis (MS), 58, 209–220, 349, 351
    autoimmune hypothesis of MS, 210, 220
    autoimmnity in, 209–220
    genetic analysis in autoimmunity, 59–60
    immunotherapies, 216
    linkage analysis report, 58
    multiple facets of multiple sclerosis, 210–217
        autoimmunity from immunotherapies of MS, 216–217
        B cell- or antibody-mediated autoimmunity, evidence for, 215–216
        clinical spectrum of, 210–211
        pathological spectrum of, 211–212
        T cell mediated autoimmunity, evidence for, 212–215
    organ-specific autoimmune diseases, 407
    predominant in women, 182
    protective autoimmunity, 220
    triggers for autoimmune reactions in MS patients, 216–219
        autoimmune reactions caused by a defect in immune regulation, 217–218
        autoimmune reactions caused by infections, 218–219
    treatment with IFN γ, 340–341
Multiple valvular lesions (MVLs) in RHD patients, 117

Multiplex families, 56
    of the various autoimmune diseases, 58
Mutation(s)
    found in *OCTNl* was a missense substitution, 61
    G–C transversion, 61
    in genes encoding sarcomeric proteins, 5
    missense substitution, 61
Myalgia, 333
Myasthenia gravis (MG), 101–110, 151, 155, 159, 350, 352
    acetylcholine receptor, 102
    cytokines in
        lymphotoxin α and β, 102
        TNF-α, 102
    experimental autoimmune MG, 103
    pathogenesis, role of the thymus, 152
    patients
        germinal center (GC) hyperplasia, 152
        thymomas, 152
    predominant in women, 182
    thymic pathogenesis of, 158–159
    thymomas, 154
Myocardial inflammation, 2
Myocarditis, 1
    autoimmune, 1–11
    common cause of, 1
    EAM, 1–2, *see* experimental autoimmune myocarditis
    experimental models of, 2–3
        cardiac myosin-induced autoimmune myocarditis, 2–3
        coxsackievirus B3 (CB3)-induced autoimmune myocarditis, 2
        peptide-induced myocarditis, 3
    innate immune system and myocarditis, 5–11
        chemokines and myocarditis, 10–11
        complement and myocarditis, 6–8
        cytokines and myocarditis, 8–10
        NK cells and myocarditis, 8
    mouse genotype, impact on prevalence and severity of, 4–5
Myocyte damage due to viral cytotoxicity, 2
Myoid cells, 153–156, 160
Myosin
    fragments capable of inducing myocarditis and valvulitis, 123
Myosin-induced myocarditis, 2

N
NARAC
    consortium in the US focused on European Americans and erosive arthritis, 60

Index

Natural killer (NK) cell, 357
NEMO protein, 19
Nephropathy, 340
Nephrotoxic nephritis, 33–34
  effect of CRP, 34
  role of Fcγ receptors, 34
Neutropenia, 341
Nicotinic acetylcholine receptor
  in demonstration of ADCC, 109
  on the neuromuscular junction, 101
  role of inflammatory mechanisms, 102
  symptoms of, 101
NK cell
  and myocarditis, 8
  reconstitution of recipient, 358
NKT cells, 44–45
  and autoimmune type 1 diabetes (T1D), 43–50
  future directions, 49–50
  iNkt cells in the pathogenesis of T1D, role of, 45–49
    iNKT cell activation induces protection against T1D, 47–49
    iNKT cell deficiency and T1D, 45–47
  type 1 diabetes (T1D), 44, 58–59
Non-Hodgkin's lymphoma, 350
Nordic multiplex families with SLE, 63

O

Oral tolerance, 370, 372–374
  induction by feeding B27PD or PDSAg, 367–369
Orchiectomy, 184
Osteoclasts, formation and activation, 334

P

Pedigree disequilibrium test (PDT), 57
Pemphaxin, 130
Pemphigus vulgaris (PV)
  active animal model of pemphigus vulgaris, 135
  autoantibody reactivity against desmogleins, 129–131
  autoimmunity against desmogleins in, 127–136
  autoreactive T lymphocytes in pemphigus, 132–134
  clinical phenotype of Pemphigus vulgaris, 128
  epidemiology of pemphigus and association with HLA class II alleles, 128
  passive animal models of pemphigus vulgaris, 135
  pathogenesis of pemphigus, 128–129
  regulatory T lymphocyes in pemphigus, 134–135
Pentraxins
  binding to apoptotic cells, 27
Peptide-induced myocarditis, 3
Peptides
  gliadins, 141, 143
  gluten, 141
  glutenins, 141
Peptidylarginine deiminase type 4, an enzyme involved in citrullination, RA, 60
Peripheral blood mononuclear cells (PBMCs), 105, 183
  culturing of, 190
  from SLE patients, 189
Peripheral muscle mononuclear cells
  density of β2-AR on, 106
Plasma C4 protein levels
  determinants of, 89
Plasma lipoproteins, 393
  low-density lipoproteins (LDLs), 393
Plasminogen, 168, 175
Polychondritis, relapsing, 355
Polymorphic disease alleles, 56
Polymorphism(s), 55–57
  genetic mutations involved in complex diseases, 55
  insulin gene, 58
  restriction fragment length polymorphism (RFLP), 103
Polymyositis, 340
  in patients with chronic hepatitis C, 340
Polyneuritis, 339
Polyubiquitination, 19
Pregnancy and flares of lupus, 183
Preproinsulin, key autoantigen in T1D pathogenesis, 59
Primary progressive multiple sclerosis (PPMS), 210
Proinflammatory and anti-inflammatory cytokines
  IL-1, TNF, IL-4, IL-6, and IL-10
  T cell and/or B cell activation by cytokines, 102
Proinflammatory cytokines
  methotrexate, leflunomide, 331
Proteases
  synthesis of, 331
Psoriasis, 63, 340, 349–350
PV-associated HLA class II alleles, 132

R

Regulatory cells
  CD4⁺ CD25⁺, 252, 255

Relapsing-remitting multiple sclerosis (RRMS), 210
RF/RHD
  molecular basis of autoimmunity in, 116
  patients, T cell receptor (TCR) analysis and cytokine production, 116
Rheumatic fever (RF)
  clinical signs and symptoms of, 115
  occurrence of erythema marginatum, 116
  production, TLR 9 as a target for regulation of, 21–22
  streptococci groups (A, B, C, F, and G), 116
  throat group A streptococcal (GAS) infection, 115
Rheumatic heart disease (RHD), 115
  animal models, 122–123
  antigen-presenting cells (APCs) cytokines, 122
  generation of streptococcal peptides, 122
  genetic susceptibility, 116–118
  mitral valve regurgitation (MVR) in RHD patients, 117
  molecular basis of autoimmune reactions, 115–123
  molecular mimicry and RF/RHD, 118–122
    humoral and cellular immune responses interface in RF/RHD, 121
    T cell receptor usage, 121–122
    the cellular immune response, 119–121
    the humoral immune response, 119
  multiple valvular lesions (MVLs) in RHD patients, 117
  the etiopathogenic agent, *Streptococcus pyogenes*, 116–117
Rheumatoid arthritis (RA), 21, 58, 63, 341, 349–352, 407
  genetic analysis in autoimmunity, 60–61
  linkage analysis report, 58
  predominant in women, 182
  TNF-α antibodies in the treatment of, 103f
Rheumatoid synovitis
  increase of angiogenesis, 333
Rotavirus, 366, 371

S
Scleroderma, 181, 351
Secondary progressive multiple sclerosis (SPMS), 210
Self-antigens (Ags), 227
Self-antigens, 181
  apoptosis, anergy or tolerance, 348
  autoantibodies against, 182
  limiting levels of, 157–158
  presentation of, 348
Self-DNA
  CpG sequences trigger autoantibody production, 20–21
  immunostimulatory, 22
Serum amyloid P component (SAP), 27
  and anti-nuclear antibodies, 31
  and chromatin clearance, 30
  and nephritis, 31
  interaction with chromatin, 30
  uptake of apoptotic cells, 31
Severe aplastic anemia (SAA), 349
Single-nucleotide polymorphisms (SNPs)
  analysis of a dense map of, 60
  close characterization of, 58
  gene encoding (32-AR), 106
  promoter SNPs -1772 (C/T) and -1661 (A/G), 108
Sjögren's syndrome, 181, 229, 336, 339, 349, 352
Skin
  against desmocollins, 129
  against desmoglein 3, 127–129, 133
  autoimmune disease pemphigus vulgaris (PV), 127
SLE, 21, 63, 85–86, 89–95, 105, 227, 358
  antibodies associated with arterial and venous thrombosis in, 401
  disease activity index (SLEDAI), 354
  gene, associated in familial RA in Europeans, 61
  risk factor for, 183
SLE-like syndrome associated with increase in MZ B cells, 230
Spleen, B lymphopoesis in, 187
Splicing mutation in *CTLA4*, 63
Spondyloarthropathy, 340
Streptococci groups (A, B, C, F, and G), 116
Streptococcus pneumoniae, 402
Streptococcus pyogenes, 116–117
  surface adhesion proteins, in the external cell wall of
    M, 117, 110–111, 121–122
    T, 117
    R, 117
  throat infection by, 115
Supramolecular activation complex (SMAC), 249
Susceptibility to myocarditis, 3
Sydenham's chorea, clinical signs and symptoms of, 115
Synoviocytes, 331, 334
Synovitis
  inflammatory reaction, 333

Index

Systemic autoimmunity
   B cell receptor-mediated signal transduction, negative regulators, 234–237
   B cell receptor-mediated signaling checkpoints, 230
   B lymphocytes participate in both innate and adaptive immunity, 228
   crippled B lymphocyte signaling checkpoints in, 227–239
   critical regulators of B cell receptor signaling, 231–233
   critical role of B cells in autoimmunity, 229–230
   disrupted B cell signaling pathways in human autoimmunity, 237–239
   disrupted T cell receptor signaling pathways in, 245–256
   signaling abnormalities of T cells in
     early, 252–253
     intermediate and late, 254–255
   signaling pathways in T cells, 246–250
   T cell signaling abnormalities in systemic autoimmune disease, 250–256
     signaling abnormalities in antigen-presenting cells and autoimmune disease, 250–251
     signaling abnormalities in T cells and autoimmune disease, 252
Systemic lupus erythematosus (SLE), 58, 69, 181, 245, 256, 349–350, 400
   accumulation of self-antigens in, 279–286
   altered pattern of tyrosine phosphorylation and calcium responses, 264–265
   antigen clearance and autoimmunity in DNASE1-deficient patients, 281–282
     effect of autoreactivity, 282
     gene mutation and clinical features, 281–282
     laboratory findings, 282
   defective clearance of self-antigens in SLE, 282–286
     clearance of self-antigens as therapeutic strategy, 286
     evidence from knockout mice, 282–284
     mechanisms of accumulation of self-antigens in SLE, 284–285
   effect of CRP on, 31
   genetic analysis in autoimmunity, 62–63
   immune cell signaling and gene transcription in, 263–275
   levels of CRP, 31
   linkage analysis report, 58
   mechanisms of increased TCR/CD3-mediated $[Ca^{2+}]$ response in SLE T cells, 268–270

   altered composition and dynamics of lipid rafts, 269–270
   FcRγ chain substitutes for defective ζ chain, 268–269
   protein kinase A function, 271–272
   regulation of transcription determines interleukin 2 deficiency in SLE T cells, 272–274
   T cell in human lupus, 280–281
   TCR ζ chain deficiency, 265–268
     impaired posttranslational functions, 267
     impaired TCR ζ chain gene transcription, 266
     impaired translation and posttranscription events, 266–267
     oxidative stress, 267
     role of IFNγ, 268
Systemic sclerosis (SSc), 351–353
   regimen-related pulmonary toxicity, 353
   improvement of skin score and functional status, 353

T

T cell(s), 365–368, 370–371
   γδ T cells isolated from spleens, 368
   αβ T cells, 368
   activated, autoaggressive, 365
   activation by IL-15, 337
   activation of autoreactive, 158
   autoreactive, 197
   costimulatory factors
     CD28 and CTLA-4, 102
     CTLA-4, negative regulator for T cell activation, 106
   HLA peptide–specific, 366
   immigrants self-reactive, 158
   in lymphoid follicles, 366
   receptor (TCR), 366
   receptor (TCR) chains
     overexpression of selected gene families, 102
   receptor excision circles (TRECs) in T cells, 358
   role of regulatory T cells, 348
   stimulatory gluten peptides
     and Homologs in the Hordeins of barley, 144
     characteristics of a selected group, 143
   Tregs, regulatory T cells, 407
T lymphocytes, autoreactive, in pemphigus, 132–134
T regulatory (Treg) cells
   $CD4^+$, $CD25^+$, 134
   Tr1 and Th3 cells, 134
T1D, 63

T1D pathogenesis, preproinsulin, 59
Tandem mass spectrometry, 142–143
TCR rearrangement diversity recovery of, 357
TGFβ-activating kinase (TAK1), 19–20
Th1 cells, 365
Thrombocytopenia, 339, 341
Thrombocytosis, 333
Thymectomy, 152
Thymic epithelial cells (TECs), neuromuscular AchRs on, 154–156
Thymic hyperplasia, 152, 154
Thymic medullary epithelial cells, 156–157, 160
Thymic pathology, 158
Thymic tolerance, 63
Thymoma lymphocytes, 152
Thymomas, MG, 154
Thymus, 151–157
  role in MG pathogenesis, 152–153
  and central immune tolerance, 157
  and T cell trafficking, 157–158
  peripheral T cell entry and activation in, 158–160
Thyroglobulin (Tg) as an autoantigen, 197
Thyroiditis, 181, 340, 350
  CD4+ T cells as mediators of induced resistance, 199–201
    effect of cytokines on CD4+ regulatory T cell induction and function, 200–201
    protection from EAT induction by elevating the circulatory thyroglobulin level, 199–200
  CD4+CD25+ T cell as peripheral barrier to autoimmune thyroiditis, 204
  CD25 expression on CD4+ T cells in induced resistance, 201–203
    abrogation of established tolerance by CD4+CD25+ T cell depletion, 201–203
    interference with CD4+CD25+ regulatory T cell function by cross-linking TNFR family molecules, 203
  MHC class II gene control of susceptibility, 198–199
  murine autoimmune, 197–206
  T cell regulation and MHC restriction, 204–205
  Treg cell, MHC class II gene control of susceptibility in autoimmune, 197–206
Thyroid-stimulating hormone (TSH), 197
TID, 61
TLR 9, *see* Toll-like receptor 9
TNF receptor(s)
  in apoptosis controlled by Fas, 331

type I (p55-TNF-R), 330
type II (p75-TNF-R), 330
TNFα
  and lupus, 335
  and lymphotoxin α (LTα), 330
  antibodies to
    in treatment of rheumatoid arthritis, 103
  contribution to apoptosis, 336
  importance in
    cell-cell interactions, 331
  in lymph node hypertrophy, 336
  inhibition in multiple sclerosis, 336
  inhibitors on the incidence of lymphomas, 336
  its receptors, 330–331
  lack of efficacy in CD, 334
  local and systemic effects of inhibition, 332–334
  mode of action of specific inhibitors, 331–332
  side effects of inhibitors, 335–336
  treatment of RA and CD, 336
  types of anti-TNFα antibodies
    adalimumab, 332
    etanercept, 332–333, 338
    infliximab, 332–334, 338
TNF-converting enzyme (TACE), 331
TNF receptor-associated factor 6 (TRAF–6), 19
TNFα-independent inflammatory pathway, 338
Tolerance
  breaking of, 251
  in autoreactive T cells, 249
  maintenance of, 252–254
  oral, 370, 372–374
    induction by feeding B27PD or PDSAg, 367–369
  rupture of, 254
  to autoantigens, 245
Tolerogen, 366, 372–374
Toll-like receptor (TLR) 9 and autoimmunity, 17–22
  CpG sequences in self-DNA trigger autoantibody production, 20–21
  TLR 9 and the immunostimulatory effects of bacterial DNA, 18
  TLR 9 as a target for regulating RF production, 21–22
  TLR 9 and intracellular signaling, 18–20
  TLRs as receptors for pathogen-associated molecules, 17–18
Toll-like receptors (TLRs), 228
Total body irradiation (TBI), 351
Transmission disequilibrium test, TDT, 57
Treatment with oral peptide B27PD, 372–373

Index

Treg cells influence on susceptibility in autoimmune thyroiditis, 197–206
Tregs, regulatory T cells, 407
TRM risk, 355
Tuberculosis, frequency in RA, 335
Tumor necrosis factor (TNF)-α, 407
Tumor-infiltrating lymphocytes (TILs), 340
Type 1 diabetes (T1D), 44, 58–59, 407
Type 1 diabetes (T1D) mellitus
   linkage analysis report, 58
Type 1 diabetes (T1D), NKT cells and autoimmune, 43–50

U
Ulcerative colitis, 182, 349
   identification of mutations in the *CARD15* gene, 61
   genetic analysis in autoimmunity, 61

Uveitis, 365–374

V
Vaccination with autoantigens for treating autoimmune disease, 407
Variable number of tandem repeat (VNTR), IL–4, 105
Variable numbers of an 86-bp tandem repeat (VNTR), 103
Vasculitis, 349, 355
Viral hepatitis
   IFNα for the treatment of, 339
Virus-mediated damage, 2
Visual acuity, 373–374

W
Wegener's granulomatosis, 105